S0-BUA-661

FUNDAMENTAL CIRCUIT ANALYSIS

FUNDAMENTAL CIRCUIT ANALYSIS

JOHN W. TONTSCH
Los Angeles Pierce College

THEODORE F. BOGART, JR., P.E.
University of Southern Mississippi

E. CHARLES ALVAREZ
Los Angeles Pierce College

SCIENCE RESEARCH ASSOCIATES, INC.
Chicago, Palo Alto, Toronto, Henley-on-Thames, Sydney
A Subsidiary of IBM

Acquisition Editor Alan Lowe
Project Editor Judith Fillmore
Cover Designer Barbara Ravizza
Illustrator John Foster

Reprinted with revisions 1983. Originally published as *Fundamental Circuit Analysis* by E. Charles Alvarez and John W. Tontsch.

Printed in the United States of America

Library of Congress Cataloging in Publication Data

Tontsch, John W.
Fundamental circuit analysis.

Revision of: Fundamental circuit analysis / E. Charles Alvarez, John W. Tontsch.
Includes index.
1. Electronic circuit design. 2. Electronic circuit design—Data processing. I. Bogart, Theodore F. II. Alvarez, E. Charles. Fundamental circuit analysis. III. Title.
TK7867.T65 1982 621.3815'3'02854 82–19173
ISBN 0–574–21570–0

10 9 8 7 6 5 4 3 2 1

PREFACE

Fundamental Circuit Analysis is an electrical fundamentals text for electronics and computer technology programs. This text is designed to prepare students for today's world of digital and computer electronics. Included, for example, are matrix theory and the solutions of equations expressed in matrix form; Bartlett's network theorem used for analyzing symmetrical integrated and filter circuits; and ladder theory, because of its applicability to digital-to-analog conversion. The traditional theorems are not neglected but explored in new ways, such as the application of Millman's theorem to analog summing circuits.

This text takes the unique approach of teaching BASIC language programming and integrating routine computer-assisted problem-solving on an optional basis throughout. This philosophy is in keeping with the current trend in both industry and institutions of higher learning to provide access to "computer power" via microcomputers or on-line terminals. Inherent in these ongoing technical advances is the necessity for technicians (using microcomputers, computer terminals, or programmable calculators) to master at least essential programming techniques if they are to take advantage of the benefits offered by these "tools." This textbook is a pioneer in that approach.

Sophisticated "string" techniques, which are better employed by professional programmers for report generation, are omitted here. Only those elementary statements oriented toward mathematical solutions are discussed. Reasonably advanced programming can be mastered in a short amount of time, thereby continuing to give major emphasis to the basic electrical theory. At no time have the authors attempted to emphasize programming at the expense of circuit theory, but rather to integrate into the theory the use of the computer as a modern tool of circuit interpretation. No prerequisites are necessary for this phase of the text.

One of our major objectives in preparing this revised printing was to ensure compatibility of all BASIC language material with the more popular microcomputers now available. Appendix C lists system-command differences for various micros. Appendix D shows how to use and solve simultaneous equations with Microcomputer BASIC, so that various matrix operations can be performed. Necessary technical and typographical corrections have also been incorporated into this revision, and selected passages have been rewritten in response to user suggestions. The authors and publisher would like to thank the following users for their valuable feedback:

Charles Garay, Electronics Technology Institute
Lou Maze, Triton College
Tom Oliver, Los Angeles Pierce College
Bert Sheffield, Adjunct, Trenton State College

The mathematical rigor of the text has been designed to coincide with a standard course in algebra and trigonometry. The required mathematics progresses from simple linear equations to quadratic functions and j operators. The delta notation (denoting change) is occasionally used to illustrate laws but does not require an understanding of the calculus.

Alternating current theory is presented with an emphasis on the methods of solving circuits. Rather than presenting all the possible variations of series and parallel combinations, the authors have chosen to develop an *approach* to the solution that will apply to any ac circuit. Network theorems are generously used in developing complex ac network analysis.

For those who desire three-phase power concepts in their course, an inexpensive paperback supplement entitled *Polyphase Systems with Computer Solutions* is available from the publisher (order number 13-4530). The language used in this supplement is BASIC-PLUS, because of the programming "power" required for most polyphase solutions.

A related dc/ac lab manual, developed by coauthor T. F. Bogart and entitled *Experiments for Electrical Circuit Analysis with BASIC Programming* (order number 13-4565), is now available to support this and other electrical theory texts. Each electrical experiment presented is accompanied by a BASIC language lesson and exercise.

The primary objective of both this text and the lab manual is to effectively prepare the student for a changing industry environment increasingly oriented toward digital electronics and computer techniques.

CONTENTS

PART I

Many questions arise when a student begins the study of electronics. Some of these questions are personal in nature; others are technical. In the former group, a student may say to himself: "I am going to spend several years of study for a degree in electronics. Is there a future?" To answer this question, consider the vast spectrum of current electronic applications. In that area of specialization referred to as *discrete* or *digital electronics*, applications in the computer field alone can range from sophisticated pocket calculators (of the type currently used by many students) to computerized photoenhancement of distant planetary pictures telemetered back to Earth. In that area of specialization referred to as *analog* or *communication type electronics*, many new commercial areas are developing. One such area of far-reaching significance includes the electronic circuits used in conjunction with fiber optics, a new medium replacing coaxial cable and copper wiring for transmission of telephone and TV signals. There is not room enough here to state all of the many exciting new areas in which electronics finds application. To answer the question previously posed, there is indeed a future in electronics.

The second area in which many students initially seek answers is technical in nature. What constitutes electronics? Is it difficult? At what level should I start? All of these questions are typical for the new entrant. Their answers now bring us to PART I of this textbook. In fact, these questions may best be answered in reverse order.

Where do I start? Obviously, with a good foundation. PART I starts with the fundamental concepts and laws governing scientific notation which, incidentally, is universally applicable throughout all scientific disciplines. It continues by discussing such fundamental ideas as:

1. When you speak of a 12-volt car battery, what is meant by a "volt"?
2. When the tag on the back of a TV set says "rated at 3.5 amps," what exactly is meant by an amp?
3. What is resistance?

These and other similar concepts form the foundation upon which the more complex principles are built.

Is it difficult? Not by applying good study habits right from the beginning. Read the material carefully and study the examples. Use the answers to problems supplied at the end of the textbook as a check against your work, not as a quick "cop-out" to handing in last night's assignment.

What constitutes electronics? This concept and the laws governing measurable electronic quantities are covered in the sequential chapters. The important first step is to build that foundation upon which comprehensions must lie. PART I of this text constitutes that foundation. Welcome to your future.

CHAPTER 1

NOTATIONS, SYSTEMS, AND CONVERSIONS

1.1 SCIENTIFIC NOTATION

The most common usage of mathematics, that of manipulating numbers, is an everyday occurrence. It may consist of averaging student test scores, balancing one's checking account, paying for and receiving change from purchases, etc. The normal magnitude of these everyday encountered numbers is usually within an easily recognizable range. In many occupations, however, one can encounter numbers in magnitudes not so easily recognizable. For example, persons working with federal or other governmental budgets will work in number magnitudes usually beyond the average everyday range. An even more illustrative example involves persons working in sciences such as astronomy. A typical number might look like

5,874,000,000,000 miles

This is the approximate distance in miles that light travels in one year, a frequent measurement of length in the science of astronomy. The field of electronics also has examples of large numbers. Frequency, measured in units called *hertz* (Hz), might look like

160,250,000 Hz

In order to simplify the reading (or writing) of numbers of unusual magnitudes, a system of number notations has been developed. It is called *scientific notation.* As is to be expected, expression in scientific notation involves the application of simple rules to reduce the number in question to a more recognizable magnitude. For numbers greater than 1, the rule is as follows: *Move the decimal point to the left until the value of the number is between 1 and 10. Multiply this resultant by 10 raised to a positive power equal to the number of places the decimal point was shifted leftward.*

If the decimal point in the above example of one light year's distance is moved to the left until the value represents a number between 1 and 10, then it can be seen that the decimal point must be shifted 12 places. To retain the original value of the number, we must now multiply the resultant 5.874 times 10 raised to a power equaling the number of places the decimal point was shifted. Since the decimal point has been moved 12 positions to the left, the correct notation of this example is 5.874×10^{12}. By the same analogy, if the frequency example were written in scientific notation, its value could be expressed as 1.6025×10^{8}.

Example 1.1 Rewrite each of the following numbers in scientific notation:

a. $3{,}256 = 3.256 \times 10^3$
b. $1{,}750{,}000 = 1.75 \times 10^6$
c. $2{,}115{,}000{,}000 = 2.115 \times 10^9$
d. $42.7 = 4.27 \times 10^1$

Numbers can also frequently occur in extremely minute quantities. For example, the definition of *capacitance* (an electronic term to be studied in subsequent chapters) is measured in a unit called the *farad.* As the reader will learn, a farad is a rather large unit of measurement. If defined in this unit, a typical capacitor value might look like

0.00000000015 farad

Another example would be to write the charge of an electron in terms of the unit definition of charge, namely, the *coulomb.* The electron charge would be written as follows

0.00000000000000000016 coulomb

As can be seen, very small numbers written in this manner also present some difficulty in grasping their value. To write a number whose value is less than 1 in scientific notation, the rule is as follows: *Move the decimal point to the right until the value of the number is between 1 and 10. Multiply the resultant by 10 raised to a negative power equal to the number of places the decimal point was shifted rightward.*

If the decimal point in the above value of capacitance is to be moved to the right until the value represents a number between 1 and 10, then it can be seen that the decimal point must be shifted 10 places. To retain the original value of the number, we must now multiply the resultant 1.5 times 10 raised to a power equaling the number of places the decimal point was shifted. Since the decimal point has been moved 10 positions to the right, the correct notation of the above example is 1.5×10^{-10}. By the same analogy, if the charge on the electron were written in scientific notation, its value would be expressed as 1.6×10^{-19}.

Example 1.2 Rewrite each of the following numbers in scientific notation:

a. $0.00126 = 1.26 \times 10^{-3}$
b. $0.169 = 1.69 \times 10^{-1}$
c. $0.0000025 = 2.5 \times 10^{-6}$
d. $0.000200 = 2 \times 10^{-4}$

As another application illustrating how much easier it is to read scientific notation, suppose it is desired to recognize which of the following two numbers is the larger:

0.00000000000123

or

0.000000000000567

On the surface it appears that the significant digits of the second number, namely 567, would be larger than the significant digits of the former number, namely 123. However, when expressed in scientific notation, the two numbers appear as 1.23×10^{-12} and 5.67×10^{-13}. The powers of 10 of the second number indicated that the decimal point is actually further to the left of the first significant digit than is the case with the former number, and hence the second number is actually the smaller of the two.

As a further aid in making it easier to express numbers, a prefix is frequently used in conjunction with the unit designation. Just as commas are used after every third zero in very large numbers, so it is that each prefix is representative of powers of 10 measured in units of 3. Refer to Table 1.1. Notice that in starting with the smallest prefix (pico) and progressing upward toward the largest (tera), each successive prefix is one thousand times larger than its predecessor. Since 1000 represents three zeros, it follows that all exponents of 10 assigned a prefix name are divisible by 3. This rule applies when the power is greater than $|3|$. Ten raised to the power ± 1 is called *deca* and *deci* respectively, and 10 raised to the power ± 2 is called *hecto* and *centi* respectively. For example, 10^{-5} and 10^8 do not have assigned prefix names. Also take note that 10^0 equals unity. For that matter, any coefficient except zero, when raised to the power of 0, has a value of unity. If we are to utilize these abbreviations, a slight modification will have to be made to the methodology of

TABLE 1.1

Number	Power of 10	Prefix	Abbreviation
1,000,000,000,000	10^{12}	tera	T
1,000,000,000	10^9	giga	G
1,000,000	10^6	mega	M
1,000	10^3	kilo	k or K
1	10^0	—	—
0.001	10^{-3}	milli	m
0.000001	10^{-6}	micro	μ
0.000000001	10^{-9}	nano	n
0.000000000001	10^{-12}	pico	p

writing a number in scientific notation. *Move the decimal so that the resulting number lies between 1 and 1000 and the resultant displacement of the decimal is divisible by 3.* The following examples illustrate the point in question.

Example 1.3 Convert the following into a number between 1 and 1000 with the appropriate prefix. (Note: The following units of measurement are electronic terms which will be studied in subsequent chapters.)

a. 1,625,000 ohms =
   1.625×10^6 ohms =
   1.625 megohms (MΩ)

b. 32,700 Hz =
   32.7×10^3 Hz =
   32.7 kilohertz (kHz)

c. 0.0152 henry =
   15.2×10^{-3} henry =
   15.2 millihenries (mH)

d. 0.000000000125 farad =
   125×10^{-12} farad =
   125 picofarad (pF)

1.2 LAWS OF EXPONENTS

All formulas involve some mathematical relationship between variables. Since many of these variables will be expressed in units of powers of 10, the student will need to know the laws that govern the handling of these powers to arrive at the correct results. Suppose that in the formula

$$X_L = 2\pi f L$$

f is expressed as 5×10^6 and L is expressed as 0.2×10^{-3}. Multiplying the coefficients together presents no problem. The question now is, How does one handle the exponents associated with the powers of 10? The generalized rule involving the product of powers may be expressed as follows:

$$\boxed{a^m \times a^n = a^{m+n}} \qquad (1.1)$$

As can be seen, multiplication requires that exponents be added. It should also be noted that there is no restriction as to the value of a. Syntactically,

a is referred to as the *base*, and must be the same in each term of the equation. For example,

$$2^3 \times 2^2 = 2^5$$

In the above example, f times L would be

$$5 \times 10^6 \times 0.2 \times 10^{-3} =$$
$$(5 \times 0.2) \times (10^6 \times 10^{-3}) = 1 \times 10^3$$

In short, *the coefficients are multiplied together and the exponents of the base are added together.* A few illustrations will help to clarify the point.

Example 1.4 Simplify the following products:

a. $3 \times 10^3 \times 4 \times 10^{-1} =$
$(3 \times 4) \times (10^3 \times 10^{-1}) =$
12×10^2

b. $4.1 \times 10^9 \times 3 \times 10^{-9} =$
$(4.1 \times 3) \times (10^9 \times 10^{-9}) =$
$12.3 \times 10^0 =$
12.3

c. $8 \times 10^{-5} \times 4 \times 10^{-4} =$
$(8 \times 4) \times (10^{-5} \times 10^{-4}) =$
32×10^{-9}

The remaining rules involving exponents are as follows:

$$\frac{a^m}{a^n} = a^{m-n} \tag{1.2}$$

$$(a^m)^n = a^{mn} \tag{1.3}$$

$$\sqrt[n]{a^m} = a^{m/n} \tag{1.4}$$

Example 1.5 Simplify the following examples:

Ref. Eqs.

a. $\dfrac{4 \times 10^{-2}}{2 \times 10^3} = \left(\dfrac{4}{2}\right) \times \left(\dfrac{10^{-2}}{10^3}\right) = 2 \times 10^{-2-3} =$
2×10^{-5} (1.2)

b. $(5 \times 10^2)^3 = (5)^3 \times (10^2)^3 = 125 \times 10^{2 \times 3} =$
125×10^6 (1.3)

c. $\sqrt[3]{8^6} = 8^{6/3} = 8^2 = 64$ (1.4)

d. $\sqrt[2]{64 \times 10^{-8}} = \sqrt[2]{64} \times 10^{-8/2} =$
8×10^{-4} (1.4)

e. $\dfrac{12 \times 10^3 \times 6 \times 10^{-2}}{3 \times 10^4} = \left(\dfrac{12 \times 6}{3}\right) \times \left(\dfrac{10^3 \times 10^{-2}}{10^4}\right) =$
$24 \times 10^{3+(-2)+(-4)} = 24 \times 10^{-3}$ (1.1, 1.2)

f. $\sqrt[2]{3^3 \times 10^4} = (3^3)^{1/2} \times 10^{4/2} =$

$27^{1/2} \times 10^2 = 100\sqrt{27}$ (1.4)

g. $(2^2 \times 2^4)^3 = (2^6)^3 = 2^{18}$ (1.1, 1.3)

h. $\sqrt[2]{(1000)^2 \times 10^2} = \sqrt[2]{10^6 \times 10^2} =$

$\sqrt[2]{10^8} = 10^4$ (1.1, 1.3, 1.4)

i. $(3 \times 10^2 \times 4 \times 10^3)^2 = (3 \times 4)^2 \times (10^2 \times 10^3)^2 =$
$(12)^2 \times (10^5)^2 = 144 \times 10^{10}$ (1.1, 1.3)

Observe that if the number under the radical sign of Eq. 1.4 is a number with an implied exponent of 1, that is, of the form $\sqrt[n]{A}$, then an alternate method to write such an expression is

$$\boxed{\sqrt[n]{A} = A^{1/n}} \qquad (1.5)$$

Example 1.6 Express each radical as a fractional exponent.

a. $\sqrt[2]{21} = 21^{1/2}$

b. $\sqrt[3]{37} = 37^{1/3}$

c. $\sqrt[2]{42^3} = 42^{3/2}$

d. $\sqrt[4]{17^2 \times 10^3} = 17^{2/4} \times 10^{3/4}$

e. $\sqrt[4]{0.125^3 \times 10^8} = 0.125^{3/4} \times 10^{8/4} =$
$0.125^{3/4} \times 10^2$

f. $\sqrt[3]{17^2} \times \sqrt[3]{17^4} =$
$17^{2/3} \times 17^{4/3} = 17^{2/3+4/3} =$
$17^{6/3} = 17^2$

g. $\sqrt[n]{a^2} \times \sqrt[m]{a^3} = a^{2/n} \times a^{3/m} = a^{2/n + 3/m}$
$= a^{(2m + 3n)/mn}$

Caution should be taken if adding or subtracting numbers in powers of 10. The procedure in cases of this nature would be to manipulate the coefficient(s) of the terms until the powers of 10 of all terms are identical.

The coefficients can then have the indicated mathematical functions performed, and the resultant is multiplied by the common power of 10. The following examples illustrate this point.

Example 1.7 Simplify the following:

a. $14 \times 10^2 + 3 \times 10^3 =$
$1.4 \times 10^3 + 3 \times 10^3 =$
4.4×10^3

b. $6 \times 10^{-3} - 241 \times 10^{-5} =$
$6 \times 10^{-3} - 2.41 \times 10^{-3} =$
3.59×10^{-3}

c. $5.21 \times 10^4 - 17{,}800 + 0.63 \times 10^5 =$
$5.21 \times 10^4 - 1.78 \times 10^4 + 6.3 \times 10^4 =$
9.73×10^4

1.3 UNIT SYSTEMS

The reader has unquestionably heard of units in systems of measurement other than the English system. For example, the speedometer of many foreign cars is calibrated in kilometers per hour as opposed to the American car which is calibrated in miles per hour. The name given the former system of measure is the *Metric system.** There are two versions of this system: MKS and CGS. The major difference between the two versions is only in the scale of magnitude of each of the measurable units. The MKS system is more closely aligned in the values of its units of measurement to the English system than is the CGS system. Generally speaking, the CGS version is better adapted to the measurement of very small quantities than is the MKS version. The CGS, however, possesses the same advantages as MKS in that its units are easily converted as multiples of 10.

Table 1.2 lists the basic measurable units in each of the three systems. Notice how easy it is to remember the names of measurable units in

*The international standard system of units is called the *SI* system (abbreviated from the French: *système international*). This is the system of units that the United States and other countries are slowly adopting; it is popularly referred to as "the Metric system." For our purposes in this text, we will find that it differs very little from the MKS system. The SI system assigns to certain MKS quantities some new names (e.g., *Tesla*, *Siemen*) that we may use interchangeably with their MKS equivalents.

TABLE 1.2 Unit Systems

Dimension	English	MKS	CGS
Length	foot (ft)	meter (m)	centimeter (cm)
Mass	slug (32.2 lb)	kilogram (kg)	gram (g)
Time	second (s)	second	second
Force	pound (lb)	newton (N)	dyne
Energy (force × distance)	foot-pound (ft-lb)	joule (newton-meter)	erg (dyne-centimeter)

Metric. MKS stands for meter, kilogram, second. CGS stands for centimeter, gram, second. The only common unit between the three systems is the unit of time, namely, the second. Table 1.3 gives a relative comparison in the magnitudes of different measurable units between the MKS system and the English system.

The rationale behind the growing popularity of the Metric system becomes obvious when one considers the awkwardness of numbers in the English system versus that of the Metric system. For example, as one advances up the scale of length in the English system, it will take 12 inches to equal a foot, 3 feet to equal a yard, and 1760 yards to equal a mile. Certainly, such conversion numbers are not at all convenient. A comparison of the Metric unit of length illustrates the ease of conversion: 10 millimeters equals 1 centimeter, 10 centimeters equals 1 decimeter, 10 decimeters equals 1 meter, etc. Even the names in the Metric system are more rational. From the previous section the reader has learned that using the prefix *milli* before the unit *meter* implies $\frac{1}{1000}$ of a meter, which of course is the correct relationship. The prefix *centi* before the unit *meter* implies $\frac{1}{100}$ of the standard unit of length. The same reasoning exists in the other prefixes of the Metric system, such as deci, kilo, etc. It is little wonder that most of the major industrial countries have already gone to the Metric system. Current legislation in the United States calls for a gradual transition to the Metric system over the next decade. For this reason, most of the examples and problems in this section will deal with Metric conversions.

1.4 UNIT CONVERSIONS

There are two types of conversions that the reader will study. The first involves changes of prefix. For example, it is easier both to read and write 65 pF rather than .000065 μF. It is apparent in both cases that capacitance is still expressed within the basic unit of the farad. Only the prefix preceding farad is changed to make the number easier to read and express verbally (65 picofarads as opposed to 65 millionths of a microfarad).

TABLE 1.3 Comparison between Metric and English Units

English unit	equals	MKS unit
1 in		0.0254 m or 2.54 cm
1 ft		0.3048 m or 30.48 cm
1 yd		0.9144 m or 91.44 cm
1 mi		1.6094 km
1 lb		0.4536 kg 4.45 N
1 slug (32.2 lb)		14.6 kg
1 ft^2		0.0929 m^2
1 ft^3		0.0283 m^3
1 gal		3.7853 liters
1 qt		0.9463 liter
MKS unit	**equals**	**English unit**
1 meter		39.37 in 3.2808 ft 1.0936 yd
1 km		0.62137 mi
1 kg		2.2046 lb
1 m^2		10.7629 ft^2
1 m^3		35.3144 ft^3
1 liter		0.2642 gal 1.0567 qt
Miscellaneous Conversions		
1 newton (MKS force) = 10^5 dyne (CGS force)		
1 joule (MKS energy) = 10^7 ergs (CGS energy)		
Temperature F° = (9/5)C° + 32°		F° = degrees Fahrenheit
Temperature C° = (5/9)(F° − 32°)		C° = degrees Celsius

Note: Prior to 1948, the Celsius temperature scale was called "centigrade." In 1948, the Ninth General Conference on Weights and Measurements changed the name to Celsius, in honor of the scientist instrumental in developing the scale.

The key to conversions is so simple, it is frequently overlooked. It really involves nothing more than multiplying the unit to be converted by 1. It would not help to multiply by numeric 1, meaning unity, but rather to multiply by the equivalent of 1. Any ratio in which the numerator and denominator are equal equates to 1. Every student knows that any number, let us arbitrarily say 5.4, divided by itself is unity. This basic truth is also valid for measurable units such as length, time, etc., not just purely dimensionless numbers. For example, since there are one million picofarads (the smaller unit) in one microfarad (the larger unit), then the ratio of 10^6 picofarads over one microfarad must be 1.

$$\frac{10^6 \text{ pF}}{1 \ \mu\text{F}} = 1$$

Expanding this idea to the original illustration in this section, that is, to convert 0.000065 μF to pF, produces the following:

$$0.000065\ \mu\text{F} \times 1 \text{ is the same as}$$

$$0.000065\ \mu\text{F} \times \frac{10^6\ \text{pF}}{1\ \mu\text{F}} = 65\ \text{pF}$$

The key in determining the equivalent of 1, that is, the conversion ratio, is to place the unit desired in the numerator, and the unit to be changed in the denominator. As an example, to change milliamperes to microamperes, the conversion factor should be

$$\frac{10^3\ \mu\text{A}\ \text{(unit desired in numerator)}}{1\ \text{mA}\ \text{(unit to be changed in denominator)}}$$

Table 1.4 lists several common conversions that illustrate the basic approach.

Example 1.8 Perform the indicated prefix conversions.

a. Convert 0.0176 mA to μA.

$$0.0176\ \text{mA} \times \frac{10^3\ \mu\text{A}}{1\ \text{mA}} = 0.0176 \times 10^3\ \mu\text{A} = 17.6\ \mu\text{A}$$

TABLE 1.4 Conversions of Prefix

Changing from	to	Multiply by
micro (10^{-6})	milli (10^{-3})	$\frac{1\ \text{milli}}{10^3\ \text{micro}}$
micro (10^{-6})	unity	$\frac{1}{10^6\ \text{micro}}$
milli (10^{-3})	micro (10^{-6})	$\frac{10^3\ \text{micro}}{1\ \text{milli}}$
milli (10^{-3})	unity	$\frac{1}{10^3\ \text{milli}}$
milli (10^{-3})	kilo (10^3)	$\frac{1\ \text{kilo}}{10^6\ \text{milli}}$
unity	mega	$\frac{1\ \text{mega}}{10^6}$
unity	kilo	$\frac{1\ \text{kilo}}{10^3}$
unity	milli	$\frac{10^3\ \text{milli}}{1}$
unity	micro	$\frac{10^6\ \text{micro}}{1}$

b. Convert 1,560,000 Hz to kHz.

$$1{,}560{,}000 \text{ Hz} \times \frac{1 \text{ kHz}}{10^3 \text{ Hz}} = 1560 \text{ kHz}$$

c. Convert 0.000024 henry (H) to μH.
(Note: Henry is the unit measurement of inductance.)

$$0.000024 \text{ H} \times \frac{10^6 \ \mu\text{H}}{1 \text{ H}} = 24 \ \mu\text{H}$$

d. Convert 1,500,000 ohms to megohms.

$$1{,}500{,}000 \text{ ohms} \times \frac{1 \text{ M}\Omega}{10^6 \text{ ohms}} = 1.5 \text{ M}\Omega$$

e. Convert 0.000000150 ampere to nanoamperes.

$$0.000000150 \text{ A} \times \frac{10^9 \text{ nA}}{1 \text{ A}} = 150 \text{ nA}$$

The second area of study in making conversions involves changes between systems. An example would be changing the unit of length in the English system (mile) to the unit of length in the MKS system (kilometer). At this point, it is important to understand that all units in the final answer must exist within the same system. For example, to measure energy in a unit called newton-mile would not be acceptable, because a newton is associated with the MKS system and mile is the basic unit of length in the English system.

As in the prefix conversions, the coefficients of the units to be changed are multiplied by the equivalent of 1, so structured that the numerator of the ratio alters the solution toward the desired result and the denominator of the ratio cancels the unwanted unit. Ratios within the same system are already familiar to the reader. Within the English system, an example would be 12 inches equals 1 foot; within the MKS system, an example would be 100 cm equals 1 meter. Table 1.3 is useful to obtain conversions between systems. The following examples will illustrate this procedure.

Example 1.9 Convert 60 feet per second to millimeters per minute.

Step 1: $$\frac{60 \text{ ft}}{\text{s}} \times \frac{60 \text{ s}}{1 \text{ min}} = \frac{3600 \text{ ft}}{\text{min}}$$

Step 2: $$\frac{3600 \text{ ft}}{\text{min}} \times \frac{12 \text{ in}}{\text{ft}} = \frac{43{,}200 \text{ in}}{\text{min}}$$

Step 3: $$\frac{43{,}200 \text{ in}}{\text{min}} \times \frac{2.54 \text{ cm}}{\text{in}} = \frac{109{,}728 \text{ cm}}{\text{min}}$$

$$\text{Step 4:} \quad \frac{109{,}728\ \text{cm}}{\text{min}} \times \frac{10\ \text{mm}}{1\ \text{cm}} = \frac{1.09728 \times 10^6\ \text{mm}}{\text{min}}$$

Now in practice, all these steps are performed as one operation, instead of four distinct operations.

$$\frac{60\ \text{ft}}{\text{s}} \times \frac{60\ \text{s}}{\text{min}} \times \frac{12\ \text{in}}{\text{ft}} \times \frac{2.54\ \text{cm}}{\text{in}} \times \frac{10\ \text{mm}}{\text{cm}} = \frac{1.09728 \times 10^6\ \text{mm}}{\text{min}} \approx \frac{1.1 \times 10^6\ \text{mm}}{\text{min}}$$

Example 1.10 Convert 65 miles per hour to kilometers per hour.

$$\frac{65\ \text{mi}}{\text{h}} \times \frac{1.609\ \text{km}}{\text{mi}} = \frac{104.6\ \text{km}}{\text{h}}$$

Example 1.11 Convert 0.85 gallon to liters.

$$0.85\ \text{gal} \times \frac{3.7853\ \text{liters}}{1\ \text{gal}} = 3.218\ \text{liters}$$

Example 1.12 Convert 2.85 cubic yards to cubic meters

$$2.85\ \text{yd}^3 \times \frac{27\ \text{ft}^3}{1\ \text{yd}^3} \times \frac{0.0283\ \text{m}^3}{\text{ft}^3} = 2.178\ \text{m}^3$$

SUMMARY

Scientific notation simplifies the writing and recognition of numbers of unusual magnitudes. To express a large number as a value between 1 and 10, move the decimal to the left until this value is reached. Multiply the resultant by 10 raised to a positive power equal in magnitude to the number of places shifted leftward. To express a small number as a value between 1 and 10, move the decimal to the right until this value is reached. Multiply the resultant by 10 raised to a negative power equal in magnitude to the number of places shifted rightward.

Prefix notations are a further aid in expressing number magnitudes. If a number can be written with a power of 10 equal to a prefix notation, then this notation may be substituted for the power of 10 when expressing the number magnitude.

When mathematically manipulating numbers expressed in scientific notation, the laws of exponents are utilized. They are more frequently applied when the base of the number is 10. Their applicability, however, is valid for any base numbering system.

It is often necessary to convert unit prefixes or to convert units between different systems of measurement. The procedure in either case is to multiply the undesired unit by the "equivalent" of unity. Choose the multiplier so that in forming the product, the unwanted terms cancel out and are replaced with the desired terms.

PROBLEMS

Reference Section 1.1

1. Rewrite the following numbers as a number between 1 and 10 times the appropriate power of 10:
   a. 0.000156
   b. 0.000000235
   c. 1685
   d. 0.152
   e. 3,650,000
   f. 0.0123
   g. 12
   h. 12.4
   i. 0.1
   j. 10
   k. 100
   l. 10.1
2. Express each of the following as a number between 1 and 1000 and use the appropriate prefix:
   a. 1,875,000 Hz
   b. 0.000050 farad
   c. 0.0025 henry
   d. 0.00000007 ampere
   e. 1400 watts
   f. 1000 ohms
   g. 1,000,000 ohms
   h. 0.1 ampere
   i. 11,100 watts
   j. 0.0001 henry

Reference Section 1.2

3. Perform the indicated operations and express in scientific notation.
   a. $2 \times 10^2 \times 6.1 \times 10^4$
   b. $3.2 \times 10^{-2} \times 4.6 \times 10^{-5} \times 1.7$
   c. $(3.94 \times 10^4)/(1.76 \times 10^{-2})$
   d. $(4.78 \times 10^6)(8.62 \times 10^{-3})/(3.1 \times 10^{-4})$
   e. $(10{,}000 \times 6.2 \times 10^{-2})/(3.1 \times 10^6)$
   f. $[(0.0025 \times 10^3)/(1.61 \times 10^{-2})]^2$
   g. $(89.6 \times 10^6)^{1/2}$

h. $[(23.4 \times 10^{-2})/4.6]^{1/2}$
i. $(0.000008)^3$
j. $[(0.000064)^{1/2}]^{1/3}$
k. $[(1000)^3 \times 10^{-9}]^{1/2}$
l. $[(0.125)^{1/3} \times 60 \times 10^{-3}]/(30 \times 10^{-3})$
m. $(a^{1/2} \times a^{2/3})/a^{1/3}$
n. $((0.0002)^3)^2$

Reference Section 1.4

4. Make the following prefix conversions:
   a. 0.000015 F = __________ μF = __________ pF
   b. 1,650,000 ohms __________ kΩ = __________ MΩ
   c. 85.6 kHz = __________ Hz = __________ MHz
   d. 1635 MHz = __________ GHz = __________ kHz
   e. 4650 nA = __________ μA = __________ mA
   f. 0.015 mH = __________ H = __________ μH
5. Make the following system conversions:
   a. 60 mi/h = __________ km/h
   b. 42 yd/min = __________ cm/s
   c. 3 ft^2 = __________ cm^2
   d. 2.5 ft^3 = __________ m^3
   e. 2.5 gal = __________ liters
   f. 8.4 yd = __________ decimeters
   g. 0.735 slug = __________ grams
   h. 245 yd/min = __________ cm/h
6. Energy is measured in force times distance. Refer to Tables 1.2 and 1.3 and determine the conversion factor between ft-lb and joules.
7. 45° Celsius is what temperature in degrees Fahrenheit?
8. Convert 87.6×10^8 ergs to ft-lb.
9. -62.4°F is what temperature in degrees Celsius?
10. At absolute zero temperature (that point at which molecular activity theoretically ceases), the following temperatures exist:

$$°C = -273.15$$

$$°F = -459.67$$

    As temperature increases, a point is reached in which both °C and °F are equal. What is this temperature point?
11. The basic equation for the linear distance transversed by a moving object is $s = vt$, where s = distance, v = velocity, and t = time. What is the distance traveled in meters in 0.15 second if an object travels 10 miles per minute?
12. A rectangular object covers 6.25 square feet. If one side is 0.85 meter long, what is the length of the other side in meters?

CHAPTER 2

ATOMS, AMPERES, AND VOLTS

2.1 STRUCTURE OF THE ATOM

During your study of electronics, you will frequently encounter three words which are extremely important in determining the characteristics exhibited by various circuit elements. These words are *conductors*, *semiconductors*, and *insulators*. The explanation as to what determines each of these properties is traceable to the smallest known masses, the particles that constitute the atom. This section will discuss some of the more basic concepts of the atom, and how they contribute to specific characteristics in electronic circuits. It is equally important that you understand the properties of the atom if you are to grasp the more fundamental concepts of the ampere and volt to be discussed later in this chapter.

Perhaps your earliest concept of the atom is a picture of a miniature solar system, analogous to the planets revolving about the sun. Within the atom, this is equivalent to the electrons revolving about the nucleus. In a similar manner, just as planets revolve in orbits that have different radii from the sun, so it is that electrons revolve in different orbits about their nucleus. Except for the simplest of all elements, hydrogen, the nucleus of an atom consists of *protons*, which are positive charged particles, and *neutrons*, which are neutral in charge. The number of protons and neutrons that constitute a given nucleus is determined by the atomic number and atomic weight of the element in question. *To retain electrical neutrality within the atom, electrons are negative in charge and equal in number to the number of positive protons in the nucleus.* Figure 2.1 illustrates this concept. For the first two elements, namely, hydrogen and helium, notice that the number of electrons is equal to the number of protons. This concept, of course, is true for all neutral atoms.

As just stated, there are two types of numbers associated with an atom. The first is its atomic number, or its position on the periodic table, and the second is the atomic weight of the atom. Appendix G contains the periodic chart of elements. This chart utilizes the standard chemical abbreviations of the elements. Appendix H has been added to assist in identifying the more formal element name associated with each chemical abbreviation. In addition to listing the elements in alphabetical order for ease of location, Appendix H also charts the electron shell distribution of each element. The importance of this distribution in determining the characteristics of elements will be discussed in subsequent paragraphs. Note that hydrogen is given atomic number 1. This is because it has one

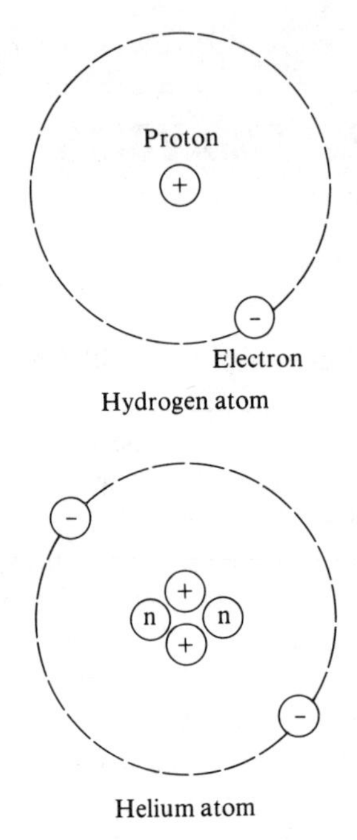

Figure 2.1 Atomic Structure of Hydrogen and Helium Atoms

positive charge in its nucleus, a proton, and not because it has one electron in its outer shell. As you will learn later, an ion (non-neutral atom) does not possess this characteristic, therefore atomic numbers refer to the number of protons in a nucleus. The element *helium* is given the atomic number 2 because it has two protons in its nucleus. This pattern, as it progresses through the periodic table, can be summarized as follows: *The atomic number of an atom indicates the number of protons in its nucleus.*

The second important number, *the atomic weight of an atom, indicates the total number of protons plus neutrons in the nucleus.* For example, the designation of the more common form of the ninety-second element (uranium) is U238. The U is the abbreviation for uranium and 238 indicates the atomic weight of that atom. Simple arithmetic then suggests the following: There are 92 protons and 92 electrons, and the difference between 238 and 92 results in 146 neutrons. Perhaps you have already heard of another rather famous atom of uranium, namely, U235. The fact that this is a uranium atom still suggests that there are 92 protons and 92 electrons; however, there are 235 minus 92 or 143 neutrons. U238 occurs in the natural state much more frequently than does U235. Hence, U235 is called an *isotope* of the uranium atom. Generally speaking, *an isotope is an atom which has the same number of protons (and electrons) as the more common element version, but it has a different number of neutrons.* Most elements have atoms which are slightly different in the number of neutrons which constitute their nuclei, and hence most elements do have isotopes.

There are some interesting relationships involving the masses of the three basic particles and their relative distances within the atom. These relationships are significant in that they contribute toward those properties which cause some elements to be conductors and others insulators. The following mass relationship exists between the three basic particles of any atom:

$$\text{Mass of neutron} = 1674.7 \times 10^{-30}\ \text{kg}$$
$$\text{Mass of proton} = 1674.4 \times 10^{-30}\ \text{kg}$$
$$\text{Mass of electron} = 0.9108 \times 10^{-30}\ \text{kg}$$

For all practical purposes, the mass of the neutron and the mass of the proton can be considered equal. But notice how many times greater this mass is than that of the electron.

$$\frac{\text{Mass of neutron}}{\text{Mass of electron}} = \frac{1674.7 \times 10^{-30}\ \text{kg}}{0.9108 \times 10^{-30}\ \text{kg}} = 1838.7$$

This number shows that the orbiting electron contributes very little to the mass of an atom.

Another interesting relationship of comparable significance is the fact that the radius of a proton is approximately equal to

$$1.2 \times 10^{-15}\ \text{m}$$

The very first orbiting shell, however, is at a distance of

$$0.529 \times 10^{-10}\ \text{m}$$

from the nucleus. The ratio of these two numbers produces the following result:

$$\frac{\text{Radius of first shell}}{\text{Radius of proton}} = \frac{0.529 \times 10^{-10}\ \text{m}}{1.2 \times 10^{-15}\ \text{m}} \approx 44{,}000$$

Considering these concepts, one conjures up the following incredible picture of an atom. It would be analogous (in mass) to a marble rapidly revolving around the nucleus consisting of a bowling ball, at a distance having a radius of about 10 km. It should be apparent, then, that virtually all the mobility within a crystal structure (an organized pattern of atomic alignments) lies with the orbiting electron, and that the cubic space occupied by an atom is almost totally void of mass.

It was stated earlier that the electrons revolving about an atomic nucleus have different orbits. Although each major orbit can be subdivided, for our analysis, we will think only in terms of the major orbits, as groups of electrons tend to revolve in relatively well defined radii from the nucleus. These orbits are designated *K*, *L*, *M*, *N*, *O*, *P*, and *Q*.

Figure 2.2 is a representation of the copper atom. It has atomic number 29; therefore, there are 29 positive proton charges in its nucleus and 29 negative electrons revolving in orbits about the nucleus. Notice the number of electrons in each orbit. If instead of letters, we assign numbers to the orbital shells, where $K = 1$, $L = 2$, etc., then an orbit "fills" according to the formula $2n^2$, where n is the number of the shell. In shell *K* (orbit 1) the total number of electrons is $2(1)^2 = 2$. In a like manner, the *L* or second orbital shell is filled when the number of electrons equals $2(2)^2 = 8$. The third orbital shell fills with $2(3)^2 = 18$ electrons. In the copper example, there is only one electron left to begin filling the fourth shell. This is because 28 electrons $(2 + 8 + 18)$ of copper's 29 electrons are used to fill the first three shells. (It is appropriate at this point to indicate that the concept of each progressively higher numbered element adding one more electron to a shell until it is filled is valid only for the first three shells, namely *K*, *L*, and *M*. Beginning with shell *N*, the interaction of atomic particles produces a somewhat more complex scheme for the placement of each additional electron. The significance of Appendix H to determine the number of electrons in an element's outermost shell should now be apparent.) For copper, the one singular electron re-

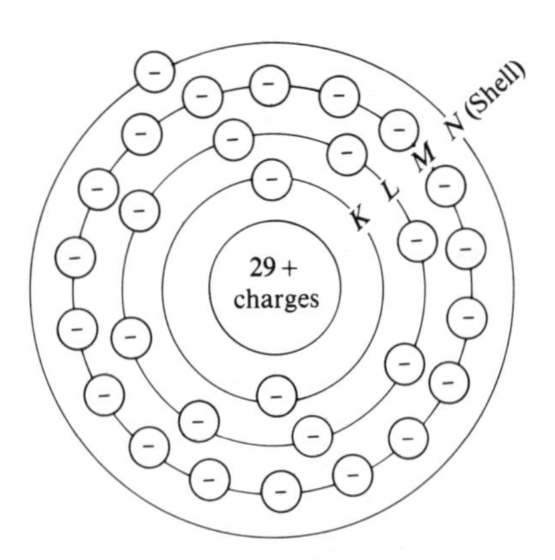

Figure 2.2 Copper Atom

maining in the outermost shell is what constitutes one favorable factor as to why it is a good conductor.

An examination of Appendix H will show that silver also has one electron in the *O* shell, and that gold has one electron in the *P* shell. As you may already know, all three elements are excellent conductors. This concept, however, will be explored in more detail in the next section. For now, let it suffice to summarize these important concepts:

1. The nuclei of atoms are rigidly locked within a given stable crystalline structure.
2. The extremely light and highly mobile electrons possess the ability to transfer charge.
3. The outermost shell of an atom, or more specifically, the number of electrons in that shell determines whether it is a conductor, semiconductor, or insulator. This shell is known as the *valence* shell. Generally speaking, a filled outer shell results in an inert element such as helium, neon, argon, krypton, etc. Refer again to the last column of the Periodic Table in Appendix G. It is when the outer shell is not filled that desirable characteristics can result.

2.2 CONDUCTORS, SEMICONDUCTORS, AND INSULATORS

In the previous section it was seen that atoms had valence electrons in orbits at different radii from the nucleus. It is also recalled that the orbiting electrons were negative in charge, whereas the nucleus possessed a net positive charge. The reader is already familiar with the expression "like charges repel, unlike charges attract." The positive protons in the nucleus of an atom exert an attractive force on the orbiting negative electrons. Coulomb's law expresses the magnitude of this attraction as follows:

$$F = \frac{KQ_1Q_2}{r^2} \tag{2.1}$$

F is measured in newtons.
K is a constant equal to 9×10^9.
Q_1, Q_2 is the charge in coulombs.
r is the distance in meters measured between the centers of the charges.

One coulomb of charge is defined as the charge represented by 6.24×10^{18} electrons; therefore, one electron possesses a quantity charge of $1/(6.24 \times 10^{18})$ or 1.6×10^{-19} coulombs.

Example 2.1 Calculate the force of attraction between the particles of a hydrogen atom.

$$Q_1 = Q_2 = 1.6 \times 10^{-19} \text{ coulomb}$$

$$r = 0.529 \times 10^{-10} \text{ m}$$

$$F = \frac{9 \times 10^9 \times 1.6 \times 10^{-19} \times 1.6 \times 10^{-19}}{(0.529 \times 10^{-10})^2}$$

$$= 82.33 \times 10^{-9} \text{ newton}$$

Example 2.2 A proton and an electron exert an attracting force of 125×10^{-9} newton. What is their separation?

$$|Q_1| = |Q_2| = 1.6 \times 10^{-19} \text{ coulomb}$$

$$F = 125 \times 10^{-9} \text{ newton}$$

$$r^2 = \frac{KQ_1Q_2}{F} = \frac{K|Q|^2}{F}$$

$$r = Q\sqrt{\frac{K}{F}} = 1.6 \times 10^{-19}\sqrt{\frac{9 \times 10^9}{125 \times 10^{-9}}}$$

$$= 1.6 \times 10^{-19} \times 2.68 \times 10^8$$

$$= 4.29 \times 10^{-11} \text{ meter}$$

The valence electrons, by virtue of their distance from the nucleus, are at the highest energy level within the atom. If the valence electrons can receive enough external energy, they can overcome this force of attraction and break free of the parent atom. Figure 2.3 depicts an example

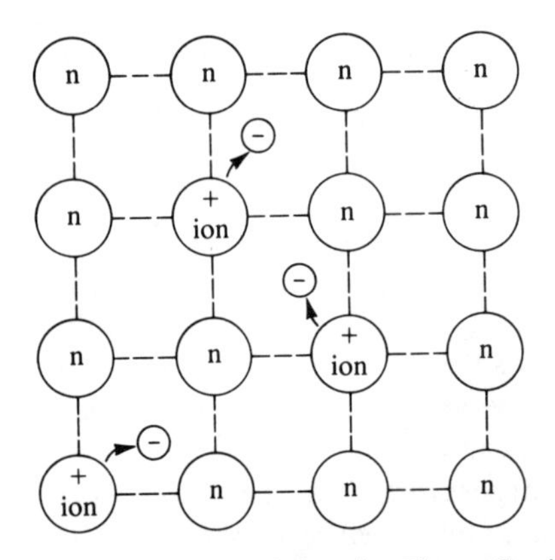

Figure 2.3 Example of Negative Charge Carriers

of free electrons moving within the crystalline structure of several atoms. The atom from which an electron is lost now has one less negative charge than it had in the neutral state. This atom is now a positive ion, and as such, could attract another electron moving too closely to it through the crystalline structure. There results then, in a crystalline structure, a situation in which some atoms are losing valence electrons which drift throughout the structure, and other positive ions are capturing electrons which had previously broken free of their parent atom. A constant exchange between atoms is always occurring. Some atoms are losing their valence electrons and other positive ions are recapturing these electrons. At any given time, however, there is a net statistical number of free electrons unattached to a parent atom.

In the atoms of copper, silver, and gold, the one valence electron farthest removed from the nucleus is attracted with the least amount of force. Additionally, this electron is at the highest energy level. The combination of being both the most weakly held and at the same time possessing the highest energy level gives such elements properties such that they become excellent conductors. For the above mentioned elements, conditions are so ideal that virtually every atom in the crystal structure contributes its valence electron to become a charge carrier. To give some idea of the availability of the charge carriers in copper, consider that one cubic centimeter contains approximately 8.4×10^{22} atoms, and hence virtually the same number of free electrons. This number can be determined by the following equation:

$$n = \frac{D \times N_0}{M} \qquad (2.2)$$

n = number of atoms/cm^3
D is the density of one cubic centimeter Cu = 8.9 grams
N_0 = Avogadro's constant = 6.02×10^{23} atoms/mole
M is the atomic weight of Cu = 63.5 grams/cm^3

Therefore,

$$n = \frac{8.9 \times 6.02 \times 10^{23}}{63.5} = 8.4 \times 10^{22} \text{ atoms/cm}^3$$

It can be concluded that good conductors readily give up their valence electron under normal conditions. The usual source of energy to free a valence electron is room temperature. If the theoretical value of absolute zero temperature is reached ($-273°$C), no electron has sufficient energy to become a charge carrier.

The foregoing discussion suggests that crystalline structures which have a strong attraction to their valence electrons cannot provide enough charge carriers to become good conductors. This is exactly the case, and such materials are referred to as *insulators*. It would take so much energy to free the valence electrons of an insulator that probable deformation and/or destruction of the material would occur before electrons were freed. The most commonly encountered insulators are rubber (which frequently encases electrical cords and house wiring) and glass or mica used as standoffs on high power lines.

Unquestionably, one of the most interesting developments of modern day electronics is the semiconductor. Recall that in Section 2.1 it was stated that some elements were extremely stable. One was helium, with a filled K shell of two electrons. All other inert elements contain eight electrons in their valence shell: the L shell for neon, the M shell for argon, the N shell for krypton, etc. The laws describing atomic physics seem to provide a high degree of stability to any element that can achieve eight electrons in its valence shell. The Periodic Table and Appendix H both reveal that there are two elements ideally suited to be semiconductors: silicon (Si) of atomic number 14 and germanium (Ge) of atomic number 32. Simple arithmetic shows that there are four valence electrons in the M shell of Si and four valence electrons in the N shell of Ge. While this in itself does not meet the highly stable requirement of eight valence electrons, the same effect is achieved if a pure crystalline structure of either element is produced. Figure 2.4 shows that, through a method

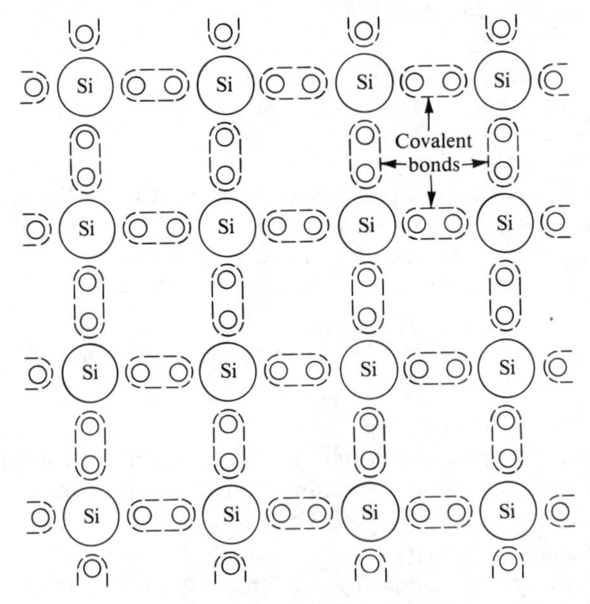

Figure 2.4 Covalent Bonding of Silicon (or Germanium) Crystal

called *covalent bonding*, each atom not only has its own four electrons, but effectively shares an electron with each of its neighbors.

Since the elements of Ge and Si "effectively" have eight electrons, a crystalline structure composed of these elements would be an excellent insulator, since nature does not readily break up a combination equivalent to eight valence electrons. Through manufacturing techniques, however, properties of this material can be altered by *doping* the pure crystalline structure with impurities. If arsenic (As), which has atomic number 33, is introduced into the crystalline structure, one charge carrier is freed. This is explained as follows. The *N* shell of arsenic has five electrons. Since only four electrons are used in the process of covalent bonding, the unneeded electron becomes loosely held by the arsenic nucleus and readily breaks free to drift through the crystalline structure. The significance of this is that during the manufacturing process, the degree of conductivity desired can be artificially manufactured by controlling the concentration of impurities. Figure 2.5(a) illustrates this point in a silicon crystal. In this illustration, the one impurity atom (arsenic) contributes one electron carrier.

To give some idea of the doping concentration, it is usually on the order of one impurity atom per one million silicon atoms. It is also possible to introduce an impurity called gallium (Ga) of atomic number 31. The same effect, but opposite charge carrier, is produced as was the case for arsenic. Since gallium has only three valence electrons, the covalent bonding phenomenon results in seven electrons and the appearance of a *hole*. The hole in reality is actually the absence of the desired eighth electron. Figure 2.5(b) illustrates the hole concept. Since any electron drifting through the structure is readily captured by the hole, the hole produces the same effect as though it were a positive charge.

In summary, then, nature has provided that elements like copper, silver or gold easily relinquish their valence electron to become good conductors. If materials can be produced which have a strong bonding of their valence electrons, such as rubber or mica, good insulators result. The semiconductor is the resultant of a controlled manufacturing process which will produce a charge carrier for each impurity atom introduced into the otherwise pure crystalline structure. It possesses properties of conductivity between those of conductors and insulators.

2.3 THE AMPERE

It has been shown that a good conductor like copper has a significant number of free electrons available to drift through the crystalline structure. Barring the application of any external forces, there is no organized

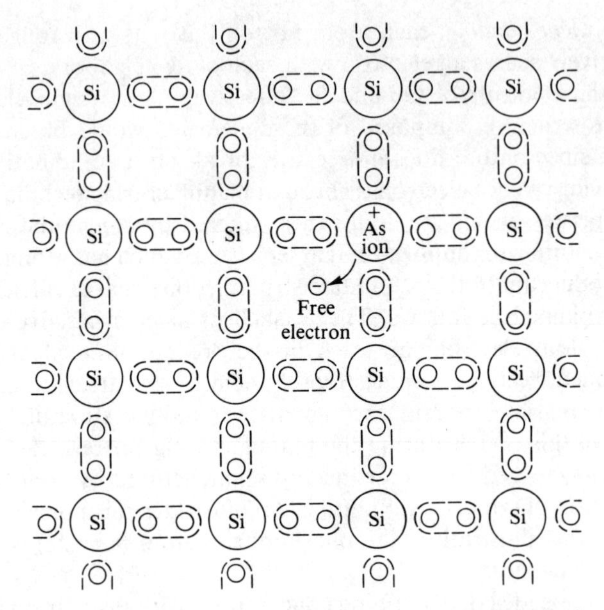

Figure 2.5(a) Doping of Semiconductor Material with Arsenic

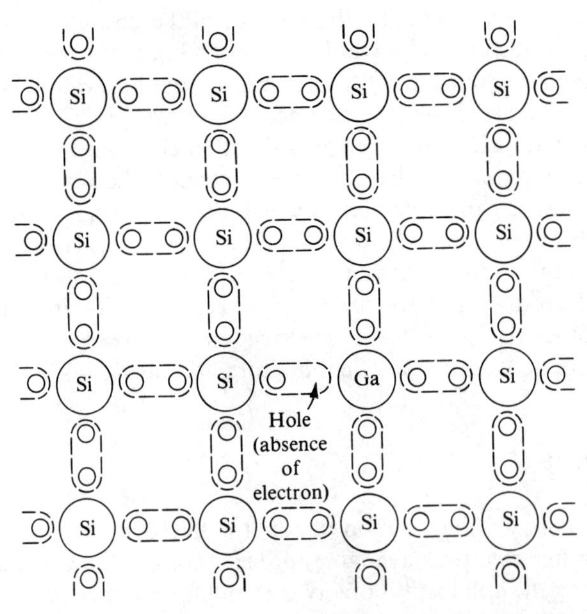

Figure 2.5(b) Doping of Semiconductor Material with Gallium

direction to the drift of these electrons. In short, there is as much likelihood that a given number of electrons might be moving in a general upward direction as in a general downward direction; or conversely, there is as much likelihood that atoms will be drifting in a general direction to the right as to the left. Figure 2.6 depicts this random movement of negative charges in a piece of copper wire.

Now if an external force is applied to this copper, such as the source of voltage illustrated in Figure 2.7, then a general direction of movement results among the free electrons. In this figure, there is a general drift away from the negative terminal and toward the positive terminal of the external source. Once again, this is in keeping with the concept of repulsion between like charges and attraction between the unlike charges. In this latter situation, there is a net migration of charge as compared to the net cancellation of charge movement illustrated in Figure 2.6.

A convenient unit has been adopted to measure the magnitude of this drift. If a hypothetical cross-sectional cut were made of the copper wire, and the number of electrons drifting past this cross section in 1 second amounted to 1 coulomb of charge, then this would be called 1 ampere of current. (See Figure 2.8.)

$$\boxed{1 \text{ ampere} = \frac{1 \text{ coulomb}}{1 \text{ second}}} \tag{2.3}$$

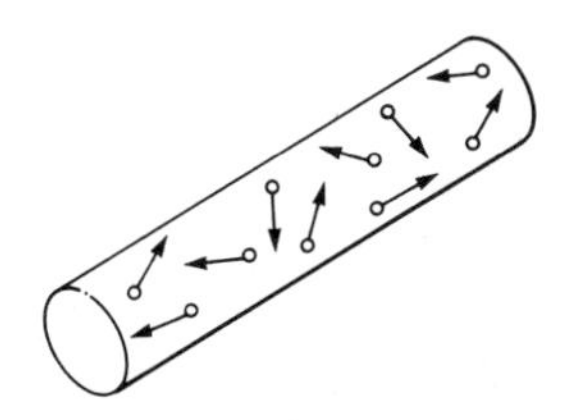

Figure 2.6 Random Motion of Free Negative Charge Carriers

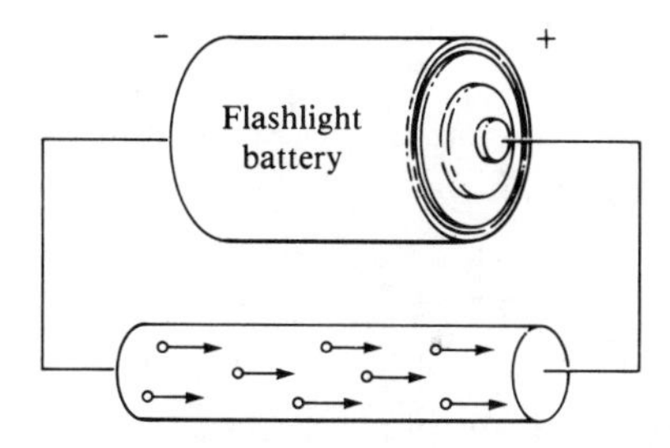

Figure 2.7 Directed Motion of Free Negative Charge Carriers

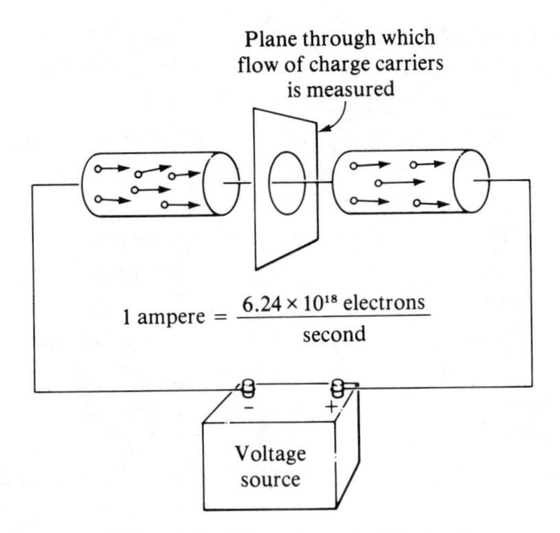

Figure 2.8 Measurement of Current

In its most fundamental form, 1 ampere could be expressed as 6.24×10^{18} electrons drifting past this cross-sectional area per second. This rather large number of electrons results from the fact that if each electron has a charge of 1.6×10^{-19} coulomb, then it would take 6.24 billion billion (6.24×10^{18}) electrons to add up to 1 coulomb of charge.

In summary, an ampere is a measurement of the migration of a given number of electrons and/or charge passing a given point per second. The letter I has been reserved by the electronics industry to represent current.

Example 2.3 How much current is represented by 3.12×10^{18} electrons moving past a given point in 1 second?

$$\frac{3.12 \times 10^{18} \text{ electrons}}{6.24 \times 10^{18} \text{ electrons}} = \frac{1}{2} \text{ coulomb}$$

$$I = \frac{\frac{1}{2} \text{ coulomb}}{1 \text{ second}} = \frac{1}{2} \text{ ampere}$$

Example 2.4 What is the current if $\frac{1}{4}$ coulomb of charge moves past a given point in 5 seconds?

$$I = \frac{\frac{1}{4} \text{ coulomb}}{5 \text{ seconds}} = \frac{1}{20} \text{ ampere} = 0.05 \text{ ampere}$$

Example 2.5 How much current is represented by 1.56×10^{15} electrons moving past a given point in 2 seconds?

$$\frac{1.56 \times 10^{15} \text{ electrons}}{6.24 \times 10^{18} \text{ electrons}} = 0.25 \times 10^{-3} \text{ coulomb}$$

$$I = \frac{0.25 \times 10^{-3} \text{ coulomb}}{2 \text{ seconds}} = 0.125 \text{ mA}$$

It should be kept in mind that although there is a net directional drift to the charge carriers when an external force is applied, the velocity of this movement is extremely slow. As the electrons are moving through a piece of copper wire, constant collision with other atoms in the crystal structure tends to impede their forward motion. In a typical size house wire, it would take several minutes for an electron to transverse a length of 1 meter. Even though the electron drift itself is relatively slow, the reader should not confuse this speed with the velocity with which the electric field effect transverses the wire. The latter moves with nearly the speed of light. It is analogous to applying pressure to a hose already full of water. Although the water itself may leave the end of the hose relatively slowly, the transmission of the water pressure through the hose is much more rapid.

2.4 VOLTAGE

Reference has been made to input pressure in a water hose pushing all the molecules of water through the hose, and of course, supplying the input with more water. By the same analogy, a form of electrical pressure must be established if electrons are to be moved through the copper wire. The same pressure source must also supply electrons to the input of the circuit. The unit of measurement given to this "electrical pump" is *voltage*. This unit is heard almost daily; probably the most common exposure is the 12-volt battery on most cars. The magnitude of the voltage is representative of the source's ability to do work. Work is calculated by multiplying the magnitude of force applied over the distance of the application. For example, to lift a heavy object and move it any distance certainly requires work.

Visualize a stationary negative charge and another movable negative charge, such as an electron. Start at a point where the force of repulsion between the two charges is negligible; for example, at a point an infinite distance away. Moving the electron toward the negative charge would require the application of force to overcome this repulsion. Suppose the electron is moved to point A as in Figure 2.9. At point A, the electron possesses potential energy, because if released, it would move away from the fixed charge by a force equal to a magnitude as calculated by Coulomb's law. If now the electron is moved to point B, closer to the

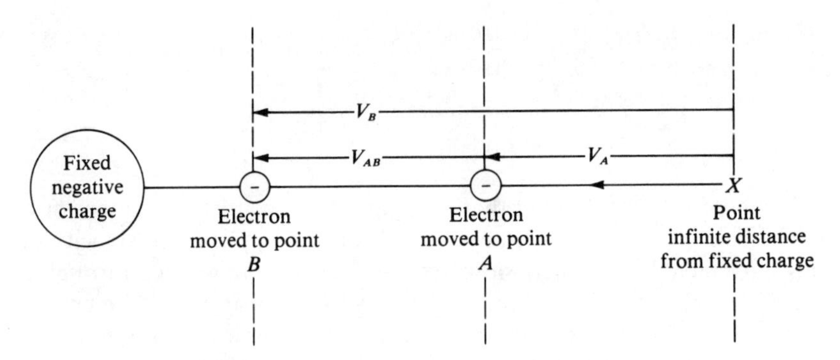

Figure 2.9 Voltage Difference between Two Points

fixed charge, it possesses an even greater potential. If released, it would move away from the fixed charge with a greater force than it possessed at point A. The difference in potential between points A and B is a measure of the voltage pressure that can move this electron. If it had required 1 joule of work to move 1 coulomb of charge from point A to point B, then the voltage difference would have been 1 volt.

$$V_B - V_A = V_{AB} \text{ (volts)} = \frac{\text{work (joules)}}{\text{charge (coulombs)}} \tag{2.4}$$

Example 2.6 What is the voltage difference between two points if ¼ joule of work is expended to move a charge of ½ coulomb?

$$V = \frac{\frac{1}{4}\text{ joule}}{\frac{1}{2}\text{ coulomb}} = \frac{1}{2}\text{ volt}$$

Example 2.7 A 12-volt car battery supplies 200 amperes for 1 second to start the motor. How much work is expended?

$$\begin{aligned} 200\text{ amperes} &= 200\text{ coulombs/second} \\ W = VQ &= 12\text{ volts} \times 200\text{ coulombs} \\ &= 2400\text{ joules} \end{aligned}$$

There are essentially two forms that voltage can take: *alternating* (variable) and *direct* (constant). The former version will be discussed in later chapters. For now, discussion will be restricted to direct or battery supplied voltages. A battery has two terminals: one marked *positive* and the other marked *negative*. The determination of each terminal's polarity is the result of a chemical reaction. The metals that constitute the positive terminal of a battery react chemically with the electrolyte solution in the battery in such a way that valence electrons are stripped from the crystal structure of the terminal. The absence of these electrons, which would otherwise constitute an equilibrium force to an atom's protons, leaves the terminal in a net positive state. It can be concluded, therefore, that being positive is the result of an atom having lost its equalizing negative valence electrons. Recall that this is the definition of an ion. Conversely, the negative terminal chemically reacts with the battery's electrolytic solution in such a way that *holes* in the terminal's valence structure are filled with *freed* electrons, resulting in a net negative charge for that terminal. There exists between these two terminals an imbalance. One terminal has too many electrons (negative) and the other terminal has too few electrons (positive). The degree of this disparity is a measure of the battery's voltage rating, and hence its ability to do work.

From a practical point, to extract energy from a battery it will be necessary to insert a *load* (a device which receives the battery's energy) between the positive and negative terminals of the battery. A bridge is now completed, such that the excess electrons on the negative terminal have a path through which they may travel to the positive terminal (which is deficient of electrons). The electrons set into motion between the negative and positive terminals by the *electric pump* must also pass through the inserted load. As the positive terminal receives electrons from the negative terminal, the chemical reactions within the battery continue to "steal" these electrons back from the positive terminal and pass them over to the negative terminal through the electrolytic solution. What results, then, is a chemical reaction to establish a disparity of charge. Copper wire (when connecting a load between the terminals) provides a bridge to alleviate this disparity.

Figure 2.10 illustrates a car battery which, when connected into the ignition system via the ignition switch, can provide sufficient electrons to work at "turning over" the motor and starting the car. This concept of a voltage source (battery) supplying current (electrons) to a load (car's electrical system) will be more thoroughly investigated in later sections. The reader must first understand the remaining missing link; the concept of *resistance*, which is to be investigated in the next chapter.

SUMMARY

Electrical neutrality exists within an atom when the number of protons (positive charges within the nucleus) equals the number of electrons (orbiting negative charges). Neutrons, also a part of an atom's nucleus, do not possess an electrical charge, but do contribute to the total atomic weight (mass) of an atom. The total atomic weight of an atom is the sum

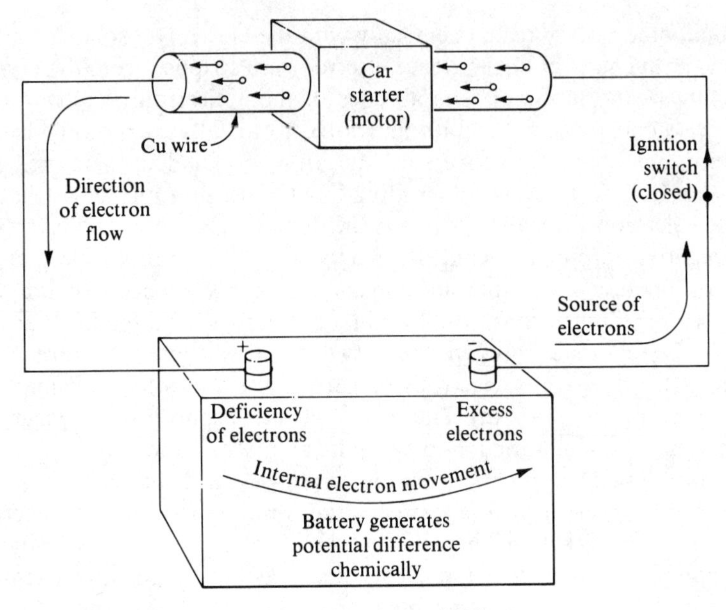

Figure 2.10 Current and Voltage

of the neutrons plus protons in its nucleus. The atomic number of an atom is equal to the number of protons in the atom's nucleus. An isotope of an element has the same number of protons but a different number of neutrons than the more commonly encountered atom.

The nucleus of an atom constitutes virtually all its mass. Nuclei of atoms are rigidly locked within a crystalline structure. Valence electrons contribute to a material's characteristics, such as whether it is suitable as a conductor, semiconductor, or insulator. A conductor easily gives up its valence electrons to become charge carriers. An insulator does not readily release valence electrons to contribute to charge migration. Semiconductors display characteristics between those of conductors and insulators.

Coulomb's law calculates the electrical force between charges. It is directly proportional to the charge magnitudes and inversely proportional to the square of their separation.

The unit of measurement of current is the ampere. It represents a charge migration of 1 coulomb past a given point in 1 second. The magnitude of the electrical pressure that causes charge migration is measured in volts. It would require 1 joule of work to move 1 coulomb of charge through a potential difference of 1 volt.

PROBLEMS

Reference Section 2.1

1. Arsenic is used as a dopant in semiconductors. It has atomic number 33 and atomic weight 75.
   a. How many electrons, protons, and neutrons does it contain?
   b. Identify the shell and the number of electrons in each shell.
2. Cobalt is atomic number 27 and atomic weight 59. What is the mass of the nucleus? What percentage of the total mass of this atom is constituted in the nucleus?
3. An isotope of hydrogen is deuterium of atomic weight 2. Identify the particles and calculate the mass.
4. What percentage of the mass of a hydrogen atom is contained in its nucleus?
5. What is the mass of a silicon atom of atomic weight 28?

Reference Section 2.2

6. Calculate the force of repulsion in newtons between two protons separated by 10^{-10} cm.
7. What is the distance between a proton and electron if their attractive force is 15×10^{-9} newton?
8. A charge of 6.5×10^{-18} coulomb is separated from another charge by 10^{-8} meter. If the force between them is 10^{-6} dyne, what is the magnitude of the second charge?
9. A copper wire is 10 meters long and has a diameter of 0.35 mm. How many free electrons are in the wire?
10. A copper wire is 20 feet long and contains 50.5×10^{22} free electrons. What is the radius of the wire in centimeters?
11. The density of silver (Ag) is 10.5 grams/cm^3. The atomic weight of silver is 107.87 gm/cm^3. How many free electrons are there in 1 cubic centimeter of Ag?

Reference Section 2.3

12. What current is represented by 9.32×10^{17} electrons flowing past a point in 0.682 second?
13. How many electrons per second are represented by a current flow of 2.1 amperes?
14. How much time does it take a current flow of 1.6 amperes to transfer 0.125 coulomb of charge?
15. How many electrons are represented if a total current of 18.75 mA moves past a point in 2.5 seconds?

Reference Section 2.4

16. How much work must be expended to move 0.75 coulomb through a potential difference of 20 volts?
17. Moving a negative charge through a potential difference of 20 volts required the expenditure of 0.25 joule of work. How many electrons does this charge represent?
18. A 12-volt car battery supplies 2750 joules to start the motor. How much charge is supplied by the battery? How many electrons does this represent?
19. The work done to move 0.56 coulomb was 2.1 joules. What is the voltage potential difference between the starting and ending points of the move?
20. A force of 3.75 newtons was applied to move 1.9 coulombs through a voltage difference of 5 volts. How many centimeters was the charge displaced?

CHAPTER 3

RESISTANCE AND CONDUCTANCE

3.1 RESISTANCE AND RESISTIVITY; CONDUCTANCE AND CONDUCTIVITY

The fundamental principle upon which electronic circuit analysis is built is the transferance of energy from a source (such as a battery) to a device (load). This transferance is via charge migration: from the source, through the device, and back to the source again. The last chapter dealt with charge migration (current measured in amperes) and energy potential (voltage). This chapter will discuss the remaining factors which determine the magnitude of the current through a device. It is essential for the reader to thoroughly understand these concepts if he is to form a foundation for subsequent, more complex circuit analysis.

Suppose a circuit is constructed as in Figure 3.1. It consists of a battery and two rods, one of copper and one of carbon, both of which are connected in *parallel* across the battery terminals. Let it further be supposed that the physical dimensions of the rods are absolutely identical, that is, the length and cross-sectional area and temperature of both rods are identical. Connecting these rods across the 12-volt battery produces a current through each rod. If a measuring device records the currents I_1 and I_2, it would be found that I_1 is significantly greater than I_2. A question then arises: All other conditions being equal, why is there a difference in the currents? The only obvious answer is that the materials of which the rods are made are different. The atomic structure of each rod is different, and the number of charge carriers that each rod can contribute to the flow of current is different.

Within both rods, the number of free carriers and the frequency of the collisions of these carriers with the atoms locked within the crystal are different. Every element or alloy represents some opposition to the net migration of charge carriers, and hence represents an opposition to the flow of current through the material comprising the rod. Based upon their particular chemical composition, all materials exhibit this property of *resistivity*. This characteristic, for a fixed temperature, is wholly dependent upon the material. For most practical purposes, resistivity is a measurement taken at room temperature, considered to be 20° Celsius.

To explain the concept of resistivity, and how it relates to a conductor's opposition to current flow, an analogy will be drawn between an electrical circuit and a hydraulic circuit. Assume a conductor is connected between the terminals of a voltage source, and a water line is connected

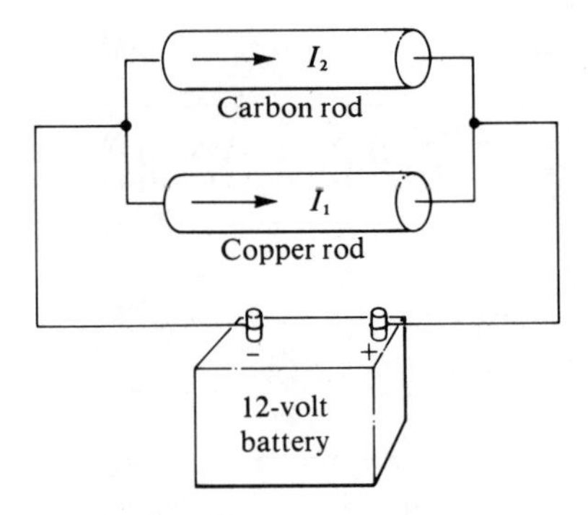

Figure 3.1 Resistivity

to a water pump. If a water pump of a given force is suddenly required to pump water through a line which has doubled in length, then the water pressure per unit length of the line must proportionately decrease. Otherwise, to maintain the same pressure would have required a pump having twice the capacity of its predecessor. What this is saying, then, is that the pressure distributed along the water line is proportionate to the capacity of the pump and inversely proportionate to the length of the water line. By a similar analogy in the electrical circuit, if we represent the voltage by the symbol V, the length of the conductor by ℓ, and the electric field strength distributed along the line by $\mathscr{E}$ (epsilon), then

$$\mathscr{E} = \frac{V}{\ell} \tag{3.1}$$

Another concept to consider within an electrical circuit is that of density. Within the two rods of Figure 3.1, the copper rod contributes more charge carriers than the carbon rod. Since both rods are the same size, the density of the carriers in the copper rod is greater. It would take a much greater cross-sectional area for the carbon rod to contribute as many carriers to the flow of current as the copper rod. Expressed in an equation:

$$J = \frac{I}{A} \tag{3.2}$$

where J is the current density, I is the current, and A is the cross-sectional area.

For a given material, resistivity, symbolized by the Greek letter ρ, is a measurement of the ratio of the intensity of the electric field to the density of the current, or mathematically:

$$\rho = \frac{\mathscr{E}}{J} \tag{3.3}$$

It should be noted that ρ is a characteristic of the material and is independent of the physical dimensions of the material. To assure this independence of the dimensions, ρ is calculated in a rather unusual unit of size, namely, ohms per meter cube. Note that the word cube, as used in this measurement, is not an exponent, but rather implies the shape of the element under which the measurement is taken. In short, if a block of the material for which the measurement was made was 1 meter on a side, then it would not matter across which dimension of parallel surfaces the measurements were taken; the resistivity would be a constant. As a result, ρ is sometimes called *specific resistance*, and for a given temperature, is a constant of the material being measured. Although ρ is a constant for a material, and is therefore independent of the dimensions of the material, tables are frequently constructed in which values of ρ are more applicable to the practical dimensions of current carriers. Therefore, for most practical conductors, resistivity is measured in ohms per meter length, times the cross-sectional area of the conductor measured in millimeters squared.

$$\rho(\text{units}) = \frac{\text{ohm} \cdot \text{mm}^2}{\text{m}}$$

Table 3.1 lists the values of the resistivity of several conductors. Observe that silver, copper, and gold exhibit the lowest opposition to current flow of all the elements listed. Table 3.2 is included to show the immense difference between the resistivity of conductors and that of insulators. Note that in Table 3.2 the unit of resistivity is the ohm-meter. To obtain a comparison of the ratio of resistivity of an insulator to a conductor, it will first be necessary to convert Table 3.2 into the same units of measurement as are utilized in Table 3.1.

TABLE 3.1 Resistivity of Various Conductive Elements (Metric System)

Conductive element	Resistivity at 20°C $\frac{\text{ohm} \cdot \text{mm}^2}{\text{m}}(10^{-2})$
Aluminum	2.7
Copper	1.72
Gold	2.21
Lead	20.7
Magnesium	4.6
Manganin	44.0
Mercury	95.8
Nickel	7.8
Platinum	10.8
Silver	1.59
Zinc	5.8

TABLE 3.2 Resistivity of Various Insulators

Insulators	Resistivity at 20°C ohm-m
Glass, crown type	4×10^7
Porcelain	3×10^{12}
Rubber	2×10^{13}
Shellac, natural	1×10^{14}

Example 3.1 Convert the resistivity of porcelain in Table 3.2 to the (ohm-mm^2)/m unit as is used in Table 3.1.

From Table 3.2.

$$\rho(\text{porcelain}) = 3 \times 10^{12} \text{ ohm} \cdot \text{m} \times \frac{10^6 \text{ mm}^2}{\text{m}^2} = 3 \times 10^{18} \frac{\text{ohm} \cdot \text{mm}^2}{\text{m}}$$

Example 3.2 How many times greater is the resistivity of porcelain (an insulator) compared to the resistivity of copper (a conductor)?

From Example 3.1,

$$\rho(\text{porcelain}) = 3 \times 10^{18} \frac{\text{ohm} \cdot \text{mm}^2}{\text{m}}$$

From Table 3.1,

$$\rho(\text{copper}) = 1.72 \times 10^{-2} \frac{\text{ohm} \cdot \text{mm}^2}{\text{m}}$$

Therefore,

$$\frac{\rho(\text{porcelain})}{\rho(\text{copper})} = \frac{3 \times 10^{18}}{1.72 \times 10^{-2}} = 1.74 \times 10^{20}$$

In the preceding paragraphs, it was stated that both rods in Figure 3.1 were of the same length, cross-sectional area, and temperature. Common reasoning suggests that a material's total opposition to current flow must also be a function of its physical dimensions. This total opposition is called *resistance*. It is identified by R and is measured in units called *ohms*. The symbol for ohms is Ω (omega), which, for a fixed value of ρ, also takes into account a conductor's length and cross-sectional area. To tie all these terms together, rewrite Eq. (3.3) as follows:

$$\rho = \frac{\mathscr{E}}{J} = \frac{V/\ell \text{ (ref. Eq. 3.1)}}{I/A \text{ (ref. Eq. 3.2)}} = \frac{V}{I}\left(\frac{A}{\ell}\right) \tag{3.4}$$

In chapter 5, it will be shown that total resistance R is related to the voltage and current by the equation

$$R = \frac{V}{I} \tag{3.5}$$

Substituting Eq. (3.5) into Eq. (3.4) results in

$$\rho = \frac{RA}{\ell} \tag{3.6}$$

Solving for R;

$$\boxed{R = \frac{\rho \ell}{A}} \tag{3.7}$$

Total resistance, therefore, is directly proportional to the material used (ρ) and the length of the conductor (ℓ). It is inversely proportional to the cross-sectional area (A).

Example 3.3 What is the resistance of a copper wire whose length is 10 meters and whose cross-sectional area is 0.1 mm^2?

From Table 3.1,

$$\rho = 1.72 \times 10^{-2}$$
$$\ell = 10 \text{ meter}$$
$$A = 0.1 \text{ mm}^2$$

Therefore,

$$R = \frac{1.72 \times 10^{-2} \times 10}{0.1} = 1.72\ \Omega$$

Example 3.4 Assume a copper bar of 0.05 centimeter on a side. What would be the length of the bar if its resistance is 12.5 ohms?

$$0.05 \text{ cm} \times \frac{10 \text{ mm}}{\text{cm}} = 0.5 \text{ mm}$$

$$A = (0.5 \text{ mm})^2 = 0.25 \text{ mm}^2$$

$$\ell = \frac{RA}{\rho} = \frac{12.5 \times 0.25}{1.72 \times 10^{-2}} = 1.82 \times 10^2$$

$$= 182 \text{ m}$$

Example 3.5 What is the resistance of 1000 meters of aluminum wire having a diameter of 0.2 mm?

$$A = \pi r^2 = \frac{\pi d^2}{4} = \frac{3.1416 \times (0.2)^2}{4} = 3.1416 \times 10^{-2}$$

$$\rho = 2.7 \times 10^{-2}$$

$$R = \frac{2.7 \times 10^{-2} \times 10^3}{3.1416 \times 10^{-2}} = 0.86 \times 10^3 = 860 \text{ ohms}$$

It is sometimes desirable to know the *ease* rather than the *opposition* which a material presents to the flow of current. The unit that describes this characteristic is called *conductivity*. It is assigned the symbol sigma (σ) and is the opposite of resistivity. Its unit of measurement is siemens per meter (S/m), formerly the mho/meter. (Note that even the spelling of mho is the opposite of ohm.)

$$\sigma(\text{conductivity}) = \frac{1}{\rho(\text{resistivity})} \tag{3.8}$$

By the same analogy, the reciprocal of resistance is called *conductance*. It is designated by the letter G. The usefulness of the latter term will become more apparent in subsequent chapters which discuss parallel circuits.

$$G = \frac{1}{R} = \frac{1}{\rho\ell/A} = \frac{A}{\rho\ell} = \frac{1}{\rho}\frac{A}{\ell}$$

$$\boxed{G = \frac{\sigma A}{\ell}} \tag{3.8a}$$

Example 3.6 What is the conductance of magnesium wire of 1500 meters length and 0.025 centimeter diameter?

$$d = 0.025 \text{ cm} = 0.25 \text{ mm}$$

$$A = \frac{\pi d^2}{4} = \frac{\pi(0.0625)}{4} = 0.049 \text{ mm}^2$$

$$G = \frac{1}{R} = \frac{1}{\rho\ell/A} = \frac{A}{\rho\ell}$$

$$= \frac{0.049}{(4.6 \times 10^{-2}) \times (1.5 \times 10^3)} = 0.0071 \times 10^{-1} = 710\ \mu\text{S}$$

3.2 CIRCULAR MILS AND SQUARE MILS

In the previous section, the measurement of the resistance of a conductor was based on values of ρ, ℓ, and A as developed in the Metric system. Since the English system of measurements uses different units, it is to be expected that a table of resistivity of conductors in the English system would have different values than in the Metric system. Table 3.3 lists the English system of values of resistivity in units called ohm-CM/ft. The terms *ohm* and *ft* are familiar and have already been discussed. The

TABLE 3.3 Resistivity of Various Conductive Elements (English System)

Conductive elements	Resistivity at 20°C ohm-CM/ft
Aluminum	16.24
Copper	10.35
Gold	13.29
Lead	124.5
Magnesium	27.67
Mercury	576.3
Nickel	46.92
Platinum	64.97
Silver	9.56
Zinc	34.89

designation *CM** stands for circular mil, and is, as will be shown, a term created by definition. CM is defined as the area occupied by a circle 1 mil ($\frac{1}{1000}$ of an inch) in diameter. Do not confuse this as being 1 square mil. The difference between the two is illustrated in Figure 3.2. This figure clearly shows that the area of a square, 1 mil on a side, is greater than 1 CM. If all conductors were circular, this unit of resistivity would be very convenient. If, however, a conductor is noncircular, such as a copper bar, it becomes necessary to find a relationship between square mils (SM) and circular mils (CM) if Table 3.3 is to be utilized to compute resistance.

The units of CM are used in Table 3.3 because it is very easy to find A_{CM} for a conductor having a circular cross-section, as all electrical wire does. Since 1 CM is the cross-sectional area of a circle having diameter 1 mil, then a circle of diameter D would have D^2 times as much area. (The area of a circle, $A = \pi D^2/4$, is proportional to the square of its diameter.) For example, a circle having diameter 2 mils would have area $2^2(1) = 4$ CM. In short,

$$A_{CM} = D^2 \text{ CM} \tag{3.9}$$

where D is the diameter of the circle expressed in mils.

Mathematically, the actual area (A_{SM}) in square mils (SM) of a circle of diameter D equates as

$$A_{SM} = \frac{\pi}{4} D^2 \text{ SM} \tag{3.10}$$

*Take care to distinguish between upper- and lowercase letters to avoid confusion with centimeters.

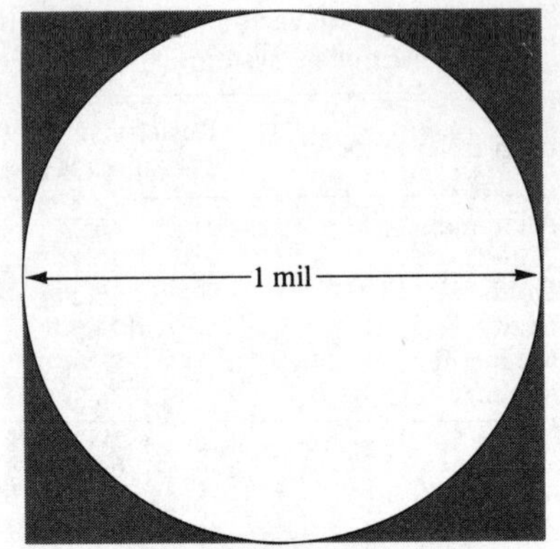

Shaded area is difference
between one square mil and
one circular mil

Figure 3.2 Circular Mil vs Square Mil

The ratio of A_{SM} to A_{CM} is

$$\frac{A_{SM}}{A_{CM}} = \frac{\frac{\pi}{4}D^2}{D^2} = \frac{\pi}{4}$$

$$\therefore \quad A_{CM} = \frac{A_{SM}}{\frac{\pi}{4}} = \left(\frac{4}{\pi} \times A_{SM}\right) \text{CM} \tag{3.11}$$

Equation (3.11) suggests the following: In the English system, so long as conductors are circular, Table 3.3 may be utilized to compute resistance wherein the area of a conductor is calculated to be the square of the diameter in mils. If conductors are not circular (such as rectangular or square), then the square area must be converted to CM before Table 3.3 is utilized. The following examples will illustrate this point.

Example 3.7 What is the resistance of copper wire 4 mils in diameter and 1 mile long?

From Table 3.3,

$$\rho(\text{Cu}) = 10.35 \text{ ohm-CM/ft}$$

$$A_{CM} = 4^2 = 16 \text{ CM}$$

$$R = \frac{\rho \ell}{A} = \frac{10.35 \times 5280}{16} = 3.4 \text{ k}\Omega$$

Example 3.8 A circular wire 100 yards long and 5 mils in diameter has a resistance of 114.7 ohms. What material is the wire made of?

$$A_{CM} = 25 \text{ CM}, 100 \text{ yd} = 300 \text{ ft}$$

$$\therefore \quad \rho = \frac{RA}{\ell} = \frac{114.7 \times 25}{300} = 9.56 \text{ ohm-CM/ft}$$

From Table 3.3, the material is silver.

Example 3.9 A copper bus bar is 3 ft. long. Its cross sectional area is rectangular in shape and has dimensions of $\frac{1}{4}$ inch by 3 inches. What is its resistance?

$$A_{SM} = 250 \text{ mils} \times 3000 \text{ mils} = 75 \times 10^4 \text{ mils}^2$$

$$A_{CM} = \frac{4}{\pi} \times 75 \times 10^4 = 95.49 \times 10^4 \text{ CM}$$

$$R = \frac{\rho\ell}{A_{CM}} = \frac{10.35 \times 3}{94.49 \times 10^4} = 0.0000329 \ \Omega$$

3.3 INFERRED ZERO TEMPERATURE AND TEMPERATURE COEFFICIENT

Sections 3.1 and 3.2 made frequent reference to room temperature when taking resistance measurements and when speaking of the resistivity of a material. Although there are a few noted exceptions like carbon, the resistance of a conductor increases with increased temperature. In the basic equation of resistance,

$$R = \frac{\rho\ell}{A}$$

increases in temperature do not appreciably affect the length or area of a conductor. Even if it did, there would tend to be an offsetting effect, since length is in the numerator and area is in the denominator, and thus the ratio would tend to remain reasonably constant. If resistance increases with temperature (again note that there are a few exceptions), it must be because resistivity increases with temperature. Expressed mathematically,

$$\rho = f(T) \tag{3.12}$$

that is, resistivity is a function of temperature.

To understand this relationship, consider the graph of Figure 3.3. Suppose it is determined that a conductor of copper has a resistance of R_1 ohms at a temperature T_1. If the temperature is raised to T_2 degrees, the resistance of the copper conductor increases to R_2 ohms. Between a rather extended range of degrees, T_3 through T_4, this increase in resistance vs temperature is linear. If discussion is restricted to within this range (which is a practical limitation of temperatures), the mathematical discussion is greatly simplified.

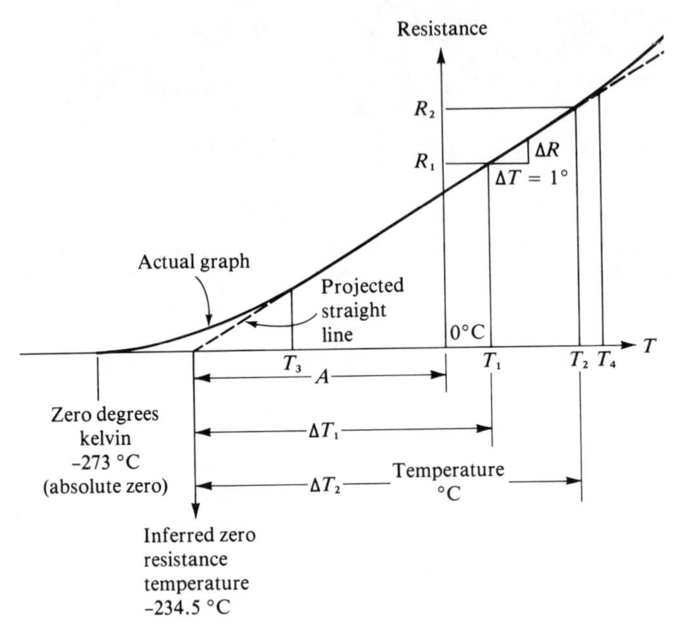

Figure 3.3 Resistance vs Temperature for Annealed Copper

The straight line portion of the graph, extended until it intersects the temperature axis, is called the *inferred zero resistance point.* All conductors and/or materials have a unique inferred zero resistance point. Table 3.4 lists several such values for commonly encountered materials. Notice that some materials such as nichrome, constantan, and manganin, or some elements such as mercury, have inferred zero resistance points lower in temperature than absolute zero. This merely means that the

TABLE 3.4 Inferred Zero Resistance Temperature Intercept Points

Material	Inferred zero resistance temperature (−°C)
Aluminum	236
Brass	480
Carbon	Negative slope, no −°C intersection
Copper	234.5
Constantan	125,000
Gold	274.0
Iron	180.0
Lead	224.0
Manganin	94,000
Mercury	1100
Nichrome	2480
Nickel	147.0
Silver	243.0
Tin	218.0
Tungsten	204.0

slope of the linear portion of their graph is very flat; and if this linear portion were extended until it intersected the temperature axis, then the point of intersection would be as listed in Table 3.4. In actual practice, materials exhibit zero resistance at a temperature of absolute zero. Figure 3.4, representing an actual graph for mercury, shows how abruptly these changes can take place. The important point to remember, however, is that within normally encountered temperature ranges, the straight line approach is highly accurate.

From Figure 3.3, $\Delta T_1 = 234.5 + T_1$ and $\Delta T_2 = 234.5 + T_2$. By similar triangles,

$$\frac{\Delta T_1}{R_1} = \frac{\Delta T_2}{R_2} \tag{3.13}$$

Substituting the equivalent values of ΔT_1 and ΔT_2 into Eq. (3.13) results in

$$\frac{234.5 + T_1}{R_1} = \frac{234.5 + T_2}{R_2} \tag{3.14}$$

Solving for R_2,

$$R_2 = R_1 \left(\frac{234.5 + T_2}{234.5 + T_1} \right) \tag{3.15}$$

This equation is good only for annealed copper, because of the constant 234.5. In general, if the distance from the inferred zero point to the origin is called A, then the resistance (R_2) at a different temperature (T_2) is related to the reference resistance (R_1) at the reference temperature (T_1) by the expression

$$\boxed{R_2 = R_1 \left(\frac{A + T_2}{A + T_1} \right)} \tag{3.16}$$

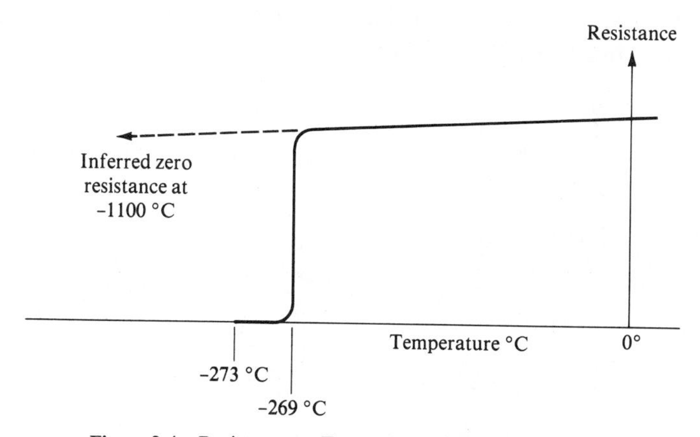

Figure 3.4 Resistance vs Temperature Graph for Mercury

Note that A is really nothing more than the inferred zero temperature of the material in question.

Example 3.10 The resistance of an annealed copper conductor is 25 ohms at 20°C. What is its resistance at 70°C?

Using Eq. (3.16),

$$R_1 = 25\ \Omega$$
$$T_1 = 20°\text{C}$$
$$T_2 = 70°\text{C}$$
$$R_2 = 25\left(\frac{234.5 + 70}{234.5 + 20}\right)$$
$$= 25\left(\frac{304.5}{254.5}\right) = 29.91\ \Omega$$

Example 3.11 An annealed copper conductor has a resistance of 15 ohms at 35°C. What is its resistance at −40°C?

$$R_1 = 15\ \Omega$$
$$T_1 = 35°\text{C}$$
$$T_2 = -40°\text{C}$$
$$R_2 = 15\left(\frac{234.5 + (-40)}{234.5 + 35}\right)$$
$$= 15\left(\frac{194.5}{269.5}\right) = 10.83\ \Omega$$

Example 3.12 A tungsten bar has a resistance of 8 ohms at 20°C. At what temperature will the resistance have increased 10%?

$$T_2 = \frac{R_2}{R_1}(A + T_1) - A$$
$$R_1 = 8\ \Omega$$
$$R_2 = 8.8\ \Omega = R_1 + 10\%$$
$$T_1 = 20°\text{C}$$
$$A = 204°\text{C}$$
$$T_2 = \frac{8.8}{8}(204 + 20) - 204 = 1.1(224) - 204$$
$$= 42.4°\text{C}$$

An alternate method exists to determine the resistance of a material at temperatures other than the reference temperature. Refer to temperature T_1 on Figure 3.3. Note that if a change in temperature (ΔT) takes place for a positive slope of the straight line, there will be a corresponding change in resistance (ΔR). There is defined a unit of measurement called *temperature coefficient of resistivity*, designated α (alpha). It measures the rate at which resistance changes with temperature, divided by the resistance of the material at the temperature at which the measurement is taken.

$$\alpha_T \equiv \frac{\Delta R}{\Delta T}\left(\frac{1}{R_T}\right) \qquad (3.17)$$

Since discussion is restricted to the straight line portion of the graphs, let ΔT equal a change of 1 degree as in Figure 3.3. Then,

$$\alpha_T = \frac{\Delta R}{R_T}$$

The student should note that α_T is not a constant, even for a given conductor. On a straight line graph, each ΔT of 1 degree will produce the same ΔR. The resistance at any given temperature, however, is changing; therefore α_T will always have the same numerator but a changing denominator. Given a conductor with a positive slope for resistance vs temperature, α_T is greater at lower temperatures because R_T is smaller and decreases with increasing temperature. A legitimate question at this point is, "What good is the temperature coefficient if it changes, even for the same conductor?" The answer is that a temperature coefficient table may be constructed for materials at the most normally encountered temperature, and all resistance variations referenced to this norm.

A typical room operating temperature is represented by 20°C. Having a reference point at α_{20} alleviates the need to know the inferred zero resistance temperature to calculate a new resistance due to temperature variations. Table 3.5 lists α_{20} for the same materials as in Table 3.4.

Rewrite Eq. (3.16) as follows:

$$\frac{R_2}{R_1} = \frac{(A + T_2) + (T_1 - T_1)}{A + T_1}$$

$$= \frac{A + T_1}{A + T_1} + \frac{T_2 - T_1}{A + T_1}$$

$$\frac{R_2}{R_1} = 1 + \frac{\Delta T}{A + T_1} = 1 + \left(\frac{1}{A + T_1}\right)\Delta T \qquad (3.18)$$

where ΔT is the difference in temperature ($T_2 - T_1$).

TABLE 3.5 Temperature Coefficients at 20°C

Material	Temperature coefficient α_{20} (10^{-3})
Aluminum	3.9
Brass	2.0
Carbon	−0.5
Copper	3.93
Constantan	0.008
Gold	3.4
Iron	5.0
Lead	4.1
Manganin	0.006
Mercury	0.89
Nichrome	0.4
Nickel	6.0
Silver	3.8
Tin	4.2
Tungsten	4.5

An alternate expression will now be found for the term $1/(A + T_1)$ in Eq. (3.18).

From Figure 3.3, similar triangles produce

$$\frac{A + T_1}{R_1} = \frac{A + T_2}{R_2} \tag{3.19}$$

Solve Eq. (3.19) for A,

$$A = \frac{R_1 T_2 - R_2 T_1}{R_2 - R_1} \tag{3.20}$$

Substituting Eq. (3.20) into the term $1/(A + T_1)$ results in

$$\frac{R_2 - R_1}{R_1(T_2 - T_1)} = \frac{\Delta R}{\Delta T}\left(\frac{1}{R_1}\right)$$

But this is the definition of α (Eq. 3.17). Therefore,

$$\frac{1}{A + T_1} = \alpha$$

Substituting this simplification into Eq. (3.18) results in

$$\frac{R_2}{R_1} = 1 + \alpha \Delta T$$

$$\boxed{R_2 = R_1(1 + \alpha \Delta T)} \tag{3.21}$$

The new resistance at a new temperature can be found knowing the resistance at 20°C and the α_{20} temperature coefficient.

Example 3.13 A copper conductor has a resistance of 12 ohms at 20°C. What is its resistance at 55°C?

$$\begin{aligned} \Delta T &= 55^\circ\text{C} - 20^\circ\text{C} = 35^\circ\text{C} \\ R_1 &= 12\ \Omega \\ \alpha_{20} &= 3.93 \times 10^{-3} \\ R_2 &= 12[1 + (3.93 \times 10^{-3}) \times 35] \\ &= 13.65\ \Omega \end{aligned}$$

Example 3.14 A conductor made of nickel has a resistance of 18 ohms at 20°C. What is its resistance at −30°C?

$$\begin{aligned} \Delta T &= -30^\circ\text{C} - 20^\circ\text{C} = -50^\circ\text{C} \\ R_1 &= 18\ \Omega \\ \alpha_{20} &= 6.0 \times 10^{-3} \\ R_2 &= 18[1 + (6.0 \times 10^{-3}) \times (-50)] \\ &= 12.6\ \Omega \end{aligned}$$

Example 3.15 An aluminum conductor has a certain resistance at 20°C. At what temperature would that resistance have increased 15%?

Let X = resistance at reference = R_1

Let $1.15X$ = original resistance + 15% = R_2

$$\begin{aligned} \frac{R_2}{R_1} &= 1 + \alpha\Delta T \\ 1.15 &= 1 + (3.9 \times 10^{-3})\Delta T \\ \Delta T &= \frac{0.15}{3.9 \times 10^{-3}} = \frac{150}{3.9} = 38.46^\circ\text{C} \\ T_2 &= 20^\circ\text{C} + \Delta T \\ &= 58.46^\circ\text{C} \end{aligned}$$

3.4 WIRE RESISTANCE

The most common wire shape is the standard circular conductor. To bring some standardization to the manufacture of conductors, wire sizes are specified by gauge numbers. Table 3.6 lists the standard American Wire Gauge (AWG) numbers, their diameter in mils, and their resistance at two different temperatures in both the Metric and English systems.

Notice that as the gauge number increases, the size of the wire decreases. Observe the resistance of the smallest wire (gauge #40) in either system of units and at either 25°C or 65°C. Counting up three gauge

TABLE 3.6 Resistance of Standard Annealed Copper Wire

Gauge number	Diameter, mils	Ohms per 1000 meters		Ohms per 1000 ft	
		25°C	65°C	25°C	65°C
0000	460	0.164	0.190	0.050	0.058
000	410	0.207	0.240	0.063	0.073
00	365	0.262	0.302	0.080	0.092
0	325	0.328	0.381	0.100	0.116
1	289	0.413	0.479	0.126	0.146
2	258	0.522	0.604	0.159	0.184
3	229	0.659	0.761	0.201	0.232
4	204	0.830	0.958	0.253	0.292
5	182	1.05	1.21	0.319	0.369
6	162	1.32	1.53	0.403	0.465
7	144	1.67	1.92	0.508	0.586
8	128	2.10	2.42	0.641	0.739
9	114	2.65	3.06	0.808	0.932
10	102	3.35	3.87	1.02	1.18
11	91	4.20	4.86	1.28	1.48
12	81	5.31	6.14	1.62	1.87
13	72	6.69	7.74	2.04	2.36
14	64	8.46	9.74	2.58	2.97
15	57	10.66	12.30	3.25	3.75
16	51	13.42	15.52	4.09	4.73
17	45	16.93	19.55	5.16	5.96
18	40	21.36	24.64	6.51	7.51
19	36	26.94	31.10	8.21	9.48
20	32	34.12	39.04	10.4	11.9
21	28.5	42.98	49.54	13.1	15.1
22	25.3	54.13	62.34	16.5	19.0
23	22.6	68.24	78.74	20.8	24.0
24	20.1	85.96	99.08	26.2	30.2
25	17.9	108.3	125.0	33.0	38.1
26	15.9	136.5	157.5	41.6	48.0
27	14.2	172.2	198.8	52.5	60.6
28	12.6	217.2	250.7	66.2	76.4
29	11.3	273.6	315.9	83.4	96.3
30	10.0	344.5	397.0	105	121
31	8.9	436.4	502.0	133	153
32	8.0	547.9	633.2	167	193
33	7.1	692.3	797.2	211	243
34	6.3	872.7	1007	266	307
35	5.6	1099	1270	335	387
36	5.0	1388	1601	423	488
37	4.5	1749	2018	533	615
38	4.0	2208	2546	673	776
39	3.5	2782	3212	848	979
40	3.1	3510	4035	1070	1230

numbers (to # 37), notice that the resistance is approximately one-half the value of gauge #40. This pattern repeats itself every three gauge numbers. For any given column, the measurements of resistance are for the same length and resistivity. The area of the wire, therefore, must double if resistance is to decrease by one-half. Calculating the area every three gauge numbers verifies this fact. The pattern in the wire tables, then, is that starting with gauge #40 and progressing toward gauge #0000, the resistance decreases by one-half every third number.

It should be observed that for any given wire size, in either the Metric or English systems, the resistance is higher at 65°C than at 25°C. This is to be expected, as copper has a positive temperature coefficient.

Example 3.16 What is the resistance of 700 meters of No. 20 annealed copper wire at 25°C?

From Table 3.6, R (1000 meters #20 at 25°C) = 34.12 ohms. Therefore,

$$R = \frac{700}{1000} \times 34.12\ \Omega = 23.88\ \Omega$$

Example 3.17 Determine the temperature coefficient of annealed copper wire at 65°C.

Choose any convenient wire size, such as AWG #18. From Table 3.6,

$$D = 40 \text{ mils}$$

$$\therefore \quad A_{CM} = D^2 = 1600 \text{ CM}$$

From Table 3.3,

$$\rho_{20} = 10.35$$

$$\therefore \quad R(1000 \text{ ft at } 20°\text{C}) = \frac{\rho \ell}{A_{CM}} = \frac{10.35 \times 10^3}{1.6 \times 10^3} = 6.47\ \Omega$$

Table 3.6 lists $R_{65°C}$ for 1000 ft of #18 annealed copper wire as 7.51 ohms. Therefore,

$$R_1 = 7.51\ \Omega, \quad T_1 = 65°\text{C}$$

$$R_2 = 6.47\ \Omega, \quad T_2 = 20°\text{C}$$

$$\Delta T = T_2 - T_1 = 20°\text{C} - 65°\text{C} = -45°\text{C}$$

From Eq. (3.21),

$$R = R_1(1 + \alpha \Delta T)$$

$$\alpha_{65} \Delta T = \frac{R_2}{R_1} - 1 = \frac{6.47}{7.51} - 1 = -0.138$$

$$\alpha_{65} = \frac{-0.138}{\Delta T} = \frac{-0.138}{-45} = 3.08 \times 10^{-3}$$

3.5 RESISTORS AND COLOR CODES

The resistance of wire is distributed uniformly along its length. The voltage distribution, therefore, is also uniform. In an electrical circuit, it is impractical to use voltage differences in this fashion. To achieve a significant voltage difference within a small physical space, *resistors* are purposely manufactured using materials of much higher resistivity than the conductors connecting them into the circuit. This is analogous to a long, large-diameter water line having a short length of small-diameter pipe fitted into the line. A disproportionate amount of the water pump's pressure is expended across this "reduction," per unit length, as compared to an equal length of the larger diameter pipe. So it is, then, that in an electrical circuit, resistors have a disproportionate percentage of the circuit's total voltage impressed across them as compared to the low resistance of the wire conductors.

Resistors come in a large variety of shapes and sizes. The most common form encountered is tubular in shape and is usually composed of a carbon composition. Protruding from each end is a low resistance copper wire used for connecting the resistor into an electrical circuit. Figure 3.5 illustrates the principles discussed in the preceding paragraph. Notice the addition of "color bands" on the resistor in Figure 3.5. If it is impractical to print the resistance value on the resistor, then a set of color coded bands is painted on one end. Always read the sequence of colors starting from the end of the resistor and progressing toward its middle. Table 3.7 specifies the values assigned to each color band.

The first two color bands represent the first two significant digits, respectively. The third color band indicates the number of zeros following

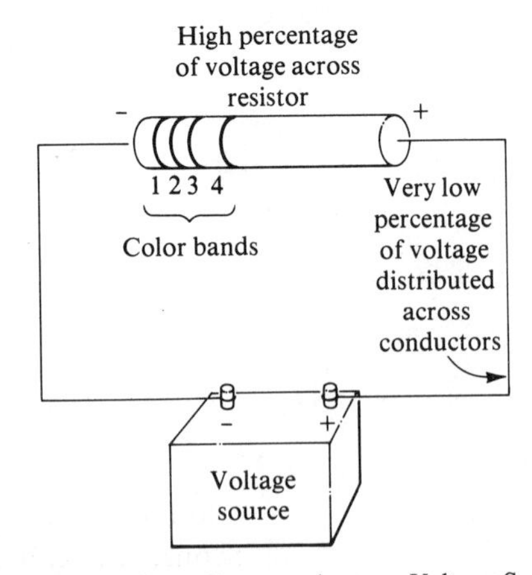

Figure 3.5 Resistor Connected across Voltage Source

the significant digits. The fourth color band may or may not be present. If there is a fourth color band, gold indicates 5% tolerance and silver indicates 10% tolerance. Absence of the fourth color band indicates 20% tolerance.

Example 3.18 What resistance value and tolerance are indicated by a color sequence of red, red, orange, silver?

$$\begin{aligned} \text{First color (red)} &= 2 \\ \text{Second color (red)} &= 2 \\ \text{Third color (orange)} &= 3 \text{ zeros} \\ \text{Fourth color (silver)} &= \pm 10\% \end{aligned}$$

$$R = 22{,}000 \pm 10\% = (22{,}000 \pm 2200) \text{ ohms}$$

Example 13.18 indicates that R may range from 19,800 ohms to 24,200 ohms, with an optimum value of 22,000 ohms. The smaller the tolerance (that is, the more precision the resistance), the more expensive is the resistor. In most applications, extensive precision is not warranted, hence tolerance in the value of resistors should be accepted as commonplace.

Example 3.19 What is the range of values a resistor could have if coded brown, black, red?

Brown	Black	Red	(No color band)
1	0	00	±20%

$$R = (1000 \pm 20\%) \text{ ohms}$$

R ranges from 800 to 1200 ohms.

TABLE 3.7 Resistor Color Codes

Decimal value	Color
0	Black
1	Brown
2	Red
3	Orange
4	Yellow
5	Green
6	Blue
7	Violet
8	Gray
9	White
Tolerance percentage	
5%	Gold
10%	Silver

3.6 LINEAR AND NONLINEAR RESISTORS

The carbon composition resistor illustrated in Figure 3.5 displays fixed resistance characteristics, assuming a fixed temperature. If this resistor were connected to a variable voltage source, the graph of the current through the resistor would be a linear function of the voltage. Figure 3.6 shows that when the voltage is zero (that is, when there is no electrical pressure to move charge carriers), the current is zero. As voltage is increased, every equal incremental increase in voltage produces a corresponding equal incremental increase in current. Graphing such a function results in a straight line, hence this is called a *linear graph.* The graph progresses upward and to the right, hence, the *slope* of the line is positive.

Some devices do not exhibit a linear resistance. Figure 3.7 shows a typical current-voltage graph for a thermistor. A thermistor is a device whose temperature coefficient is highly dependent upon the temperature. The graph shows that equal incremental changes in voltage do not corre-

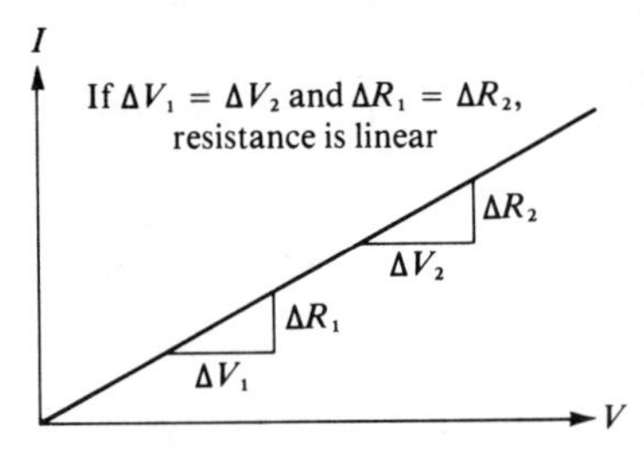

Figure 3.6 Graph of Linear Resistance

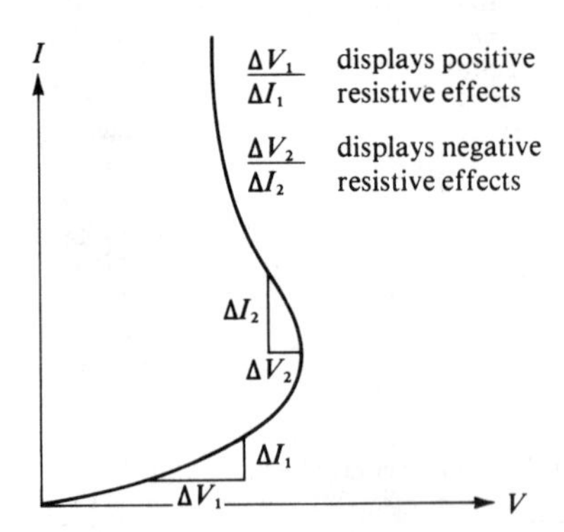

Figure 3.7 Nonlinear Resistive Characteristics of a Thermistor

spond to equal incremental changes in current. In fact, different values of current can produce the same voltage drop across the device. Resistive devices like thermistors exhibit nonlinear characteristics.

This book will deal primarily with linear resistors. Nonlinear resistive devices are left for subsequent, more advanced courses.

SUMMARY

All materials exhibit an opposition to current. The property that describes this characteristic is called *resistivity*. At a given reference temperature, usually 20°C, each material has a fixed value of resistivity. Conductors have very low values; insulators have very high values.

The unit of measurement of resistivity in the Metric system is $(\text{ohm} \cdot \text{mm}^2)/\text{m}$; in the English system it is $(\text{ohm} \cdot \text{CM})/\text{ft}$. If other than circular conductors are used in the English system, it will become necessary to convert their square cross-sectional area to CM before calculating resistance.

The unit of resistance is the ohm. It is directly proportional to a material's resistivity and length and inversely proportional to its cross-sectional area. The reciprocal of resistance is conductance. It is measured in units called siemens.

The resistivity, and hence the resistance of most materials, increases with temperature. A notable exception is carbon. The value of resistance of a material at a temperature different from a reference temperature can be calculated, provided one knows the value of resistance at the reference temperature and either the material's inferred zero temperature or its temperature coefficient.

Tables have been calculated that list the diameter and resistance of virtually the entire range of wire sizes. They are usually tabulated in ohms per thousand meters for Metric notation or ohms per thousand feet for English notation. Resistor magnitudes are indicated by color band codes. Linear resistors display straight line characteristics on their current-voltage graphs.

PROBLEMS

Reference Section 3.1

1. What is the resistance of copper wire 20 meters long if it has a diameter of 0.025 cm?
2. Find the resistance of a silver wire 12.5 meters long and having a diameter of 0.02 inches.

3. Aluminum wire is 42 meters long and has a resistance of 1 ohm. What is its radius in inches?
4. One hundred feet of aluminum wire has a resistance of 12 ohms. What is its diameter in millimeters?
5. The diameter of magnesium wire is 0.01 inch. How long must it be to have a resistance of 100 ohms?
6. A certain conductor has a radius of 0.9 mm. It is 100 meters long. What is the probable material if the resistance of the conductor is 1.06 ohms?
7. Copper wire 1200 ft long has a diameter of 0.04 inches. What is its conductance?
8. The conductance of a spool of nickel wire 180 ft long is 43,500 micro-siemens. What is its radius in mm?
9. What would be the dimensions of a magnesium cube of 0.001 ohm resistance?

Reference Section 3.2

10. The cross-sectional dimensions of a bus bar are $\frac{3}{8}$ inch by $2\frac{1}{4}$ inch. How many CM is this equivalent to?
11. The cross-sectional area of a square bus bar is equivalent to 10^5 CM. What are the dimensions of the cross-sectional area in inches?
12. A square copper bar 0.15 inch on a side is 2 ft long. What is its resistance?
13. A square aluminum bar is 0.8 mm on a side. If it is 2.3 yards long, what is its resistance?
14. A square copper wire 0.05 inch on a side is 6 ft long. It is desired to replace this with a circular copper wire 12 ft long. What is the radius in mils of the circular copper conductor if the resistance is to remain the same?
15. What is the conversion factor to convert ohm-CM/ft to ohm-mm^2/m?
16. What is the conversion factor to convert ohm-mm^2/m to ohm-CM/ft?
17. Manganin is listed in Table 3.1 as 44×10^{-2} ohm-mm^2/m. Convert this to the English system of ohm-CM/ft.

Reference Section 3.3

18. The resistance of a copper rod is 120 ohms at 20°C. What is its resistance at 55°C?
19. If a hypothetical element had a resistance of 8 ohms at 20°C and 9 ohms at 68°C, what would be its inferred zero resistance temperature point?
20. A silver conductor has a resistance of 12.5 ohms at 20°C. At what temperature will the resistance have increased 15%?

21. What would be the percentage of change in resistance of a length of manganin wire if temperature increases from 20°C to 120°C?
22. If a column of mercury has a resistance of 0.5 ohm at 20°C, at what temperature would its resistance be 0.51 ohm?
23. A hypothetical material has a resistance of 12.7 ohms at 20°C and a resistance of 14.5 ohms at 85°C. What is its α_{20} temperature coefficient?
24. The final value of resistance of nichrome wire is 2.86 ohms at 75°C. What was its resistance at 20°C?

Reference Section 3.4

25. What is the resistance of 725 ft of #18 copper wire at 25°C?
26. How many meters long is #25 copper wire if its resistance is 1150 ohms at 25°C?
27. What is the area in mm^2 of #28 copper wire?

Reference Section 3.5

28. What is the color code of 220 kΩ, 10% tolerance?
29. What is the color code for an 80 ohm resistor, 20% tolerance?
30. What is the color code for a 670 kΩ, 5% tolerance resistor?
31. What are the upper and lower values of resistance for the resistor of problem 3.30?
32. What is the color code for a 1 ohm resistor, ±20%?

PART II

Part I of this text has built the foundations necessary for circuit analysis. The reader now has knowledge of the three basic building blocks of modern circuits: voltage, current, and resistance. Circuits will be constructed, and reactions to variations of input parameters will be analyzed.

To further modernize the approach taken throughout this text, the concepts of programming will be studied. It is possible that no computer, or even that no terminal, is available at the reader's educational institution. It is highly probable, however, that some students already possess "programmable" pocket calculators. It is certainly likely, in light of the recent history of the calculator market, that the frequency of utilization of these newer devices will significantly increase. It would, therefore, be most beneficial for the reader to study the following sections on computer analysis, as he will become familiar with this subject for subsequent applications following his studies. If it is still desired to omit these sections, that is easily accomplished, as they will be designated "optional." A more abbreviated but yet nondetrimental analysis will result, because subsequent topics will be presented in the more traditional approach and then embellished with optional computer analysis examples.

To provide a foundation for this more advanced form of instruction, it will be necessary to defer circuits until Chapter 5. Chapter 4 will introduce the reader to how a program is written and entered into a computer (or programmable calculator) for solution. Following this introductory chapter on programming, applications of the interacting principles of voltage, current, and resistance will resume.

CHAPTER 4

COMPUTER PROGRAMMING OF ELECTRONIC PROBLEMS

4.1 INTRODUCTION

With the advent of low-cost pocket calculators, it is not an infrequent sight to see students in college classrooms using their own electronic minicalculators. In many instances, this is adequate to determine an immediate numerical solution to a mathematical problem. Nonprogrammable pocket calculators, however, are limited in ability to provide hardcopy results of a series of calculations, or to store resident programs for later instantaneous recall. The newer, programmable pocket calculators are becoming increasingly popular, as they permit the solution of more sophisticated problems. They do, however, require a logical application of programming steps before a solution can be reached. Although programmable calculators have a significant advantage over the nonprogrammable types, most still lack the ability to store library subroutines for easy application to problem solutions. Fortunately, the same technology that produced pocket calculators has made large-scale computers much more accessible, via locally installed terminals, and has produced the now widely available "personal computer," or microcomputer. While eliminating the deficiencies inherent in a small calculator, it is still necessary for the student to apply simple programming steps if these sophisticated "tools" are to be utilized.

Projected trends in employment indicate an ever-increasing interrelationship between the technical knowledge of electronic circuits on the one hand, and the ability of the technician or engineer to readily program computers to solve everyday problems. The question is not whether it is appropriate for the electronics student to learn programming, but rather at what point in his study this should be undertaken. It seems most logical that the two areas of study, namely, the electronics circuit theory and the ability to program problems for a solution, should be undertaken simultaneously. This textbook, throughout this and the following chapters, will familiarize the electronics student with one of the most powerful and popular languages currently available. By the end of Part II of this text, the reader should be able to program reasonably complex problems because of the easy, conversational nature of the language to be discussed.

It should not be construed that this introductory level to programming techniques will make the reader a proficient, scientific programmer.

Rather, the reader will be able to readily use the instructions to be discussed to write programs at a terminal connected to a computer system and receive accurate solutions. Since this book is intended primarily as a course in the study of the fundamentals of dc and ac electronics, and not as a thorough in-depth introduction to programming techniques, this discussion will be limited to the more basic statements necessary to solve simple electronic problems.

4.2 THE PROGRAMMING LANGUAGE

When a person says that he is going to study a language, the typical reaction is to think of a spoken language like English, Spanish, French, etc. All these languages basically perform the same function, namely, to provide a means of communication. They differ only in the rules relating to word pronunciation and grammatical structure. Programming languages, like spoken languages, also provide a means of communication. They have names like FORTRAN, PL/1, APL, BASIC, etc. Again, like spoken languages, they differ only in syntactical and structural rules. A natural question arises: What programming language should be utilized? As stated in the Preface, it becomes obvious that there is an inherent transference of concepts and methodologies regardless of the language chosen. The choice, then, should be a language that is easy to comprehend (conversational in nature), oriented toward projected future trends, and more universally accessible for study as opposed to being structured for a special device. BASIC was chosen because it is both English-language-oriented in its command statements and is also in keeping with current industrial trends such as timesharing (many isolated programmers, each simultaneously using remote terminals to share powerful computing systems).

Broadly speaking, two kinds of computers are commonly used for programming in the BASIC language. These are the microcomputer and the large, fast, and typically expensive computer that can be programmed using any one of a large number of different languages. This second type of computer is usually located in a special room or building to which programmers do not have access. Instead, the programmer communicates with the computer from a remote location, using a *video terminal* connected to the computer through a telephone line. The terminal has a keyboard, much like a typewriter keyboard, that the programmer uses to send programming instructions to the computer. The terminal also has a screen for the computer to display results. Figure 4.1 shows a typical terminal.

Installations of the type just described are usually found in large universities and corporations. A typical system may have dozens of terminals capable of communicating with the same computer. In fact, by a process

Figure 4.1 Typical Input Terminal (Manufacturer: Applied Digital Data Systems.)

known as *time-sharing*, many programmers may apparently be using the same computer at the same time. Actually, the computer is so fast and has such a large memory that it can allocate short periods of time to each user in turn, responding to each only as much as necessary to accommodate what the user has programmed since the last turn. For example, if the time allocated to each user is $^1/_{60}$ second and if there are 12 users on the system, then each user is serviced every $^1/_5$ second. To gain access to the computer in this kind of system, the programmer must go through a "log-on" procedure by typing in certain key words or numbers, usually including an account number. The computer will then recognize the programmer as an authorized user and a record can be kept of the amount of computer time to be charged to the account.

The other type of computer widely used for BASIC programming is the *microcomputer*. Designed around an integrated-circuit chip called a microprocessor, the microcomputer is much smaller and much less expensive than the time-shared computer—characteristics that account for its increasing popularity in homes, businesses, schools, and laboratories in recent years. The typical microcomputer cannot be programmed in as wide a variety of languages as a large computer, but virtually every micro-computer is programmable in BASIC. Furthermore, the microcomputer cannot generally be programmed by more than one user at a time—it is strictly a "one-on-one" situation. Many microcomputers look very much like the video terminals described previously; they have keyboards that the programmer uses to enter programs into the computer and video screens on which results are displayed. The computer itself is so small that it is packaged within the terminal enclosure. Figure 4.2 shows a typical microcomputer.

Microcomputer installations can also be connected to optional accessories, such as auxiliary memory ("floppy disks") to expand the capability to store programs and data and printers to obtain "hard copy" of programs and results. The procedure for gaining access to a microcomputer for BASIC programming purposes varies according to the manufacturer, but it is usually simpler than for a large computer and it does not generally require the use of an account number.

4.3 VERSIONS OF BASIC

Many aspects of the BASIC programming language are shared by all computers that can be programmed in BASIC; there are differences, however. Different computers have different *versions* of BASIC; in other words, a BASIC program written for one type of computer may not work on another type, or may not even work on another computer of the same type but of different manufacture. The differences among versions are most often (but not always) related to the *power* that one version has as compared to another—the ability of one version to perform certain kinds of computations that another version cannot perform. For example, versions of BASIC are given names such as "extended BASIC," "BASIC-PLUS," "tiny BASIC," and "integer BASIC" (which cannot do computations that involve fractions or decimal points). A program written in a "weaker" version can often be run on a computer with a more powerful version, but not vice versa.

The differences among versions are particularly noticeable when comparing the BASIC used in large, time-shared installations with that used in microcomputers; the former is generally much more powerful. For example, very few microcomputer versions of BASIC are capable of handling matrix operations (which we will discuss in a later chapter), while most versions used in large systems have that capability. In this text, we will assume a powerful version of BASIC (BASIC-PLUS) but, in those examples where we rely on the extra power of the language, we will also describe alternate programming techniques that could be used with most microcomputer versions of BASIC.

Figure 4.2 Typical Microcomputer (Photo courtesy Tandy Corporation)

One important feature of all versions of BASIC is the ability to permit *interactive programming* of a computer. This kind of programming allows the user to write a program so that the computer will generate and display intermediate results during processing of the program and also pause at selected times to allow the user to enter new data. Unlike some languages, BASIC was specifically designed for this kind of spontaneous communication between the user and the computer. BASIC is an example of a language that is controlled in the computer by another program, called an *interpreter*, which gives it the very useful ability to inform the programmer immediately when a mistake is made in the way that a programming instruction is entered. This is accomplished by displaying an *error message* on the video screen.

Most computers, including microcomputers, have *system commands* that the programmer must use to gain access to a certain language, store programs in the computer's memory, retrieve programs already stored in memory, modify (edit) existing programs, or perform similar tasks. These system commands are not part of the BASIC language, but the programmer must know them and be able to apply them to use a particular computer. The commands vary widely depending on the manufacturer, so students must begin their study of BASIC by becoming familiar with the system commands appropriate for the type of computer being used. Manufacturers always provide a *programming manual*, which should be consulted for this kind of information. In Appendix C, we summarize a number of the important system commands for several currently popular computers.

4.4 INTRODUCTION TO PROGRAMMING

For the benefit of new programmers approaching their first programming experience, there are four phases in programming a computer:

1. Writing the computer program.
2. Entering the program to the computer system.
3. Testing and debugging the program.
4. Running the finished program.

BASIC-PLUS is the language in which the user writes programs for input to a computer system. Input of the completed program is generally performed from the terminal keyboard. Figure 4.1 (page 63) pictures a video terminal (also known as a cathode ray tube or CRT terminal) that can be used for program entry in a time-shared system. A programmer communicating directly with a computer is said to be operating *on-line*.

Programs can also be entered into a computer system through other various peripheral devices, such as paper tape readers, magnetic tape, punched cards, etc.; however, the initial creation of a program is usually performed on-line to the computer from the terminal keyboard.

Ideally, a program runs correctly as written; but in practice, this is seldom the case. A program can contain simple typing mistakes or complex logical errors. Typing and syntactical errors are detected as the program is typed at the terminal and appropriate error messages are printed. BASIC-PLUS also evaluates the entire program for commonly made errors and generates messages which explain the mistakes to the user. Program errors are corrected on-line from the terminal keyboard.

The testing and debugging process is continued until the program appears to execute correctly. This is a good time to explain to the new user that a computer program only does what the programmer has written. The calculations performed by the computer are not necessarily those that will produce the correct results. In order to obtain correct results from a computer, the user must write a program which is not only free of detectable errors, but one which correctly analyzes his problem.

4.5 LINE NUMBERS

Before the program can be properly written, the writer (programmer) must understand the ground-rules of the language that he will use to communicate with the computer. Just as there are rules and symbols to correctly express a mathematical equation, it follows that there are precise rules of formatting so that the monitor will understand the programmer's statements. This subdivision will explain those formatting rules and the symbolic representations necessary to initiate a computer instruction.

Each BASIC program line is preceded by a line number. Line numbers:

1. Indicate the order in which statements are normally evaluated.
2. Enable the normal order of evaluation to be changed; that is, the execution of the program can branch or loop through designated statements (this is explained further in the sections on the GOTO, GOSUB, and IF-THEN statements.
3. Enhance program debugging by permitting modification of any specified line without affecting any other portion of the program.

The total line numbers allowed depends on the version of BASIC being used, but it is typically more than 30,000. BASIC maintains programs in line number sequence, rather than the order in which lines are entered into the system. It is good programming practice to number lines in increments of 5 or 10 when first writing a program, to allow for insertion of forgotten or additional lines when debugging the program.

When a program is executed, the BASIC processor evaluates the statements in the order of their line numbers, starting with the smallest line number and going to the largest.

4.6 STATEMENTS

Each line number is followed by a BASIC statement. These statements are frequently one word but may be as many as three related words. BASIC statements identify the type of operation to be performed and how to treat the data (if any) that follows the word(s). In some versions of BASIC, more than one statement can be written on a single line, as long as each statement (except the last) is terminated with a special symbol. Several versions use a colon for this symbol. Thus, only the first statement on a linc can (and must) have a line number. For example:

```
10 INPUT R1, R2, R3
```

is a single-statement line, while:

```
20 LET P = P + 1: PRINT I1, E2,P4: IF P = 20E - 3 GOTO 10
```

is a multiple-statement line containing three statements: a LET, a PRINT, and an IF-GOTO statement.*

* Note: In this and all programs, be sure to observe the difference between the letter "I" and the number "1."

As the programmer continues to enter a line into the computer, his typing appears across the face of the CRT terminal (such as in Figure 4.1). It would be extremely rare if a single statement exceeded the length of one line (typically 80 characters), but if so, the programmer may wish to break the statement into two or more independent lines. Whether this be the case or whether the programmer merely wishes to terminate a single- or multiple-statement line as it nears the end of the CRT terminal, either is accomplished by depressing the RETURN key. This informs the computer that the current line statement(s) is terminated. If more statements are used in the total program, they are usually followed by line numbers numerically higher than the one just entered.

4.7 SPACES AND TABS

In the programming examples that follow throughout this text, the reader will note the use of *spaces* or *tabs*. The use of the space bar or tab key has no effect on the program. They are used to make statements easier to read. For example, each of the following three statements will calculate inductive reactance (designated here as $X3$):

```
10 LET X3 = 2 * PI * F3 * L3

20 LETX3=2*PI*F3*L3

30 LETX3    =2*    PI*F3*    L 3
```

Certainly, because of the use of the space, statement 10 is easier to read than statements 20 or 30.

Tabs are similar to spaces, except that one depression of the tab key can cause a predetermined displacement, usually equal to many spaces. As stated above, their only function is to make the program easier to read. An example follows:

```
10 FOR K=1 TO 3
20        FOR I=1 TO 10
30                    FOR J=1 TO 10
40                    A(I,J) = K/(I+J-1)+A(I,J)
50                    NEXT J
60        NEXT I
70 NEXT K
```

4.8 VARIABLES

A variable is a data item whose value can be changed by the program. A numeric variable is denoted by a single letter or by a letter followed by a single digit. Thus BASIC interprets $E8$ as a variable, along with A, X, $N5$, $L0$, and 01.

Variables are assigned values by LET, INPUT, and READ statements. The value assigned to a variable does not change until the next time a LET, INPUT, or READ statement is encountered that contains a new value for that variable or when the variable is incremented by a FOR statement. (These conditions are explained further in later sections.) In most versions of BASIC, all variables are set equal to zero before program execution. It is only necessary to assign a value to a variable when an initial value other than zero is required. However, good programming practice would be to set variables equal to (0) wherever necessary. This practice ensures that later changes or additions will not misinterpret values.

4.9 NUMBERS

Numbers, called *numeric constants* because they retain a constant value throughout a program, can be positive or negative. Numeric constants are written using decimal notation, as follows:

+2
−3.675
1234.56
−123456
−.000001

The following are not acceptable numeric constants in BASIC:

$$\frac{14}{3} \qquad \sqrt{7}$$

However, BASIC can find the decimal expansion of those two mathematical formulas as shown below:

$\frac{14}{3}$ is expressed as 14/3

$\sqrt{7}$ is expressed as SQR(7)

These formats are explained in later sections.

Scientific notation allows further flexibility in number representation. Numeric constants can be written using the letter *E* to indicate "times ten to the power," thus:

.000123456	can be written in BASIC as	123.456E−6
1234560000.	can be written in BASIC as	123456E4
−12345678900.	can be written in BASIC as	−1.23456789E10

The E format representation of numbers is very flexible since a number such as .001 can be written as $1E-3$, $.01E-1$, $100E-5$, or any number of ways. If more than six digits are generated during any computation, the result of that computation is automatically printed in E format. (If the exponent is negative, a minus sign is printed after the E; if the exponent is positive, a space is printed: $1E-4$; $1E\ 4$.)

The combination $E7$, however, is not a constant, but a variable. The term $1E7$ is used to indicate that 1 is multiplied by 10^7.

4.10 RELATIONAL SYMBOLS

The following relational symbols exist with BASIC. Their application and meaning are best understood when using the IF-THEN statement.

Mathematical symbol	*BASIC symbol*	*Example*	*Meaning*
=	=	A=B	*A* is equal to *B*
<	<	A<B	*A* is less than *B*
≤	<=	A<=B	*A* is less than or equal to *B*
>	>	A>B	*A* is greater than *B*
≥	>=	A>=B	*A* is greater than or equal to *B*
≠	<>	A<>B	*A* is not equal to *B*

Although it is not available on microcomputer versions of BASIC, some large system versions have an additional operator called *approximately equal* (= =). The term "approximately equal to" means that the two quantities look the same when printed. Within the computer, floating-point numbers can differ by a miniscule amount in the last decimal place but still be considered equal for all practical purposes. This last decimal place within the computer does not always cause two numbers to have different values when printed. Numbers are carried internally at greater than 6 digits of precision, but are rounded to 6 digits for output or an (≈) comparison. Thus, two numbers identical when rounded to 6 digits of precision are approximately equal, whereas two numbers equal to the internally carried limits of precision are truly equal (=).

4.11 OPERATORS

Arithmetic Operator

BASIC automatically performs the mathematical operations of addition, subtraction, multiplication, division, and exponentiation. Formulas to be evaluated are represented in a format similar to standard mathematical notation. Five arithmetic operators are used to write such formulas; they are as follows:

Operator	*Example*	*Meaning*
+	A+B	Add *B* to *A*
−	A−B	Subtract *B* from *A*
*	A*B	Multiply *A* by *B*
/	A/B	Divide *A* by *B*
↑	A↑B	Calculate *A* to the *B* power, A^B

The symbol used for exponentiation (raising to a power) varies, depending on the version of BASIC being used. Many versions, including several microcomputer versions, use the symbol ** instead of ↑. Another symbol used in microcomputer versions (including Apple and Atari) is ∧. In this text, we will use both the ** and ↑ symbols.

When more than one operation is to be performed in a single formula, as is most often the case, rules are observed as to the precedence of the above operators. The arithmetic operations are performed in the following sequence, with (a) having the highest precedence:

a. Any formula within parentheses is evaluated before the parenthesized quantity is used in further computations. Where parentheses are nested, as follows:

   (A+(B*(D↑2)))

   the innermost parenthetical quantity is calculated first.

b. In the absence of parentheses in a formula, BASIC performs operations in the following sequence: (1) exponentiation, (2) multiplication and division, and (3) addition and subtraction

c. In the absence of parentheses in a formula involving more than one operation on the same level in (b) above, the operations are performed left to right, in the order that the formula is written. For example:

   A/B/C is evaluated as (A/B)/C

   A*B/C is evaluated as (A*B)/C

The expression A+B*C↑D is evaluated sequentially as follows: (1) *C* is raised to the *D* power, (2) The result of the first operation is multiplied by *B*, and (3) The result of the previous operation is added to *A*.

Parentheses are used to indicate any other order of evaluation. For example, if the product of *B* and *C* is to be raised to the *D* power, the expression would look as follows:

```
A+(B*C)↑D
```

If it is desired to multiply the quantity $A+B$ by C to the D power:

```
(A+B)*C↑D
```

The user is encouraged to use parentheses even where they are not strictly required in order to make expressions easier to read. Ambiguities can exist only in the programmer's mind; the computer always performs the operations as we have explained.

Logical Operators

Although BASIC recognizes a number of logical operators, this portion of the text will limit discussion to the three most commonly encountered: AND, OR, and NOT operators. Additional logical operators will be discussed in Chapter 7. Logical operators are used in IF-THEN statements where some condition is tested to determine subsequent operations. For this discussion, *A* and *B* are relational expressions having only *true* and *false* values. The above logical operators are defined as follows:

Operator	*Example*	*Meaning*
AND	A AND B	The logical product of *A* and *B*. *A AND B* has the value *true* only if *A* and *B* are both true, and has the value *false* if either *A* or *B* is false.
OR	A OR B	The logical sum of *A* and *B*. *A OR B* has the value *true* if either *A* or *B* is true, and has the value *false* only if both *A* and *B* are false.
NOT	NOT A	The logical negative of *A*. If *A* is true, *NOT A* is false.

Logical operators may at first seem difficult to comprehend. The reader will find, however, that they are extremely powerful statements. Keep in mind that logical operators are not used to perform mathematical calculations. They are used to test conditions; then, based upon the conclusion reached by that test, decisions can be made as to which portion of the program the computer should branch in order to complete the program. As an example, suppose it is desired to make a test of the following circuit condition: If the power dissipated by resistor 3 is less than 20 milliwatts OR the power dissipated by resistor 5 exceeds 50 milliwatts, then the next instruction is at statement 100. This test is expressed as follows:

```
25 IF P3 <2ØE-3 OR P5> 5ØE-3 THEN GOTO 1ØØ
```

If either stipulated power condition is *true*, then the logical test is true (by virtue of the definition of the OR test) and the program transfers to statement 100.

4.12 EXPRESSIONS

As stated earlier, a statement tells the computer the nature of what it is to do. When all elements associated with the statement are combined, this collection is referred to as an *expression*. In addition to the statement itself, an expression is usually composed of numbers and/or variables separated by arithmetic and/or logical operators. The following are examples of expressions:

Arithmetic

```
1Ø LET E3 = 25
```

which translates to "voltage source 3 is initially set equal to 25 volts."

```
2Ø LET X = 1/(2*PI*F*C)
```

which translates to "capacitive reactance equals the reciprocal of the product of two times Pi times frequency times capacitance."

Logical

```
3Ø IF P3 < 15E-3 AND P1 > 2ØE-3 GOTO 9Ø
```

which translates to "if the power in resistor 3 is less than 15 milliwatts AND the power in resistor 1 is greater than 20 milliwatts, then GOTO statement 90, ELSE GOTO the next sequential statement."

4.13 ELEMENTARY BASIC STATEMENTS

This section describes the simplest forms of the more elementary BASIC statements. These statements are sufficient, by themselves, for the solution of most problems. Once they are mastered, the user can investigate the more advanced applications of these statements and the additional statements and features explained in Chapter 8.

The reader should understand that any problem which can be solved with the more advanced techniques can also be solved with the simpler statements, although the solution may not be as efficient. As long as the user understands the details of his problem, he can represent it in BASIC on a number of levels ranging from the simple to the sophisticated.

Certain documentation conventions are used throughout this chapter to clarify examples of BASIC syntax. Each BASIC statement is described at least once in general terms using the following conventions:

a. Items in italics (*formula, variable,* etc) are supplied by the user according to rules explained in the text. Items in capital letters (LET, IF, THEN, etc.) must appear exactly as shown because they form the vocabulary of the BASIC language.

b. The term *line number* used in examples indicates that any line number is valid.

c. Angle brackets indicate essential elements of the statement or command being described. For example:

$$\textit{line number}\ \text{LET}\quad \langle \textit{variable} \rangle = \langle \textit{expression} \rangle$$

d. Square brackets indicate a choice of one element among two or more possibilities. For example:

$$\textit{line number}\ \text{IF}\ \langle \textit{expression} \rangle \begin{bmatrix} \text{THEN}\ \langle \textit{statement} \rangle \\ \text{THEN}\ \langle \textit{line number} \rangle \\ \text{GOTO}\ \langle \textit{line number} \rangle \end{bmatrix}$$

e. Braces indicate an optional statement element or a choice of one element among two or more optional elements:

$$\textit{line number}\ \text{IF}\ \langle \textit{expression} \rangle \begin{bmatrix} \text{THEN}\ \langle \textit{statement} \rangle \\ \text{THEN}\ \langle \textit{line number} \rangle \\ \text{GOTO}\ \langle \textit{line number} \rangle \end{bmatrix} \begin{Bmatrix} \text{ELSE}\ \langle \textit{statement} \rangle \\ \text{ELSE}\ \langle \textit{line number} \rangle \end{Bmatrix}$$

Remarks and Comments

It is often desirable to insert notes and messages within a user program. Such data as the name and purpose of the program, how to use it, how certain parts of the program work, and expected results at various points are useful to have present in the program for ready reference by anyone using that program. Comments can be inserted into a user program by:

a. the REMARK statement, and
b. use of a special character (depending on the version of BASIC).

The word *REMARK* is abbreviated to REM for typing convenience, and the message itself can contain any printing characters on the keyboard. BASIC completely ignores anything on a line following the letters REM. Typical REM statements are shown below:

```
1Ø REM – THIS PROGRAM COMPUTES THE POWER
11 REM – IN A LOAD RESISTOR
```

The special character mentioned in (b) above is used to insert a remark on the same line as another statement. Some versions use ! for this purpose, and others use \. Many microcomputer versions use the combination :REM, where the colon simply denotes the beginning of another statement on the same line and REM begins a REMARK statement. Following are two examples using the ! character:

```
125 LET F=1/(2*PI*SQR(L*C))    !CALCULATE RESONANT FREQUENCY
13Ø PRINT F2-F1                !PRINT BANDWIDTH
```

Messages in REMARK statements are generally called *remarks*; those after the exclamation mark, *comments*. Remarks and comments are printed when the user program is listed, but they do not affect program execution.

LET Statement

The LET statement assigns a numeric value to a variable. Each LET statement is of the form:

line number LET ⟨*variable*⟩ = ⟨*expression*⟩

This statement does not indicate algebraic equality, but performs the calculations within the expression (if any) and assigns the numeric value to the indicated variable. For example:

```
1Ø LET R  =R+1E3         ! INCREMENT RESISTOR R BY 1K
2Ø LET W2=(A4-X↑3)*(Z-A/B)
```

In line 10, the old value of R is increased by 1kΩ and becomes the new value of R. In line 20, the formula on the right-hand side is evaluated and the numeric value assigned to $W2$.

The LET statement can be a simple numerical assignment, such as

```
5Ø LET A=47E3    ! THIS IS THE INITIAL VALUE OF RESISTOR A
```

or require the evaluation of a formula so long that it is continued on the next line.

Many versions of BASIC, including BASIC-PLUS and most microcomputer versions, allow the user to omit the word LET from the LET statement. You may find it easier to type

```
1Ø X = 12*(S + 7)
```

than

```
1Ø LET X=12*(S+7)
```

This is a convenience and does not alter the effect of the statement.

In versions of BASIC that permit multiple statements on a line, more than one LET statement can appear on a line; for example:

```
10 X = 44: Y = X↑2 + Y1: B2 = 3.5*A
```

Some powerful versions of BASIC (but not most microcomputer versions) allow the user to assign a value to multiple variables in the same statement. For example,

```
10 LET X,Y,Z = 5.7
```

causes each of the three variables to be set equal to 5.7.

Programmed Input and Output

This section describes the techniques used in performing BASIC program I/O (an abbreviation for the term *input/output*, which includes the processes by which data is brought into and sent out of the computer). The most elementary forms of the READ, DATA, PRINT, and INPUT statements are presented here so that the user is able to create simple BASIC programs.

Using the LET statement and the following executable statements, the user can easily write a BASIC program. If he should want to try his program, these simple I/O statements provide a means of obtaining tangible output.

Read-Data Statements READ and DATA statements are used to enter information into the user program during execution. A READ statement is used to assign to the listed variables those values which are obtained from a DATA statement. Neither statement is used without the other.

A READ statement is of the form:

line number READ ⟨*variable list*⟩

A DATA statement is of the form:

line number DATA ⟨*value list*⟩

A READ statement causes the variables listed to be assigned sequential values in the collection of DATA statements. Before the program is run, BASIC takes all DATA statements in the order they appear and creates a data block. Each time a READ statement is encountered in the program, the data block supplies the next value.

READ and DATA statements appear as follows:

```
150 READ X,Y,Z,X1,Y2,Q9
330 DATA 4,2,1.7
340 DATA 6.73E-3, -174.321, 3.1415927
```

The assignments performed by line 150 are as follows:

```
 X=4
 Y=2
 Z=1.7
X1=6.73E-3
Y2=-174.321
Q9=3.1415927
```

Since data must be read before it can be used in a program, READ statements normally occur near the beginning of a program. The location of DATA statements is arbitrary, as long as they occur in the correct order. A good practice is to collect all DATA statements near the end of the program.

Input Statement The second way to input data to a program is with the INPUT statement. There is a distinct advantage to using this form of data entry as opposed to READ-DATA statements. If a program is going to be reused many times and only the values of the variables are changing, it is usually more convenient to type in each new set of data than to rewrite the entire DATA statement.

The INPUT statement is of the form:

line number INPUT ⟨*list*⟩

Suppose the computer is processing the program, and a point is reached where it is necessary for the programmer to supply the values of variables before the processing can be completed. At this point in the program, when the INPUT statement is reached, the processor stops, types a (?), and waits for the programmer to supply the new variable data before continuing. For example:

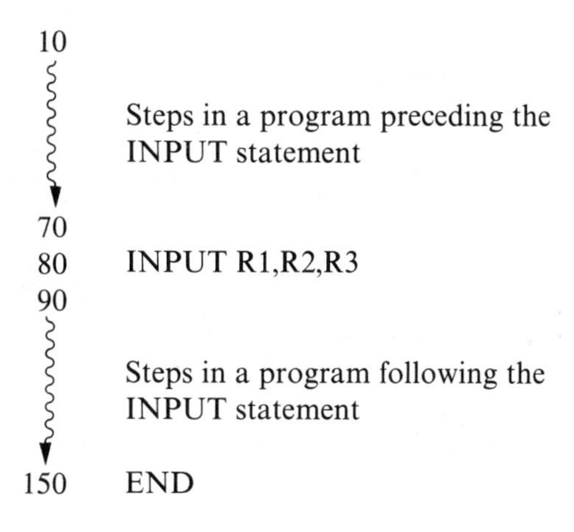

In this sample program, the processor is executing the program statements between lines 10 and 70. In order to complete the program (to line 150), values of resistors 1, 2, and 3 will be needed. When the processor reaches line 80, it stops, prints a (?), and waits for the requested input data before continuing. If the values of the three resistors are 47, 22, and 1.5 kΩ, respectively, the programmer would type in

```
47E3, 22E3, 1.5E3
```

and then depress the RETURN key. The processor would immediately resume the program at line 90. Between this reentry point and the program END (line 150), calculations involving resistors 1, 2, and 3 would have the values assigned by the programmer.

It should be apparent that if this particular program were continuously reused, and the resistive values constantly changing, it would be simpler to merely key-in the new value than to retype each DATA statement.

Print Statement The PRINT statement is used to obtain results (output) from the computer. Each time a PRINT statement is executed, the output it generates is displayed on the CRT of the terminal or microcomputer. To obtain hard copy, a printer is required. In some installations that include a printer, output is automatically printed in response to PRINT statements, either in addition to or in lieu of a CRT display. Other installations require the use of special system commands to actuate the printer. Figure 4.3 shows a typical I/O device with a keyboard for input and a printer for output, but no CRT. When such a device is used for input, all characters typed in from the keyboard are printed as they are typed, so the device serves the dual purpose of displaying input and providing hard copy of output.

The general format of the PRINT statement is:

line number PRINT {*list*}

where the list can contain expressions, text strings, or both. As the braces indicate, the list is optional. Used alone, the PRINT statement,

```
25 PRINT
```

causes a blank line to be printed (a carriage return/line feed operation is performed).

PRINT statements can be used to perform calculations and print results. Any expression within the list is evaluated before a value is printed. Consider the following program:

```
LIST
10 LET A=1: LET B=2: LET C=3+A
20 PRINT
30 PRINT A+B+C

RUN

 7
```

The command LIST in this example causes the computer to print out the program in the same form it was entered (including any remarks). The command RUN causes the computer to execute the program and to print any output generated by PRINT statements (the number 7, in this example).

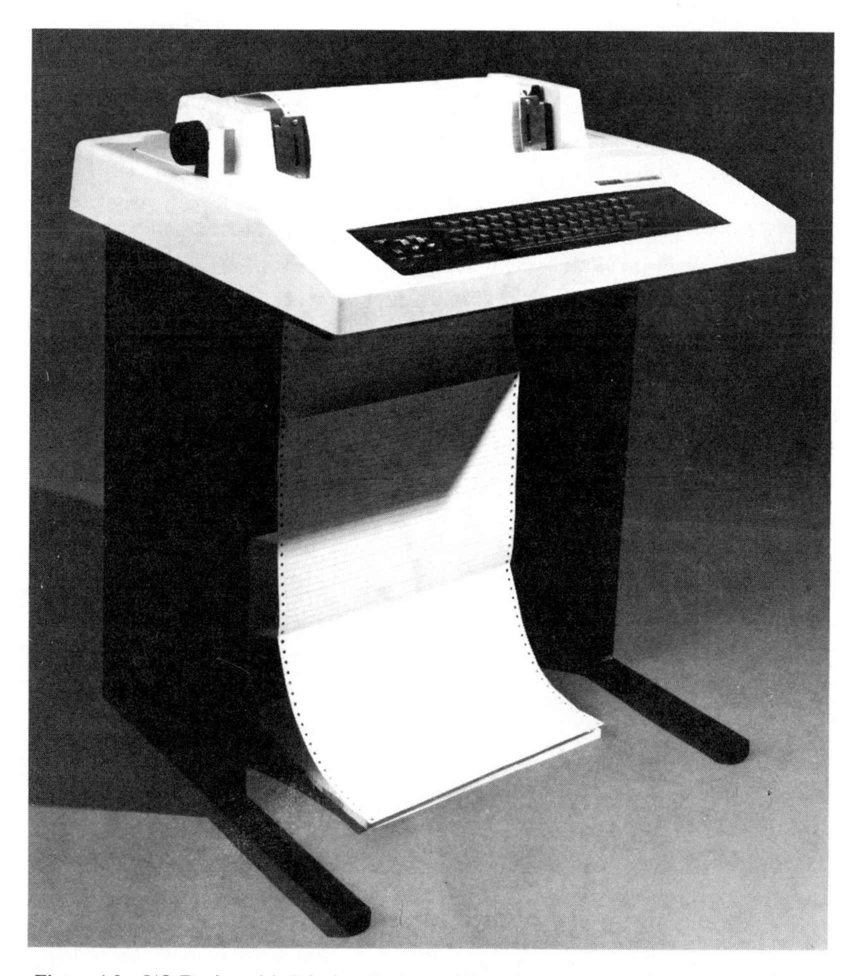

Figure 4.3 I/O Device with Printing Features (Manufacturer: Digital Equipment Corp.)

All numbers are printed in the form:

[space/−] ⟨number⟩ ⟨space⟩

More than one PRINT statement can appear on a line if multiple-statement lines are permitted. For example,

```
1Ø A=1: PRINT A: A=A+5: PRINT: PRINT A
```

would cause the following to be printed on the terminal when executed:

```
RUN
 1

 6
```

Notice that a carriage return/line feed is performed at the end of each PRINT statement. Thus, the first PRINT statement causes a 1 and a carriage return/line feed to be output, the second PRINT statement is responsible for the blank line, and the third PRINT statement causes a 6 and another carriage return/line feed to be output.

BASIC considers the terminal printer to be divided into 5 zones of 14 spaces each. Terminals with greater than 83 columns have additional print zones in units of 14 spaces. When an item in a PRINT statement is followed by a comma, the next value to be printed appears in the next available print zone. For example:

```
1Ø LET A=3: LET B=2
2Ø PRINT A, B, A+B,A*B,A-B,B-A
```

When the preceding lines are executed, the following is printed:

```
 3             2             5             6             1
-1
```

Notice that the sixth element in the PRINT list is printed as the first entry on a new line, since a 72-character line has 5 print zones.

Two commas together in a PRINT statement cause a print zone to be skipped. For example:

```
LIST
1Ø LET A=1: LET B=2
2Ø PRINT A,B,,A+B

RUN
 1             2                            3
```

If the last item in a PRINT statement is followed by a comma, no carriage return/line feed is output, and the next value to be printed (by a later PRINT statement) appears in the next available print zone. For example:

```
LIST
10 A=1:B=2:C=3
20 PRINT A,:PRINT B: PRINT C

RUN
 1           2
 3
```

If a tighter packing of printed values is desired, the semicolon character can be used in place of the comma. A semicolon causes no further spaces to be output. A comma causes the print head to move at least one space to the next print zone or possibly perform a carriage return/line feed. The following example shows the effects of the semicolon and comma.

```
LIST
10 LET A=1: B=2: C=3
20 PRINT A;B;C;
30 PRINT A+1;B+1;C+1
40 PRINT A,B,C

RUN
 1  2  3  2  3  4
 1           2           3
```

The PRINT statement can be used to print a message, either alone or together with the evaluation and printing of numeric values. Characters are indicated for printing by enclosing them in single or double quotation marks (therefore each type of quotation mark can only be printed if surrounded by the other type of quotation mark). For example:

```
LIST
10 PRINT "POWER IN LOAD RESISTOR IS"
20 PRINT ' "POWER IN LOAD RESISTOR IS" '

RUN
POWER IN LOAD RESISTOR IS
"POWER IN LOAD RESISTOR IS"
```

In another example, consider the following line:

```
40 PRINT "AVERAGE GRADE IS";X
```

which prints the following (where X is equal to 83.4):

```
AVERAGE GRADE IS 83.4
```

When a character string is printed, only the characters between the quotes appear; no leading or trailing spaces are added. Leading and trailing spaces can be added within the quotation marks using the keyboard space bar; spaces appear in the printout exactly as they are typed within the quotation marks.

When a comma separates a text string from another PRINT list item, the item is printed at the beginning of the next available print zone.

Semicolons separating text strings from other items are ignored. Thus, the previous example could be expressed as:

```
40 PRINT "AVERAGE GRADE IS" X
```

and the same printout would result. A comma or semicolon appearing as the last item of a PRINT list always suppresses the carriage return/line feed operation.

The following example demonstrates the use of the formatting characters (,) and (;) with text strings:

```
120 PRINT "STUDENT NUMBER"X,"GRADE ="G;"AVE. ="A;
130 PRINT "NO. IN CLASS ="N
```

could cause the following to be printed (assuming calculations were done prior to the line 130):

```
STUDENT NUMBER 119      GRADE = 87 AVE. = 85 NO. IN CLASS = 26
```

The spacing of printed output can also be controlled, using the TAB function. Available in most versions of BASIC, this function allows the user to specify precisely where output should be printed. When a TAB (n) statement is executed, the next data item printed will be in column n (counting from zero). TAB specifications are written with the PRINT statement, as illustrated in the following example:

```
PRINT X TAB(5) Y TAB(11) "END"
```

In this example, the value of Y will be printed, beginning in column 5, and the word "END" will be printed, beginning in column 11.

Stop and End Statements

The STOP and END statements are used to terminate program execution. The END statement must be the last statement of a program in many versions of BASIC, but this is optional in many microcomputer versions. The STOP statement can occur several times throughout a single program, with conditional jumps determining the actual end of the program. The END statement is of the form:

line number END

The line number of the END statement should be the largest line number in the program, since any lines having line numbers greater than that of the END statement are not executed.

The STOP statement is of the form:

line number STOP

When a STOP statement is executed, a message is displayed to tell the user the line number where the STOP occurred. The form of the message depends on the computer system being used. Typical examples are:

```
STOP AT LINE 1.n.                    (BASIC-PLUS)
READY
BREAK IN 1.n.                        (TRS-80 and Commodore)
READY
STOPPED AT LINE 1.n.
```

1.n. is the line number where the STOP was executed. Typing CONTINUE at this point (CONT in most microcomputers) causes execution to resume at the statement following STOP.

In some systems, execution of an END statement also causes a message, such as READY, to be displayed. This signals that the execution of a program has been terminated or completed, and BASIC is able to accept further input. The execution of an END statement also closes all files in a BASIC program.

4.14 PROGRAMMING EXAMPLES

The following examples illustrate how to write a program to solve electrical problems and obtain a printout of the results. Note the logical sequence of the statements.

Example 4.1 Using Eq. (2.1), write a program in BASIC to calculate the force of repulsion between two electrons separated by 1.75×10^{-10} meter.

```
10      LET Q1 = 1.6E−19:  Q2 = 1.6E−19:  K = 9E9:  R = 1.75E−10
20      F = K*Q1*Q2/R↑2
30      PRINT "FORCE IN NEWTONS IS" F
40      END

FORCE IN NEWTONS IS 7.52328E−09
```

Example 4.2 Write a program in BASIC to calculate the current represented by 1.56×10^{18} electrons moving past a given point in 2.55 seconds.

```
LIST
10      REM PROGRAM TO CALCULATE CURRENT
20      LET E1 = 1.56E18:  E2 = 6.24E18
25      LET T = 2.55
30      LET Q = E1/E2:  I = Q/T
40      PRINT "CURRENT IN MA IS" I*1E3
50      END

RUN
CURRENT IN MA IS 98.0392

READY
```

Example 4.3 Read the following resistor values into the computer: 1 kΩ, 3.3 kΩ, 2.2 kΩ, 4.7 kΩ, 6.8 kΩ. Print out all five resistors according to the following formats:

a. in a vertical column
b. in a horizontal row
c. in a diagonal line

```
LIST
10      REM     THIS PROGRAM SHOWS PRINTING TECHNIQUES
20      READ    R1,R2,R3,R4,R5
30      DATA    1E3,3.3E3,2.2E3,4.7E3,6.8E3
40      PRINT R1: PRINT R2: PRINT R3: PRINT R4: PRINT R5
45      PRINT:PRINT
50      PRINT R1,R2,R3,R4,R5
55      PRINT:PRINT
60      PRINT R1: PRINT ,R2: PRINT ,,R3: PRINT ,,,R4: PRINT ,,,,R5
70      END

RUN
 1000
 3300
 2200
 4700
 6800

 1000          3300          2200          4700          6800

 1000
               3300
                             2200
                                           4700
                                                         6800
```

Example 4.4 Calculate the range of values of each of the following resistors (22 kΩ ± 10%, 4.7 kΩ ± 5%, 68 kΩ ± 20%). Print out three columns containing *R*(low), *R*, and *R*(high). Label the columns as stated.

```
LIST
10      REM     THIS PROGRAM CALCULATES RESISTOR RANGE
20      R1 = 22E3
25      R2 = 4.7E3
30      R3 = 68E3
40      PRINT" R(LOW)               R                  R(HIGH)"
50      PRINT R1 - .1*R1,R1,R1 + .1*R1
60      PRINT R2 - .05*R2,R2,R2 + .05*R2
70      PRINT R3 - .2*R3,R3,R3 + .2*R3
80      END

RUN
 R(LOW)         R             R(HIGH)
 19800          22000         24200
 4465           4700          4935
 54400          68000         81600
```

Example 4.5 Given $A = 2.71$, $B = 13.8$, $C = 1.115$, and $D = 6.175$, calculate and print out the following results:

$$K_1 = \frac{AB - C}{D}$$

$$K_2 = \frac{AB}{D} - C$$

$$K_3 = \left(\frac{A}{D}\right)^B - C$$

$$K_4 = \frac{A^B - C}{D}$$

```
LIST
10      READ A,B,C,D
20      DATA 2.71, 13.8, 1.115, 6.175
30      K1 = (A*B - C)/D: K2 = (A*B)/D - C: K3 = (A/D)↑B - C
31      K4 = (A↑B - C)/D
40      PRINT"K1 = "K1
50      PRINT"K2 = "K2
60      PRINT"K3 = "K3
70      PRINT"K4 = "K4
80      END

RUN
K1 = 5.87579
K2 = 4.94136
K3 = -1.11499
K4 = 152876
```

SUMMARY

Computer programming of technical problems has advanced to a near conversational mode. A constant interchange occurs between the machine and the programmer, facilitating correction of programming errors on a real-time basis. Timesharing techniques permit multiple users to share systems having more powerful capabilities than are usually available to single, dedicated users.

The BASIC programming language permits mixing both fixed point and floating point operands. All BASIC statements must be numbered. The computer processes the program in an ascending, numerical sequence. An order of priority is established in computing arithmetic operations.

Several methods are used to input data. The READ, DATA statements are used together in the program to initialize variables. The INPUT statement requests a response from the user during program execution. The PRINT statement is the most frequently encountered statement to output data. In addition to the calculations, the PRINT statement facilitates formatting the output. The END statement is the final statement in a program. The STOP statement may be utilized to temporarily halt proceedings. Program execution is resumed by typing CONTINUE.

PROBLEMS

Reference Section 4.13

Write a program in BASIC to do the following:

1. Print out the letters BASIC horizontally, vertically, and diagonally.
2. Given $R_1 = 4.7\,\text{k}\Omega$, $R_2 = 680$ ohms, and $R_3 = 2.2\,\text{k}\Omega$, calculate and print out the values of A, B, C and D for the following equations:

$$A = R_1 \times R_2$$
$$B = (R_1/R_2) \times R_3$$
$$C = (R_1 - R_2)/R_3$$
$$D = \frac{(R_1)(R_2) + (R_1 - R_3)}{(R_2)(R_3)}$$

3. Given $A = 17.15$, $B = -3.61$, and $C = 2.8$, calculate and print out the solution of D for any given value of X:

$$D = AX^2 + BX + C$$

   Hint: Use the READ, DATA statement for A, B, and C. Use the INPUT statement for X.
4. Given $A = 2.34$, $B = 1.78$, and $C = 1.15$. Write a program to evaluate and print out the following solutions of T:

$$T_1 = A \times BC - B^A C$$
$$T_2 = (AC)^B - (C + B)^A$$
$$T_3 = A + 2B^C + A^B C$$

5. Calculate and print out the number of free electrons in 100 meters of copper wire having a diameter of 0.175 mm.
6. Calculate and print out the radius in mils of magnesium wire that is 150 feet long and has a resistance of 1.52 ohms.
7. Calculate and print out the dimensions in inches of a side of a square copper bar that is 3 feet long and has a resistance of .00000135 ohm.
8. Calculate and print out the resistance of a copper rod at 55°C if its resistance was 67.5 ohms at 20°C.
9. Calculate and print out the percentage of change of manganin wire resistance if the temperature decreases from 20°C to −65°C.

CHAPTER 5

OHM'S LAW, POWER, AND EFFICIENCY

5.1 SCHEMATIC SYMBOLS

In the preceding chapters, voltage, current, and resistance were illustrated pictorially. In diagramming electrical circuits, however, it is impractical to represent circuit elements this way. The electronics industry has adopted instead a set of simple symbols to represent these circuit elements. Figure 5.1(a) shows the symbol for a source of voltage. Notice that the longer of the two vertical lines represents the positive terminal, and the shorter the negative terminal. In circuit diagrams a series of voltage cells is sometimes stacked together as in Figure 5.1(b). In practice, each cell of a battery produces approximately 1.5 volts. Figure 5.1(b) would depict a 6-volt battery source, since four cells have been cascaded (connected in succession). As voltage sources increase in magnitude, even this kind of diagram becomes impractical. To alleviate the difficulty, this text will use the two simplified voltage schematics of Figures 5.1(c) and 5.1(d). The magnitude of the voltage will be written next to the symbol.

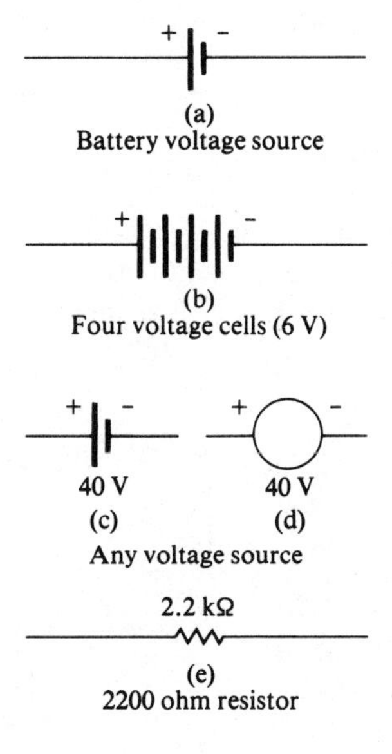

Figure 5.1 Electronic Symbols

Resistance is symbolized by a series of jagged lines, as shown in Figure 5.1(e). The magnitude of the resistance is written next to the symbol. Figure 5.1(e) indicates that the resistor has a magnitude of 2.2 kilohms (2200 ohms).

When a voltage source and a resistance are connected by conductive wiring, the circuit shown in Figure 5.2 results. Note that the current (I) is shown merely as an arrow indicating the direction in which current is presumed to flow (see Section 5.2). Current is shown flowing counterclockwise in Figure 5.2. The direction of electron flow is from the negative terminal of the 20 V source, through the 10 kΩ resistor, to the positive terminal. Note that a voltage difference is indicated across the resistor. Electron flow develops a potential difference across resistors, in the direction of the flow, from negative polarity to positive polarity.

For all practical purposes, the wire connecting the resistor to the voltage source is considered to be of negligible resistance; therefore, all the voltage pressure of the source is utilized in pushing current through the resistor. In the circuit of Figure 5.2, all 20 volts of the source are *dropped* across the 10 kΩ resistor; hence V_{AB} equals 20 V. The convention used in subscripted notation is that the first subscripted letter is the reference polarity to which the second subscripted letter is compared. If point B is more positive than point A, the polarity is positive. If point B is more negative than point A, the polarity is negative. Henceforth, the source will be designated simply as V (volts). Voltage drops across elements that consume the energy of the source will generally be labeled with subscripted letters.

5.2 DIRECTION OF CURRENT FLOW

Up to this point, we have portrayed freed valence electrons as the carriers of current. Current has been shown flowing from negative to positive in the external circuit and from positive to negative in the voltage source itself.

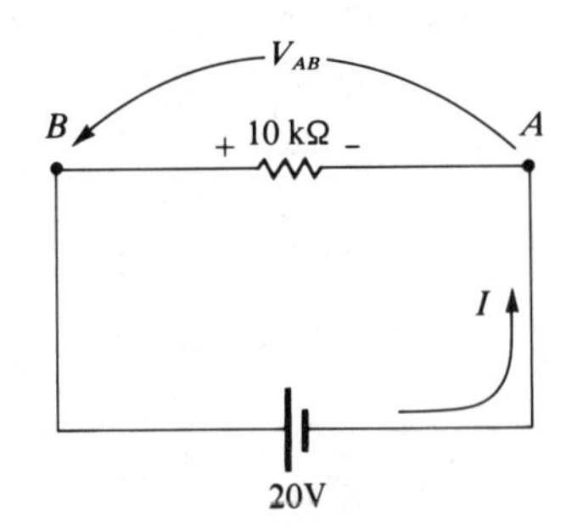

Figure 5.2 Circuit Illustrating Voltage, Current, and Resistance

Historically, the earliest experiments on the nature of electricity preceded our present knowledge of atomic structure. Current was known to be a migration of charge, but the direction of migration was not known. An arbitrary decision was made to consider the direction of current flow to be from positive to negative. In consequence, a pattern was established: texts on electronic theory portrayed current as flowing from positive to negative through the external circuit. We have adopted this "conventional" current flow in this text, since it forms a basis for the later study of solid-state devices.

Figure 5.3 illustrates both *electron* current flow and *conventional* current flow. Furthermore, it shows that there is no difference in the voltage polarity developed across the resistance. In Figure 5.3(a), electron current flow is from point A to point B, leaving a "−" to "+" voltage drop. V_{AB} is positive, since point B is positive with respect to point A. In Figure 5.3(b), conventional current flow is from point B to point A, leaving a "+" to "−" voltage drop. However, V_{AB} is still positive, because point B is still positive with respect to point A. In either figure, V_{BA} is negative (because point A is negative with respect to point B) and equals $-V_{AB}$.

5.3 OHM'S LAW

Earlier analogies were constructed between sources of pressure, such as water pumps and batteries, between the length and diameter of water lines and the resistance of circuit elements, and between the quantity of water moved and the net migration of charge. In the water analogy, the total quantity of water moved must be proportional to the pressure exerted by the pump and inversely proportional to the total opposition

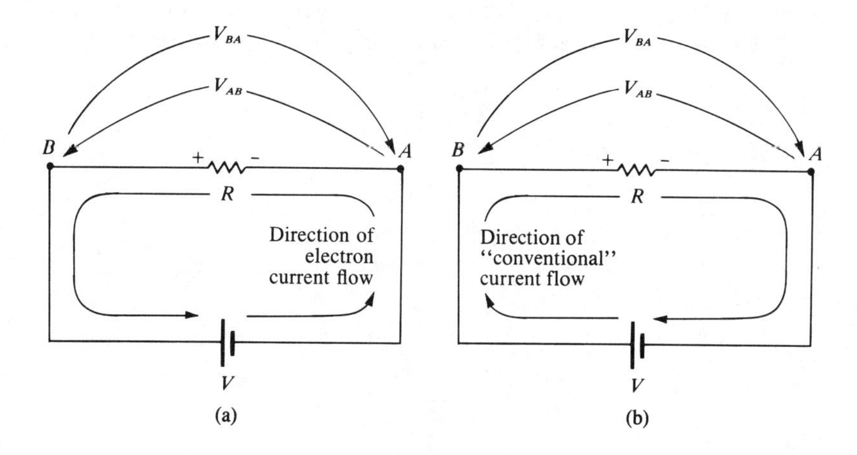

Figure 5.3 Electron Current Flow vs Conventional Current Flow

offered by the water lines. This same principle may be extended to the electrical circuit. The total amount of charge moved per second (I) is directly proportional to the magnitude of the voltage source (V) and inversely proportional to the total resistance of the circuit (R).

Ohm's law states a relationship among these three quantities. If 1 volt is impressed across 1 ohm of resistance, the resultant current flow will be 1 ampere. Expressed mathematically,

$$\boxed{I = \frac{V}{R}} \tag{5.1}$$

This equation can be rearranged to express resistance in terms of voltage and current:

$$\boxed{R = \frac{V}{I}} \tag{5.2}$$

or to express voltage in terms of current and resistance:

$$\boxed{V = IR} \tag{5.3}$$

Given any two of the three quantities, the unknown quantity can always be determined.

The relationships just given imply a linear resistance—that is, there is no current flow through the resistor when the voltage is zero, and the current increases in direct proportion to any increase in voltage. Figure 5.4 illustrates such a relationship for two resistors, R_1 and R_2. If the source voltage is increased from zero to V, the current through R_1 increases from zero to I_1 and the current through R_2 increases from zero to I_2. The increase through R_1 and R_2 is linear: every increase in voltage

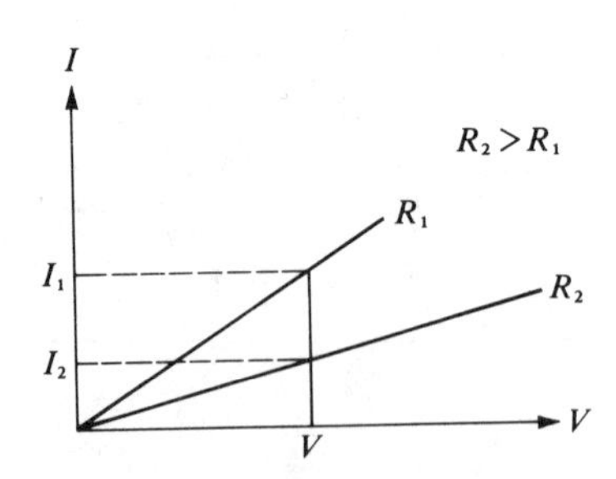

Figure 5.4 Linear Resistors

produces a proportional increase in current. Notice that for any given value of voltage, the current through R_1 is greater than the current through R_2. Since current is inversely proportional to resistance, R_1 must be smaller in magnitude than R_2. The slope of a resistance line on an $I-V$ graph indicates its magnitude. The closer the resistance line is to the voltage axis, the greater is the magnitude of the resistance.

Example 5.1 Determine the magnitude of I in Figure 5.5. Note that conventional current flow is assumed.

$$I = \frac{V}{R} = \frac{20 \text{ V}}{100 \ \Omega} = 0.2 \text{ ampere}$$

Example 5.2 Determine the magnitude of I in Figure 5.6.

$$I = \frac{V}{R} = \frac{8 \text{ V}}{2.2 \text{ k}\Omega} = \frac{8}{2.2 \times 10^3}$$

$$= 3.64 \text{ mA}$$

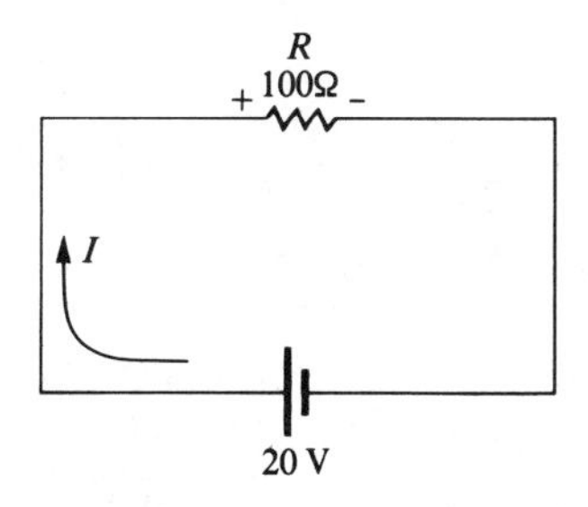

Figure 5.5 Reference Example 5.1

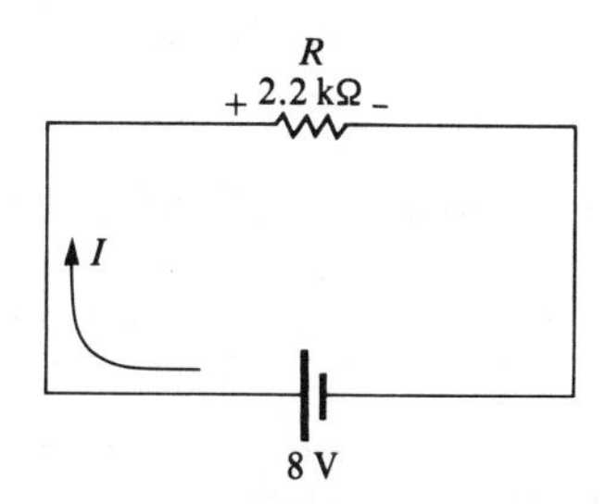

Figure 5.6 Reference Example 5.2

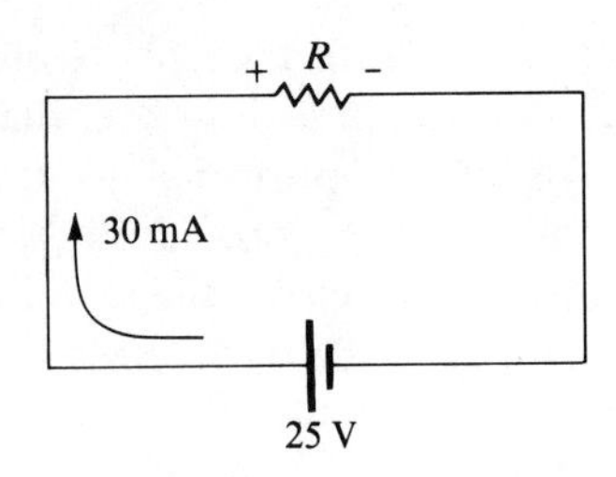

Figure 5.7 Reference Example 5.3

When volts are divided by R, expressed in kilohms, the answer is in milliamperes.

$$I = \frac{\text{V}}{R \times 10^3} = I \times 10^{-3} = I \text{ mA}$$

Example 5.3 Determine the magnitude of R in Figure 5.7.

$$R = \frac{V}{I} = \frac{25 \text{ V}}{30 \times 10^{-3}} = \frac{5}{6} \times 10^3 \text{ ohms} = 833 \text{ ohms}$$

Example 5.4 What value of resistance should be connected across a 12 V source if 25 μA is to flow?

$$R = \frac{V}{I} = \frac{12 \text{ V}}{25 \times 10^{-6} \text{ A}} = \frac{12}{25} \times 10^6 \text{ ohms}$$

$$= 0.48 \text{ M}\Omega = 480 \text{ k}\Omega$$

Example 5.5 How much voltage is dropped across a 2.2 kΩ resistor if 27.5 mA of current flows through the resistor?

$$V = IR = 27.5 \times 10^{-3} \text{ A} \times 2.2 \times 10^3 \text{ ohms}$$

$$= 27.5 \times 2.2 \times 10^0 \text{ V} = 60.5 \text{ volts}$$

Example 5.6 Determine V_{AB} in Figure 5.8.
Since point A is at a higher potential than point B, V_{AB} is negative.

$$V_{AB} = -IR$$

$$= -(7.5 \text{ mA} \times 3.3 \text{ k}\Omega) = -(7.5 \times 3.3) \text{ volt} = -24.75 \text{ volt}$$

5.4 POWER

When force is applied over a distance—for example, in lifting a heavy weight—*energy* is expended and *work* is performed. The amount of work

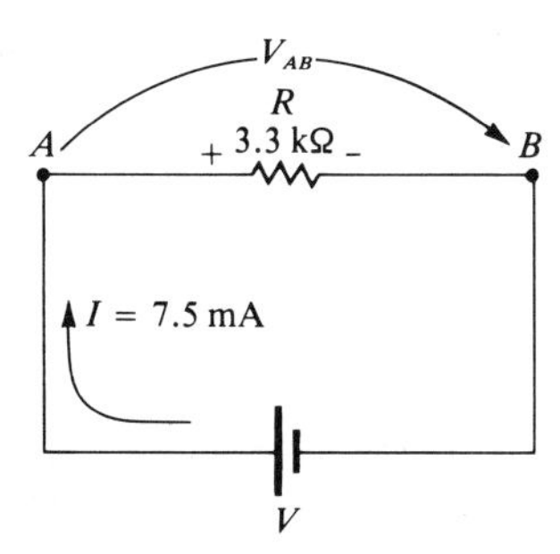

Figure 5.8 Reference Example 5.6

done depends on the amount of force and the distance through which it is applied. In the MKS system, the unit of work is the *newton-meter*, or *joule*. If one person can perform a given amount of work in a short time, whereas it takes another person a longer time to do the same amount of work, we say that the first person has more *power*. Power, then, relates work to the time in which it can be done. Since the unit of work is the joule and the unit of time is the second, power is measured in *joules per second*, or *watts*:

$$\boxed{P = \frac{W(\text{newton-meter})}{t(\text{second})} = \frac{\text{joule}}{\text{second}} \equiv \text{watt}} \tag{5.4}$$

The watt is a familiar unit: everyone has heard it in connection with household appliances and light bulbs. It expresses the rate at which an electrical device converts energy. Equations for power are typically expressed in terms of current, voltage, and resistance. From Eq. (2.4),

$$W = QV \tag{5.5}$$

Dividing both sides of Eq. (5.5) by t,

$$\frac{W}{t} = P = \frac{QV}{t} = \left(\frac{Q}{t}\right)V \tag{5.6}$$

According to Eq. (2.3), $I = Q/t$. Therefore, Eq. (5.6) may be rewritten as

$$P = IV \tag{5.7}$$

This equation is in a form more convenient for ordinary use.

Example 5.7 How much power is consumed by a toaster plugged into a standard household outlet of 120 V, if the current rating is 5 A?

$$P = IV = 120\text{ V} \times 5\text{ A} = 600\text{ watts}$$

Example 5.8 How much power is consumed by the resistor of Figure 5.9?

$$I = \frac{V}{R} = \frac{20\text{ V}}{10\text{ k}\Omega} = 2\text{ mA}$$

$$P = IV = 2\text{ mA} \times 20\text{ V} = 40\text{ mW}$$

Instead of expressing power in terms of voltage and current, we can derive alternate formulas relating power to current and resistance or to voltage and resistance. Equation (5.1) expresses the current through a resistor in relation to the voltage and resistance.

$$I = V/R$$

Substituting into Eq. (5.7) results in

$$P = IV = \left(\frac{V}{R}\right)V = \frac{V^2}{R} \tag{5.8}$$

Alternately, we can use Eq. (5.3), which relates the voltage across a resistor to the current and resistance:

$$V = IR$$

Substituting into Eq. (5.7) results in

$$P = IV = I(IR) = I^2R \tag{5.9}$$

The three equations relating current, voltage, and resistance to power are summarized as follows:

$$\boxed{P = IV} \tag{5.7}$$

$$\boxed{P = V^2/R} \tag{5.8}$$

$$\boxed{P = I^2R} \tag{5.9}$$

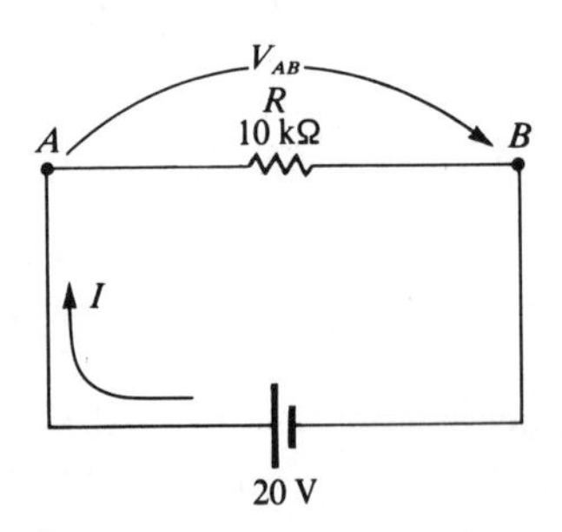

Figure 5.9 Reference Example 5.8

Which equation to use depends on what two quantities are known. If current and voltage are known, use Eq. (5.7). If voltage and resistance are known, use Eq. (5.8). If current and resistance are known, use Eq. (5.9).

Example 5.9 How much power is consumed by the resistor in Figure 5.10?

$$P = V^2/R = \frac{(-30)^2\,\text{V}}{4.7\ \text{k}\Omega} = \frac{900}{4.7} \times 10^{-3}\ \text{watt}$$
$$= 191.49\ \text{mW}$$

Example 5.10 How much power is consumed by resistor R in Figure 5.11?

$$P = I^2R = (25 \times 10^{-3}\ \text{A})^2 \times (22 \times 10^3)\ \text{ohms}$$
$$= (625 \times 10^{-6}) \times (22 \times 10^3)\ \text{watts}$$
$$= (0.625 \times 10^{-3}) \times (22 \times 10^3)\ \text{watts}$$
$$= 13.75\ \text{watts}$$

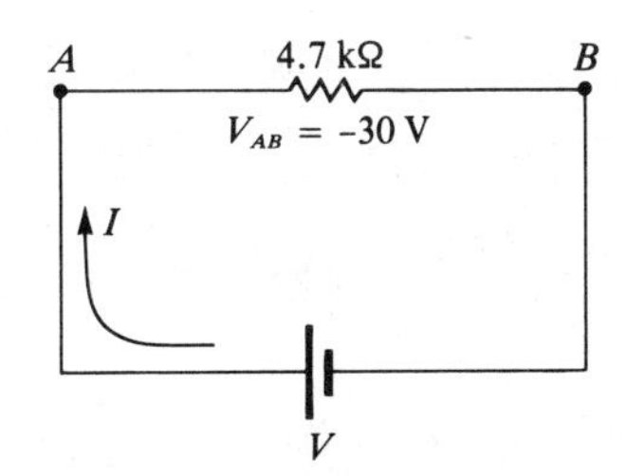

Figure 5.10 Reference Example 5.9

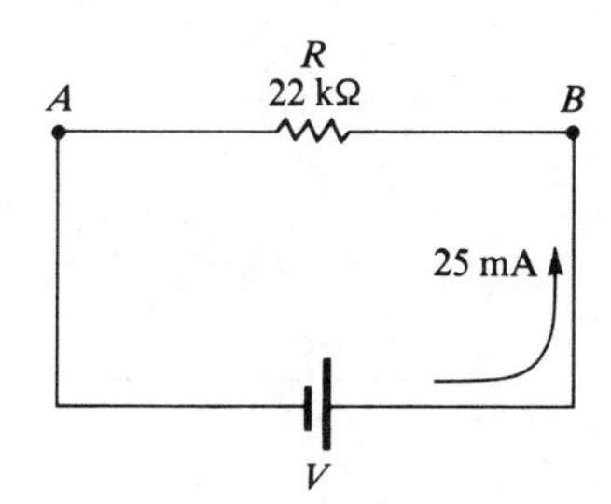

Figure 5.11 Reference Example 5.10

Example 5.11 What is the value of the resistance R in Figure 5.12?

$$P = I^2R$$

$$R = P/I^2 = \frac{23.5\ \text{W}}{(45 \times 10^{-3}\ \text{A})^2} = \frac{23.5}{2025 \times 10^{-6}}$$

$$= \frac{23.5}{2.025} \times 10^3\ \text{ohms}$$

$$= 11.6\ \text{k}\Omega$$

Example 5.12 What is the value of the voltage V_{AB} in Figure 5.13?

$$P = (V_{AB})^2/R$$

$$V_{AB} = \sqrt{PR} = \sqrt{100 \times 10^{-3} \times 5.6 \times 10^3}$$

$$= \sqrt{560} = 23.7\ \text{volts}$$

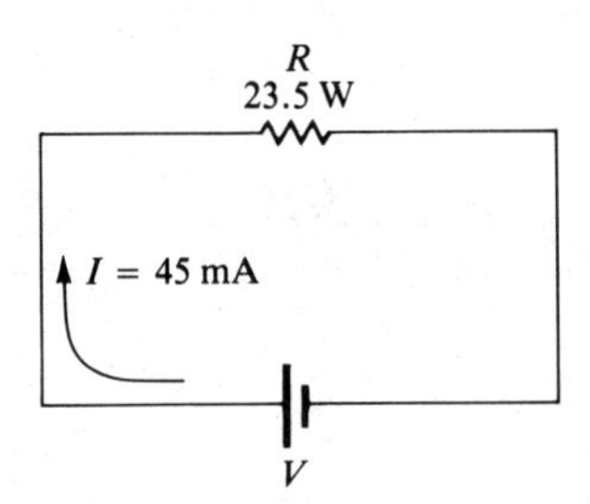

Figure 5.12 Reference Example 5.11

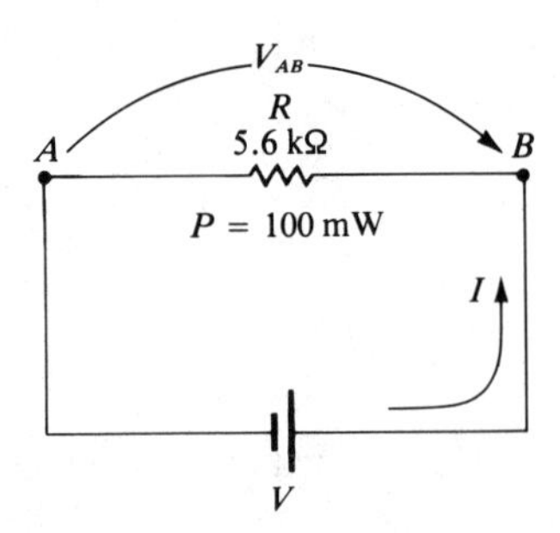

Figure 5.13 Reference Example 5.12

5.5 ENERGY

The most common exposure to energy is the monthly or bimonthly electric bill. When an electrical device is plugged into a standard household outlet, the electric company furnishes the voltage (electrical pressure) and the current to drive the device. What the electric company is selling, then, is the energy to drive the device.

We can rearrange Eq. (5.4).

$$\text{Energy (joules)} = P(\text{watts}) \times t(\text{seconds}) \qquad (5.10)$$

Energy, therefore, is a measurement of power times the lapse of time over which it is consumed. What your monthly electric bill indicates is the product of power times how long that power was used. This seems reasonable, since it would not be fair for one person to pay the same amount for using 1000 W of energy for 1 minute as compared to another person who used 1000 W for 1 hour. One's total electric bill is not just a measurement of how many watts were used, but also how long this use was taking place. Now in practice, a watt-second is an extremely small amount of energy; therefore, a more convenient unit is the measurement of 1000 watts (1 kW) consumed for 1 hour.

$$\text{Energy (kWh)} = \frac{\text{power (watts)} \times \text{time (hours)}}{1000} \qquad (5.10a)$$

Households have *watt-hr meters* installed by the electric company to measure this total consumption. The calibration of this meter also takes into consideration the time over which power is consumed. The numbers on the dials indicate the new value of kilowatthour, and when the previous month's reading is subtracted from this value, then the current month's total kilowatthour consumption is known.

Example 5.13 A TV is played for 4 hours. If it is rated as 2.5 A at 120 V, how much cost is incurred if the rate is 4.5 cents per kWh?

$$W\text{ (kWh)} = \frac{2.5 \times 120 \times 4}{1000} = 1.2\text{ kWh}$$

$$\text{Cost} = \frac{4.5¢}{\text{kWh}} \times 1.2\text{ kWh} = 5.4¢$$

Example 5.14 How long must a 100 W light bulb burn to cost 10 cents, if the rate is 4.5 cents/kWh?

A 100 W bulb, burning for 10 hours, consumes 1000 W or 1 kW of power. Therefore,

$$\text{Cost} = 4.5¢ \text{ for 10 hours of use}$$

By ratios,

$$\frac{10 \text{ hours}}{4.5¢} = \frac{X \text{ hours}}{10¢}$$

$$X \text{ (hours)} = \frac{10 \times 10}{4.5} = 22.22 \text{ hours}$$

Work or energy was initially expressed as joules in the MKS system and ft-lb in the English system. Both are measurements relating force and distance. As such, they are essentially mechanical measurements of power. Just as the watt is the electrical unit of power, *horsepower* is the mechanical equivalent, and it is commonly used to rate automobile engines and electric motors. To determine a relationship between horsepower and watts, the following equation is needed:

$$1 \text{ hp} = \frac{33{,}000 \text{ ft-lb}}{\text{minute}} = 550 \frac{\text{ft-lb}}{\text{second}} \tag{5.11}$$

This expression was determined by James Watt when comparing the output power of an engine to the output power of a workhorse.

Example 5.15 Convert 1 hp to watts.

$$1 \text{ hp} = \frac{550 \text{ ft-lb}}{\text{second}} \times \frac{1 \text{ meter}}{3.28 \text{ ft}} \times \frac{4.45 \text{ newtons}}{\text{lb}} = 746 \frac{\text{N-meter}}{\text{second}}$$

By definition, (newton-meter/second) = (joule/second) = watt. Therefore

$$\boxed{1 \text{ hp} = 746 \text{ W}} \tag{5.12}$$

Example 5.16 What is the cost of running a $2\frac{1}{2}$ hp motor for 24 hours at 4.5 cents/kWh?

$$2.5 \text{ hp} \times 746 \frac{\text{W}}{\text{hp}} = 1.865 \text{ kW}$$

$$1.865 \text{ kW} \times 24 \text{ hours} \times \frac{4.5¢}{\text{kWh}} = \$2.014$$

Example 5.17 An electric motor is rated at 1 hp. If it is connected across 203 V, what is its current rating?

$$1 \text{ hp} = 746 \text{ W} = IV = I\,(203)$$

$$I = \frac{746}{203} = 3.67 \text{ amperes}$$

Example 5.18 Assume a motor to be 100% efficient. If it is connected across 120 V and delivers 1500 ft-lb/s of power, what is its current rating?

$$\frac{1500 \text{ ft-lb/s}}{550 \dfrac{\text{ft-lb/s}}{\text{hp}}} = 2.727 \text{ hp}$$

$$2.727 \text{ hp} \times \frac{746 \text{ W}}{\text{hp}} = 2034 \text{ watts}$$

$$I = \frac{P}{V} = \frac{2034}{120} = 16.95 \text{ amperes}$$

5.6 EFFICIENCY

Example 5.18 assumed a motor to be 100% efficient. In reality, no device is 100% efficient. Friction, along with other types of mechanical and electrical imperfections, results in all devices producing less output power than was originally supplied to drive the device. The efficiency of a device, therefore, is effectively a measurement of how well the device delivers power compared to how much power it consumed. Expressed mathematically,

$$\text{Efficiency} = \frac{\text{power out}}{\text{power in}}$$

The symbol for efficiency is lower case Greek eta.

$$\boxed{\eta \text{ (efficiency)} = \frac{P_o}{P_i}} \qquad (5.13)$$

Example 5.19 A motor delivers 1 hp of power. It draws 7 A at 120 V. What is its efficiency?

$$P_i = 7 \times 120 = 840 \text{ watts}$$

$$P_o = 1 \text{ hp} = 746 \text{ watts}$$

$$\therefore \quad \eta = \frac{P_o}{P_i} = \frac{746}{840} = 0.8881$$

To convert to percent, multiply by 100.

$$\eta = 88.81\%$$

Example 5.20 A motor draws 12 A at 203 V. It delivers 3 hp. What is its efficiency?

$$P_o = 3\text{ hp} = 3 \times 746\text{ W} = 2238\text{ watts}$$
$$P_i = IV = 12 \times 203 = 2436\text{ watts}$$
$$\eta = \frac{P_o}{P_i} = \frac{2238}{2436} = 0.9187 = 91.87\%$$

Example 5.21 A motor is 85% efficient. It delivers 2 hp when connected across a 120 V line. What current does it draw?

$$\eta = \frac{P_o}{P_i} = 0.85 = \frac{2 \times 746}{I \times 120}$$
$$I = \frac{1492}{120 \times 0.85} = 14.63\text{ amperes}$$

Suppose a series of devices is connected together, such as in Figure 5.14. If the input power is 1 kW, and device 1 is 80% efficient, its output power is 0.8 kW. This is the input power to device 2. In turn, if device 2 is 85% efficient, it can only deliver 85% of what it receives as its input power. In this example, device 2 has 0.8 kW as input and not 1 kW, which is the input to the overall system. System output power, then, is 85% of 80% of 1 kW, for a net output of 0.68 kW. Generally speaking, if a series of devices is connected together, the final output percentage is equal to the product of the percentages of each device in the chain.

$$\boxed{P_o = (\eta_1 \eta_2 \cdots \eta_n) P_i} \tag{5.14}$$

Example 5.22 A generator supplies voltage and current to a motor. The generator is 90% efficient and the motor is 80% efficient. If the input to the generator is mechanically equal to 1750 ft-lb/s, what is the final output power rated in horsepower?

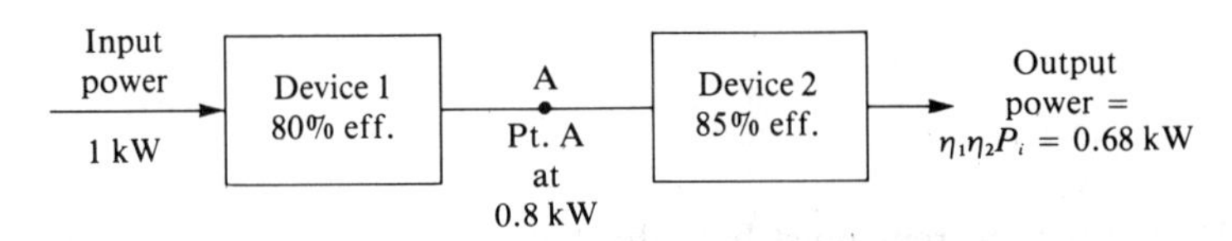

Figure 5.14 Efficiency of Devices in Series

$$\frac{1750 \text{ ft-lb/s}}{550 \dfrac{\text{ft-lb/s}}{\text{hp}}} = 3.182 \text{ hp}$$

$$\begin{aligned} P_o &= \eta_1 \times \eta_2 \times P_i \\ &= 0.90 \times 0.80 \times 3.182 \\ &= 2.291 \text{ hp} \end{aligned}$$

SUMMARY

The electronics industry has established symbols to depict various circuit elements. Additionally, subscripts can be utilized to establish a point of reference to which another point is compared. Generally speaking, the first subscript is the reference point and the second subscript is the point which is compared to this reference.

Ohm's law represents the basic equation that describes quantitative circuit analysis. It relates the magnitude of current through a device to the voltage pressure and total resistance within the circuit. There are currently two accepted theories concerning the direction that current can be assumed to flow. Conventional current flow is from the positive terminal of the source, through the resistive elements comprising a closed circuit, and to the negative terminal of the source. Electron current flow is opposite to the assumed direction of conventional current. Regardless of the direction chosen, voltage polarity and magnitude are consistent.

Power is a measurement of the work performed per unit length of time. The basic unit of energy is the kilowatt-hour. The cost of energy is the product of the total energy consumed times the cost per kWh. The relationship between mechanical power (measured in horsepower) and electrical power (measured in watts) is 1 hp = 746 watts.

Efficiency is the ratio of the power delivered by a device to the power the device consumes. This ratio is always less than unity. If a series of devices is cascaded, the total efficiency of the system is the product of the efficiencies of each device.

PROBLEMS

Reference Section 5.3

1. What is the voltage dropped across a 3.3 kΩ resistor if 12 mA of current flows through it?
2. How much current must flow through a 330 kΩ resistor if 16.5 V is to be developed across the resistor?

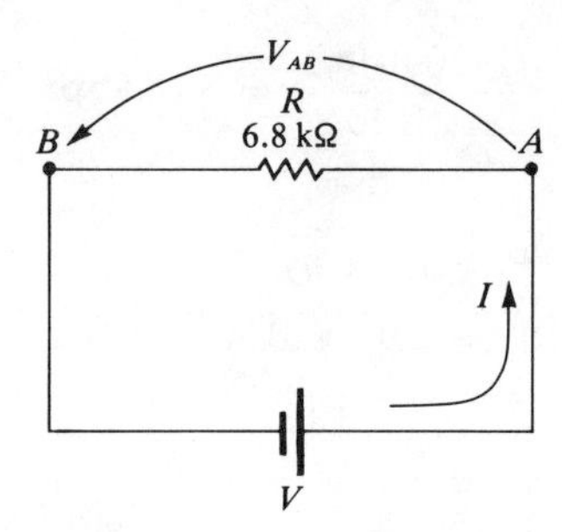

Figure 5.15 Reference Problem 4

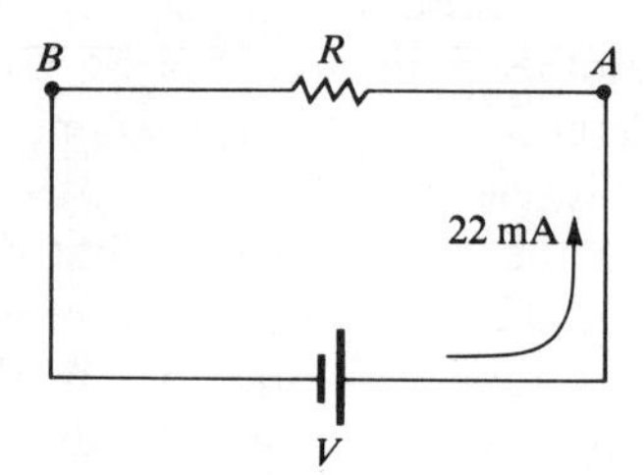

Figure 5.16 Reference Problem 5

3. What value of R would drop 13.85 V if 16 mA were flowing through it?
4. V_{AB} in Figure 5.15 is -32 V. What is the magnitude of current through R?
5. What is the value of R in Figure 5.16 if $V_{BA} = 30$ V?
6. What is the current magnitude if 22.5 V is impressed across a resistor of brown, black, orange color bands?
7. What is the maximum current that can flow in the circuit of Problem 6? What is the minimum current?
8. A circuit has a voltage source of 20 V. Allowable current tolerance is 18 mA to 22 mA. What should be the color bands on the resistor?

Reference Section 5.4

9. How much power is dissipated by R in Figure 5.17?
10. If the power dissipated by a resistor is 700 mW, what current flows if $R = 6.8$ kΩ?
11. What is the voltage across a 47 kΩ resistor if it dissipates 2 W of power?
12. The power dissipated by a resistor is 1.8 W. The voltage across the resistor is 10 V. What is the magnitude of the current?

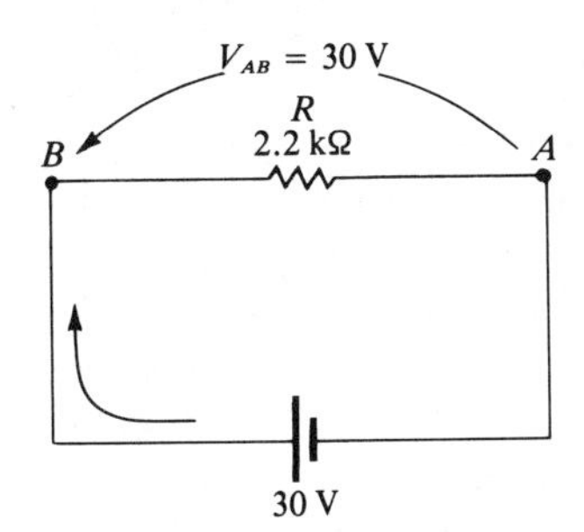

Figure 5.17 Reference Problem 9

13. A device converts 2400 joules of energy over a timespan of 1 minute. How many watts are dissipated by the device per second?

Reference Section 5.5

14. A 60 W light bulb burns for 10 hours a day, 7 days a week. If the rate for power is 4.7¢ per kWh, how much does it cost to burn this light for a year?
15. A device draws 4.8 A at 120 V. How much does it cost to run the device for 12 hours if the rate is 4.6¢ per kWh?
16. A force of 14,000 newton-meters is applied over a period of 2.5 minutes. Convert this to a horsepower rating.
17. A force of 18,750 ft-lb is applied over a period of 6 minutes. Convert this to watts.
18. A device draws 1.85 A when connected across 203 V. It costs 24¢ to run the device for 10 hours. What is the cost per kWh?
19. A $3\frac{1}{2}$ hp motor is connected across 120 volts. What is the current rating of the motor?
20. The power rating of a device is 720 ft-lb/s. If it is connected across 203 V, what is its current rating?
21. What is the rating in newton-meters/minute of a 2.2 hp motor?
22. If the cost per kWh is 4.7¢, how much would it cost to run a device for 24 hours if the device delivered 18,750 N · m/s of power?
23. A soldering iron is rated at 250 W. If power costs 4.2¢ per kWh, how much does it cost per year to heat the iron 12 hours per day, 5 days a week?
24. An oscilloscope is turned on for 60 hours per week. When connected across a 120 V source, it was calculated that the cost of operating the instrument would be $3.15 per week. How much current does the oscilloscope draw if the cost is 4.5¢ per kWh?

Reference Section 5.6

25. A motor delivers 1.5 hp. It is 92% efficient. How much does it cost to run the motor for 24 hours if the rate is 3.9¢ per kWh?
26. It costs $0.60 to run a $\frac{3}{4}$ hp motor for 18 hours. If the efficiency is 87%, what is the rate per kWh?
27. Device number one is 78% efficient. It is cascaded to device number two, which is 81% efficient. If the input is 2.5 A at 120 V, what horsepower is available at the output?
28. Two devices of equal efficiency are cascaded. If the input power is 3 hp and the output power is 1.8 hp, what is the efficiency of each device?
29. Three devices are cascaded. Each device is 90% efficient. The output power is 0.68 hp. If the input is connected across 203 V, what is the input current?

Optional Computer Problems

30. Write a program in BASIC to accept any input value of horsepower. Calculate and print out the equivalent value in watts and foot-pounds per second.
31. Write a program in BASIC to input any value of power in kW delivered to a system. The system consists of two devices cascaded together: device 1 is 87.5% efficient and device 2 is 91.25% efficient. Calculate and print out the output power rating of each device in horsepower.

CHAPTER 6

SERIES AND PARALLEL CIRCUITS

6.1 CURRENT AND RESISTANCE: SERIES CIRCUITS

Frequently, in practical circuits, more than one resistor or other device is involved within a closed circuit. Assume that, initially, resistor R is connected across a voltage source, as in Figure 6.1(a). The electrical symbolism is illustrated in Figure 6.1(b). The resistor has a basic length ℓ, a resistivity ρ, and a cross-sectional area A. By Eq. (3.7),

$$R = \frac{\rho \ell}{A}$$

which, for the resistor in Figure 6.1, is 10 kΩ.

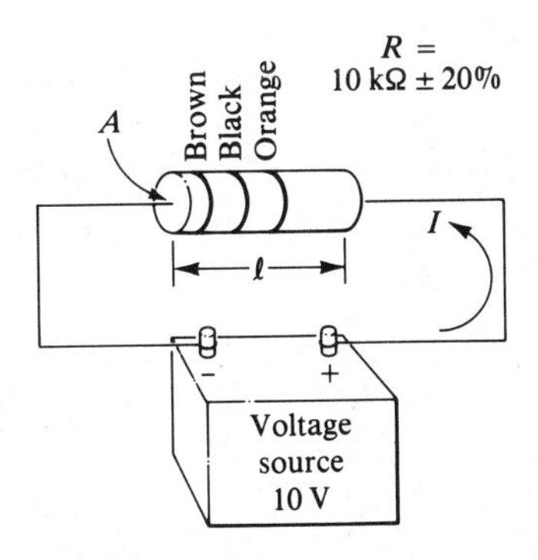

Figure 6.1(a) Closed Resistive Circuit

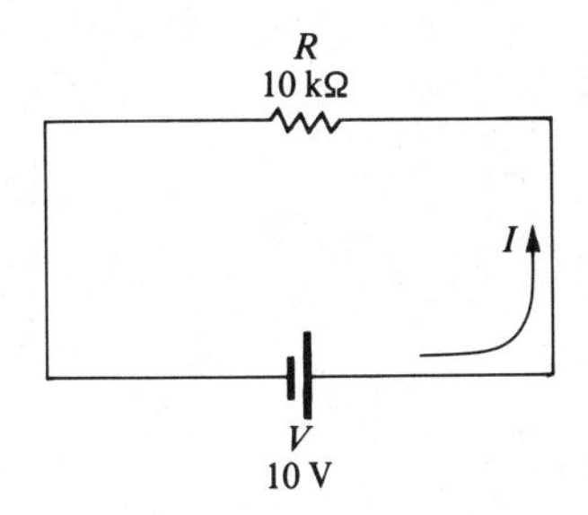

Figure 6.1(b) Electrical Equivalent of Figure 6.1(a)

What would happen if R were divided into two equal parts? All factors would remain constant, except ℓ, which would be one-half its former length. Since resistance is directly proportional to length, each half of the former R would be equal to one-half its former value. Figure 6.2 would result, with each half equal to 5 kΩ.

Reason suggests that the current I in Figure 6.1 must equal the I in Figure 6.2, since the resistance is the same in both figures. If this is so, then the sum of R_1 and R_2 in Figure 6.2 must equal the original R in Figure 6.1. Expressed mathematically, in this closed series circuit,

$$R_T = R_1 + R_2 \tag{6.1}$$

By definition, *a series circuit means that there is only one path through which current may flow.* In Figure 6.2, all of the I must flow through both R_1 and R_2 in moving from the positive terminal of V to the negative terminal. Implicit in this statement is the fact that current is the same at all points in the series circuit. To assume otherwise would be analogous to saying that water flowing through a pipe "bunches up" at selected points. If the number of electrons per second flowing past a cross-sectional

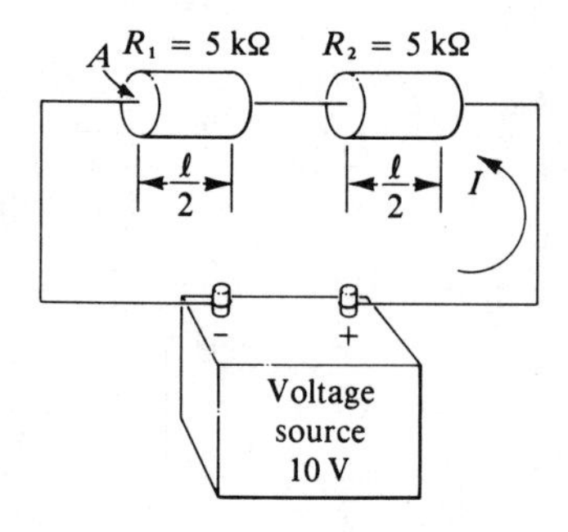

Figure 6.2(a) Resistor of Figure 6.1 Halved

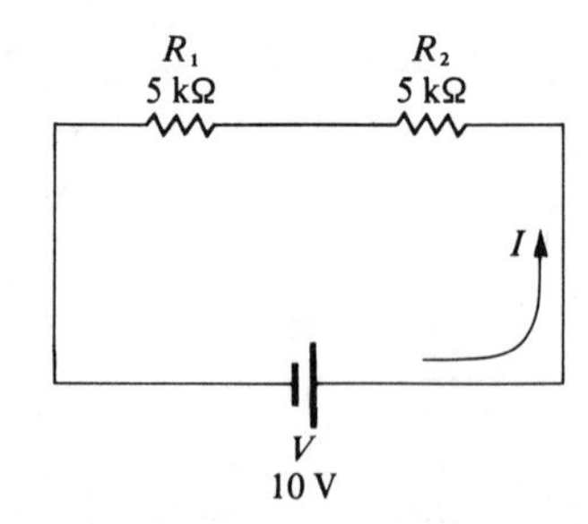

Figure 6.2(b) Electrical Equivalent of Figure 6.2(a)

area in an electrical circuit were counted, it would be the same at any point in the circuit.

In summary, the following two important conclusions are established for a series circuit:

1. The total resistance "seen" by the source in a series circuit is the sum of all the individual resistances.
2. The current flowing in a series circuit is the same at all points in the circuit.

Example 6.1 What is the total resistance of Figure 6.3?

$$\begin{aligned} R_T &= R_1 + R_2 + R_3 \\ &= 10\text{ k}\Omega + 5\text{ k}\Omega + 8.7\text{ k}\Omega \\ &= 23.7\text{ k}\Omega \end{aligned}$$

Example 6.2 What is the total current of Figure 6.3?

$$I_T = \frac{V}{R_T} = \frac{20\text{ V}}{23.7\text{ k}\Omega} = 0.844\text{ mA}$$

Example 6.3 What is the total current of Figure 6.4?
First, find the total resistance.

$$\begin{aligned} R_T &= R_1 + R_2 + R_3 + R_4 \\ &= 10\text{ k}\Omega + 8.5\text{ k}\Omega + 1.2\text{ k}\Omega + 3.7\text{ k}\Omega \\ &= 23.4\text{ k}\Omega \end{aligned}$$

The total current may now be found using Ohm's law.

$$I_T = \frac{V}{R_T} = \frac{50\text{ V}}{23.4\text{ k}\Omega} = 2.137\text{ mA}$$

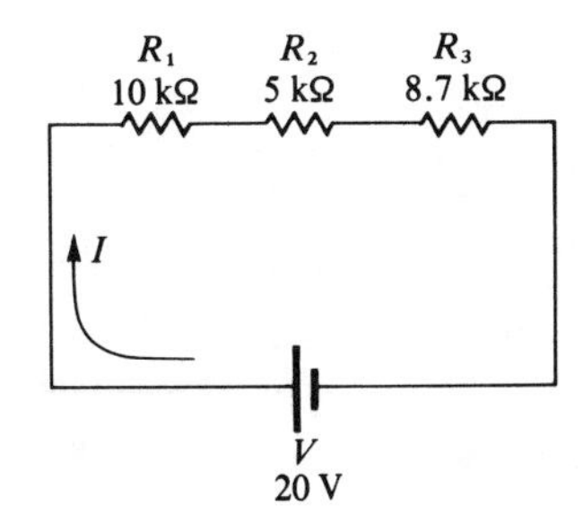

Figure 6.3 Reference Examples 6.1 and 6.2

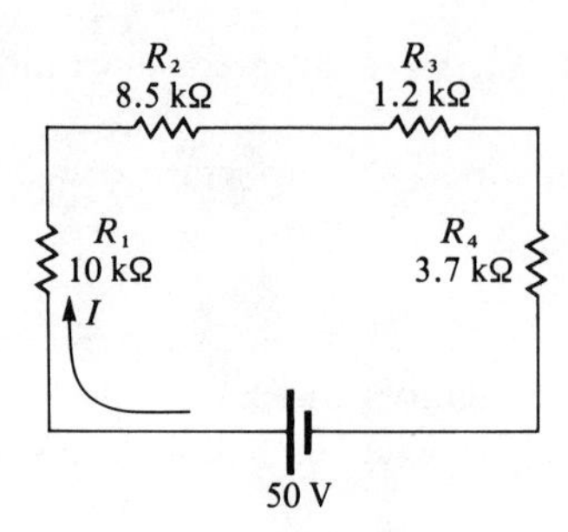

Figure 6.4 Reference Example 6.3

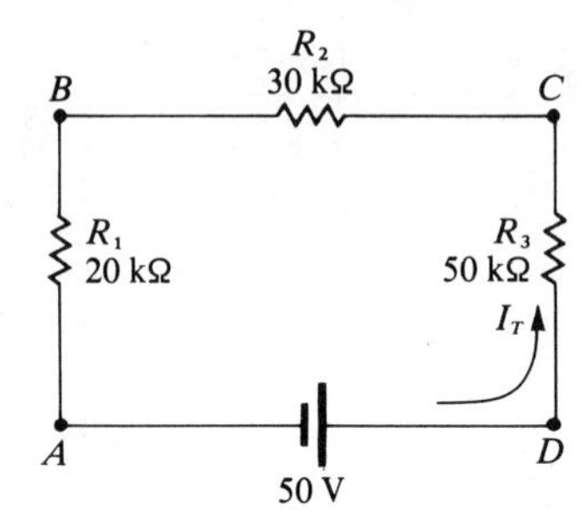

Figure 6.5 Reference Example 6.4

Example 6.4 What is the total current of Figure 6.5?

$$R_T = R_1 + R_2 + R_3$$
$$= 20\text{ k}\Omega + 30\text{ k}\Omega + 50\text{ k}\Omega = 100\text{ k}\Omega$$
$$I_T = \frac{V}{R_T} = \frac{50\text{ V}}{100\text{ k}\Omega} = 0.5\text{ mA}$$

Example 6.5 What are the voltages V_{BA}, V_{CB}, V_{DC} in Figure 6.6?

$$R_T = 35\text{ k}\Omega + 45\text{ k}\Omega + 70\text{ k}\Omega = 150\text{ k}\Omega$$
$$I_T = \frac{50\text{ V}}{150\text{ k}\Omega} = \tfrac{1}{3}\text{ mA}$$
$$V_{BA} = \tfrac{1}{3}\text{ mA} \times 35\text{ k}\Omega = 11\tfrac{2}{3}\text{ volts}$$
$$V_{CB} = \tfrac{1}{3}\text{ mA} \times 45\text{ k}\Omega = 15\text{ volts}$$
$$V_{DC} = \tfrac{1}{3}\text{ mA} \times 70\text{ k}\Omega = 23\tfrac{1}{3}\text{ volts}$$

In this example, all voltages are positive because voltages use (−) as reference.

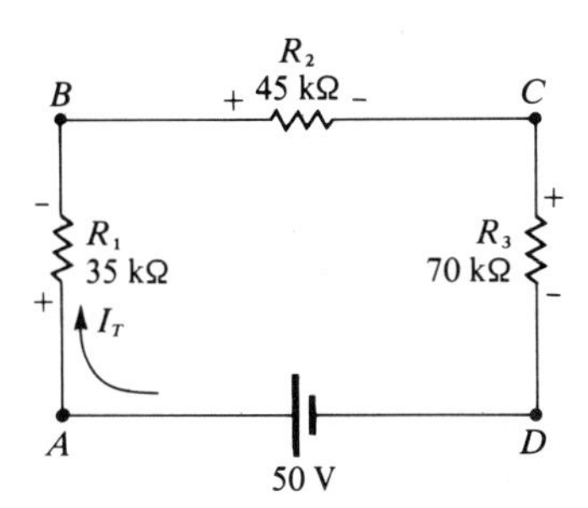

Figure 6.6 Reference Examples 6.5 and 6.7

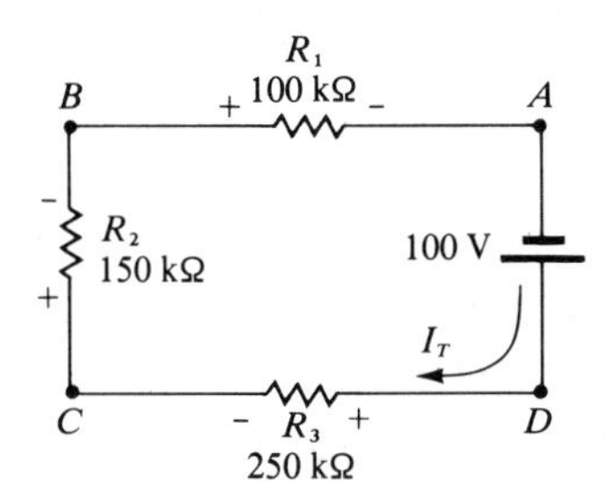

Figure 6.7 Reference Examples 6.6 and 6.8

Example 6.6 What is the sum of the voltages across each resistor from point A to point D of Figure 6.7?

$$R_T = 100\text{ k}\Omega + 150\text{ k}\Omega + 250\text{ k}\Omega = 500\text{ k}\Omega$$

$$I_T = \frac{V}{R_T} = \frac{100\text{ V}}{500\text{ k}\Omega} = 0.2\text{ mA}$$

$$V_{AB} = 0.2\text{ mA} \times 100\text{ k}\Omega = 20\text{ V}$$

$$V_{BC} = 0.2\text{ mA} \times 150\text{ k}\Omega = 30\text{ V}$$

$$V_{CD} = 0.2\text{ mA} \times 250\text{ k}\Omega = 50\text{ V}$$

$$\therefore \quad V_{AD} = V_{AB} + V_{BC} + V_{CD}$$

$$= 20\text{ V} + 30\text{ V} + 50\text{ V} = 100\text{ V}$$

6.2 KIRCHHOFF'S VOLTAGE LAW

Example 6.6 shows that the sum of the voltage "drops" from point A to point D is equal to the source voltage. This is the essence of Kirchhoff's voltage law, which states that:

"The algebraic sum of the voltages in a closed loop is equal to zero."

$$\sum \text{V (in a closed loop)} = 0 \text{ volts} \tag{6.2}$$

Note that in Example 6.6 the voltage V_{AB} is from a negative polarity to a positive polarity, and hence is +20 V. By the same reasoning, V_{BC} is positive 30 V and V_{CD} is positive 50 V. These sums, counterclockwise from point A to point D in Figure 6.7, add up to positive 100 V. To complete the closed loop the circuit must be continued from point D to point A, which is the voltage across the source. Point D, however, is the positive terminal of the source and point A is the negative terminal of the source. In progressing from point D to point A, V_{DA} decreases in potential by 100 V. The sum of the voltages in this closed loop is zero.

$$V_{AB} + V_{BC} + V_{CD} + V_{DA} = 0 \text{ volts}$$

$$20 \text{ V} + 30 \text{ V} + 50 \text{ V} - 100 \text{ V} = 0 \text{ volts}$$

Example 6.7 Verify Kirchhoff's voltage law in Figure 6.6. Voltages across each resistor were determined in Example 6.5.

$$V_{DC} + V_{CB} + V_{BA} + V_{AD} = 0 \text{ volts}$$

$$23\tfrac{1}{3} \text{ V} + 15 \text{ V} + 11\tfrac{2}{3} \text{ V} - 50 \text{ V} = 0 \text{ volts}$$

Example 6.8 Assuming the voltage supply in Figure 6.7 is doubled, verify Kirchhoff's voltage law.

$$R_T = 500 \text{ k}\Omega \text{ (from Example 6.6)}$$

$$I_T = \frac{200 \text{ V}}{500 \text{ k}\Omega} = 0.4 \text{ mA}$$

$$V_{AB} = I_T R_1 = 0.4 \text{ mA} \times 100 \text{ k}\Omega = 40 \text{ V}$$

$$V_{BC} = I_T R_2 = 0.4 \text{ mA} \times 150 \text{ k}\Omega = 60 \text{ V}$$

$$V_{CD} = I_T R_3 = 0.4 \text{ mA} \times 250 \text{ k}\Omega = 100 \text{ V}$$

Verifying Kirchhoff's voltage law, we have:

$$V_{AB} + V_{BC} + V_{CD} + V_{DA} = 0 \text{ volts}$$

$$40 \text{ V} + 60 \text{ V} + 100 \text{ V} - 200 \text{ V} = 0 \text{ volts}$$

6.3 PROPORTIONAL VOLTAGE RATIO

Suppose it is desired to know the voltage across a single element in a series circuit. The initial approach to a problem of this nature is to determine the total resistance, then calculate the total current. Since this current is the same through all elements in the series, the voltage across any single element may now be calculated.

Example 6.9 Calculate the voltage across R_2 in the circuit of Figure 6.8.

$$R_T = 10\ \text{k}\Omega + 15\ \text{k}\Omega + 75\ \text{k}\Omega = 100\ \text{k}\Omega$$

$$I_T = \frac{50\ \text{V}}{100\ \text{k}\Omega} = 0.5\ \text{mA}$$

This is the current through all elements in the series circuit, therefore,

$$V_{BC} = I_T R_2 = 0.5\ \text{mA} \times 15\ \text{k}\Omega = 7.5\ \text{V}$$

Although this approach produces the desired answer, there is a more rapid method to determine the solution without first calculating current.

From the expression

$$V = IR$$

it is seen that voltage is directly proportional to both current and resistance. In a series circuit, I is the same in all elements. This being the case, the voltage dropped is directly proportional to the resistance across which it is developed. To express the relationship between a specific resistive voltage (V_n), the total source voltage (V_T), a specific resistance (R_n), and the total resistance (R_T), refer to Figure 6.9. Suppose it is desired to determine V_2.

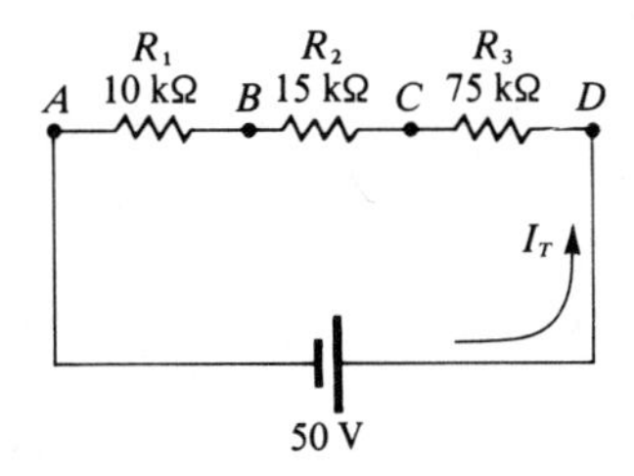

Figure 6.8 Reference Example 6.9

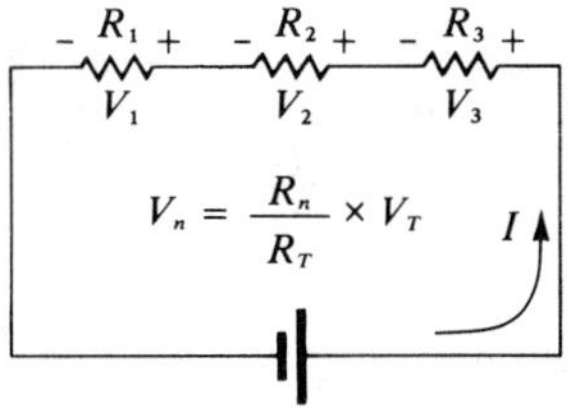

$$V_n = \frac{R_n}{R_T} \times V_T$$

Figure 6.9 Relationship of V_n to V_I

$$V_2 = IR_2$$
$$= \left(\frac{V_T}{R_T}\right)R_2$$
$$= \frac{R_2}{R_T} \times V_T$$

In general terms,

$$\boxed{V_n = \frac{R_n}{R_T} \times V_T} \tag{6.3}$$

Note that V_n is expressed without reference to current. All the information that is needed is to know the ratio of the resistance in question to the total series circuit resistance. This ratio is then multiplied by the source voltage.

Example 6.10 What is V_{BC} in Figure 6.10?

$$V_{BC} = \frac{12\ \text{k}\Omega}{10\ \text{k}\Omega + 12\ \text{k}\Omega + 13\ \text{k}\Omega} \times 45\ \text{V}$$

$$V_{BC} = \frac{12}{35} \times 45\ \text{V} = 15.43\ \text{V}$$

Example 6.11 Determine all the voltages in Figure 6.11. Verify Kirchhoff's voltage law.

$$R_T = 20\ \text{k}\Omega + 30\ \text{k}\Omega + 85\ \text{k}\Omega + 65\ \text{k}\Omega = 200\ \text{k}\Omega$$

$$V_{R_1} = \frac{20}{200} \times 100 = 10\ \text{V}$$

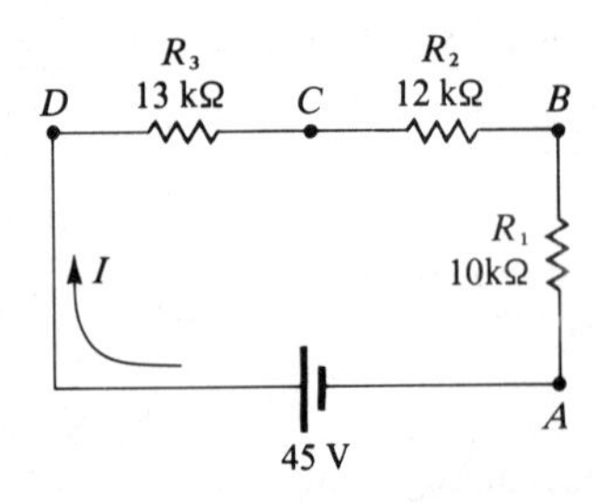

Figure 6.10 Reference Example 6.10

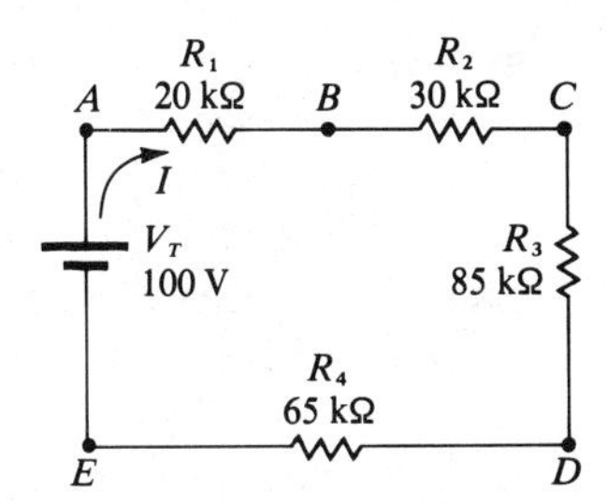

Figure 6.11 Reference Example 6.11 and 6.12

$$V_{R_2} = \frac{30}{200} \times 100 = 15 \text{ V}$$

$$V_{R_3} = \frac{85}{200} \times 100 = 42.5 \text{ V}$$

$$V_{R_4} = \frac{65}{200} \times 100 = 32.5 \text{ V}$$

Kirchhoff's voltage law states that the algebraic sum of voltages in a closed loop equals zero. Progressing counterclockwise from point E,

$$V_{ED} + V_{DC} + V_{CB} + V_{BA} + V_{AE} = 0 \text{ volts}$$
$$32.5 \text{ V} + 42.5 \text{ V} + 15 \text{ V} + 10 \text{ V} - 100 \text{ V} = 0 \text{ volts}$$

Example 6.12 Determine V_{CA} in Figure 6.11.

$$V_{CA} = \frac{R_1 + R_2}{R_T} \times V_T$$

$$= \frac{50}{200} \times 100 \text{ V} = 25 \text{ V}$$

6.4 VOLTAGE DIVIDERS

When a series of resistors is connected across a source for purposes of developing specific voltage values, this is known as a *voltage divider*. Within most electronic devices, various values of voltages are usually required to drive the different elements comprising the device. For example, in the modern pocket calculators, the voltage requirements to illuminate the light indicator segments can be different from the voltage requirements of the IC chips that comprise the calculator.

Figure 6.12(a) represents a voltage divider. The source V_T is shown as a battery, but it could just as easily represent the output of an ac voltage converted to a dc level. The important concept is that V_T has a steady dc value and that a reference point is established. The reference voltage in Figure 6.12(a) is ground. Since the negative terminal of V_T is connected to the reference (point *D*), all other voltages can be expressed in relation to this reference.

Another way to represent Figure 6.12(a) is illustrated by the solid lines of Figure 6.12(b). The dashed portion is assumed to be a source V_T with its negative terminal tied to reference ground and its positive terminal tied to point *A*. The fact that many circuits in this and subsequent chapters will not show the battery symbol does not mean the circuit is open. Current through the resistors completes a closed loop through the source V_T. The following examples will clarify these concepts.

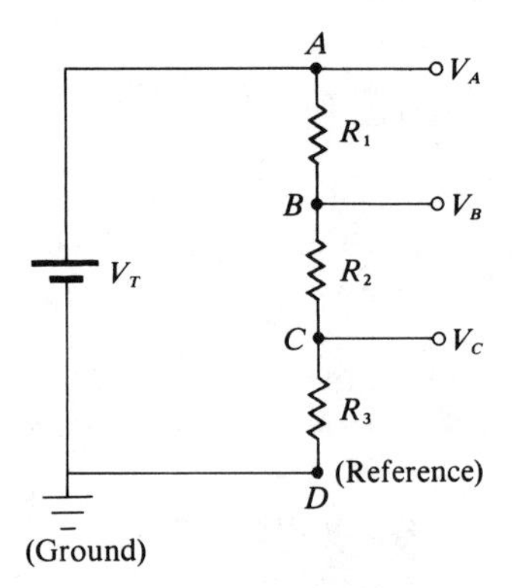

Figure 6.12(a) Voltage Divider

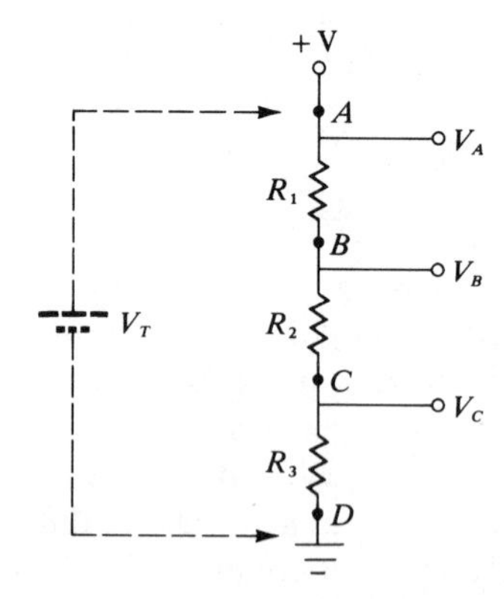

Figure 6.12(b) Simplified Representation of a Voltage Divider

Example 6.13 What are the values of V_A, V_B, and V_C in Figure 6.13? Note that the voltages required are actually V_{DA}, V_{DB}, and V_{DC}. Since D is the ground reference for all cases, it may be dropped from the notation.

$$R_T = 5\ \text{k}\Omega + 15\ \text{k}\Omega + 55\ \text{k}\Omega = 75\ \text{k}\Omega$$

$$V_A = \frac{5}{75} \times 50\ \text{V} = 3.33\ \text{volts}$$

$$V_B = \frac{5 + 15}{75} \times 50\ \text{V} = 13.33\ \text{volts}$$

$V_C = 50$ V (V_C is tied to the positive terminal of the source)

Example 6.14 What are V_{AB} and V_{AC} in Figure 6.13? In this example, the voltages at points B and C are desired in reference to the voltage at point A, not in reference to ground.

$$V_{AB} = \frac{15}{75} \times 50\ \text{V} = 10\ \text{volts}$$

$$V_{AC} = \frac{R_{AB} + R_{AC}}{R_T} \times V_T = \frac{70}{75} \times 50\ \text{V} = 46.67\ \text{volts}$$

Notice that the value of voltage at a given point depends upon its point of reference. In Example 6.13, V_B, with reference to ground, was 13.33 V. With reference to point A, however, it is only 10 V. It should be clear now why it is necessary to establish a reference point when expressing voltage magnitude.

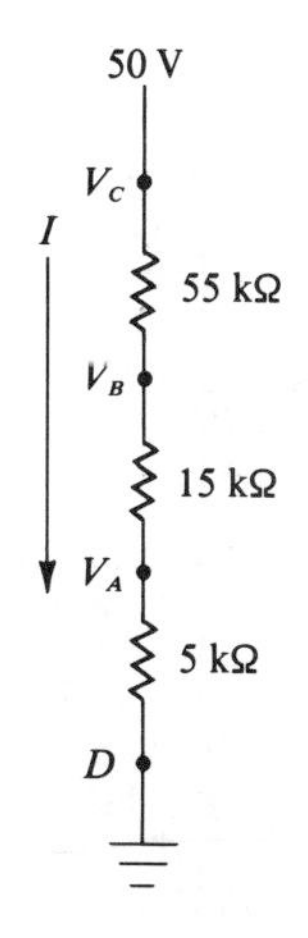

Figure 6.13 Reference Examples 6.13 and 6.14

If the reference point is not ground, then the actual reference value must be accounted for in determining the voltage at a given point in a voltage divider. In general,

$$V_n = V_{\text{ref}} + \frac{R_n}{R_T}(\Delta \text{V}) \tag{6.4}$$

Comparing this to Eq. (6.3),

$$V_n = 0 + \frac{R_n}{R_T}(V_T)$$

it is seen that V_{ref} was considered "ground," and the term ΔV (difference in voltage) replaces the single voltage source V_T.

Example 6.15 Find V_B with respect to ground in Figure 6.14. Start at the -15 V source.

$$V_B = \underset{\text{(reference)}}{-15\text{ V}} \quad \underset{\substack{\text{(polarity}\\\text{rise)}}}{+} \quad \underset{\substack{\text{(resistance}\\\text{ratio)}}}{\frac{6\text{ k}\Omega + 5\text{ k}\Omega}{29\text{ k}\Omega}} \times \underset{\substack{\text{(voltage}\\\text{difference)}}}{55\text{ V}}$$

$$V_B = -15\text{ V} + \frac{11}{29}(55\text{ V}) = 5.86\text{ volts}$$

The principle formerly established that the assumed direction of current flow does not alter the answer is *always* valid. To further illustrate this point, consider the following example.

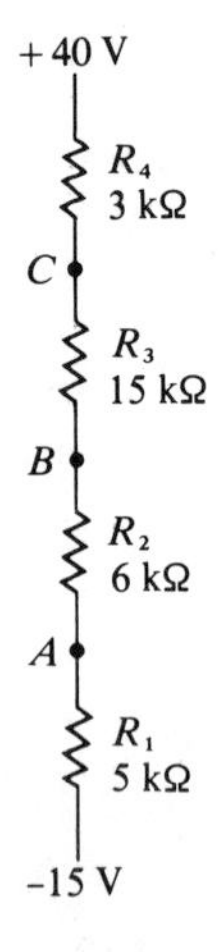

Figure 6.14 Reference Examples 6.15 through 6.18

Example 6.16 Determine V_B in Figure 6.14 beginning at the 40 V source.

$$V_B = \underset{\text{(reference)}}{40\text{ V}} \quad \underset{\substack{\text{(polarity}\\\text{drop)}}}{-} \quad \underset{\substack{\text{(resistance}\\\text{ratio)}}}{\frac{3\text{ k}\Omega + 15\text{ k}\Omega}{29\text{ k}\Omega}} \times \underset{\substack{\text{(voltage}\\\text{difference)}}}{55\text{ V}}$$

$$V_B = 40\text{ V} - \frac{18}{29}(55\text{ V}) = 5.86\text{ volts}$$

Notice that this is exactly the same answer as in Example 6.15.

Example 6.17 Find V_C in Figure 6.14 starting at the -15 V source.

$$V_C = -15\text{ V} + \frac{26\text{ k}\Omega}{29\text{ k}\Omega} \times 55\text{ V}$$

$$= 34.31\text{ volts}$$

Example 6.18 Determine V_C in Figure 6.14 beginning at the 40 V source. Compare this to the answer determined in Example 6.17.

$$V_C = 40\text{ V} - \frac{3\text{ k}\Omega}{29\text{ k}\Omega} \times 55\text{ V}$$

$$= 34.31\text{ volts}$$

6.5 VOLTAGE AND RESISTANCE: PARALLEL CIRCUITS

Section 6.1 developed relationships for resistance and current as they relate to the elements in a series circuit. In a similar fashion, it is possible to formulate rules expressing the voltage across parallel branches and the total equivalent resistance of parallel resistors. By definition, *a parallel branch means that there is more than one path through which current may flow to complete a closed loop between the terminals of a voltage source.*

The most common example of parallel current paths is the typical household wiring system illustrated in Figure 6.15. All appliances plugged into a single wiring system are in parallel. In this configuration, it can be seen that the total voltage V is impressed across each appliance. This is why most appliances have the same voltage rating, namely, 120 V (typical household voltage). From this illustration the following conclusion is reached: *the voltage is the same across all elements of a parallel circuit.*

To develop a mathematical expression for the total resistance *seen* by a voltage source for resistors in parallel, consider Figure 6.16(a). Since V must be the same across both resistors,

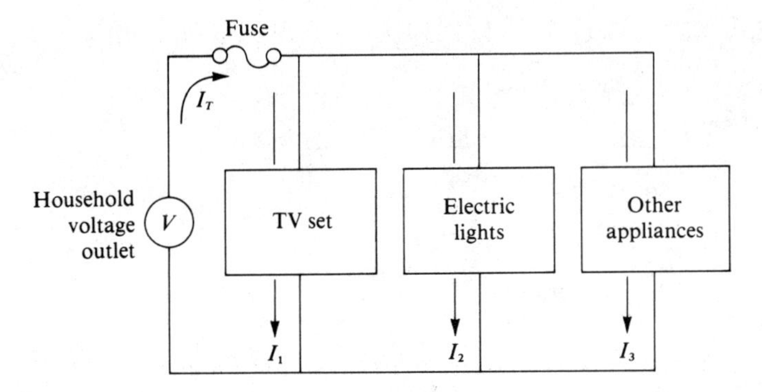

Figure 6.15 Household Appliances in Parallel

$$I_1 = \frac{V}{R_1} \quad \text{and} \quad I_2 = \frac{V}{R_2}$$

The total current delivered by V is related to the total resistance by

$$I_T = \frac{V}{R_T}$$

The current supplied by the source, however, is the current delivered to the two resistors, therefore

$$\frac{V}{R_T} = \frac{V}{R_1} + \frac{V}{R_2}$$

Dividing both sides of this equation by V results in

$$\frac{1}{R_T} = \frac{1}{R_1} + \frac{1}{R_2} \tag{6.5}$$

Solving Eq. (6.5) for R_T yields

$$R_T = \frac{1}{1/R_1 + 1/R_2} \tag{6.6}$$

In general, if there are n resistors,

$$R_T = \frac{1}{1/R_1 + 1/R_2 + \cdots + 1/R_n} \tag{6.7}$$

This states that R_T is the reciprocal of the sum of the reciprocals of each resistor. In practice, most combinations of parallel resistance occur in groups of two or three. If Eq. (6.7) is simplified for two resistors, the following results:

$$R_T = \frac{R_1 R_2}{R_1 + R_2} \tag{6.8}$$

Figure 6.16(a) Electrical Schematic of Parallel Circuit

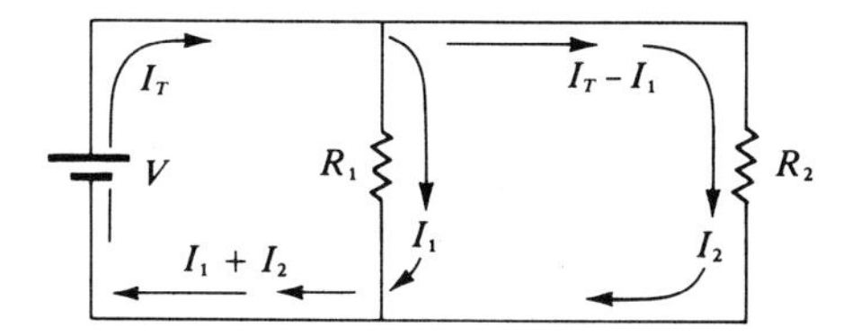

Figure 6.16(b) Relabeling Currents of Figure 6.16(a)

For three resistors, Eq. (6.7) results in

$$R_T = \frac{R_1 R_2 R_3}{R_1 R_2 + R_1 R_3 + R_2 R_3} \tag{6.9}$$

The pattern is as follows: For n number of resistors, R_T equals the quotient of a fraction in which the numerator is the product of all the resistors and the denominator is the sum of all the combinations of $(n - 1)$ of the resistors. This is usually called the *product over sum* rule. If the number of resistors exceeds three, it is easier to utilize Eq. (6.7) rather than the product over sum rule.

Example 6.19 What is the total resistance of a 20 kΩ and 30 kΩ resistor in parallel?

$$R_T = \frac{20\ \text{k}\Omega \times 30\ \text{k}\Omega}{20\ \text{k}\Omega + 30\ \text{k}\Omega} = \frac{600 \times 10^6}{50 \times 10^3} = 12 \times 10^3 = 12\ \text{k}\Omega$$

Note that this value of resistance is less than the resistance of any one of the resistors. When resistors are connected in parallel, the equivalent resistance is always less than the smallest of any of the resistors comprising the parallel network.

Example 6.20 Determine each of the resistance branch currents if a 15 kΩ, a 20 kΩ, and a 30 kΩ resistance are all connected in parallel across a 60 V source. What is the sum of these currents?

$$I_1 = \frac{60\text{ V}}{15\text{ k}\Omega} = 4\text{ mA}$$

$$I_2 = \frac{60\text{ V}}{20\text{ k}\Omega} = 3\text{ mA}$$

$$I_3 = \frac{60\text{ V}}{30\text{ k}\Omega} = 2\text{ mA}$$

$$I_1 + I_2 + I_3 = 9\text{ mA}$$

Example 6.21 Determine I_T of Example 6.20 by first computing R_T. Compare this value of I_T with the sum of the currents computed in Example 6.20.

$$\begin{aligned} R_T &= \frac{R_1 R_2 R_3}{R_1 R_2 + R_1 R_3 + R_2 R_3} \\ &= \frac{15 \times 20 \times 30}{(15 \times 20) + (15 \times 30) + (20 \times 30)}\text{ k}\Omega \\ &= \frac{9000\text{ k}\Omega}{300 + 450 + 600} = \frac{9000\text{ k}\Omega}{1350} \\ &= 6\tfrac{2}{3}\text{ k}\Omega \end{aligned}$$

Note again that R_T is smaller than any of the resistors comprising the parallel branches. From Ohm's law,

$$I_T = \frac{60\text{ V}}{6\tfrac{2}{3}\text{ k}\Omega} = \frac{60\text{ V}}{\tfrac{20}{3}\text{ k}\Omega} = 9\text{ mA}$$

This is the same as the previous calculations of the sum of currents.

In summary, the following rules apply to parallel circuits:

1. The voltage across any branch in parallel is the same.
2. The total resistance is equal to the reciprocal of the sum of the reciprocals.
3. The total resistance is less than the resistance of any single branch of the parallel network.

6.6 KIRCHHOFF'S CURRENT LAW

In addition to his law for voltages in a closed series circuit, Kirchhoff developed a law for currents in parallel. The simplicity of this law can again best be explained by another analogy to water lines. If a water main connects to a Y-shaped junction, so that the water in the main splits up between the two other lines, then the two junction water lines can only carry as much water as was supplied by the main line. In short, the water that leaves a junction can only equal as much water as entered the junction. The same holds true for current:

> "The total current entering a junction equals the sum of the currents leaving the junction."

This is Kirchhoff's current law. Recall that the essence of this law was already illustrated in Examples 6.20 and 6.21. Figure 6.16(b) algebraically labels the currents of Figure 6.16(a) to comply with Kirchhoff's current law.

Example 6.22 The current entering a junction with two branches is 16.5 mA. If the current in one branch is 8.3 mA, what is the current in the other branch?

$$I_T = I_1 + I_2$$

$$16.5 \text{ mA} = 8.3 \text{ mA} + I_2$$

$$I_2 = 16.5 \text{ mA} - 8.3 \text{ mA} = 8.2 \text{ mA}$$

To expand these examples into more practical applications, Kirchhoff's current law will be combined with the voltage divider theories of the previous sections. As the reader's knowledge of electronics expands, theories and applications will increasingly be combined to present more advanced concepts.

The preceding examples specified the values of the resistors in a voltage divider. Voltages were then calculated, based upon these values. This is analogous to putting the cart before the horse. More practically, current requirements and voltages for sections of a device are known, and what is needed is to design the voltage divider network that will supply these voltages. As a first example into design procedures, suppose that a rechargeable 9 V, 50 mA battery is the most convenient standard source for a pocket calculator. Let it further be specified that the calculator's light indicators require 7.5 V at 25 mA and that the calculator's IC chips require 3.5 V at 15 mA. Figure 6.17 illustrates these requirements.

Example 6.23 Determine the values of R_1, R_2, and R_3 in Figure 6.17.

$$V_{BA} = 9 \text{ V} - 7.5 \text{ V} = 1.5 \text{ volts}$$

$$I_{R_1} = 50 \text{ mA}$$

$$\therefore \quad R_1 = \frac{1.5 \text{ V}}{50 \text{ mA}} = 30 \text{ ohms}$$

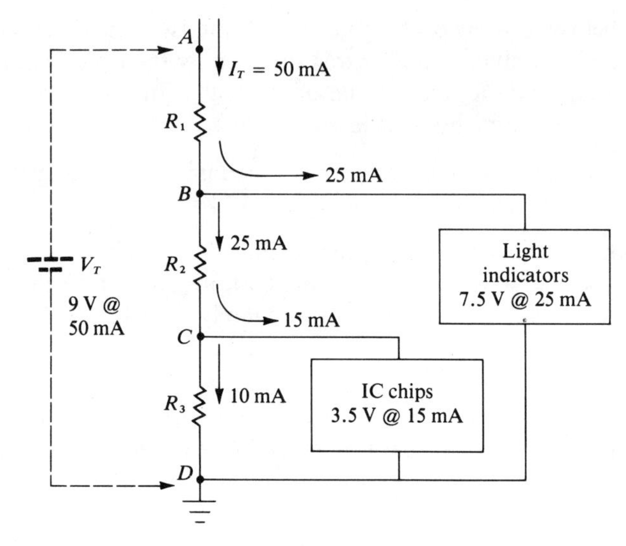

Figure 6.17 Voltage Divider Design Problem Reference Example 6.23

$$V_{CB} = 7.5\text{ V} - 3.5\text{ V} = 4\text{ volts}$$
$$I_{R_2} = 25\text{ mA}$$
$$\therefore\quad R_2 = \frac{4\text{ V}}{25\text{ mA}} = 160\text{ ohms}$$

$$V_{DC} = 3.5\text{ V} - \text{ground reference} = 3.5\text{ volts}$$
$$I_{R_3} = 10\text{ mA}$$
$$\therefore\quad R_3 = \frac{3.5\text{ V}}{10\text{ mA}} = 350\text{ ohms}$$

Example 6.24 Determine R_1, R_2, and R_3 for the voltage divider network of Figure 6.18.

$$V_{BA} = (30 - 22.5)\text{ V} = 7.5\text{ volts}$$
$$I_{R_1} = 50\text{ mA}$$
$$\therefore\quad R_1 = \frac{7.5\text{ V}}{50\text{ mA}} = 150\text{ ohms}$$

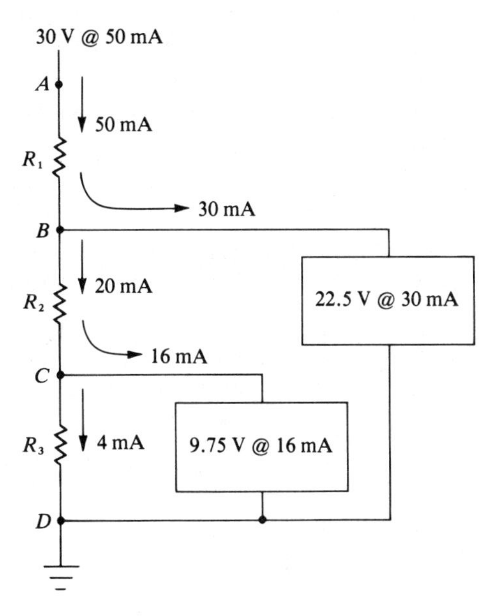

Figure 6.18 Reference Example 6.24

$$V_{CB} = (22.5 - 9.75) \text{ V} = 12.75 \text{ volts}$$

$$I_{R_2} = 20 \text{ mA}$$

$$\therefore \quad R_2 = \frac{12.75 \text{ V}}{20 \text{ mA}} = 637.5 \text{ ohms}$$

$$V_{DC} = 9.75 \text{ V} - \text{ground reference} = 9.75 \text{ volts}$$

$$I_{R_3} = 4 \text{ mA}$$

$$\therefore \quad R_3 = \frac{9.75 \text{ V}}{4 \text{ mA}} = 2.44 \text{ k}\Omega$$

6.7 INVERSE CURRENT RATIO

Previous examples of parallel networks always showed a source of voltage across the circuit. Knowing that the voltage is the same across all branches of a parallel circuit, it is easy to calculate the current through any single branch by applying Ohm's law. Suppose, however, that the voltage across a parallel network is not known; rather, what is known is the current entering the junction. Is there a way to determine the current through each branch; that is, is there a current law analogous to the proportionate voltage law?

If reference is made to Figure 6.19, I_T enters junction A, divides into I_1 and I_2, and returns to the source as I_T again. Since V_{R_1} equals V_{R_2} (parallel branches have equal voltages), then

$$I_1 R_1 = I_2 R_2$$

But

$$I_2 = I_T - I_1$$

$$\therefore \quad I_1 R_1 = (I_T - I_1)R_2$$

$$= I_T R_2 - I_1 R_2$$

$$I_1(R_1 + R_2) = I_T R_2$$

$$\boxed{I_1 = \frac{R_2}{R_1 + R_2} I_T} \tag{6.10}$$

Similarly,

$$\boxed{I_2 = \frac{R_1}{R_1 + R_2} I_T} \tag{6.11}$$

In words, the current through R_1 (which is I_1) equals the ratio of the other resistor, R_2, to the total resistance, multiplied by the total current entering the junction. The current through R_2 (which is I_2) equals the ratio of the other resistor, R_1, to the total resistance, multiplied by the total current entering the junction. This relationship is called the *inverse current law.* In general, the current through one branch of two parallel resistors is the product of the current entering the junction times the ratio of the inverse resistor to the total resistance. Be careful to remember that total resistance as used here is the sum of the two resistors and not the equivalent of two resistors in parallel.

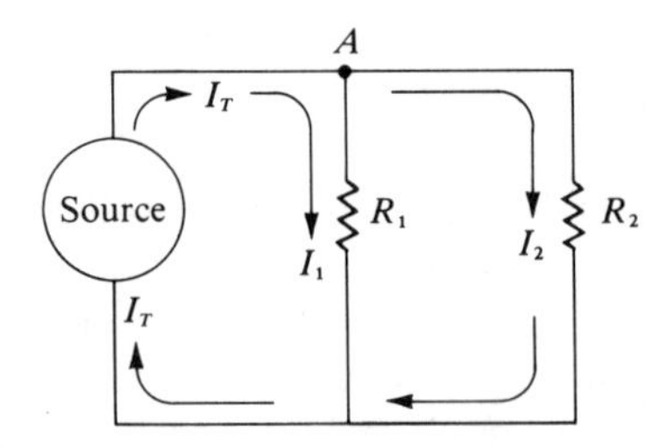

Figure 6.19 Network to Determine Inverse Current Ratio

Example 6.25 Determine I_{R_1} and I_{R_2} in Figure 6.20.

$$I_{R_1} = \frac{18\ \text{k}\Omega}{(12 + 18)\ \text{k}\Omega} \times 40\ \text{mA} = 24\ \text{mA}$$

$$I_{R_2} = \frac{12\ \text{k}\Omega}{(12 + 18)\ \text{k}\Omega} \times 40\ \text{mA} = 16\ \text{mA}$$

Example 6.26 Determine the total current in Figure 6.21, knowing only the resistors and the current through one of the resistors.

$$I_1 = I_T \times \frac{R_2}{R_1 + R_2}$$

Solving for I_T,

$$I_T = \frac{I_1(R_1 + R_2)}{R_2}$$

$$= 6.5\ \text{mA} \times \frac{55}{22} = 16.25\ \text{mA}$$

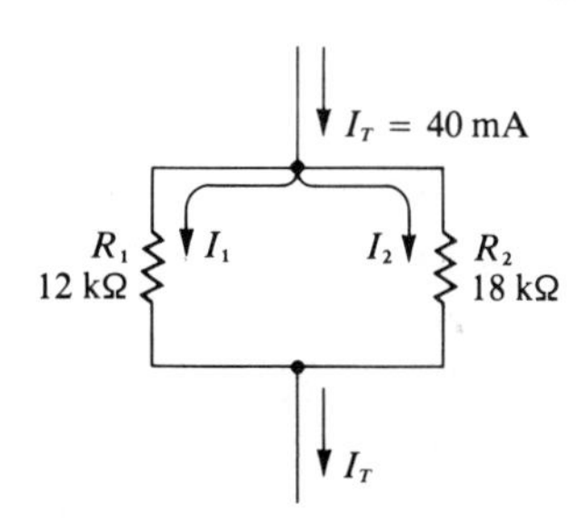

Figure 6.20 Reference Example 6.25

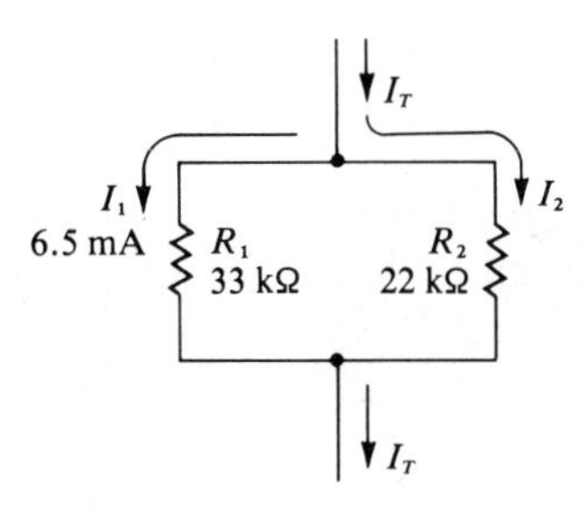

Figure 6.21 Reference Example 6.26

6.8 CONDUCTANCE OF PARALLEL RESISTORS

It has been shown that for n parallel resistors,

$$\frac{1}{R_T} = \frac{1}{R_1} + \frac{1}{R_2} + \cdots + \frac{1}{R_n} \tag{6.12}$$

Chapter 3 discussed the property of conductance (G), which by definition, was the reciprocal of resistance. It has been necessary to defer development of conductance concepts until parallel resistor analysis. Since $G = 1/R$, then for n parallel resistors,

$$\boxed{G_T = G_1 + G_2 + G_3 + \cdots + G_n} \tag{6.13}$$

Conductance in parallel is analogous to resistance in series; that is, it is additive.

Example 6.27 Using conductance, find the total current if three parallel resistors of 10 kΩ, 20 kΩ, and 50 kΩ are placed across a 40 V source.

$$G_1 = \frac{1}{10\ \text{k}\Omega} = 100\ \mu\text{S}$$

$$G_2 = \frac{1}{20\ \text{k}\Omega} = 50\ \mu\text{S}$$

$$G_3 = \frac{1}{50\ \text{k}\Omega} = 20\ \mu\text{S}$$

$$G_T = G_1 + G_2 + G_3 = (100 + 50 + 20)\ \mu\text{S} = 170\ \mu\text{S}$$

$$I_T = V_T G_T = 40\ \text{V} \times 170\ \mu\text{S}$$

$$= 6800\ \mu\text{A} = 6.8\ \text{mA}$$

Example 6.28 What is the value of resistor R_2 in Figure 6.22?

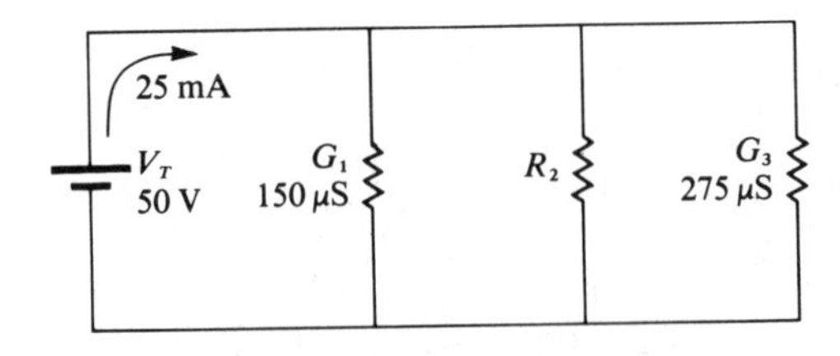

Figure 6.22 Reference Example 6.28

$$I_T = G_T V_T$$

$$G_T = \frac{I_T}{V_T} = \frac{25 \text{ mA}}{50 \text{ V}} = 500\ \mu\text{S}$$

$$G_2 = G_T - G_1 - G_3$$

$$= (500 - 150 - 275)\ \mu\text{S} = 75\ \mu\text{S}$$

$$R_2 = \frac{1}{G_2} = \frac{1}{75 \times 10^{-6} \text{ S}} = 13.33 \text{ k}\Omega$$

6.9 MULTIPLE VOLTAGE SOURCES

Up to this point in the discussion of electrical circuits, multiple resistors have been shown in series across one voltage source. But what would happen if there were multiple voltage sources in series, in lieu of or in addition to multiple series resistors? The answer is that resistors would still be numerically summed, since they exhibit no polarity; voltage sources would be algebraically summed, since they do exhibit polarity. In a series circuit, if the polarity of voltages is aligned, the net voltage is the sum of the voltages and in the direction of the original sources. If the voltages are in opposition, the net voltage is the algebraic sum of the individual sources and in the direction of the algebraically largest sum.

Example 6.29 Simplify the voltage network of Figure 6.23.

1. Between points A and B, V_1 *and* V_2 are additive; therefore they could be replaced by a single source of 25 V, with the negative terminal toward point A.

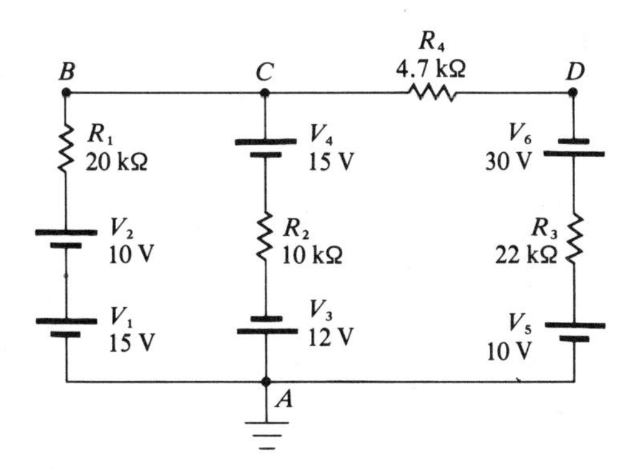

Figure 6.23 Reference Example 6.29

2. Between points A and C, the voltages are in opposition. V_3 and V_4 reduce to 3 V, with the positive terminal toward point C. For purposes of determining the current through this branch, it does not matter to which side of R_2 the equivalent source is placed. Its effect will still be the same, so long as it is connected in series with R_2 between points A and C. Recall that current is the same at all points in a series branch.

3. By the same reasoning, V_5 and V_6 reduce to a 20 V source, with the negative terminal toward point D. This resultant source is in series with R_3 connected between points A and D.

R_4, located between points C and D, is unaffected. The resultant simplification is shown in Figure 6.24.

Example 6.30 Determine the current magnitude and direction of current in Figure 6.25.

$$R_T = 12\ \text{k}\Omega + 15\ \text{k}\Omega + 6.5\ \text{k}\Omega + 1.2\ \text{k}\Omega = 34.7\ \text{k}\Omega$$

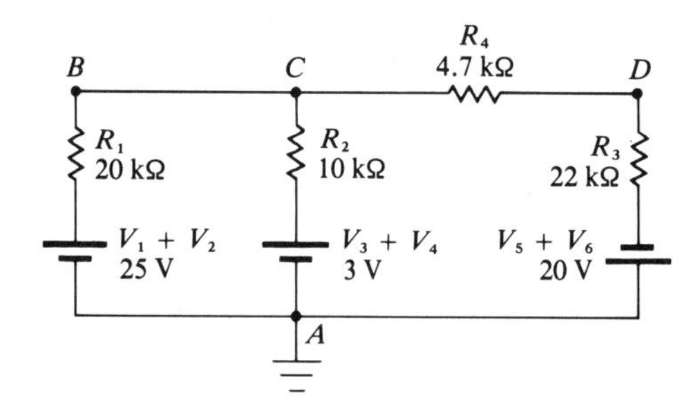

Figure 6.24 Simplified Version of Figure 6.23

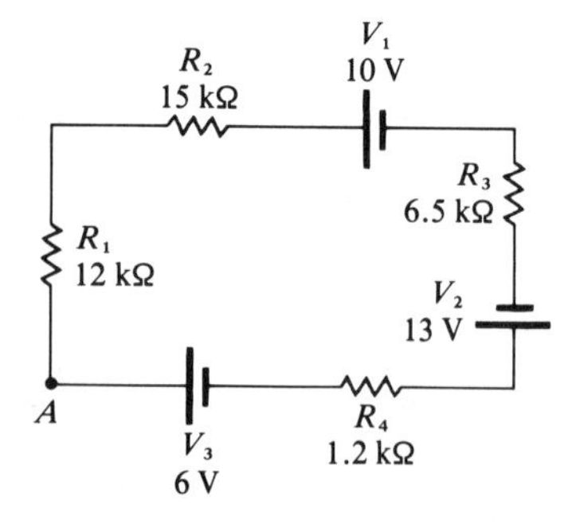

Figure 6.25 Reference Example 6.30

Starting at point A and progressing clockwise,

$$V_T = V_1 + V_2 + V_3$$
$$= -10 + 13 + 6 = 9 \text{ V (negative to positive clockwise)}$$

The simplified circuit is represented in Figure 6.26.

$$I_T \text{ (clockwise)} = \frac{9 \text{ V}}{34.7 \text{ k}\Omega} = 0.26 \text{ mA}$$

Figure 6.25 illustrates multiple voltage sources in series. A precaution must be taken for parallel arrangements:

> "Two different voltage sources may not be directly connected in parallel. Only voltage sources equal in both magnitude and polarity may be paralleled."

The rationale for this statement is that voltages across parallel branches must be equal. If two different voltage sources are paralleled, the larger of the two will discharge through the smaller, and eventually its terminal voltage will diminish to the value of the smaller. Figure 6.27(a) shows an

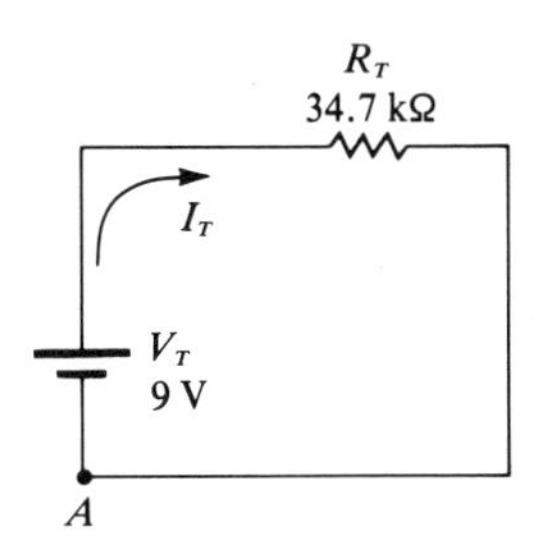

Figure 6.26 Simplified Version of Figure 6.25

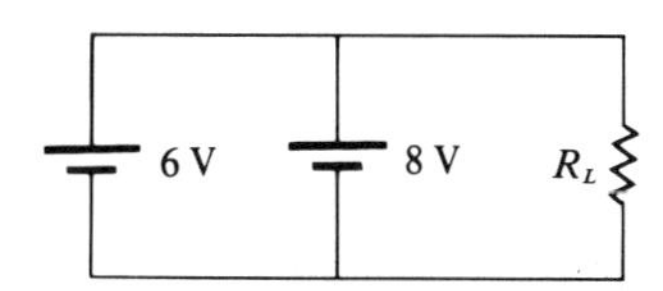

Figure 6.27(a) Invalid Voltage Connection

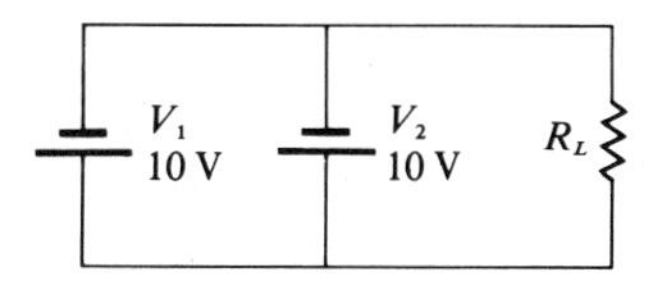

Figure 6.27(b) Valid Voltage Connection

invalid arrangement for voltage sources. Figure 6.27(b) shows a valid arrangement. This latter configuration can be useful if the current drawn by R_L is greater than that which either V_1 or V_2 can individually deliver. For example, if R_L requires 10 V at 8 mA, and if V_1 is only capable of 10 V at 4 mA, then the two parallel sources can supply the required current.

> "The current capability of equal parallel voltage sources is additive."

6.10 MULTIPLE CURRENT SOURCES

Until now, all sources of energy have been symbolized as voltage sources across a network of resistances. In addition, there exist energy sources known as *current sources*. This concept is usually somewhat more difficult for students beginning their study of electronics to grasp. Perhaps the easiest way to explain a constant current source is to describe the input/output characteristics of a transistor, especially if connected in what is referred to as the *common base* mode.

Figure 6.28 is a graph of I_C (output current) vs V_C (output voltage) of a transistor connected in the common base configuration. Terminology such as I_C, V_C, common base, etc., is not of consequence at this point, but will be discussed in subsequent courses of electronics. What is important is to conceptually understand the significance of such a graph, and hence the meaning of a constant current source.

Note that there is a very steep rise in I_C for the first few tenths of a volt of V_C. From the knee of the curve, and until a point of about 12 V, the

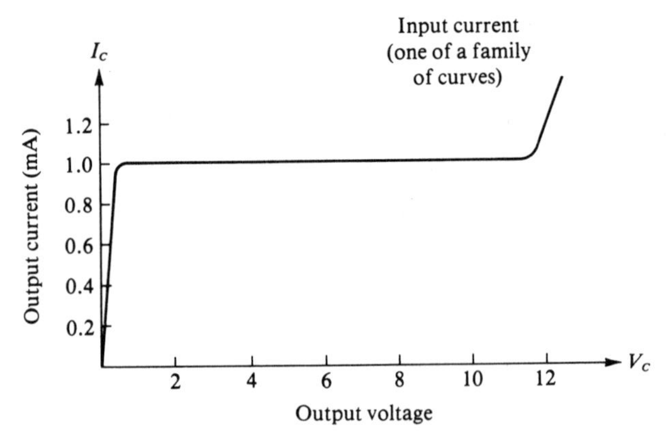

Figure 6.28 Constant Current Source

value of output current (I_C) remains constant at 1.0 mA. This means that over a relatively broad spectrum of output voltage variations, the current remains constant. Such a device is therefore appropriately called a *constant current source.* Semiconductors, such as transistors and integrated circuits, have become dominant, active elements in modern electronics. Current sources, therefore, will become more of an everyday term in subsequent electronics courses.

Symbolically, current sources will be represented as in Figure 6.29(a) and 6.29(b). The arrow indicates the direction of current flow. In Figure 6.29(a), all 2 mA of I must flow through R_1, R_2, and R_3. In Figure 6.29(b), the 8 mA separates into I_1 and I_2 at junction A, in accordance with the inverse current ratio law. The important point is this:

> "The current specified as emanating from a constant current source is the only current that may flow in the branch containing that source".

The following examples will clarify this point.

Example 6.31 How much current flows through R_4 of Figure 6.30?

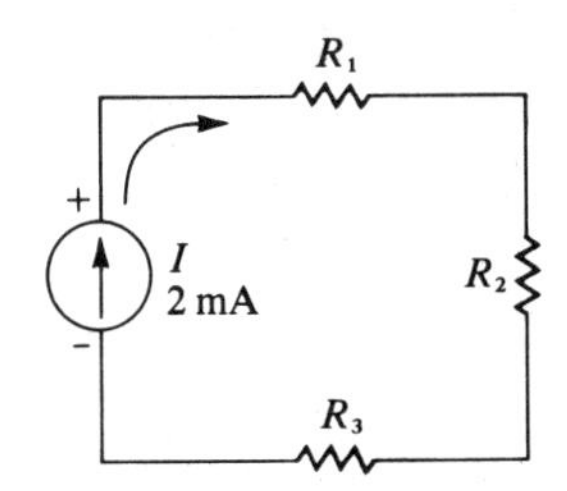

Figure 6.29(a) Constant Current Source

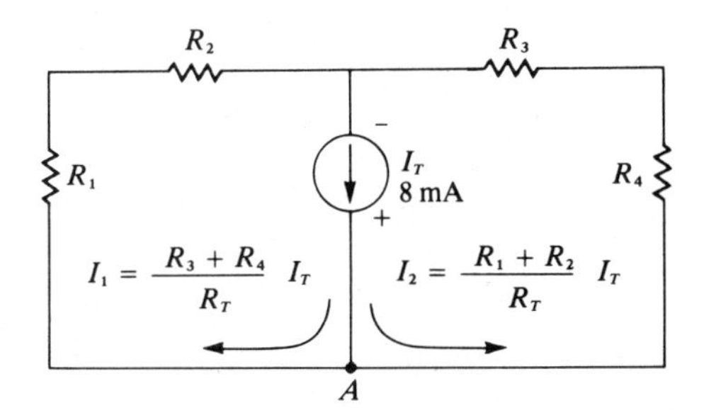

Figure 6.29(b) Current Source Dividing between Parallel Branches

From points A to B, 8.4 mA must flow. From points B to D and back to A again, 5.0 mA must flow. Therefore, 8.4 mA − 5.0 mA or 3.4 mA must flow through R_4.

Note that these values, as they divide between the two branches from the 8.4 mA generator, do not obey the inverse current ratio law. This is because I_2 "forces" a condition in which 5.0 mA must flow through branch BDA.

Example 6.32 If the current source I_2 were replaced by a short circuit in Figure 6.30, how would I_1 divide between branches BDA and BCA?

By the inverse current law,

$$I_{BDA} = \frac{10\text{ k}\Omega + 5.6\text{ k}\Omega}{10\text{ k}\Omega + 5.6\text{ k}\Omega + 5\text{ k}\Omega + 8.6\text{ k}\Omega} \times 8.4\text{ mA} = 4.49\text{ mA}$$

$$I_{BCA} = \frac{13.6\text{ k}\Omega}{(15.6 + 13.6)\text{ k}\Omega} \times 8.4\text{ mA} = 3.91\text{ mA}$$

$$I_1 = 4.49\text{ mA} + 3.91\text{ mA} = 8.4\text{ mA}$$

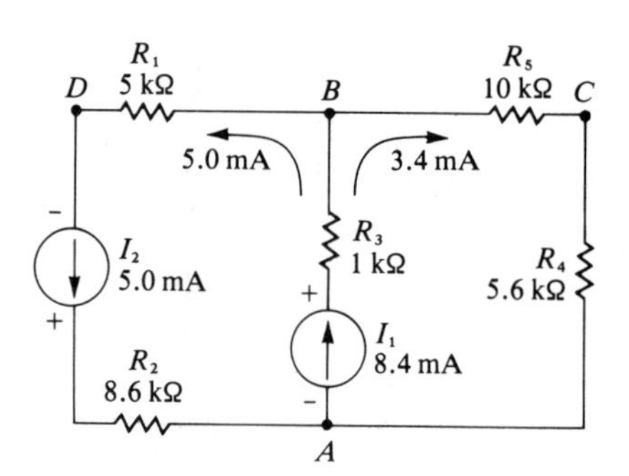

Figure 6.30 Reference Examples 6.31 and 6.32

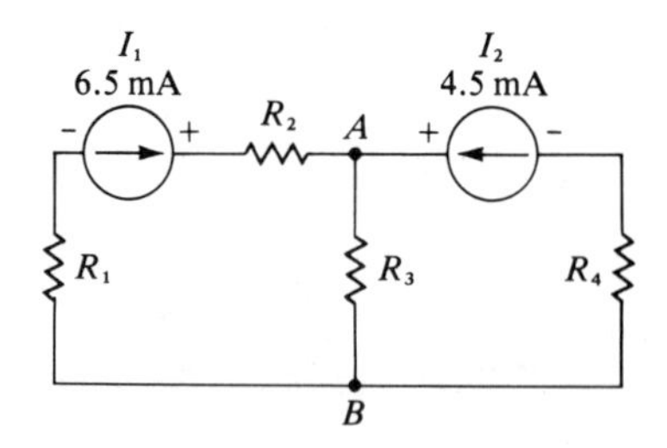

Figure 6.31 Reference Example 6.33

Example 6.33 Identify how much current flows in each resistor of Figure 6.31.

$$I_{R_1} = I_{R_2} = 6.5 \text{ mA}$$
$$I_{R_4} = 4.5 \text{ mA}$$

From junctions A to B, both I_1 and I_2 join to flow through R_3, because constant current sources may not oppose one another in series.

$$I_{R_3} = I_1 + I_2 = 6.5 \text{ mA} + 4.5 \text{ mA} = 11 \text{ mA}$$

If two or more current sources are placed in parallel, they are algebraically additive. Note that in Figure 6.32(a), the total current from source 1 reaches point A. Source 2 forces 2.5 mA away from point A, leaving a net of 1.5 mA for R_L. This 1.5 mA, along with the 2.5 mA of source 2, meets at junction B to return as 4 mA back to source 1 again. The equivalent circuit of Figure 6.32(a) is redrawn as Figure 6.32(b).

Just as parallel combinations of unequal voltage sources are invalid, there is also an invalid configuration for current sources.

> "Two current sources may not be placed in the same series branch unless they are of the same magnitude and in the same direction."

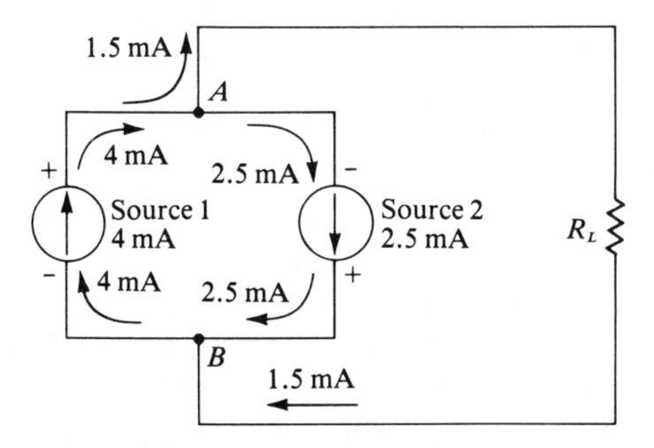

Figure 6.32(a) Two Current Sources in Parallel

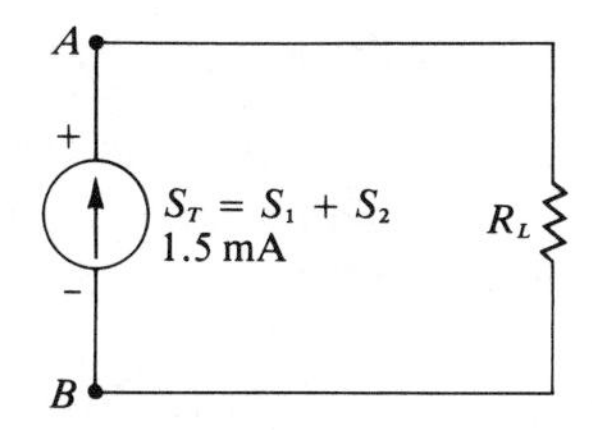

Figure 6.32(b) Equivalent Circuit of Figure 6.32(a)

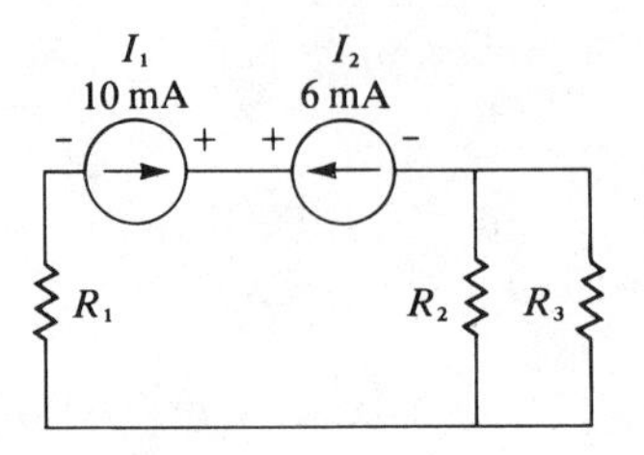

Figure 6.33(a) Invalid Current Source Connections

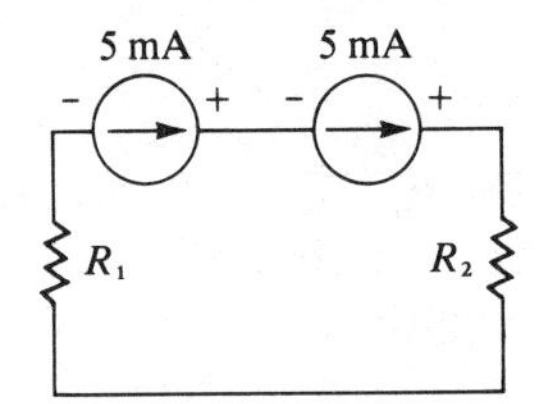

Figure 6.33(b) Valid Current Sources in Series

Figure 6.33(a) would be invalid, whereas Figure 6.33(b) is permissible. Current sources are constant and in the direction specified. For two sources to be opposing would be to force one source in a direction contrary to its electronic structure.

6.11 CONVERSION OF SOURCES

If sources are mixed within a circuit, it is frequently helpful to convert one source into the other when analyzing the circuit. Suppose it is desired to determine the current through R_2 in Figure 6.34. Part of the current is attributable to V_2 and part is due to the current generator. If the voltage source could be converted into a current generator, then I_{R_2} could easily be determined by summing the currents that each source contributes to R_2.

All voltage sources have a series resistance, even if it is only the internal resistance of the source itself. Figure 6.35 illustrates the conversion process.

> "A voltage source (V_S) with its series resistor (R_S) may be converted into a current source equal in magnitude to V_S/R_S. Resistor R_S is placed in parallel to I_S."

Example 6.34 In the voltage source of Figure 6.35(a), $R_S = 0.1$ kΩ, $R_L = 0.3$ kΩ, and $V_S = 1$ V. Compute I_{R_L} using the voltage source. Con-

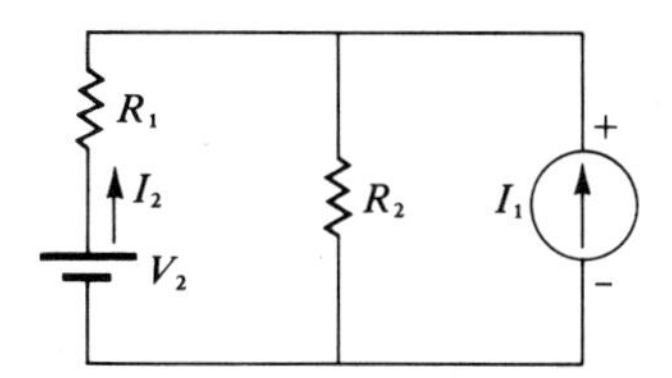

Figure 6.34 Circuit with Mixed Sources

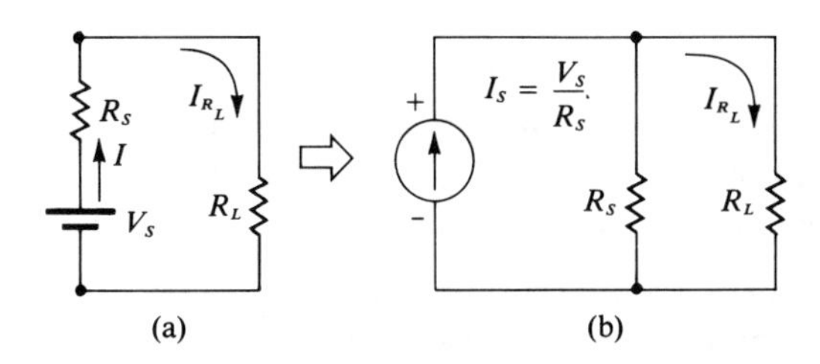

Figure 6.35 Converting Voltage Source to Current Source

vert the voltage source into a current source and verify that I_{R_L} remains constant.

a. From Figure 6.35(a),

$$R_T = R_S + R_L = 0.1\ \text{k}\Omega + 0.3\ \text{k}\Omega = 0.4\ \text{k}\Omega$$

$$I_{R_L} = \frac{1\ \text{V}}{0.4\ \text{k}\Omega} = 2.5\ \text{mA}$$

b. Convert the voltage source of Figure 6.35(a) into a current source.

$$I_S = \frac{V_S}{R_S} = \frac{1\ \text{V}}{0.1\ \text{k}\Omega} = 10\ \text{mA}$$

R_S is placed in parallel with I_S and R_L, as in Figure 6.35(b). By the inverse current ratio,

$$I_{R_L} = \frac{R_S}{R_S + R_L} \times I_S$$

$$= \frac{0.1}{0.1 + 0.3} \times 10\ \text{mA} = 2.5\ \text{mA}$$

I_{R_L} has the same value in both circuits.

To reverse the direction of source conversion, current sources can be converted into voltage sources as illustrated in Figure 6.36.

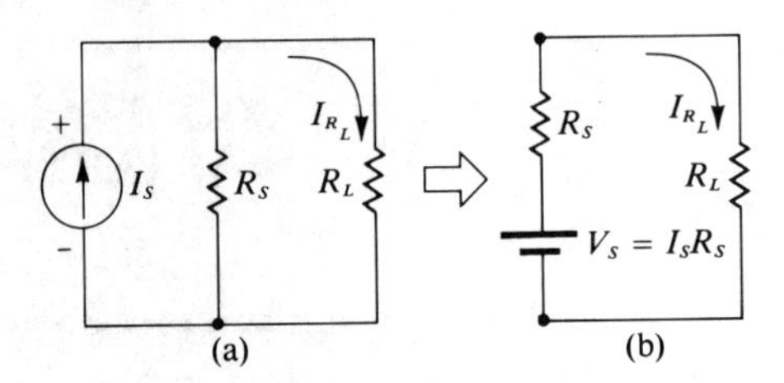

Figure 6.36 Converting a Current Source into a Voltage Source

> "To change a current source into a voltage source, V_S is the product of $I_S \times R_S$. The parallel resistor R_S is placed in series with V_S."

Always make sure when performing either conversion that the polarity of the voltage source and the arrow direction of the current source move current in the same direction.

Example 6.35 Assume that for the current source of Figure 6.36(a), $I_S = 15$ mA, $R_S = 0.3$ kΩ, and $R_L = 0.7$ kΩ. Calculate the current through R_L using the current source. Convert the current source to a voltage source and verify that I_{R_L} is unchanged.

a. Use the inverse current law for Figure 6.36(a).

$$I_{R_L} = \frac{R_S(I_S)}{R_S + R_L} = \frac{0.3}{0.3 + 0.7} \times 15 \text{ mA}$$
$$= 4.5 \text{ mA}$$

b. Convert the current source to a voltage source.

$$V_S = I_S R_S = 15 \text{ mA} \times 0.3 \text{ k}\Omega = 4.5 \text{ volts}$$

R_S is placed in series with V_S, as in Figure 6.36(b).

$$I_{R_L} = \frac{V_S}{R_S + R_L} = \frac{4.5 \text{ V}}{(0.3 + 0.7) \text{ k}\Omega} = 4.5 \text{ mA}$$

This answer agrees with I_{R_L} as determined in part "a" of this example.

Example 6.36 Determine the current through R_L in Figure 6.37.

The parallel network of the current source and R_2 will be converted into a voltage source.

$$V_2 = I_S R_2 = 1.5 \text{ mA} \times 0.8 \text{ k}\Omega = 1.2 \text{ volts}$$

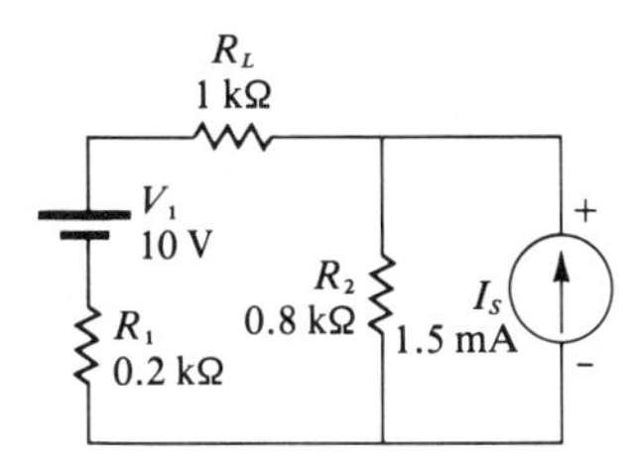

Figure 6.37 Reference Example 6.36

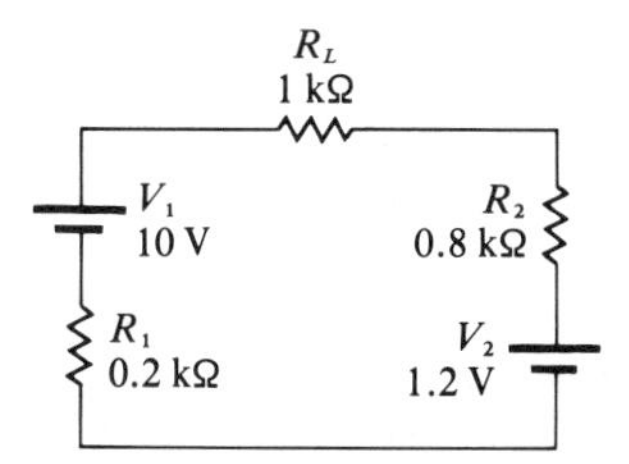

Figure 6.38 Simplified Version of Figure 6.37

Placing R_2 in series with V_2 produces the circuit shown in Figure 6.38. The two voltage sources are opposing, therefore

$$V_T = (10 - 1.2)V = 8.8 \text{ volts}$$

$$R_T = R_1 + R_2 + R_L = 2 \text{ k}\Omega$$

$$I_{R_L} = \frac{V_T}{R_T} = \frac{8.8 \text{ V}}{2 \text{ k}\Omega} = 4.4 \text{ mA}$$

6.12 SERIES/PARALLEL NETWORKS

The final step in analyzing networks is to combine series circuits with parallel circuits. Various combinations of the rules developed in this section must be employed, depending on what is given and what is desired. At times, the inverse current law is more appropriate; at other times, the proportionate voltage law is better employed to find the desired result. Experience of application is the best teacher in determining the most advantageous rule to use. The following examples will illustrate combinations of the concepts developed throughout this chapter.

Example 6.37 Determine I_T in Figure 6.39.

First, combine R_1 and R_2 into one resistor.

$$R_1 \| R_2 = \frac{20\ \text{k}\Omega \times 30\ \text{k}\Omega}{(20 + 30)\ \text{k}\Omega} = 12\ \text{k}\Omega$$

This is in series with R_3, resulting in

$$12\ \text{k}\Omega + 8\ \text{k}\Omega = 20\ \text{k}\Omega$$

Finally, this combination is in parallel with R_4.

$$R_T = \frac{20 \times 30}{50}\ \text{k}\Omega = 12\ \text{k}\Omega$$

$$\therefore \quad I_T = \frac{24\ \text{V}}{12\ \text{k}\Omega} = 2\ \text{mA}$$

Example 6.38 What is V_{CB} in Figure 6.39?

From Example 6.37, $I_T = 2$ mA flowing up to junction A. The entire network of R_1, R_2, and R_3 was previously calculated to be 20 kΩ. The circuit at this point may be replaced by Figure 6.40. From the inverse current law,

$$I_1 = \frac{30}{20 + 30} \times 2\ \text{mA} = 1.2\ \text{mA}$$

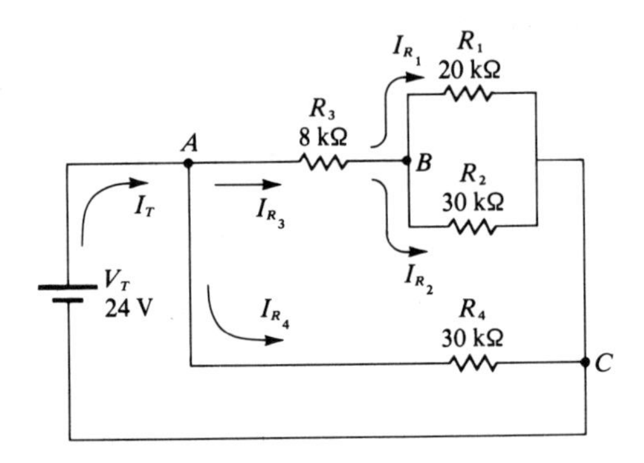

Figure 6.39 Reference Examples 6.37 and 6.38

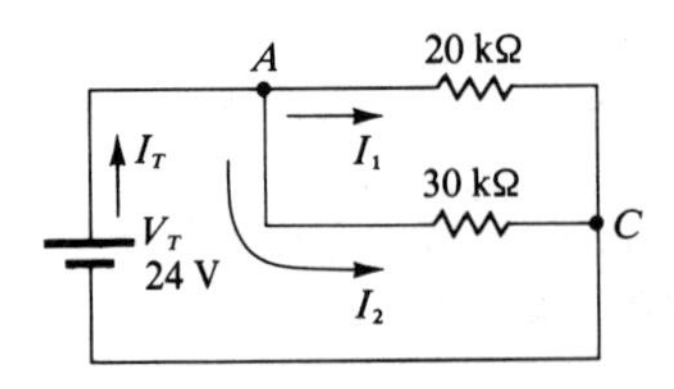

Figure 6.40 Simplified Version of Figure 6.39

This is the amount of current flowing through R_3 and up to junction B of Figure 6.39. This 1.2 mA in turn splits up between R_1 and R_2.

At this point, two different approaches could be taken to find V_{CB}. First, the inverse current ratio law could be applied to find how much current flows through either R_1 or R_2. Knowing this, Ohm's law can be used to find the voltage across R_1 or R_2, which is V_{CB}.

$$I_{R_1} = \frac{30}{20 + 30}(1.2 \text{ mA}) = 0.72 \text{ mA}$$

$$V_{CB} = V_{R_1} = I_{R_1}R_1 = 0.72 \text{ mA} \times 20 \text{ k}\Omega = 14.4 \text{ volts}$$

Alternately,

$$I_{R_2} = \frac{20}{20 + 30}(1.2 \text{ mA}) = 0.48 \text{ mA}$$

$$V_{CB} = V_{R_2} = I_{R_2}R_2 = 0.48 \text{ mA} \times 30 \text{ k}\Omega = 14.4 \text{ volts}$$

A simpler approach, requiring only one step, would be to use the proportionate voltage law. All 24 V of V_T is across terminals CA. Since R_1 and R_2 were found to equal 12 kΩ in parallel,

$$V_{CB} = \frac{12}{12 + 8}(24 \text{ V}) = 14.4 \text{ volts}$$

Notice how this example shows that several different approaches could produce the desired answer. As stated earlier, choosing the simplest method, and therefore the one least likely to result in an error, comes with practice.

Example 6.39 Find V_{EC} of Figure 6.41.

V_{EC} represents the voltage drop across resistors R_3 and R_5. R_5 has all 3.5 mA of the source current flowing through it, therefore

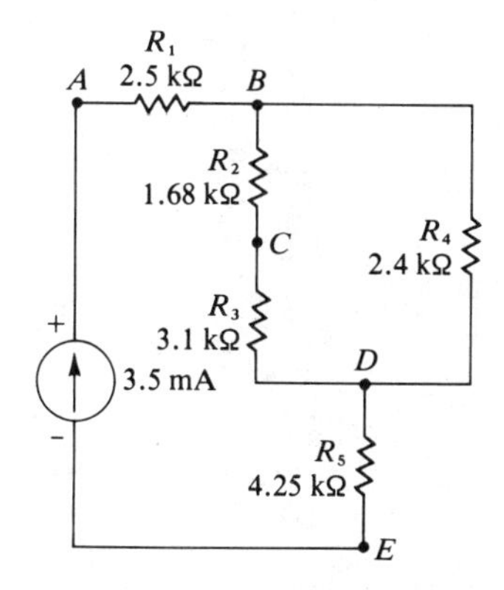

Figure 6.41 Reference Example 6.39

$$V_{R_5} = 3.5 \text{ mA} \times 4.25 \text{ k}\Omega = 14.88 \text{ volts}$$

Determining the voltage drop across R_3 is a little more difficult. That portion of the source's 3.5 mA which flows through the *BCD* branch of the parallel network must first be determined by using the inverse current law.

$$I_{(R_2+R_3)} = \frac{R_4}{(R_2 + R_3) + R_4}(3.5 \text{ mA})$$

$$I_{(R_2+R_3)} = \frac{2.4}{1.68 + 3.1 + 2.4}(3.5 \text{ mA}) = 1.17 \text{ mA}$$

$$V_{R_3} = I_{R_3}R_3 = 1.17 \text{ mA} \times 3.1 \text{ k}\Omega = 3.63 \text{ volts}$$

$$V_{EC} = V_{R_3} + V_{R_5} = 14.48 \text{ V} + 3.63 \text{ V} = 18.11 \text{ volts}$$

Example 6.40 Determine I_T, I_1, and I_2 in Figure 6.42.

The series combination branch *BCE* equals the sum of R_1 and R_2.

$$R_{BCE} = R_1 + R_2 = 4.7 \text{ k}\Omega + 3.3 \text{ k}\Omega = 8.0 \text{ k}\Omega$$

The series combination branch *BDE* equals the sum of R_3 and R_4.

$$R_{BDE} = R_3 + R_4 = 2.2 \text{ k}\Omega + 6.8 \text{ k}\Omega = 9.0 \text{ k}\Omega$$

The series combination in branch *BAE* is the source V_T plus the resistive sum of R_5 and R_6.

$$R_{BAE} = R_5 + R_6 = 1.5 \text{ k}\Omega + 1.0 \text{ k}\Omega = 2.5 \text{ k}\Omega$$

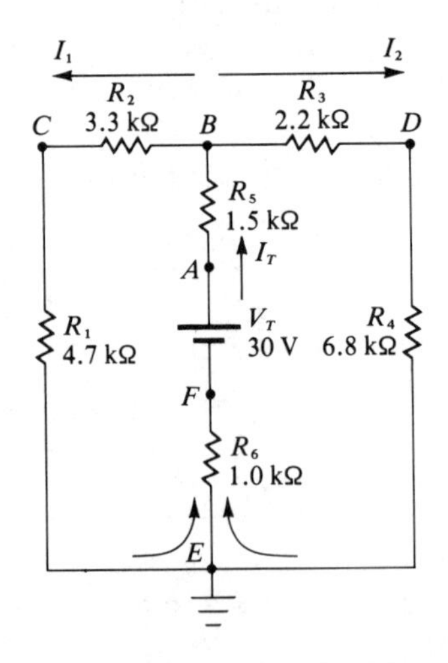

Figure 6.42 Reference Examples 6.40 and 6.41

All three branches are in parallel between point B and point E (ground reference). The equivalent circuit is redrawn as Figure 6.43.

$$\text{Let} \quad R_X = R_{BCE} \| R_{BDE} = \frac{8 \times 9}{17} \text{ k}\Omega = 4.24 \text{ k}\Omega$$

$$R_T = R_X + R_{BAE} = 4.24 \text{ k}\Omega + 2.5 \text{ k}\Omega = 6.74 \text{ k}\Omega$$

$$I_T = \frac{V_T}{R_T} = \frac{30 \text{ V}}{6.74 \text{ k}\Omega} = 4.45 \text{ mA}$$

$$I_1 = \frac{9}{17}(4.45 \text{ mA}) = 2.36 \text{ mA}$$

$$I_2 = I_T - I_1 = (4.45 - 2.36) \text{ mA} = 2.09 \text{ mA}$$

Example 6.41 Determine V_{FC} in Figure 6.42 by the following three paths:

a. $V_{FA} + V_{AB} + V_{BC}$
b. $V_{FE} + V_{EC}$
c. $V_{FE} + V_{ED} + V_{DB} + V_{BC}$

Verify that the voltage between points F and C is the same, regardless of the path taken between the two points. This is just another way of saying that voltages across parallel branches must be equal.

a. $V_{FA} = 30$ V (voltage source)
$V_{AB} = I_T R_5 = 4.45 \text{ mA} \times 1.5 \text{ k}\Omega = -6.68$ volts
$V_{BC} = I_1 R_2 = 2.36 \text{ mA} \times 3.3 \text{ k}\Omega = -7.79$ volts
$\therefore \quad V_{FC} = (30 - 6.68 - 7.79) \text{ V} = 15.53$ volts

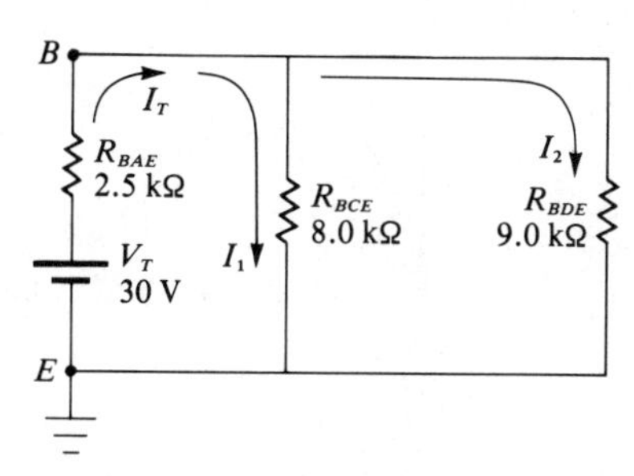

Figure 6.43 Simplified Circuit of Figure 6.42

Notice that V_{AB} and V_{BC} are negative because of the subscript conventions previously adopted, that is, the polarity of the second subscript in reference to the first subscript.

b. $V_{FE} = I_T R_6 = 4.45 \text{ mA} \times 1 \text{ k}\Omega = 4.45 \text{ volts}$
$V_{EC} = I_1 R_1 = 2.36 \text{ mA} \times 4.7 \text{ k}\Omega = 11.09 \text{ volts}$
$\therefore \quad V_{FC} = (4.45 + 11.09) \text{ V} = 15.54 \text{ volts}$

c. $V_{FE} = 4.45$ V (calculated in part "b")
$V_{ED} = I_2 R_4 = 2.09 \text{ mA} \times 6.8 \text{ k}\Omega = 14.21 \text{ volts}$
$V_{DB} = I_2 R_3 = 2.09 \text{ mA} \times 2.2 \text{ k}\Omega = 4.60 \text{ volts}$
$V_{BC} = I_1 R_2 = -7.79$ V (calculated in part "a")
$\therefore \quad V_{FC} = (4.45 + 14.21 + 4.6 - 7.79) \text{ V} = 15.47 \text{ volts}$

The minor difference in answers is only due to the roundoff of each individual solution. The important conclusion is that between any two points in an electrical circuit, the voltage must be the same regardless of which paths are used in calculating this voltage.

Example 6.42 Determine the magnitude of all the currents shown in Figure 6.44. Notice that all resistors are expressed in kilohms. It is convenient, therefore, to drop the kΩ term and automatically express all currents in mA for mathematical simplification.

$$R_4 + R_5 = 5.6 + 6.8 = 12.4$$

$$(R_4 + R_5) \| R_3 = \frac{10 \times 12.4}{10 + 12.4} = 5.54$$

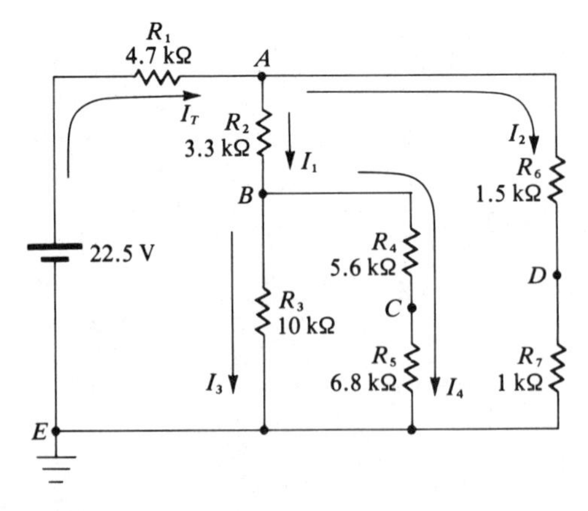

Figure 6.44 Reference Examples 6.42, 6.43, and 6.44

$$[(R_4 + R_5)\|R_3] + R_2 = 5.54 + 3.3 = 8.84 \equiv R_a$$
$$R_6 + R_7 = 1.5 + 1.0 = 2.5 \equiv R_b$$

Based on the above calculations, Figure 6.45 is the simplification of Figure 6.44.

$$R_a\|R_b = \frac{8.84 \times 2.5}{8.84 + 2.5} = \frac{22.1}{11.34} = 1.95$$
$$R_T = R_1 + (R_a\|R_b) = 4.7 + 1.95 = 6.65$$
$$I_T = \frac{V_T}{R_T} = \frac{22.5 \text{ V}}{6.65 \text{ k}\Omega} = 3.38 \text{ mA}$$
$$I_1 = \frac{R_b}{R_a + R_b}(I_T) = \frac{2.5}{11.34}(3.38) = 0.75 \text{ mA}$$
$$I_2 = I_T - I_1 = (3.38 - 0.75) \text{ mA} = 2.63 \text{ mA}$$

In turn, I_1 divides into I_3 and I_4 at junction B of Figure 6.44.

$$I_3 = \frac{(R_4 + R_5)}{R_3 + (R_4 + R_5)}(I_1) = \frac{12.4}{22.4}(0.75) \text{ mA} = 0.42 \text{ mA}$$
$$I_4 = I_1 - I_3 = (0.75 - 0.42) \text{ mA} = 0.33 \text{ mA}$$

As a check, Figure 6.44 shows that

$$I_2 + I_3 + I_4 = I_T$$
$$(2.63 + 0.42 + 0.33) \text{ mA} = 3.38 \text{ mA}$$
$$3.38 \text{ mA} = 3.38 \text{ mA}$$

Example 6.43 Using the current calculations of Example 6.42, determine V_A, V_B, V_C, and V_D (with respect to ground). Since current is in mA and resistance in kilohms, volts will automatically result.

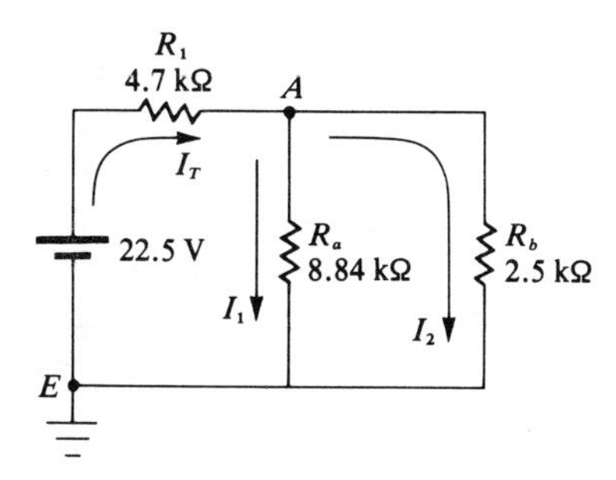

Figure 6.45 Simplification of Figure 6.44

$$V_A = \text{source voltage reference} - I_T R_1$$
$$V_A = 22.5 - (3.38 \times 4.7) = 6.61 \text{ volts}$$
$$V_B = V_A - I_1 R_2 = 6.61 - (0.75 \times 3.3) = 4.14 \text{ volts}$$
$$V_C = V_B - I_4 R_4 = 4.14 - (0.33 \times 5.6) = 2.29 \text{ volts}$$
$$V_D = I_2 R_7 = 2.63 \times 1.0 = 2.63 \text{ volts}$$

Example 6.44 Using the calculations of Example 6.43, determine V_{DA}, V_{BC}, and V_{CA}. Observe the order of the subscripts.

$$V_{DA} = V_A - V_D = 6.61 - 2.63 = 3.98 \text{ volts}$$
$$V_{BC} = V_C - V_B = 2.29 - 4.14 = -1.85 \text{ volts}$$
$$V_{CA} = V_A - V_C = 6.61 - 2.29 = 4.32 \text{ volts}$$

SUMMARY

By definition, a series circuit provides only one path for current flow. The total resistance "seen" by a source in a series circuit is the sum of all the individual resistances. The current flowing in a series circuit is the same at all points in the circuit. The sum of all voltages in a closed loop is zero.

Proportional voltage law states that the voltage across any resistance in a closed loop is equal to the ratio of that resistance to the total series resistance, multiplied by the applied voltage.

Voltage dividers are used to develop specific voltage values for the various components of a system. Individual resistance values are calculated by considering both the voltage and current requirements of the system.

By definition, a parallel branch means that there is more than one path through which current may flow to complete a closed loop between the terminals of a voltage source. The voltages across parallel branches are equal. The total resistance is less than the resistance of any single branch and is equal to the reciprocal of the sum of the reciprocals of each individual resistance.

Kirchhoff's current law states that the current entering a junction equals the sum of the currents leaving the junction. In a two-branch parallel junction, the current in one branch is the ratio of the other branch resistance to the sum of both branch resistances, multiplied by the current entering the junction.

The total conductance seen by a source in a parallel circuit is the sum of the conductances of each individual branch. This is analogous to the total resistance in a series circuit.

Circuits with multiple sources may be simplified. Current sources may be converted into voltage sources, or conversely, voltage sources may be converted into current sources. Current sources in parallel or voltage sources in series are algebraically additive. Voltage sources in parallel and current sources in series must be of the same magnitude and polarity direction.

Many circuits consist of complex combinations of both series and parallel arrangement. Analysis usually requires the application of various combinations of Kirchhoff's laws, Ohm's law, voltage ratios, and other simpler laws.

PROBLEMS

Reference Sections 6.1 through 6.3

1. Three resistors in series have values of 2.2 kΩ, 3.3 kΩ, and 6.8 kΩ. What is the total resistance?
2. Four resistors in series have values of 33 kΩ, 680 ohms, 1.2 kΩ, and 3300 ohms. What is the total resistance?
3. The sum of three resistors is 32.9 kΩ. The second resistor is twice the first and the third resistor is twice the second. What is the value of each resistor?
4. If V_T volts is impressed across the three resistors of Problem 3, what is the voltage across each resistor? Express the answer algebraically.
5. Write an algebraic expression for the total current in Figure 6.46.
6. Three resistances respectively have 12 V, 6.5 V, and 4 V developed across them. What is the applied voltage?
7. The applied voltage is 34 V. Two of three series resistances have 6.8 V and 12.35 V developed across them. How much voltage is developed across the third resistance?
8. How many volts are across R_3 in Figure 6.46?
9. What is the voltage between points A and B in Figure 6.46?
10. Determine the magnitude of V_{CB} in Figure 6.46.
11. What is V_{AC} in Figure 6.47?

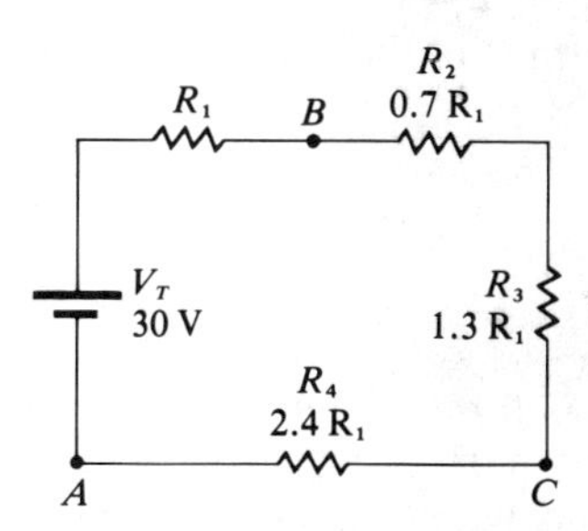

Figure 6.46 Reference Problems 5, 8, 9, and 10

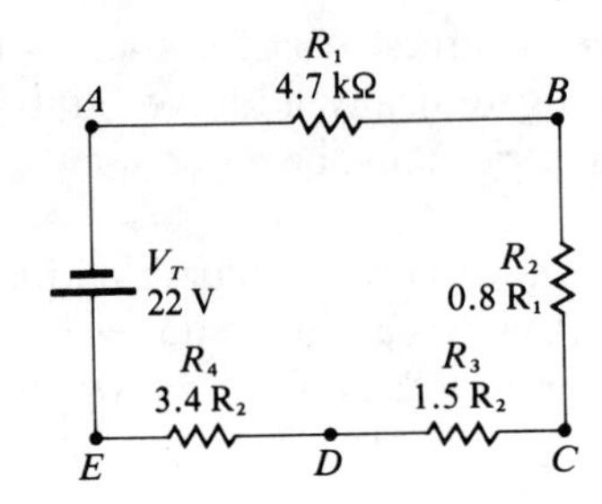

Figure 6.47 Reference Problem 11

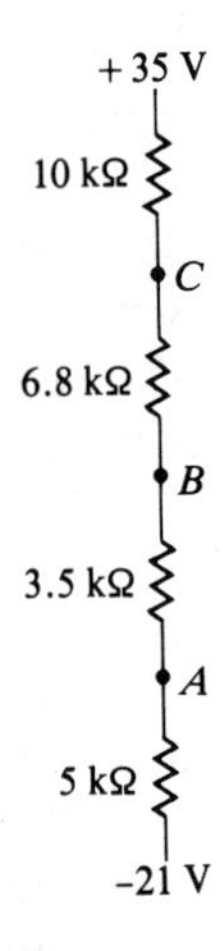

Figure 6.48 Reference Problems 12, 13, and 14

Reference Section 6.4

12. What is V_B with reference to ground in Figure 6.48?
13. What is V_{BA} in Figure 6.48?
14. What is the magnitude of the current flow in Figure 6.48?
15. It is desired that the voltage of point *B*, with reference to ground, equal 5 V in Figure 6.49. What should be the value of *R*?

Reference Section 6.5

16. It is desired to have a total resistance of 2.78 kΩ. Several resistors of higher value are available. If one is 6.8 kΩ, what value of *R* should be placed in parallel with this resistor to result in the desired R_T?

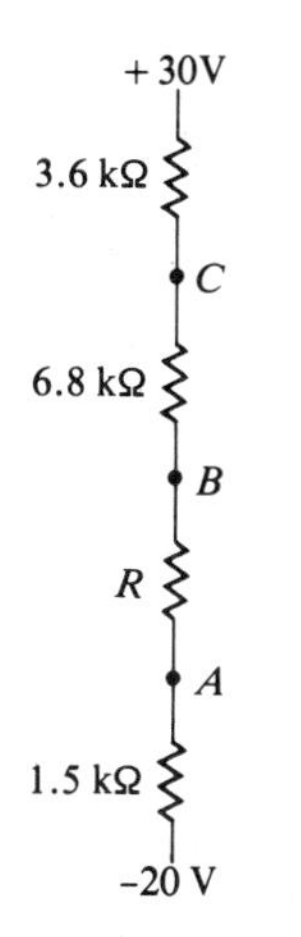

Figure 6.49 Reference Problem 15

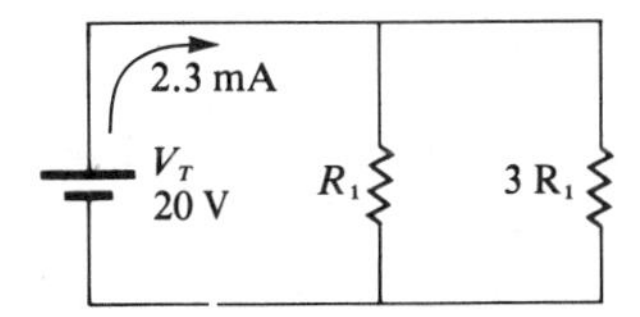

Figure 6.50 Reference Problem 18

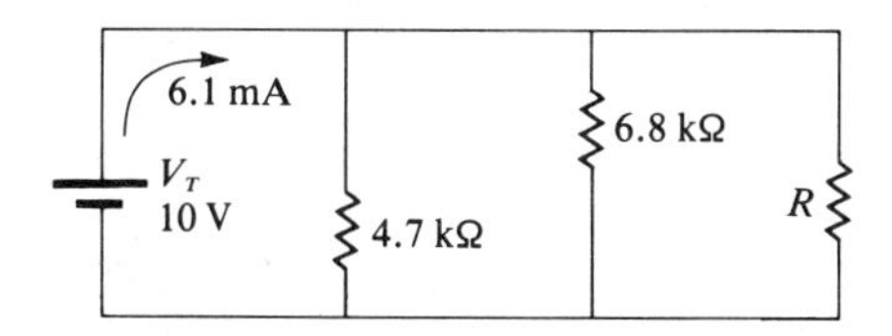

Figure 6.51 Reference Problem 19

17. What is the equivalent resistance of 2.2 kΩ, 3.3 kΩ, and 6.8 kΩ placed in parallel?
18. What is the value of R_1 in Figure 6.50?
19. What is the value of R in Figure 6.51?
20. What is the equivalent resistance of two equal resistors in parallel? What is the equivalent resistance of three equal resistors in parallel? Develop an expression for n equal resistors in parallel.

Reference Section 6.6

21. A current of 102 mA enters a junction. One of the two branches leaving the junction carries twice the current of the other branch. What are the branch currents?
22. Determine the values of R_1, R_2, and R_3 for the voltage divider network of Figure 6.52.
23. Determine the values of R_1, R_2, R_3, and R_4 for the voltage divider network of Figure 6.53.

Reference Section 6.7

24. Two resistors of 12.3 kΩ and 6.7 kΩ are connected in parallel. Using the inverse current ratio law, determine the current through each resistor if 12 mA total current enters the parallel junction.

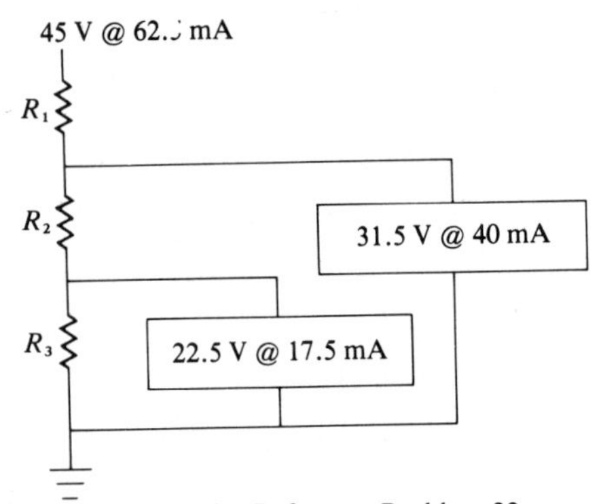

Figure 6.52 Reference Problem 22

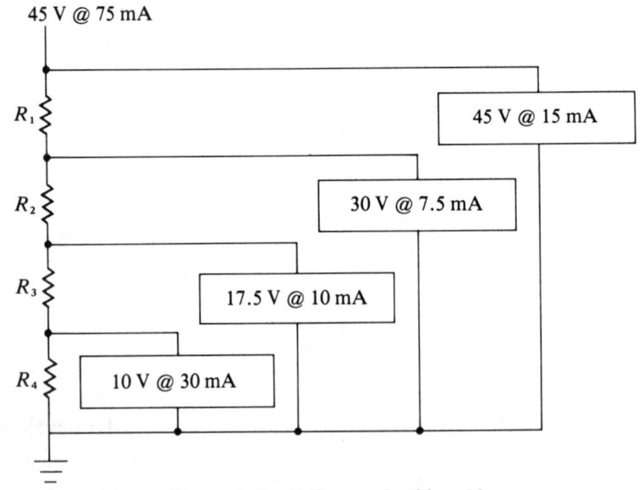

Figure 6.53 Reference Problem 23

25. A current of 250 mA enters a parallel junction of 22 kΩ, 33 kΩ, and 10 kΩ. What is the current through each resistor?
26. I_T enters the junction of three parallel resistors, R_1, R_2, and R_3. Develop an algebraic expression for the current through each resistor.

Reference Section 6.8

27. Three resistors of 470 ohms, 680 ohms, and 1.5 kΩ are in parallel. What is the total conductance of this circuit?
28. What is the value of G_2 in Figure 6.54?

Reference Section 6.9

29. Determine the current in the circuit of Figure 6.55.
30. What is V_{BE} in Figure 6.55?
31. What is V_{GC} in Figure 6.55?

Reference Section 6.10

32. What is V_{BC} in Figure 6.56?
33. What is the current through the 1.2 kΩ resistor of Figure 6.56?
34. What is the voltage across the current generators of Figure 6.56?

Reference Section 6.11

35. Determine the current through R_1 in Figure 6.57.
36. Determine the current through R_1 in Figure 6.58.

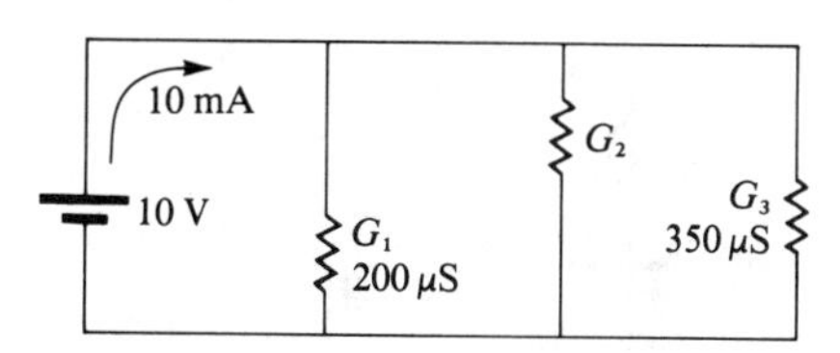

Figure 6.54 Reference Problem 28

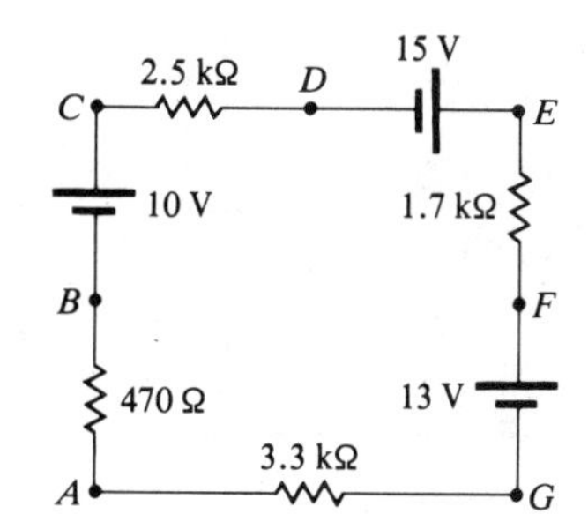

Figure 6.55 Reference Problems 29, 30, and 31

Reference Section 6.12

37. What is I_T in Figure 6.59?
38. What is I_1 in Figure 6.59?
39. What is I_2 in Figure 6.59?
40. What is V_A in Figure 6.59?
41. Determine the magnitude of the total current in Figure 6.60.
42. What is I_2 in Figure 6.60?
43. What is V_{BD} in Figure 6.60?

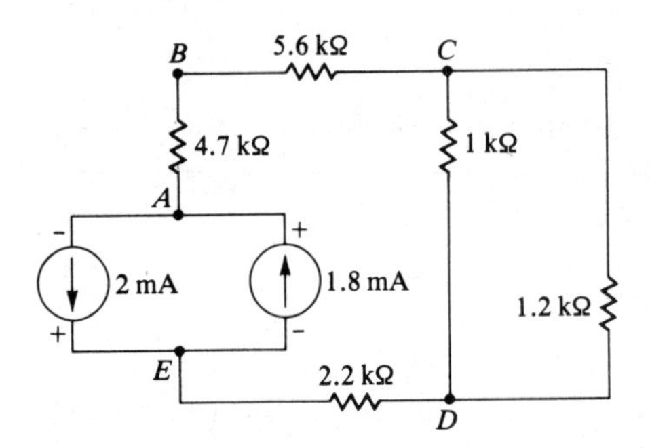

Figure 6.56 Reference Problems 32, 33, and 34

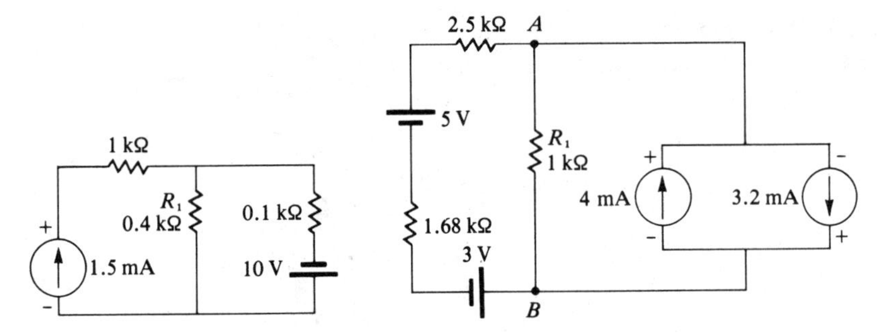

Figure 6.57 Reference Problem 35

Figure 6.58 Reference Problem 36

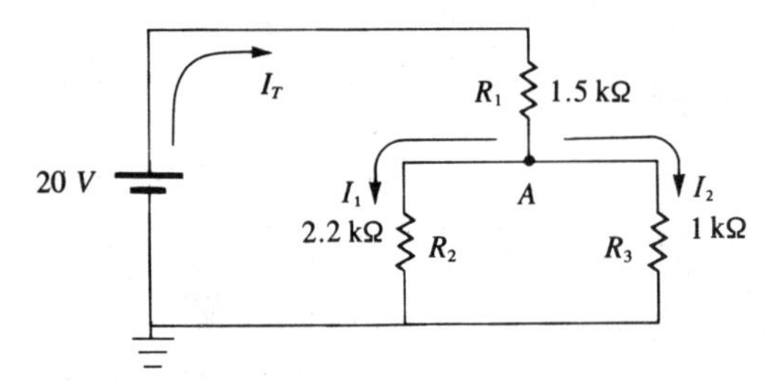

Figure 6.59 Reference Problems 37 through 40

44. What is V_{CD} in Figure 6.60?
45. Does I_4 equal I_7 in Figure 6.61?
46. Determine all currents in Figure 6.61.
47. Determine the voltage at all referenced points in Figure 6.61.
48. What is the value of V_{AC}, V_{DE}, and V_{CG} in Figure 6.61?

Optional Computer Problems

49. Write a program in BASIC to print out the voltage across each of three series resistors. The three resistor values and the total source voltage are to be input to the computer using the INPUT statement.
50. Write a program in BASIC to print out the current through each of three parallel resistors. The three resistor values and the total current entering the junction are to be input to the computer using the INPUT statement.

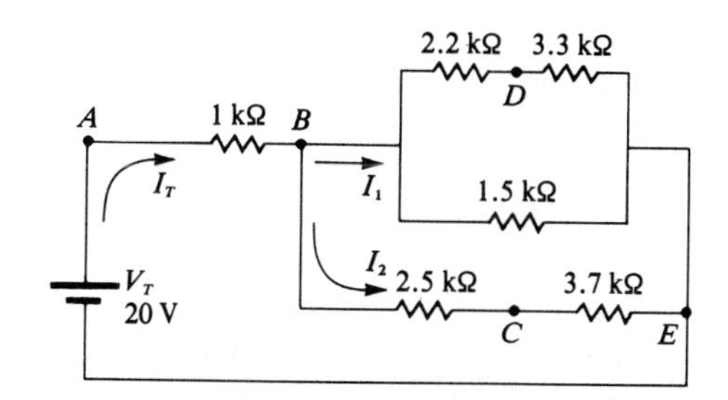

Figure 6.60 Reference Problems 41 through 44

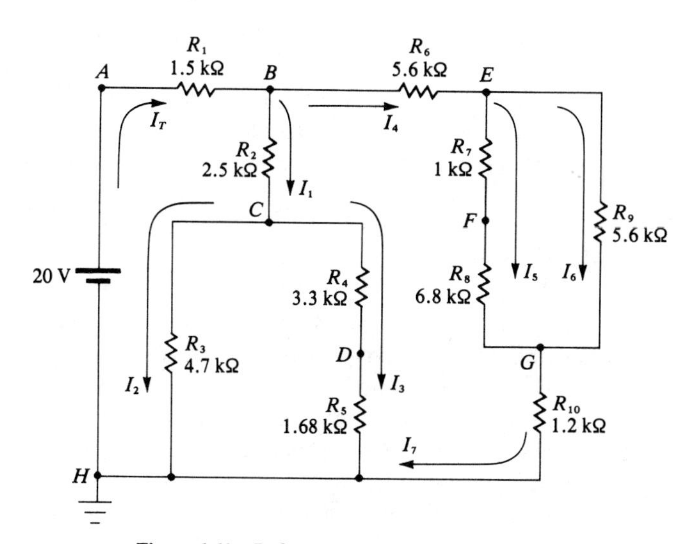

Figure 6.61 Reference Problems 45 through 48

CHAPTER 7

ADVANCED PROGRAMMING TECHNIQUES FOR ELECTRONIC PROBLEMS

7.1 INTRODUCTION

The concepts behind computer programming and several of the more elementary BASIC statements were presented in Chapter 4. These statements were sufficient to write simple programs and generate printed results. In most cases, solutions of the various assigned programming problems could be performed more rapidly with a slide rule or pocket calculator. It has been necessary to defer the study of the truly significant power capabilities of computers until sufficient background knowledge of electronic circuits was achieved. Having established such a background, it is now appropriate to progress to the more advanced programming techniques.

This chapter will explain how to program computers in such a manner that the computer itself can vary the parameters of a problem. This process facilitates the generation of hundreds, or even thousands, of solutions—one for each parameter variation. One example of the benefits such a process can produce is that the computer can continue to vary component values and calculate results until the desired answer is achieved. The time saved by freeing the technician or engineer from making so many individual calculations far outweighs the short time necessary to originally write the program.

Following a familiarization with logical and mathematical functions, this chapter will discuss advanced statements. It will conclude with a number of programming examples applied to electronic circuits.

7.2 MATHEMATICAL FUNCTIONS

Within the course of a user's programming experience, he encounters many cases where relatively common mathematical operations are performed. The results of these common operations can often be found in volumes of mathematical tables, such as sine, cosine, square root, log, etc. Since computers perform this sort of operation with speed and accuracy, such operations are built into BASIC. The user need never consult tables to obtain the value of the sine of 23° or the natural log of 144.

When such values are to be used in an expression, intrinsic functions such as:

```
SIN (23*PI/18Ø)
LOG (144)
```

are substituted.

The various mathematical functions available are detailed in Table 7.1.

TABLE 7.1 Mathematical Functions

Function code	Meaning
ABS(X)	Returns the absolute value of X
SGN(X)	Returns the sign function of X, a value of 1 preceded by the sign of X, SGN(Ø)=Ø
INT(X)	Returns the greatest integer in X which is less than or equal to X, (INT(−.5)= −1)
FIX(X)	Returns the truncated value of X, SGN(X)*INT(ABS(X)), (FIX(−.5)=Ø)
COS(X)	Returns the cosine of X in radians
SIN(X)	Returns the sine of X in radians
TAN(X)	Returns the tangent of X in radians
ATN(X)	Returns the arctangent (in radians) of X
SQR(X)	Returns the square root of X
EXP(X)	Returns the value of $e \uparrow X$, where $e = 2.71828\ldots$
LOG(X)	Returns the natural logarithm of X, $\log_e X$

Depending on the BASIC version, additional functions may be available to the programmer. These might include the common (base 10) logarithm, LOG10(X) or CLOG(X); the constant π, PI; and a function that generates random numbers, RND or RANDOM.

7.3 LOGICAL OPERATORS

The AND, OR, and NOT logical operators were first introduced in Chapter 4. Other logical operators, again available only with more advanced versions of BASIC, are the XOR, IMP, and EQV. All six have been combined into one chart and are defined on page 154. Once again, remember that

a. Logical operators are used in IF-THEN and other similar statements where some condition is tested to determine subsequent operations within the user program.
b. A and B are relational expressions having only TRUE and FALSE values.

Operator	*Example*	*Meaning*
NOT	NOT A	The logical negative of *A*. If *A* is true, NOT A is false.
AND	A AND B	The logical product of *A* and *B*. A AND B has the value TRUE only if *A* and *B* are both true, and has the value FALSE if either *A* or *B* is false.
OR	A OR B	The logical sum of *A* and *B*. A OR B has the value TRUE if either *A* or *B* is true, and has the value FALSE only if both *A* and *B* are false.
XOR	A XOR B	The logical exclusive OR of *A* and *B*. A XOR B has the value TRUE if either *A* or *B* is true but not both, and has the value FLASE otherwise.
IMP	A IMP B	The logical implication of *A* and *B*. A IMP B has the value FALSE if and only if *A* is true and *B* is false; otherwise the value is TRUE.
EQV	A EQV B	*A* is logically equivalent to *B*. A EQV B has the value TRUE if *A* and *B* are both true or both false, and has the value FALSE otherwise.

The following tables are called *truth tables* and graphically describe the results of these logical operations; both A and B are given for every possible combination of values.

A	*B*	A AND B
T	T	T
T	F	F
F	T	F
F	F	F

A	*B*	A OR B
T	T	T
T	F	T
F	T	T
F	F	F

A	*B*	A XOR B
T	T	F
T	F	T
F	T	T
F	F	F

A	*B*	A EQV B
T	T	T
T	F	F
F	T	F
F	F	T

A	*B*	A IMP B
T	T	T
T	F	F
F	T	T
F	F	T

A	NOT A
T	F
F	T

7.4 BRANCHING STATEMENTS

Unconditional Branch, GOTO Statement

The GOTO statement is used when it is desired to unconditionally transfer to some line other than the next sequential line in the program. In other words, a GOTO statement causes an immediate jump to a specified line, out of the normal consecutive line number order of execution. The general format of the statement is as follows:

line number GOTO ⟨*line number*⟩

The line number to which the program jumps can be either greater than or less than the current line number. It is thus possible to jump forward or backward within a program.

Consider the following simple example:

```
1Ø LET A=2
2Ø GOTO 5Ø
3Ø LET A=SQR(A+14)
5Ø PRINT A,A*A
```

When executed, the above lines cause the following to be printed:

```
2                 4
```

When the program encounters line 20, control transfers to line 50, line 50 is executed, and control then continues to the line following line 50. Line 30 is never executed. Any number of lines can be skipped in either direction.

When written as part of a multiple statement line, GOTO should always be the last statement on the line, since any statement following the GOTO on the same line is never executed. For example:

```
11Ø LET A=ATN(B2): PRINT A: GOTO 5Ø
```

Conditional Branch, IF-THEN and IF-GOTO Statements

The IF-THEN and IF-GOTO statements are used to transfer conditionally from the normal consecutive order of statement numbers, depending upon the truth of some mathematical relation or relations. The basic format of the IF statement is as follows:

	THEN	⟨*statement*⟩
line number IF ⟨*condition*⟩	THEN	⟨*line number*⟩
	GOTO	⟨*line number*⟩

(Note: Some versions do not permit the IF-THEN-GOTO option.)

The specified condition is tested. If the relationship is false, then control is transferred to the statement following the IF statement (the next sequentially numbered line). If the condition is true, the statement following THEN is executed, or control is transferred to the line number given after THEN or GOTO.

The deciding condition can be either a simple relational expression in which two mathematical expressions are separated by a relational operator, or a logical expression in which two relational or logical expressions are separated by a logical operator. For example:

Relational expression	Logical expression
A+2>B	A>B AND B< =SQR(C)

Both types of condition, when evaluated, are either true or false; no numeric value is associated with the results of an IF statement.

```
75 IF A*B> =B*(B+1)  THEN LET D4=D4+1
```

In the above line, the quantities A*B and B*(B+1) are compared. If the first value is greater than or equal to the second value, the variable D4 is incremented by 1. If B*(B+1) is greater than A*B, D4 is not incremented and control passes immediately to the next line following line 75.

When a line number follows the word THEN, the IF-THEN statement is the same as the IF-GOTO statement. The word THEN can be followed by any BASIC statement, including another IF statement. For example:

```
25 IF A>B THEN IF B>C THEN PRINT "A>B>C"
25 IF A>B AND B>C THEN PRINT "A>B>C"
```

The preceding two lines are logically equivalent and perform the following operation:

If *B* is both less than *A* and greater than *C*, the message

```
A>B>C
```

is printed, otherwise the line following line 25 is executed.

In the following example, the IF-GOTO statement in line 20 is used to limit the value of the variable *A* in line 10. Execution of the loop continues until the relationship A>4 is true, then immediately branches to line 55 to end the program. (A program *loop* is a series of statements which are written so that, when all of them have been executed, control transfers to the beginning of the statements. This process continues to occur until some terminal condition is reached.)

```
LIST
1Ø LET A=A+1: X=A↑2
2Ø IF A>4 GOTO 55
25 PRINT X
3Ø PRINT "VALUE OF A IS" A
4Ø GOTO 1Ø
55 END
```

When the above loop is executed, the following is printed:

```
RUN
 1
VALUE OF A IS 1
 4
```

```
VALUE OF A IS 2
 9
VALUE OF A IS 3
 16
VALUE OF A IS 4
```

(The novice BASIC programmer is advised to follow the operation of the computer through these short example programs.)

Examples of IF-THEN statements follow. The R's refer to specific resistors in an electrical circuit and the P's refer to power dissipations.

```
10   IF R1 > R2 THEN 100                 ! SIMPLE COMPARISON
20   IF P1=P2 OR P1 < P3 THEN 200
30   IF P2=P3 THEN R3=R3+4.7E3           ! ASSIGNMENT STATEMENT
40   IF P4 > 1.5 OR P5 > 2.5 THEN PRINT "POWER LIMITS EXCEEDED"
```

IF-THEN-ELSE Statement

The IF-THEN statement allows the program to transfer control to another line or execute a specified statement depending upon a stated condition. Some powerful versions of BASIC (but not most microcomputer versions, with the exception of TRS-80 Level II) permit an IF-THEN-ELSE statement.

The IF-THEN-ELSE statement is the same as the IF-THEN statement, except that rather than executing the line *following* the IF statement, another line number or statement can be specified for execution where the condition is not met. The statement is of the form:

line number IF ⟨*condition*⟩ [THEN ⟨*line number*⟩ | THEN ⟨*statement*⟩ | GOTO ⟨*line number*⟩] { ELSE ⟨*line number*⟩ | ELSE ⟨*statement*⟩ }

where the condition is defined as one of the following:

⟨*relational expression*⟩ ⟨*logical operator*⟩ ⟨*relational expression*⟩

and a relational expression is defined as:

⟨*expression*⟩ ⟨*relational operator*⟩ ⟨*expression*⟩

The relational condition is tested; if it is true, the THEN/GOTO part of the statement is executed. If the condition is false, the ELSE part of the statement is executed. Following the word ELSE is either a statement to be executed or a line number to which control is transferred.

Examples of IF-THEN-ELSE statements and their potential meanings follow:

```
75 IF P2 > P4 THEN PRINT "GREATER" ELSE PRINT "NOT GREATER"
```

IF the power dissipated by resistor 2 exceeds the power dissipated by resistor 4, then print "GREATER"; otherwise print "NOT GREATER."

```
85 IF P2+P3>=P4 OR P1<100E-3 THEN 200 ELSE R1=R1+1E3
```

IF, within a given electrical circuit, the sum of the powers dissipated by both resistors R2 and R3 is equal to or greater than the power dissipated by resistor R4 OR if the power dissipated by resistor R1 is less than 100 milliwatts, THEN go to statement number 200 ELSE if the above is not true, then increment the value of resistor R1 by 1K.

Clearly, the IF-THEN-ELSE combination can be implemented with versions of BASIC without that capability, simply by writing the statement that would follow ELSE on the line following the IF-THEN statement. For example,

```
IF A>B THEN 50 ELSE 100
```

is equivalent to

```
IF A<B THEN 50
GOTO 100
```

7.5 PROGRAM LOOPS

Loops were first mentioned in the section on the IF-THEN and IF-GOTO statements. Programs frequently involve performing certain operations a specific number of times. This is a task for which a computer is particularly well suited. With simple tasks, such as computing a list of prime numbers between 1 and 1,000,000, a computer can perform the operations and obtain correct results in a minimum amount of time. To write a loop, the programmer must ensure that the series of statements is repeated until a terminating condition is met.

Programs containing loops can be illustrated by using two versions of a program to print a table of the positive integers 1 through 100 together with the square root of each. Without a loop, the first program is 101 lines long and reads:

```
10 PRINT 1, SQR(1)
20 PRINT 2, SQR(2)
30 PRINT 3, SQR(3)
         :
         :
990 PRINT 99, SQR(99)
1000 PRINT 100, SQR(100)
1010 END
```

With the following program example, using a simple sort of loop, the same table is obtained with fewer lines:

```
1Ø LET X=1
2Ø PRINT X,SQR(X)
3Ø LET X=X+1
4Ø IF X<=1ØØ THEN 2Ø
5Ø END
```

Statement 10 assigns a value of 1 to *X*, thus setting up the initial conditions of the loop. In line 20, both 1 and its square root are printed. In line 30, *X* is incremented by 1. Line 40 asks whether *X* is still less than or equal to 100; if so, BASIC returns to print the next value of *X* and its square root. This process is repeated until the loop has been executed 100 times. After the number 100 and its square root have been printed, *X* becomes 101. The condition in line 40 is now false so control does not return to line 20, but goes to line 50 which ends the program.

All program loops have four characteristic parts:

a. *Initialization*: the conditions that must exist for the first execution of the loop (line 10 above);
b. The *body of the loop*: performance of the operation that is to be repeated (line 20 above);
c. *Modification*: some value is altered which makes each execution of the loop different from the one before and the one after (line 30 above);
d. *Termination condition*: an exit test which, when satisfied, completes the loop (line 40 above). Execution continues to the program statements following the loop (line 50 above).

FOR and NEXT Statements

The FOR statement is of the form:

line number FOR ⟨*variable*⟩ = ⟨*expression*⟩ TO ⟨*expression*⟩ {STEP ⟨*expression*⟩}

For example:

```
1Ø FOR K=2 TO 2Ø STEP 2
```

which causes program execution to cycle through the designated loop using *K* as 2, 4, 6, 8, . . . , 20 in calculations involving *K*. When *K* = 20, the loop is left behind and the program control passes to the line following the associated NEXT statement. The variable in the FOR statement, *K* in the preceding example, is known as the control variable. The expression in the FOR statement can be any acceptable BASIC expression.

The NEXT statement signals the end of the loop which began with the FOR statement. The NEXT statement is of the form:

line number NEXT ⟨*variable*⟩

where the variable is the same variable specified in the FOR statement. Together the FOR and NEXT statements describe the boundaries of the program loop. When execution encounters the NEXT statement, the computer adds the STEP expression value to the variable and checks to see if the variable is still less than or equal to the terminal expression value. When the variable exceeds the terminal expression value, control falls through the loop to the statement following the NEXT statement.

If the STEP expression is omitted from the FOR statement, +1 is the assumed value. Since +1 is a common STEP value, that portion of the statement is frequently omitted.

The expressions within the FOR statement are evaluated *once* upon initial entry to the loop. The test for completion of the loop is made prior to each execution of the loop. (If the test fails initially, the loop is never executed.)

In some versions of BASIC (but not, for example, TRS-80) the control variable can be modified within the loop. In these versions, when control falls through the loop, the control variable retains the last value used with the loop.

The following is a demonstration of a simple FOR-NEXT loop. The loop is executed 10 times; the value of *I* is 10 when control leaves the loop; and +1 is the assumed STEP value:

```
10 FOR I=1 TO 10
20 PRINT I
30 NEXT I
40 PRINT I
```

The loop itself is lines 10 through 30. The numbers 1 through 10 are printed when the loop is executed. After $I = 10$, control passes to line 40 which causes 10 to be printed again. If line 10 had been:

```
10 FOR I = 10 TO 1 STEP -1
```

the value printed by line 40 would be 1.

```
10 FOR I = 2 TO 44 STEP 2
20 LET I = 44
30 NEXT I
```

The above loop is only executed once since the value of $I = 44$ has been reached and the termination condition is satisfied.

If, however, the initial value of the variable is greater than the terminal value, the loop is not executed at all. A statement of the form:

```
10 FOR I = 20 TO 2 STEP 2
```

cannot be used to begin a loop, although a statement like the following will initialize execution of a loop properly:

```
10 FOR I=20 TO 2 STEP -2
```

For positive STEP values, the loop is executed until the control variable is greater than its final value. For negative STEP values, the loop continues until the control variable is less than its final value.

FOR loops can be nested but not overlapped. The depth of nesting depends upon the amount of user storage space available (in other words, upon the size of the user program and the amount of memory each user has available). Nesting is a programming technique in which one or more loops are completely within another loop. The field of one loop (the numbered lines from the FOR statement to the corresponding NEXT statement, inclusive) must not cross the field of another loop.

Acceptable Nesting Techniques

```
FOR I1 = 1 TO 1Ø
FOR I2 = 1 TO 1Ø
NEXT I2
FOR I3 = 1 TO 1Ø
NEXT I3
NEXT I1
```

Unacceptable Nesting Techniques

```
FOR I1 = 1 TO 1Ø
FOR I2 = 1 TO 1Ø
NEXT I1
NEXT I2
```

Three-Level Nesting

```
FOR I1 = 1 TO 1Ø
FOR I2 = 1 TO 1Ø
FOR I3 = 1 TO 1Ø
NEXT I3
FOR I4 = 1 TO 1Ø
NEXT I4
NEXT I2
NEXT I1
```

```
FOR I1 = 1 TO 1Ø
FOR I2 = 1 TO 1Ø
FOR I3 = 1 TO 1Ø
NEXT I3
FOR I4 = 1 TO 1Ø
NEXT I4
NEXT I1
NEXT I2
```

It is possible to exit from a FOR-NEXT loop without the control variable reaching the termination value. A conditional or unconditional transfer can be used to leave a loop. Control can only transfer into a loop which had been left earlier without being completed, thus ensuring that termination and STEP values are assigned.

Both FOR and NEXT statements can appear anywhere in a multiple statement line. For example:

```
1Ø FOR I=1 TO 1Ø STEP 5:  NEXT I: PRINT "I=";I
```

causes:

```
I= 6
```

to be printed when executed.

Neither the FOR nor NEXT statement can be executed conditionally in an IF statement. The following statements are *incorrect*:

```
15 IF I<>J THEN NEXT I
16 IF I=J THEN FOR I=1 TO J
```

Conditional Termination of FOR Loops

In the simple FOR-NEXT loop, the format of the FOR statement is given as:

line number FOR ⟨*variable*⟩ = ⟨*expression*⟩ TO ⟨*expression*⟩ {STEP ⟨*expression*⟩}

There are many situations in which the final value of the loop variable is not known in advance and what is really desired is to execute the loop as many times as necessary to satisfy some condition. In evaluating a function, for example, this condition might be the point at which further iterations contribute no further accuracy to the result. BASIC-PLUS provides a convenient way of specifying that a loop is to be executed until a certain condition is detected or while some condition is true. (This capability is not available in microcomputer versions of BASIC.) These statements take the following forms:

line number FOR ⟨*variable*⟩ = ⟨*expression*⟩ {STEP ⟨*expression*⟩} WHILE ⟨*relational expression*⟩

line number FOR ⟨*variable*⟩ = ⟨*expression*⟩ {STEP ⟨*expression*⟩} UNTIL ⟨*relational expression*⟩

The condition has the same structure as specified in an IF statement and can be just as elaborate, if necessary. Before the loop is executed, and at each loop iteration, the condition is tested. The iteration proceeds if the result is true (FOR-WHILE) or false (FOR-UNTIL).

The difference between a FOR loop specified with a WHILE or UNTIL and one specified with a terminal value for the loop variable is worth noting, in order to avoid potential pitfalls in the usage of each. Consider the two loops in the program below:

```
LIST
1Ø FOR I=1 TO 1Ø
15 PRINT I;
2Ø NEXT I
25 PRINT "I="I
5Ø FOR I=1 UNTIL I>1Ø
55 PRINT I;
6Ø NEXT I
65 PRINT "I="I
75 END

RUN
 1  2  3  4  5  6  7  8  9  1Ø I= 1Ø
 1  2  3  4  5  6  7  8  9  1Ø I= 11
```

Each loop prints the numbers from 1 to 10. When the loop at line 10 is done, however, the loop variable is set to the last value used (that is, 10). In the second loop beginning at line 50, the loop variable is set to the value which caused the loop to be terminated (that is, 11).

Next consider the following two loops:

```
LIST
10 X=10
20 FOR I=1 TO X
30 X=X/2: PRINT I,X
40 NEXT I
50 PRINT
60 X=10
70 FOR I=1 UNTIL I>X
80 X=X/2: PRINT I,X
90 NEXT I
95 END

RUN
 1            5
 2            2.5
 3            1.25
 4            .625
 5            .3125
 6            .15625
 7            .078125
 8            .390625E-1
 9            .195313E-1
 10           .976563E-2

 1            5
 2            2.5
```

In the case of the loop beginning with line 20, the iteration stops when *I* exceeds the initial value of *X* (that is, 10). Even though the value of *X* changes within the loop, the initial value of *X* determines the performance of the loop. In the second loop, the current value of *X* determines when the iteration ceases. Thus, after three iterations, *I* is greater than *X* in the second loop and the loop is terminated. (The STEP value, when omitted, is still assumed to be 1.)

These forms of loop control are particularly useful in iterative applications where data generated during the loop execution determines loop completion.

7.6 SUBROUTINES

The technique of looping allows the program to do a sequence of instructions a specified number of times. If the program should require that a sequence of instructions be executed several times in the course of the program, this is also possible.

A subroutine is a section of code performing some operation required at more than one point in the program. Sometimes a complicated I/O operation for a volume of data, a complex mathematical evaluation, or any number of other processes may be best performed in a subroutine.

More than one subroutine can be used in a single program, in which case they can be placed one after another at the end of the program (in line number sequence). A useful practice is to assign distinctive line numbers to subroutines; for example, if the main program uses line numbers up to 199, use 200 and 300 as the first numbers of two subroutines.

GOSUB Statement

Subroutines are usually placed physically at the end of a program before DATA statements, if any, and always before the END statement. The program begins execution and continues until it encounters a GOSUB statement of the form:

line number GOSUB ⟨*line number*⟩

where the line number following the word GOSUB is the first line number of the subroutine. Control then transfers to that line in the subroutine. For example:

```
5Ø GOSUB 2ØØ
```

Control is transferred to line 200 in the user program. The first line in the subroutine can be a remark or any executable statement.

Note that a STOP statement must be placed between the main program and the first line of a subroutine that follows it. Otherwise, when the computer has finished executing the main program, it will "fall into" the subroutine and execute it without having been instructed to do so by a GOSUB. This will cause an error message to be displayed.

RETURN Statement

Having reached the line containing a GOSUB statement, control transfers to the line indicated after GOSUB; the subroutine is processed until the computer encounters a RETURN statement of the form:

line number RETURN

which causes control to return to the statement *following* the original GOSUB statement. A subroutine always exits via a RETURN statement.

Before transferring to the subroutine, BASIC internally records the next sequential statement to be processed after the GOSUB statement; the RETURN statement is a signal to transfer control to this statement. In this way, no matter how many subroutines or how many times they are called, BASIC always knows where to go next.

Nesting Subroutines

Subroutines can be nested; that is, one subroutine can call another subroutine. If the execution of a subroutine encounters a RETURN statement, it returns control to the line following the GOSUB that called *that* subroutine. Therefore, a subroutine can call another subroutine, even itself. Subroutines can be entered at any point and can have more than one RETURN statement. It is possible to transfer to the beginning or any part of a subroutine; multiple entry points and RETURNS make a subroutine more versatile.

The maximum level of GOSUB nesting depends on the size of the user program and the amount of core storage available at the installation.

7.7 FLOWCHARTING

Flowcharting is a process frequently employed by programmers to analyze the "big picture" of a problem before beginning to code the problem. It involves pictorially structuring the logical processes or steps to be taken before writing the sequence of statements necessary for proper computer solution. Although very simple problems may not require a flowchart, their use is recommended, to make more complex problems easier.

The four most frequently encountered flowcharting symbols, illustrated in Figure 7.1, will be employed throughout this chapter.

Symbol (a) is the START and END of a flowchart. When inputting data via a READ or INPUT statement, or when outputting results via the PRINT statement, symbol (b) is used. Decisions requiring a *yes* or *no* answer use symbol (c). They are normally used in looping, and will frequently involve the FOR-NEXT statements. There are usually two branch exits from a decision symbol. All mathematical and logical functions utilize the rectangular symbol, Figure 7.1(d).

Example 7.1 Draw a flowchart to sum the square roots of all the integers from 1 to 100, inclusive. Print out the answer. Notice the use of arrows between the symbols to indicate the directions of logic flow. Also note that the direction of exits from a decision symbol is arbitrary. Directions of *yes* and *no* exits are usually chosen to facilitate flowchart construction. (See Flowchart 7.1.)

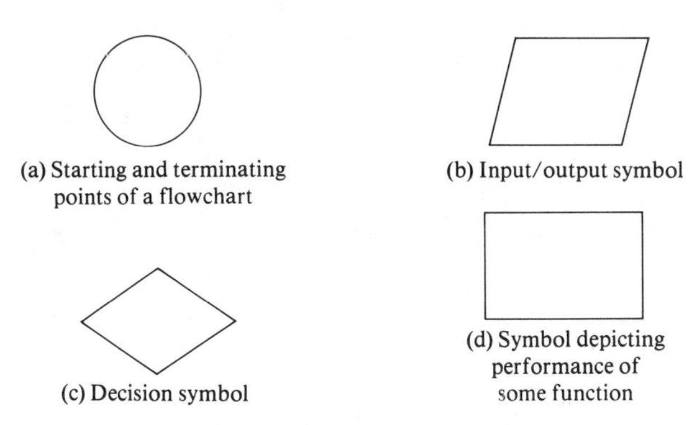

Figure 7.1 Flowchart Symbols

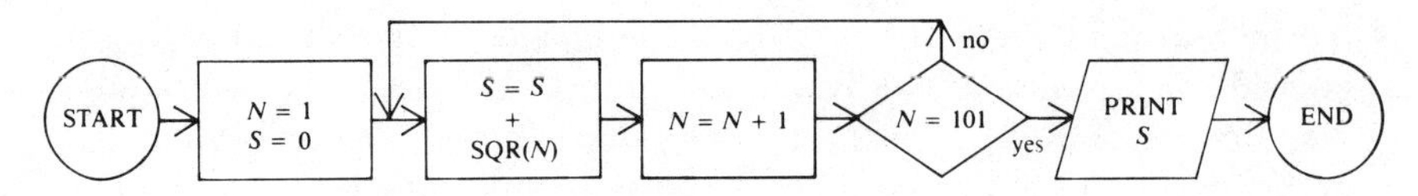

Flowchart 7.1

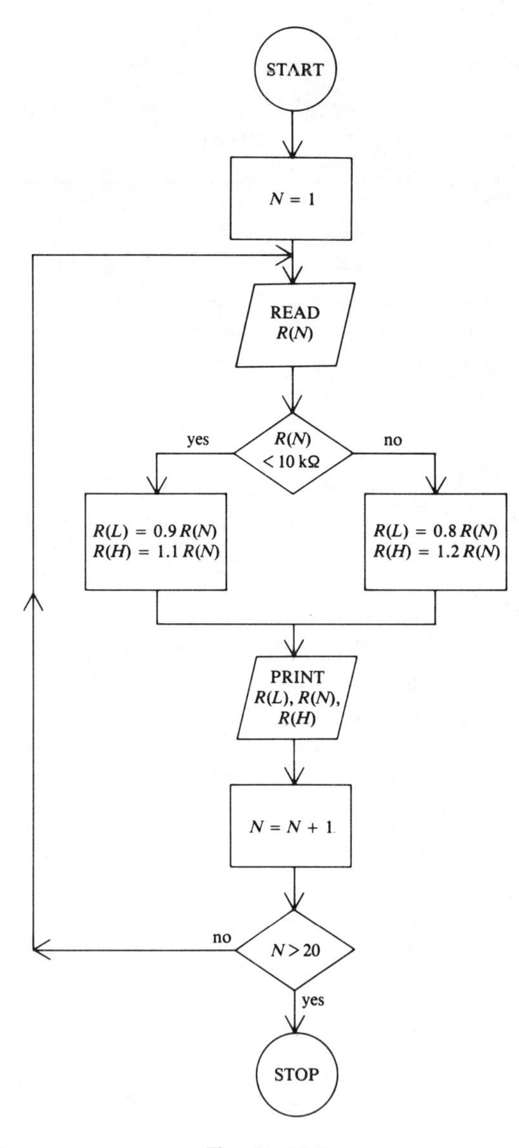

Flowchart 7.2

Example 7.2 Twenty resistor values are to be read into the computer. Resistors equal to or greater than 10 kΩ have 20% tolerance. Those resistors less than 10 kΩ have 10% tolerance. Draw a flowchart to print out R(low), R, and R(high) for each resistor. (See Flowchart 7.2.)

7.8 PROGRAMMING EXAMPLES

The following examples will combine the principles of flowcharting with the more advanced statements in using computers for the solution of electronic problems. Each programming solution is accompanied by a flowchart to help in problem analysis.

Example 7.3 (Program 1) Calculate and print out the value of current through R in Figure 7.2 for any values of R_1, R_2, and R and for any given voltage V. (See Flowchart 7.3.)

The following equations will be used in writing the program:

$$R_T = R_1 + \frac{R_2 R}{R_2 + R}$$

$$I_T = \frac{V}{R_T} = \frac{V(R_2 + R)}{R_1(R_2 + R) + R_2 R}$$

$$I = \frac{R_2}{R_2 + R} I_T = \left(\frac{R_2}{R_2 + R}\right)\left(\frac{V}{R_1 + \dfrac{R_2 R}{R_2 + R}}\right)$$

Let $$A = \frac{R_2}{R_2 + R}$$

$$\therefore \quad I = \frac{AV}{R_1 + AR}$$

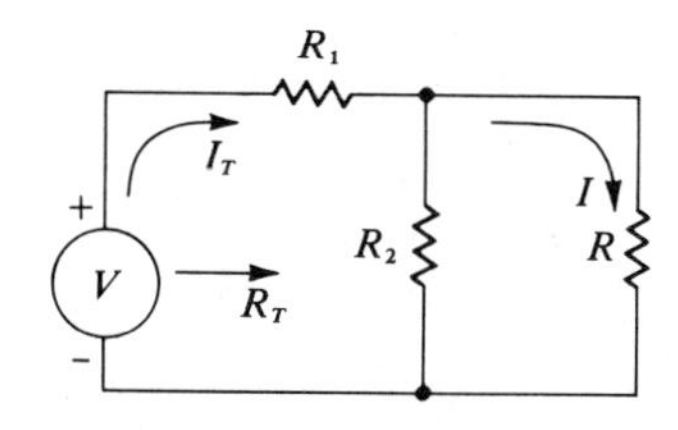

Figure 7.2 Reference Example 7.3

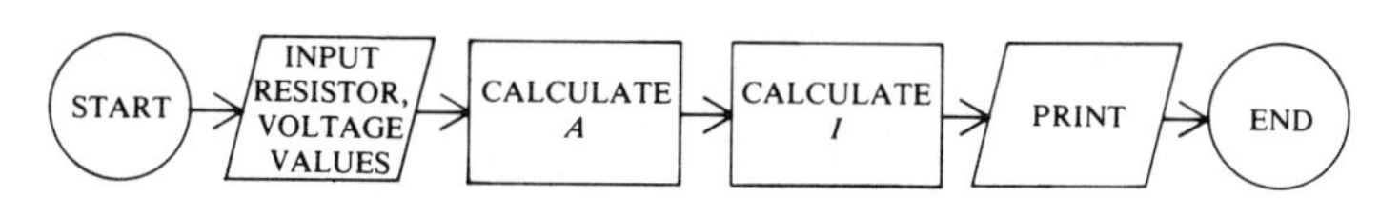

Flowchart 7.3

In this example, an arbitrary choice was made of

$$V = 20 \text{ volts}$$

$$R_1 = 10 \text{ k}\Omega$$

$$R_2 = 3.3 \text{ k}\Omega$$

$$R = 4.7 \text{ k}\Omega$$

```
LIST

10        INPUT"VALUE OF THREE RESISTORS ARE";R1,R2,R
15        INPUT"VALUE OF THE VOLTAGE SOURCE IS";V
20        LET A = R2/(R2+R)
30        LET I = A*V/(R1+A*R)
40        PRINT"VALUE OF CURRENT IS";I
50        END

RUN
VALUE OF THREE RESISTORS ARE? 1E4,3.3E3,4.7E3
VALUE OF THE VOLTAGE SOURCE IS? 20
VALUE OF CURRENT IS .691027E-3
```

Note the string inserted in statement 10 between the INPUT statement and R1, R2, R. This is permissible in many versions of BASIC.

Example 7.4 (Program 2) Calculate and print out the value of the current through R for any resistive values of R_1, R_2, R_3, and R and for any assigned voltage value of V in Figure 7.3. (See Flowchart 7.4.)

The following equations will be used in writing the program:

$$R_T = R_1 + \frac{R_2(R_3 + R)}{R_2 + R_3 + R}$$

$$I_T = \frac{V}{R_T} = \frac{V(R_2 + R_3 + R)}{R_1(R_2 + R_3 + R) + R_2(R_3 + R)}$$

$$I = \frac{R_2(I_T)}{R_2 + R_3 + R} = \frac{R_2(V)}{R_2 + R_3 + R}\left[\frac{1}{R_1 + \dfrac{R_2(R_3 + R)}{R_2 + R_3 + R}}\right]$$

$$\text{Let} \quad A = \frac{R_2}{R_2 + R_3 + R}$$

$$\text{Let} \quad B = \frac{R_2(R_3 + R)}{R_2 + R_3 + R} = A(R_3 + R)$$

$$I = \frac{VA}{R_1 + B}$$

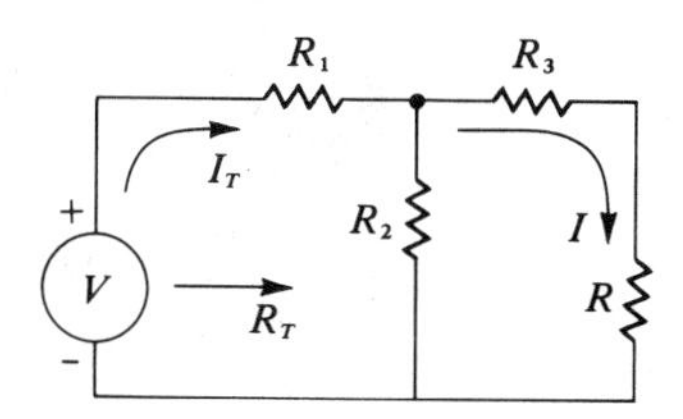

Figure 7.3 Reference Example 7.4

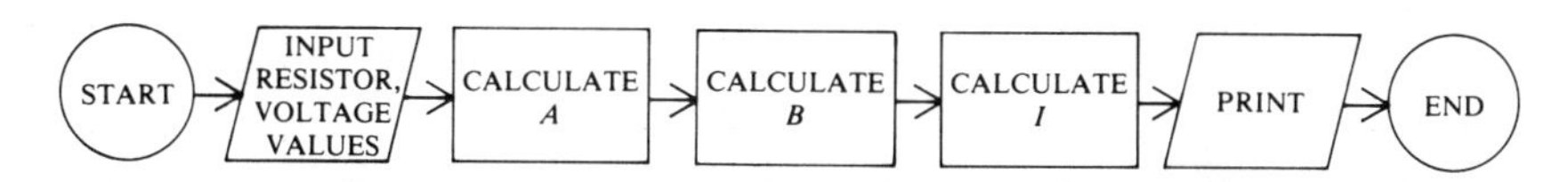

Flowchart 7.4

In this example, an arbitrary choice was made of

$$V = 30 \text{ volts}$$
$$R_1 = 6.8 \text{ k}\Omega$$
$$R_2 = 2.2 \text{ k}\Omega$$
$$R_3 = 4.7 \text{ k}\Omega$$
$$R = 1.5 \text{ k}\Omega$$

```
LIST
10      INPUT"VALUE OF FOUR RESISTORS ARE";R1,R2,R3,R
20      INPUT"VALUE OF VOLTAGE SOURCE IS";V
30      LET A = R2/(R2+R3+R)
40      LET B = A*(R3+R)
50      LET I = V*A/(R1+B)
60      PRINT"VALUE OF I IS";I
70      END

RUN
VALUE OF FOUR RESISTORS ARE? 6.8E3,2.2E3,4.7E3,1.5E3
VALUE OF VOLTAGE SOURCE IS ? 30
VALUE OF I IS .93273E-3
```

Example 7.5 (Program 3) Program the computer to print out a table consisting of R_2, the current through R_2, voltage across R_2, and the power expended in R_2 in Figure 7.4. R_2 is to vary from 1 kΩ to 10 kΩ in increments of 500 ohms. (See Flowchart 7.5.)

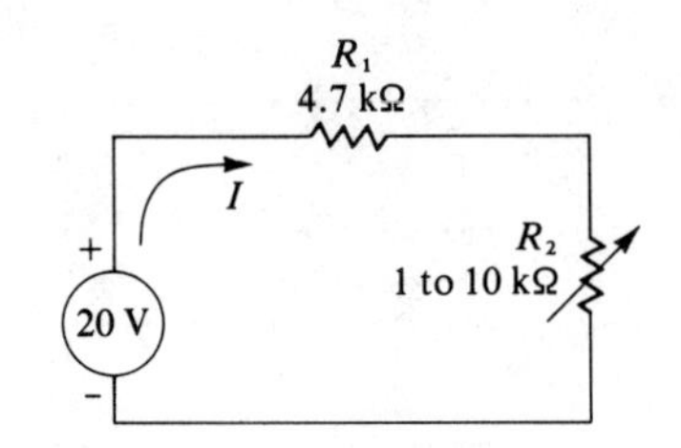

Figure 7.4 Reference Example 7.5

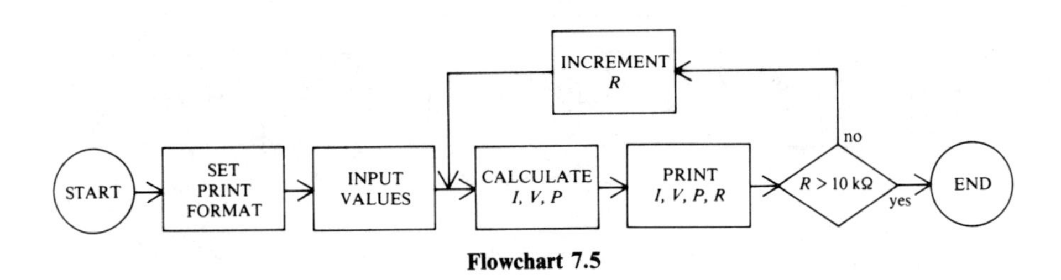

Flowchart 7.5

```
LIST
10      PRINT" R2                 I2                  V2                  P2"
30      LET V = 20:R1 = 4.7E3
40      FOR R2 = 1E3 TO 10E3 STEP .5E3
50      LET I2 = V/(R1 + R2)
55      LET V2 = (V*R2)/(R1 + R2)
60      LET P2 = (I2**2)*R2
70      PRINT R2,I2,V2,P2
80      NEXT R2
90      END

RUN

 R2            I2            V2            P2
1000          .350877E - 2  3.50877       .123115E - 1
1500          .322581E - 2  4.83871       .156087E - 1
2000          .298507E - 2  5.97015       .178213E - 1
2500          .277778E - 2  6.94444       .192901E - 1
3000          .25974E - 2   7.79221       .202395E - 1
3500          .243902E - 2  8.53659       .208209E - 1
4000          .229885E - 2  9.1954        .211389E - 1
4500          .217391E - 2  9.78261       .212665E - 1
5000          .206186E - 2  10.3093       .212562E - 1
5500          .196078E - 2  10.7843       .211457E - 1
6000          .186916E - 2  11.215        .209625E - 1
6500          .178571E - 2  11.6071       .020727
7000          .17094E - 2   11.9658       .204544E - 1
7500          .163934E - 2  12.2951       .201559E - 1
8000          .15748E - 2   12.5984       .01984
8500          .151515E - 2  12.8788       .195133E - 1
9000          .145985E - 2  13.1387       .191806E - 1
9500          .140845E - 2  13.3803       .188455E - 1
10000         .136054E - 2  13.6054       .185108E - 1
```

Example 7.6 (Program 4) Reference Figure 7.5. Print out a table which has calculated I_1, I_2, and I_3 as V varies from 10 to 24 volts in 4 volt increments and R_3 varies from 1 kΩ to 10 kΩ in 3 equal increments. (See Flowchart 7.6.)

The following equations will be used in writing the program:

$$R_T = R_4 = R_1 + \frac{R_2 R_3}{R_2 + R_3}$$

$$I_T = I_1 = \frac{V}{R_T}$$

$$I_2 = \frac{R_3 I_T}{R_2 + R_3}$$

$$I_3 = I_T - I_2 = I_1 - I_2$$

```
10    PRINT " V" TAB(6) " R3" TAB(15) " I1" ;
15    PRINT TAB(30) " I2" TAB(45) " I3"
20    PRINT
25    B=1E3: F=10E3: R1=3.3E3: R2=6.8E3
30    R=(F-B)/3
35    FOR V=12 TO 24 STEP 4
40    FOR R3=1E3 TO 10E3 STEP R
45    R4= (R2*R3)/(R2+R3)+R1
50    I1=V/R4 : I2=R3*I1/(R2+R3) : I3=I1-I2
55    PRINT V ; TAB(6) ; R3 ; TAB(15) ; I1 ;
60    PRINT TAB(30) ; I2 ; TAB(45) ; I3
65    NEXT R3
70    NEXT V
75    END
```

V	R3	I1	I2	I3
12	1000	2.87646E-03	3.68777E-04	2.50768E-03
12	4000	2.06238E-03	7.63845E-04	1.29854E-03
12	7000	1.77797E-03	9.01868E-04	8.76101E-04
12	10000	1.63318E-03	9.72132E-04	6.6105E-04
16	1000	3.83528E-03	4.91703E-04	3.34358E-03
16	4000	2.74984E-03	1.01846E-03	1.73138E-03
16	7000	2.37062E-03	1.20249E-03	1.16813E-03
16	10000	2.17758E-03	1.29618E-03	8.814E-04
20	1000	4.7941E-03	6.14628E-04	4.17947E-03
20	4000	3.4373E-03	1.27307E-03	2.16423E-03
20	7000	2.96328E-03	1.50311E-03	1.46017E-03
20	10000	2.72197E-03	1.62022E-03	1.10175E-03
24	1000	5.75292E-03	7.37554E-04	5.01537E-03
24	4000	4.12476E-03	1.52769E-03	2.59707E-03
24	7000	3.55594E-03	1.80374E-03	1.7522E-03
24	10000	3.26636E-03	1.94426E-03	1.3221E-03

Example 7.7 (Program 5) In Figure 7.6, all three resistors can vary from 1 kΩ to 10 kΩ in 1 kΩ increments. If all three resistors were mini-

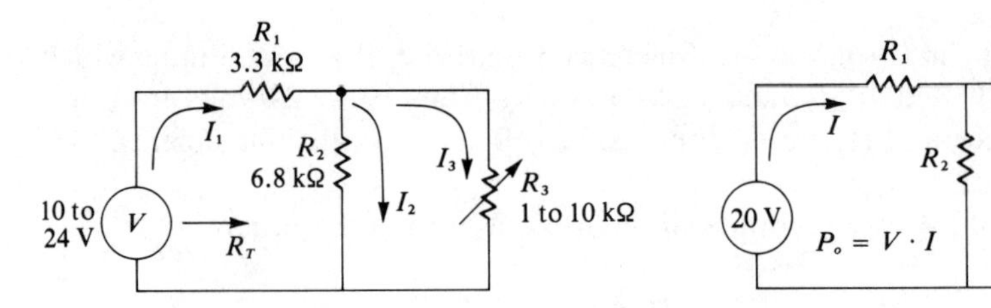

Figure 7.5 Reference Example 7.6

Figure 7.6 Reference Example 7.7

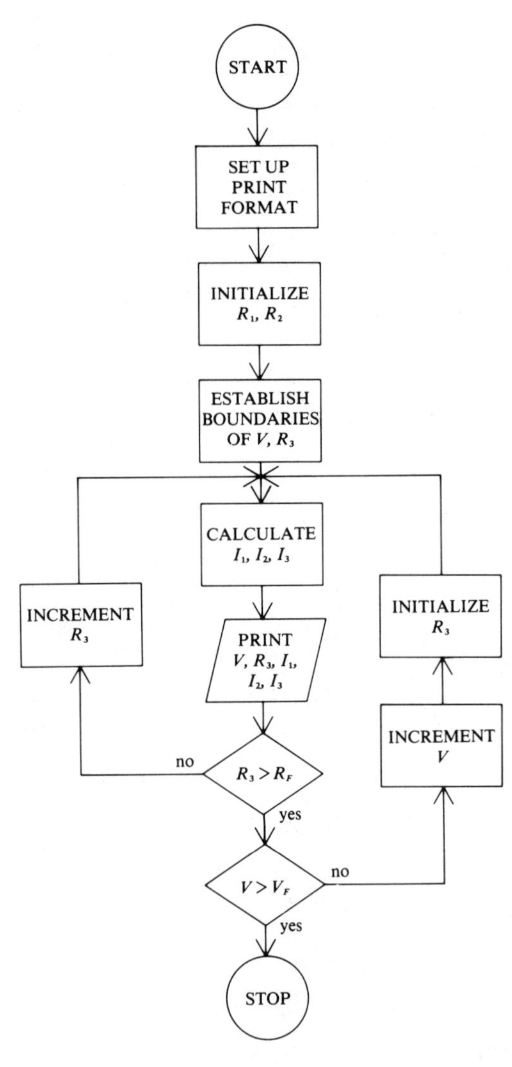

Flowchart 7.6

mum (1 kΩ), then the power supplied by the source would be maximum and equal to 266.67 mW. If all three resistors were maximum (10 kΩ), then the power supplied by the source would be minimum and equal to 26.67 mW. Write a program to print out, in tabular format, all combinations of R_1, R_2, and R_3 which would cause the source to supply between 113.55 mW and 121.75 mW of power. Also print out the power actually supplied by the source for each of the above selected combinations. (See Flowchart 7.7.)

Note the small circles labeled A and B in the flowchart diagram. This is not a new logical symbol, but rather a functional symbol that depicts connecting points. It is not desirable to cross over lines on a flowchart, nor is it possible to have lines between pages on very long flowcharts; hence the connecting symbols.

Notice that thousands of different resistive combinations exist, each requiring extensive calculations. The reader can now appreciate the computing power of a modern system and that "demarcation point" where calculators become less desirable than computers.

```
10      PRINT " R1" , " R2" , " R3" , "P(SOURCE)"
15      PRINT ,,, "  IN MW"
20      PRINT
25      FOR R1=1E3 TO 10E3 STEP 1E3
30      FOR R2=1E3 TO 10E3 STEP 1E3
35      FOR R3=1E3 TO 10E3 STEP 1E3
40      GOSUB 100
45      NEXT R3 : NEXT R2 : NEXT R1
50      STOP
100     R= R1+(R2*R3)/(R2+R3)
105     P= 4E5/R
110     IF P>=113.55 AND P<=121.75 THEN 120
115     GOTO 125
120     PRINT R1 , R2 , R3 , P
125     RETURN
130     END
```

```
 R1      R2           R3          P(SOURCE)
                                    IN MW

1000    3000         10000         120.93
1000    4000         6000          117.647
1000    5000         5000          114.286
1000    6000         4000          117.647
1000    10000        3000          120.93
2000    2000         4000          120
2000    2000         5000          116.667
2000    2000         6000          114.286
2000    3000         3000          114.286
2000    4000         2000          120
2000    5000         2000          116.667
2000    6000         2000          114.286
3000    1000         1000          114.286
```

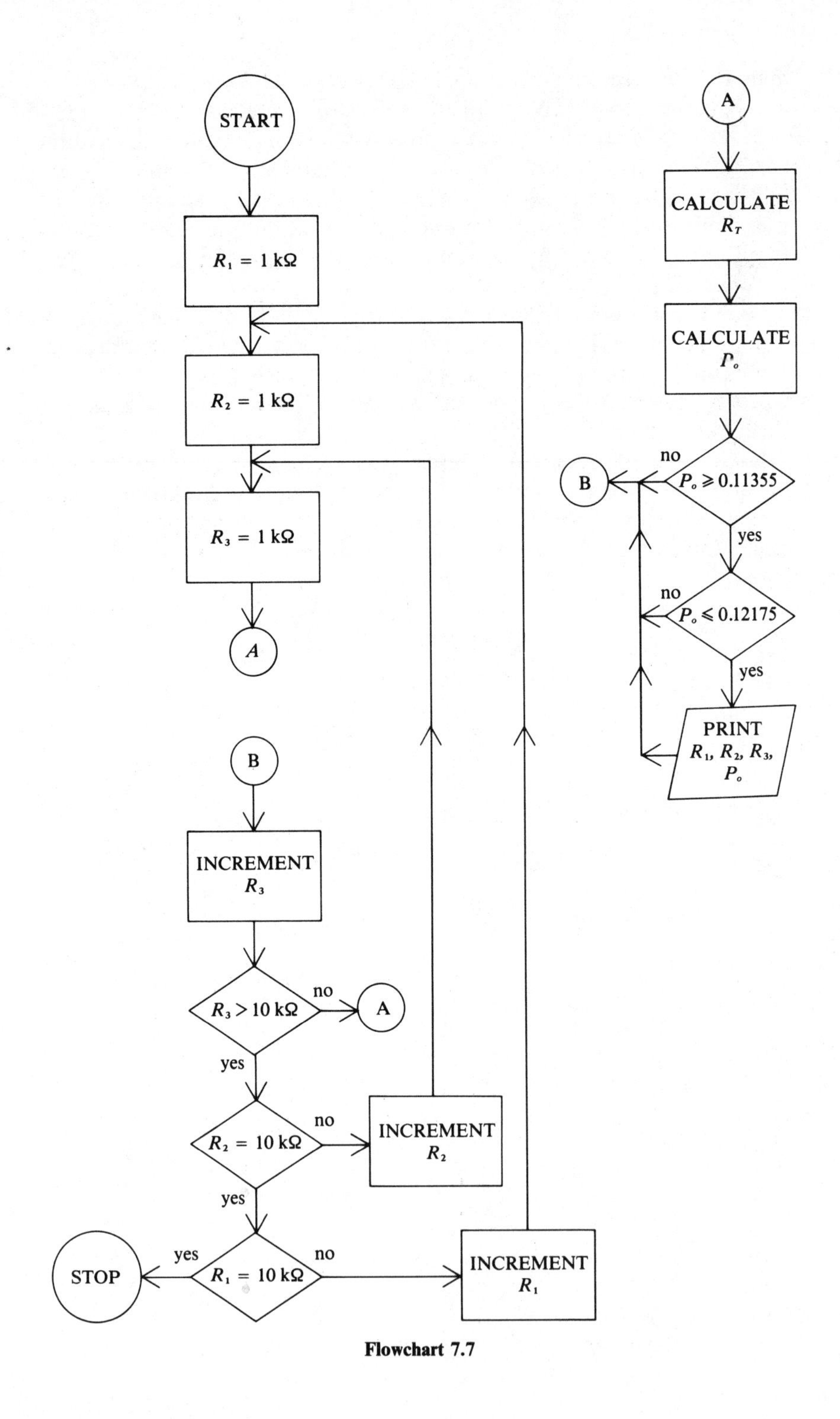

Flowchart 7.7

Example 7.8 The computer may also be used to give numerical results without programming. Using the PRINT statement without a statement number can instantly result in answers equaling the most advanced scientific pocket calculators.

Suppose an extremely complex numerical calculation like the following is needed:

$$\sqrt{\left[\epsilon^{\cos 25.6^\circ} \times \tan\left(\epsilon^{0.16}\right) \times \frac{16.3}{\pi^{1.5}}\right]^{-1.45}}$$

The following simple PRINT statement with the indicated problem produces the desired answer. The answer is printed out just below the PRINT statement.

```
PRINT SQR((EXP(COS(PI/180*25.6))*TAN(EXP(.16))*16.3/PI**1.5)**-1.45)
 .127179

READY
```

SUMMARY

For computers to be practical in solving problems, provisions must exist wherein multiple passes may be made through the program, rather than a single pass providing for only one solution. The ability of a program to branch to different instructions based on decision criteria, and to make multiple passes through the statements, provides this practicality.

Common mathematical and logical functions can be called up as routine steps in solving problems. Branching statements fall into two categories: GOTO for unconditional branching, and IF-THEN-ELSE for conditional branching.

Looping is accomplished with the FOR-NEXT statement. Looping is very useful when multiple passes are made through the program, because the computer itself is used to vary the parameters for each solution. Loops may be ended using a known terminating value, or conditionally ended utilizing the WHILE or UNTIL option.

Subroutines are normally sets of multiple statements frequently used by the program. They are usually written as a separate entity from the main program body. Subroutine entry and exit are accomplished with the GOSUB and RETURN statements.

Flowcharting looks at the overall scope of a problem before attempting to code the program. Standardized symbols are used to depict the major functions within the flowchart.

PROBLEMS

Reference Sections 7.4 through 7.8

1. Two resistors are in parallel across a 10 V source. If one resistor is 3.3 kΩ, write a program to calculate the total current if the other resistor varies from 1 kΩ to 10 kΩ in 1 kΩ increments. Print out a table of R (variable) and I (total).
2. Two resistors are in parallel across a 20 V source. Each resistor varies from 1 kΩ to 21 kΩ in 4 kΩ increments. Write a program to determine I_T, I_1, and I_2 for each combination of resistors. Print out a table of R_1, R_2, I_T, I_1, and I_2.
3. In Figure 7.7, determine V_{R_3}, I_{R_3} and P_{R_3} as R_3 varies from 1 kΩ to 10 kΩ in 1 kΩ increments. Print out a table of R_3, V_3, I_3, and P_3 for each incremental value of R_3.
4. Three resistors are in parallel. Each can vary from 1 kΩ to 10 kΩ in 0.5 kΩ increments. Write a program to determine those combinations of R_1, R_2, and R_3 resulting in an equivalent resistance of between 2810 and 2830 ohms. Print out those combinations and the equivalent resistance that each represents.
5. In Figure 7.8, V varies from 5 to 25 volts in steps of 5 volts and R_2 varies from 1 kΩ to 10 kΩ in steps of 1 kΩ. Determine those combinations of V and R_2 that result in R_3 dissipating more than 32.75 milliwatts of power.

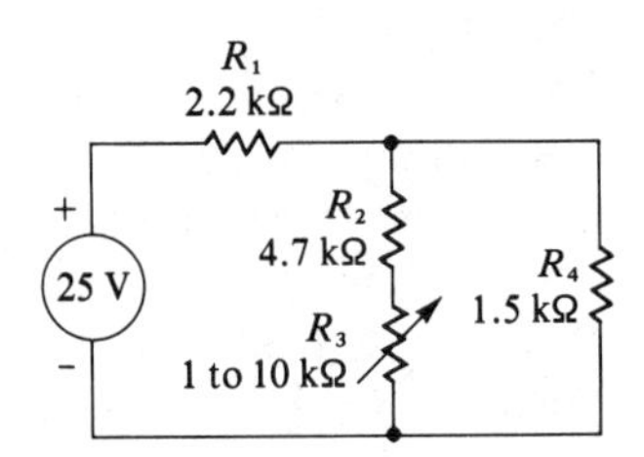

Figure 7.7 Reference Problem 3

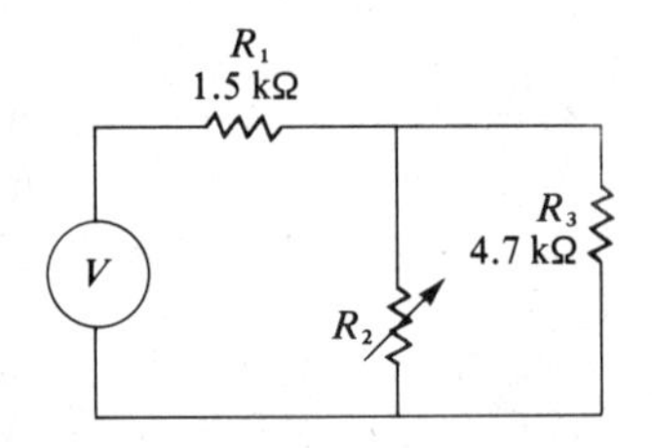

Figure 7.8 Reference Problems 5 and 6

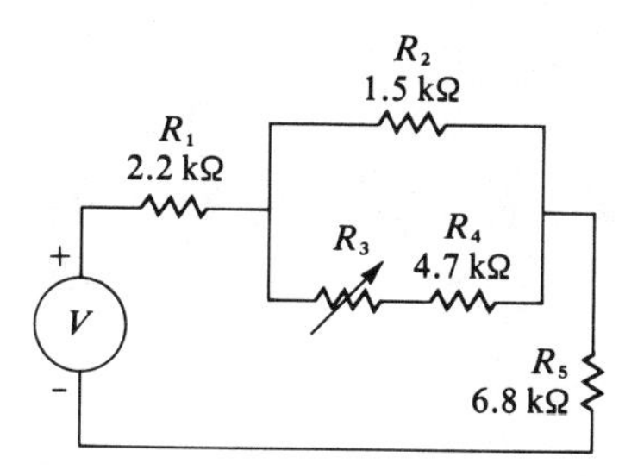

Figure 7.9 Reference Problems 7 and 8

6. In Problem 5, V ranges through 5 values and R_2 ranges through 10 values, for a total of 50 different combinations. If each combination exists for an equal interval of time, what is the average power expended by the source?
7. If V is 50 volts in Figure 7.9, what is V_{R_3} and I_{R_3} if R_3 varies from 0.2 kΩ to 2.2 kΩ in increments of 0.2 kΩ? Print out a table of R_3, V_3, and I_3.
8. If R_3 varies from 1 kΩ to 10 kΩ in 500 ohm increments in Figure 7.9, what should be each value of V so that V_3 equals 2 volts for each of R_3's values? Print out a table of R_3 and V that results in this voltage across R_3.

CHAPTER 8

DETERMINANT AND MATRIX ANALYSIS

8.1 DETERMINANTS

In electrical circuits that have clearly defined arrangements of series and parallel elements, it is relatively easy to determine current magnitude and direction through each element, as well as the voltage across each element. Many practical circuits, however, have element arrangements which are neither in series with, nor parallel to, the other elements. Such a circuit is shown in Figure 8.1. This circuit is frequently referred to as a *Wheatstone bridge.*

The current I_T, provided by the voltage source, enters junction A, where it divides between R_1 and R_3. Note that R_1 and R_3 are not in parallel with each other. Whether or not any current flows through the current meter is solely dependent upon the ratio of the four resistors in the bridge. Suppose that resistor R_1 equals R_2, and suppose further that resistor R_3 equals R_4. This being the case, no current would flow through the current meter, because by the proportionate voltage law, point B would equal $V_T/2$ and point C would also equal $V_T/2$. Since points B and C are at the same potential, the bridge would be considered balanced.

The practicality of this circuit lies in its ability to make very accurate resistance measurements of an unknown resistor (R_3) by varying R_4 until it equals the value of the unknown resistor. After balancing the bridge (no current through the meter), the value of R_3 can now be read by observing the calibrated dials of R_4. If R_3 and R_4 are not equal, the current through the meter will be in one direction or the other, depending upon whether R_4 is less than or greater than R_3.

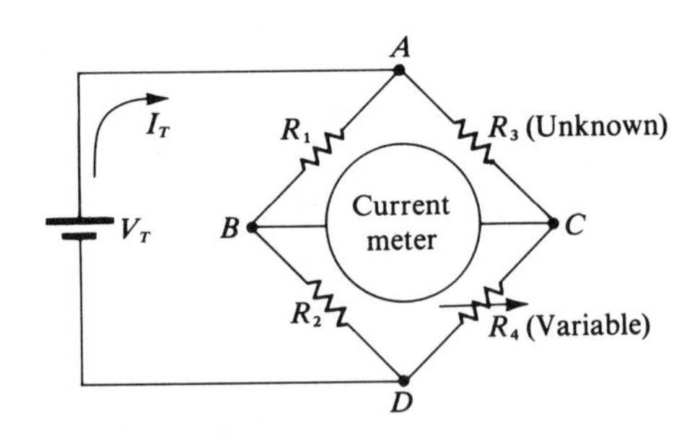

Figure 8.1 Simplified Wheatstone Bridge

In this example, the meter provided a visual reading of current magnitude and direction. If the meter were replaced by its internal resistance, or any resistance for that matter, then all visual indicators are removed. With the value of the resistors known and an unbalanced state assumed, an analytical calculation is required to determine the current between points B and C. At this point, the difficulty of solution may now become apparent. If resistors are not quite in series, nor quite in parallel, how is such a circuit analyzed? Problems of this nature necessitate a discussion of what is referred to as *loop equations* and *determinants.*

Consider the following equations with two unknown variables (a and b).

$$2a + 4b = 14 \tag{8.1}$$

$$3a + 3b = 15 \tag{8.2}$$

One method to solve these simultaneous equations is to eliminate one of the variables. If Eq. (8.1) were multiplied by 3 (the coefficient of b in Eq. (8.2)), and if Eq. 8.2 were multiplied by 4 (coefficient of b in Eq. 8.1)), then the following two equations result:

$$6a + 12b = 42 \tag{8.3}$$

$$12a + 12b = 60 \tag{8.4}$$

Subtract Eq. (8.3) from Eq. (8.4).

$$6a + 0b = 18 \tag{8.5}$$

Solve for a.

$$a = \frac{18}{6} = 3$$

Substitute this value of a into Eq. (8.2) and solve for b.

$$3(3) + 3b = 15$$

$$3b = 6$$

$$b = 2$$

The above example used specific coefficients for the two variables. It is preferred to adopt a generalized solution to the problem. The following two equations will be solved in a similar fashion, but the solution of the unknown variables (x, y) will be expressed in the general terminology of their algebraic coefficients (a, b, c, d, e, f).

$$ax + by = c \tag{8.6}$$

$$dx + ey = f \tag{8.7}$$

Multiply Eq. (8.6) by e and Eq. (8.7) by b.

$$ae\,x + be\,y = ce \tag{8.8}$$

$$bd\,x + be\,y = bf \tag{8.9}$$

Subtract Eq. (8.9) from Eq. (8.8).

$$(ae - bd)x = ce - bf$$

Solve for x.

$$\boxed{x = \frac{ce - bf}{ae - bd}} \tag{8.10}$$

In a similar fashion,

$$\boxed{y = \frac{af - cd}{ae - bd}} \tag{8.11}$$

Notice that in both solutions, the denominator is the same.

At this point, the mechanics of determinants have been established. To write a determinant,

1. Write the equations with like variables aligned, as in Eqs. (8.6) and (8.7).
2. Using Eqs. (8.6) and (8.7) as examples, write an array of coefficients in the following format.

$$x = \frac{\begin{vmatrix} c & b \\ f & e \end{vmatrix}}{\begin{vmatrix} a & b \\ d & e \end{vmatrix}} \tag{8.12}$$

$$y = \frac{\begin{vmatrix} a & c \\ d & f \end{vmatrix}}{\begin{vmatrix} a & b \\ d & e \end{vmatrix}} \tag{8.13}$$

Observe that both denominators are the same and that the coefficients are in the same order as if reading Eqs. (8.6) and (8.7) from left to right. Notice also that the numerators are arranged in such a fashion that the coefficients c and f are substituted for the coefficients of the unknown variable being solved for. This array is still in a left-to-right direction.

To complete the solutions of x and y,

3. Form the cross product of the upper left and lower right coefficients within each bracket.
4. Subtract the cross product of the lower left and upper right coefficients within each bracket.

This results in

$$x = \frac{ce - bf}{ae - bd}$$

$$y = \frac{af - cd}{ae - bd}$$

In summary, a determinant is nothing more than an organized array of numbers. The numbers are related to the coefficients of the variables and to the coefficients to which each equation is equated. The solution of the determinant, and hence the solution of the variables, is purely a mechanical operation.

Multiply in this direction:

upper left ↘ lower right

Subtract in this direction:

lower left ↗ upper right

Example 8.1 Solve the following determinant:

$$q = \frac{\begin{vmatrix} 3 & 2 \\ 6 & 1 \end{vmatrix}}{\begin{vmatrix} 1 & 4 \\ 3 & 2 \end{vmatrix}}$$

$$= \frac{(3)(1) - (6)(2)}{(1)(2) - (3)(4)} = \frac{3 - 12}{2 - 12} = \frac{-9}{-10} = 0.9$$

Example 8.2 Solve the following determinant:

$$t = \frac{\begin{vmatrix} -3 & 2 \\ -1 & 6 \end{vmatrix}}{\begin{vmatrix} 2 & -5 \\ -1 & -7 \end{vmatrix}}$$

$$= \frac{(-3)(6) - (-1)(2)}{(2)(-7) - (-1)(-5)}$$

$$= \frac{-16}{-19} = 0.842$$

Example 8.3 Using determinants, solve the following simultaneous equations:

$$4r - 5s = -32$$
$$-3r + 2s = 17$$

$$r = \frac{\begin{vmatrix} -32 & -5 \\ 17 & 2 \end{vmatrix}}{\begin{vmatrix} 4 & -5 \\ -3 & 2 \end{vmatrix}} = \frac{(-64) - (-85)}{(8) - (15)} = \frac{21}{-7} = -3$$

$$s = \frac{\begin{vmatrix} 4 & -32 \\ -3 & 17 \end{vmatrix}}{-7} = \frac{(68) - (96)}{-7} = \frac{-28}{-7} = 4$$

Notice that the denominator of the determinant for s was already solved in finding the solution of r. In general, whatever the order of the determinant, once the determinant of the denominator (D) of the first variable is solved, the value of the denominator of all other variables is established, since it is the same for all the variable quantities.

Example 8.4 Solve the following simultaneous equations using determinants:

$$y = 21 - 4x$$
$$-2x - 3 = 5y$$

First, arrange each equation into an aligned array as represented in Eqs. (8.6) and (8.7).

$$4x + y = 21$$
$$-2x - 5y = 3$$

Now solve determinants for the unknown quantities,

$$x = \frac{\begin{vmatrix} 21 & 1 \\ 3 & -5 \end{vmatrix}}{\begin{vmatrix} 4 & 1 \\ -2 & -5 \end{vmatrix}} = \frac{(-105) - (3)}{(-20) - (-2)} = \frac{-108}{-18} = 6$$

$$y = \frac{\begin{vmatrix} 4 & 21 \\ -2 & 3 \end{vmatrix}}{-18} = \frac{(12) - (-42)}{-18} = \frac{54}{-18} = -3$$

Simplification rules can also be utilized in solving determinants. Figure 8.2 illustrates the numerator of y in Example 8.4. It also indi-

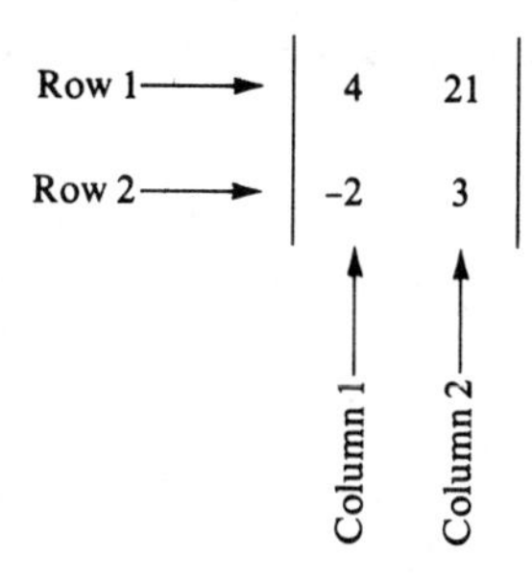

Figure 8.2 Determinant Row and Column Designation

cates both the column numbers and the row numbers of the determinant. If any column or row is evenly divisible by a coefficient, then each term of that column or row may be so divided. However, the resulting determinant must later be multiplied by these divisors, as illustrated below. (This is somewhat analogous to reducing $^{36}/_{72}$ to $^{1}/_{2}$, in that both fractions are the same, but the numbers in the latter fraction are smaller, and therefore easier to manipulate.) Column 1 is divisible by 2 and column 2 is divisible by 3. Reducing both these columns results in

$$(2)(3)\begin{vmatrix} 2 & 7 \\ -1 & 1 \end{vmatrix} = 6[(2) - (-7)] = 54$$

This is the same value that was determined in the original answer. The simplification process is best utilized when large, reducible numbers exist within the determinant. Calculating subsequent products then becomes much easier.

Example 8.5 Solve the following determinant, using simplification processes wherever possible:

$$x = \frac{\begin{vmatrix} 25 & 225 \\ 15 & 15 \end{vmatrix}}{\begin{vmatrix} 16 & 225 \\ 4 & 15 \end{vmatrix}}$$

In the numerator, column 1 is reducible by 5 and column 2 is reducible by 15. In the denominator, column 1 is reducible by 4 and column 2 is again reducible by 15.

$$x = \frac{(5)(15)\begin{vmatrix} 5 & 15 \\ 3 & 1 \end{vmatrix}}{(4)(15)\begin{vmatrix} 4 & 15 \\ 1 & 1 \end{vmatrix}}$$

The 15's cancel, and row 1 in the numerator is further reducible by 5.

$$x = \frac{(5)(5)\begin{vmatrix} 1 & 3 \\ 3 & 1 \end{vmatrix}}{(4)\begin{vmatrix} 4 & 15 \\ 1 & 1 \end{vmatrix}} = \frac{25[(1) - (9)]}{4[(4) - (15)]}$$

$$= \frac{50}{11} = 4.545$$

The preceding examples are for second-order determinants, that is, determinants solving for two unknowns. The principles apply for determinants of any order. If three equations with three unknowns are encountered, each unknown may be solved for in a like manner. For example,

$$ax + by + cz = d \tag{8.14}$$

$$ex + fy + gz = h \tag{8.15}$$

$$ix + jy + kz = \ell \tag{8.16}$$

$$x = \frac{\begin{vmatrix} d & b & c \\ h & f & g \\ \ell & j & k \end{vmatrix}}{\begin{vmatrix} a & b & c \\ e & f & g \\ i & j & k \end{vmatrix} \equiv D} \tag{8.17}$$

$$y = \frac{\begin{vmatrix} a & d & c \\ e & h & g \\ i & \ell & k \end{vmatrix}}{D} \tag{8.18}$$

$$z = \frac{\begin{vmatrix} a & b & d \\ e & f & h \\ i & j & \ell \end{vmatrix}}{D} \tag{8.19}$$

Notice that in writing all determinants, the following mechanics apply:

1. Like variables are aligned.
2. In the numerator, from left to right, the coefficients to which each equation is equated are substituted for the coefficients of the variable being solved for.
3. In the denominator, (D) is the same for all unknowns, and reading from left to right, represents the coefficients of all the variables.

In solving third-order determinants, the following mechanics must be followed:

1. All combinations of the products of three terms, from upper left to lower right, are summed.
2. From the above, all combinations of the products of three terms, from lower left to upper right, are subtracted.

Figure 8.3 illustrates all combinations of three terms from upper left to lower right. Figure 8.4 illustrates all combinations from lower left to upper right. Observe that the first subscript of each a term indicates the row. The second subscript indicates the column. This designation will become very useful when *matrix* is discussed in subsequent sections of this chapter.

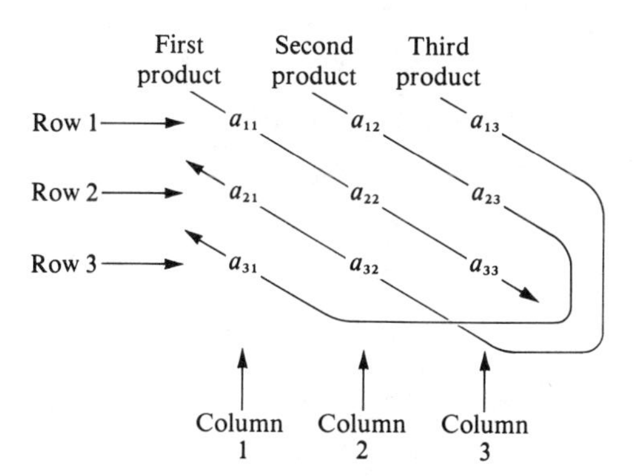

Figure 8.3 Products of All Combinations (Upper Left to Lower Right)

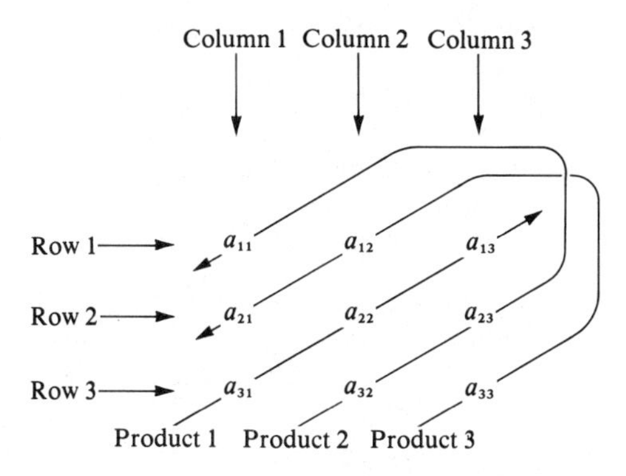

Figure 8.4 Products of All Combinations (Lower Left to Upper Right)

Example 8.6 Solve the following determinant:

$$x = \begin{vmatrix} 1 & 4 & 7 \\ 2 & 5 & 8 \\ 3 & 6 & 9 \end{vmatrix}$$

$$\begin{aligned} &= (1)(5)(9) + (4)(8)(3) + (7)(6)(2) \\ &\quad - (3)(5)(7) - (6)(8)(1) - (9)(4)(2) \\ &= 45 + 96 + 84 - 105 - 48 - 72 \\ &= 225 - 225 = 0 \end{aligned}$$

Example 8.7 Solve the following determinant, simplifying where possible:

$$y = \begin{vmatrix} 15 & 25 & 15 \\ 6 & 36 & 36 \\ 9 & 36 & 81 \end{vmatrix}$$

First, simplify the rows. Row 1 is divisible by 5, row 2 is divisible by 6, and row 3 is divisible by 9.

$$y = \underbrace{(5)(6)(9)}_{270} \begin{vmatrix} 3 & 5 & 3 \\ 1 & 6 & 6 \\ 1 & 4 & 9 \end{vmatrix}$$

Next, column 3 is divisible by 3:

$$y = \underbrace{(270)(3)}_{810} \begin{vmatrix} 3 & 5 & 1 \\ 1 & 6 & 2 \\ 1 & 4 & 3 \end{vmatrix}$$

$$\begin{aligned} &= 810(54 + 10 + 4 - 6 - 24 - 15) \\ &= 810(23) = 18{,}630 \end{aligned}$$

Example 8.8 Solve the following simultaneous equations with three unknowns:

$$\begin{aligned} 2a - 3b - c &= 9 \\ -a + 4b - 3c &= -26 \\ -3a - 4b + 6c &= 30 \end{aligned}$$

$$D \equiv \begin{vmatrix} 2 & -3 & -1 \\ -1 & 4 & -3 \\ -3 & -4 & 6 \end{vmatrix} = (48 - 27 - 4) - (12 + 24 + 18) = -37$$

$$a = \frac{\begin{vmatrix} 9 & -3 & -1 \\ -26 & 4 & -3 \\ 30 & -4 & 6 \end{vmatrix}}{D} = \frac{(216 - 104 + 270) - (-120 + 468 + 108)}{-37}$$

$$= \frac{-74}{-37} = 2$$

$$b = \frac{\begin{vmatrix} 2 & 9 & -1 \\ -1 & -26 & -3 \\ -3 & 30 & 6 \end{vmatrix}}{D} = \frac{(-312 + 30 + 81) - (-78 - 54 - 180)}{-37}$$

$$= \frac{111}{-37} = -3$$

After solving for any two unknowns, it is usually easier to substitute these values into any of the three original equations to solve for the remaining unknown. Choosing the first equation,

$$\begin{aligned} c &= 2a - 3b - 9 \\ &= 2(2) - 3(-3) - 9 = 4 \end{aligned}$$

Check by substituting the values of a, b, and c into the remaining two equations.

Eq. (2):

$$\begin{aligned} -(2) + 4(-3) - 3(4) &= -26 \\ -2 - 12 - 12 &= -26 \\ -26 &= -26 \qquad \text{(checks OK)} \end{aligned}$$

Eq. (3):

$$\begin{aligned} -3(2) - 4(-3) + 6(4) &= 30 \\ -6 + 12 + 24 &= 30 \\ 30 &= 30 \qquad \text{(checks OK)} \end{aligned}$$

Example 8.9 Solve the following simultaneous equations, simplifying where appropriate:

$$\begin{aligned} 6 - 4c &= 8a + 2b \\ -2b + 16a &= -6 - 8c \\ 6c &= 16 + 4a \end{aligned}$$

First, rearrange the equations so that like variables are aligned.

$$8a + 2b + 4c = 6 \qquad (8.20)$$

$$16a - 2b + 8c = -6 \qquad (8.21)$$

$$-4a \qquad + 6c = 16 \qquad (8.22)$$

Note the absence of a b variable in Eq. (8.22). This can work to good advantage, because if the a variable is solved, then only one unknown remains in Eq. (8.22). This is then best solved by substitution, rather than solving another determinant. In addition, the value of "zero" for the b variable in Eq. (8.22) greatly simplifies the determinant.

$$a = \frac{\begin{vmatrix} 6 & 2 & 4 \\ -6 & -2 & 8 \\ 16 & 0 & 6 \end{vmatrix}}{\begin{vmatrix} 8 & 2 & 4 \\ 16 & -2 & 8 \\ -4 & 0 & 6 \end{vmatrix}} = \frac{(2)(2)(2)\begin{vmatrix} 3 & 1 & 2 \\ -3 & -1 & 4 \\ 8 & 0 & 3 \end{vmatrix}}{(4)(2)(2)\begin{vmatrix} 2 & 1 & 2 \\ 4 & -1 & 4 \\ -1 & 0 & 3 \end{vmatrix}}$$

$$= \frac{1}{2}\left[\frac{(-9 + 32) - (-16 - 9)}{(-6 - 4) - (2 + 12)}\right]$$

$$= \frac{1}{2}\left[\frac{48}{-24}\right] = -1$$

From Eq. (8.22),

$$6c = 16 + 4a = 16 + 4(-1) = 12$$
$$c = 2$$

From Eq. (8.20),

$$2b = 6 - 4c - 8a = 6 - 8 + 8 = 6$$
$$b = 3$$

Another major simplification process to solve determinants makes use of a method called *cofactors*. Cofactors are best employed in four-order determinants or higher, or when a third order determinant has one or more zeros in the array. In keeping with the general theme of determinants, cofactors are merely a mechanical manipulation of the numbers of the array. Two rules are employed in this method.

Rule 1. A given row or column is selected, usually the one with the most zeros. In Figure 8.5, this could either be column 1 or row 2. The cofactors of each term of that row or column are then multiplied by the term for which the cofactor was created.

If column 1 is selected, then the first term (a_{11}) is 3, the second term (a_{21}) is 0, and the third term (a_{31}) is -4. Draw an imaginary line, horizontally and vertically, through each coefficient in the same row or column of the term for which the cofactor is being written. Write a two-order deter-

Column 1

$$\text{Row 2} \rightarrow \begin{vmatrix} 3 & 6 & -1 \\ 0 & 8 & -4 \\ -4 & 2 & 6 \end{vmatrix} = X$$

Figure 8.5 Selection of "Zero" Term for Cofactor

minant out of the remaining four numbers. For example, in Figure 8.6(a) the resultant cofactor of a_{11} would be

$$\begin{vmatrix} 8 & -4 \\ 2 & 6 \end{vmatrix}$$

In a like manner, from Figure 8.6(b), the cofactor of 0 (a_{21}) is

$$\begin{vmatrix} 6 & -1 \\ 2 & 6 \end{vmatrix}$$

and from Figure 8.6(c), the cofactor of -4 (a_{31}) is

$$\begin{vmatrix} 6 & -1 \\ 8 & -4 \end{vmatrix}$$

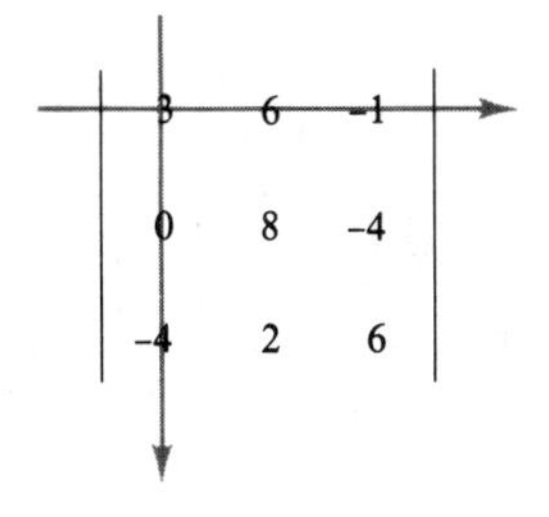

Figure 8.6(a) Cofactor of a_{11} (3)

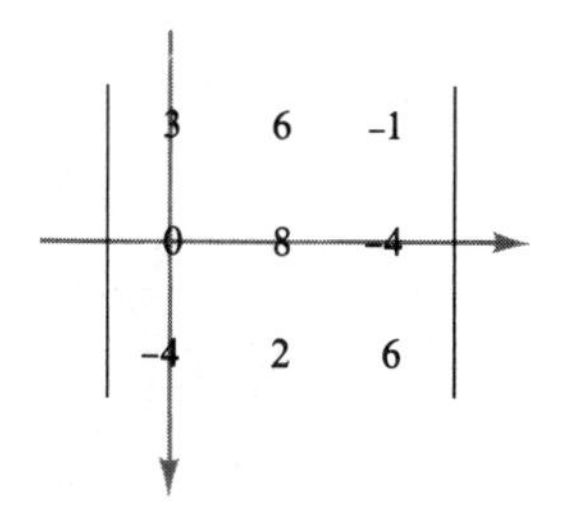

Figure 8.6(b) Cofactor of a_{21} (0)

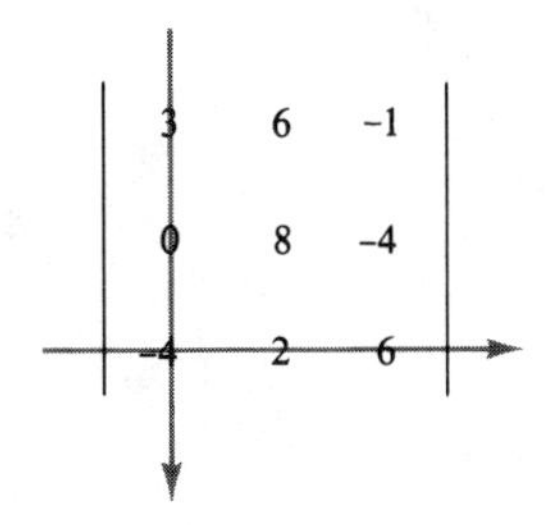

Figure 8.6(c) Cofactor of a_{31} (-4)

The solution of the determinant for "x" may now be summed as

$$x = \underbrace{(\text{sign})(3)}_{a_{11}} \underbrace{\begin{vmatrix} 8 & -4 \\ 2 & 6 \end{vmatrix}}_{\substack{\text{cofactor} \\ \text{of } a_{11}}} + \underbrace{(\text{sign})(0)}_{a_{21}} \underbrace{\begin{vmatrix} 6 & -1 \\ 2 & 6 \end{vmatrix}}_{\substack{\text{cofactor} \\ \text{of } a_{21}}} + \underbrace{(\text{sign})(-4)}_{a_{31}} \underbrace{\begin{vmatrix} 6 & -1 \\ 8 & -4 \end{vmatrix}}_{\substack{\text{cofactor} \\ \text{of } a_{31}}}$$

Three observations can now be stated:

1. The original third order determinant is reduced to the sum of a series of second order determinants.
2. The determinant cofactor preceded by zero need not be evaluated, because any quantity multiplied by zero is still zero. This is the advantage of picking the row or column with the most zeros.
3. A new *sign* quantity must be incorporated into the series.

This latter observation is essentially the second rule, namely:

Rule 2. All terms of the reduced determinant must be preceded by a *sign* according to the following, alternating pattern.

$$\begin{vmatrix} +(a_{11}) & -(a_{12}) & +(a_{13}) \\ -(a_{21}) & +(a_{22}) & -(a_{23}) \\ +(a_{31}) & -(a_{32}) & +(a_{33}) \end{vmatrix}$$

Applying rule 2 to the solution for x:

$$\begin{aligned} x &= +(3)\begin{vmatrix} 8 & -4 \\ 2 & 6 \end{vmatrix} - (0)\begin{vmatrix} 6 & -1 \\ 2 & 6 \end{vmatrix} - (4)\begin{vmatrix} 6 & -1 \\ 8 & -4 \end{vmatrix} \\ &= 3[(48) - (-8)] - 4[(-24) - (-8)] \\ &= 232 \end{aligned}$$

Example 8.10 Solve the determinant for a of Example 8.9 using cofactors. Verify that the solution is still "-1".
From Example 8.9,

$$a = \left(\frac{1}{2}\right) \frac{\begin{vmatrix} 3 & 1 & 2 \\ -3 & -1 & 4 \\ 8 & 0 & 3 \end{vmatrix}}{\begin{vmatrix} 2 & 1 & 2 \\ 4 & -1 & 4 \\ -1 & 0 & 3 \end{vmatrix}}$$

Utilizing row 3 to form cofactor terms,

$$2a = \frac{(8)\begin{vmatrix} 1 & 2 \\ -1 & 4 \end{vmatrix} - (0) + (3)\begin{vmatrix} 3 & 1 \\ -3 & -1 \end{vmatrix}}{(-1)\begin{vmatrix} 1 & 2 \\ -1 & 4 \end{vmatrix} - (0) + (3)\begin{vmatrix} 2 & 1 \\ 4 & -1 \end{vmatrix}}$$

$$= \frac{8[(4) - (-2)] + 3[(-3) - (-3)]}{-1[(4) - (-2)] + 3[(-2) - (4)]}$$

$$= \frac{48}{-24} = -2$$

$$a = -1$$

This answer agrees with the earlier calculation for a in Example 8.9.

Example 8.11 Solve the following simultaneous equations by using the most advantageous cofactors of the determinant. Complete the solution by using the most advantageous substitutions.

$$3a - 4b - c = 1 \tag{8.23}$$

$$3b + 2c = 1 \tag{8.24}$$

$$-a + 2b - c = -9 \tag{8.25}$$

Since column 1 contains a zero for variable a, this column will be used when writing cofactors.

$$a = \frac{\begin{vmatrix} 1 & -4 & -1 \\ 1 & 3 & 2 \\ -9 & 2 & -1 \end{vmatrix}}{\begin{vmatrix} 3 & -4 & -1 \\ 0 & 3 & 2 \\ -1 & 2 & -1 \end{vmatrix}}$$

$$= \frac{1\begin{vmatrix} 3 & 2 \\ 2 & -1 \end{vmatrix} - 1\begin{vmatrix} -4 & -1 \\ 2 & -1 \end{vmatrix} - 9\begin{vmatrix} -4 & -1 \\ 3 & 2 \end{vmatrix}}{3\begin{vmatrix} 3 & 2 \\ 2 & -1 \end{vmatrix} - 1\begin{vmatrix} -4 & -1 \\ 3 & 2 \end{vmatrix}}$$

$$= \frac{[-3 - 4] - [4 + 2] - 9[-8 + 3]}{3[-3 - 4] - [-8 + 3]}$$

$$= \frac{-7 - 6 + 45}{-21 + 5} = \frac{32}{-16} = -2$$

$$b = \frac{\begin{vmatrix} 3 & 1 & -1 \\ 0 & 1 & 2 \\ -1 & -9 & -1 \end{vmatrix}}{D} = \frac{3\begin{vmatrix} 1 & 2 \\ -9 & -1 \end{vmatrix} - \begin{vmatrix} 1 & -1 \\ 1 & 2 \end{vmatrix}}{-16}$$

$$= \frac{3[-1 + 18] - [2 + 1]}{16} = \frac{48}{-16} = -3$$

From Eq. (8.24),

$$2c = 1 - 3b = 1 - (3)(-3) = 10$$
$$c = 5$$

8.2 APPLICATION OF DETERMINANTS

An extensive amount of time was dedicated to the study of determinants. This ultimately resulted in a method to mechanically "crank out" solutions to simultaneous equations. But where do the equations come from so that this practical process can be applied? It is at this point that the reader now has all the background knowledge to solve the Wheatstone bridge problem posed in Section 8.1. Kirchhoff's voltage law permits writing of the necessary equations, and the methodology of determinants facilitates their solution.

The Wheatstone bridge configuration has been redrawn in Figure 8.7. There are three uniquely enclosed areas, therefore three equations must be constructed. Currents I_1, I_2, and I_3 have been arbitrarily drawn in a clockwise direction. Any direction could have been chosen, because the solution of the determinant gives the correct current magnitude. If the solution turns out to be negative, it only means that the assumed current direction must be reversed, but the magnitude is correct.

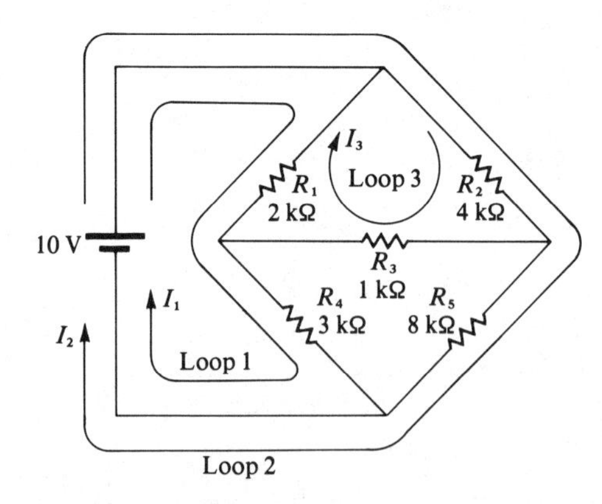

Figure 8.7 Selecting Loops for a Determinant

Care should be taken in the choice of the three loops. Each of the three choices must close back unto itself, and every element of the circuit must have at least one current flowing through it. One of many other possible selections is shown in Figure 8.8. Notice again that each element has at least one current flowing through it, and all loops close.

From Figure 8.7,

$$I_{R_1} = I_1 - I_3 \text{ (current choices are in opposite directions)} \quad (8.26)$$

$$I_{R_2} = I_2 + I_3 \text{ (current choices are in the same direction)} \quad (8.27)$$

$$I_{R_3} = I_3 \quad (8.28)$$

$$I_{R_4} = I_1 \quad (8.29)$$

$$I_{R_5} = I_2 \quad (8.30)$$

From Kirchhoff's voltage law, the sum of the voltages in a closed loop is zero. Since all resistors are in units of kΩ, current will be in units of mA, therefore the kΩ designation will be dropped to keep the equation simpler. Reference Figure 8.7:

From loop 1,

$$I_1R_1 + I_1R_4 - I_3R_1 = 10$$

$$I_1(R_1 + R_4) - I_3R_1 = 10$$

$$5I_1 + 0I_2 - 2I_3 = 10 \quad (8.31)$$

From loop 2,

$$I_2(R_2 + R_5) + I_3R_2 = 10$$

$$0I_1 + 12I_2 + 4I_3 = 10 \quad (8.32)$$

From loop 3,

$$-I_1R_1 + I_2R_2 + I_3(R_1 + R_2 + R_3) = 0$$

$$-2I_1 + 4I_2 + 7I_3 = 0 \quad (8.33)$$

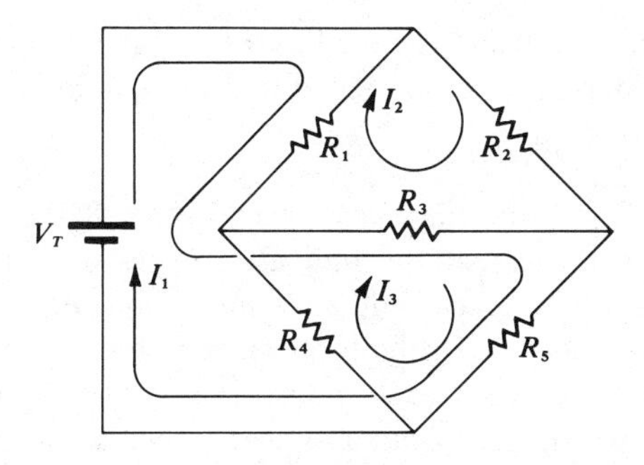

Figure 8.8 Alternate Loop Choices for Figure 8.7

Collect the three simultaneous equations together:

$$5I_1 + 0I_2 - 2I_3 = 10$$
$$0I_1 + 12I_2 + 4I_3 = 10$$
$$-2I_1 + 4I_2 + 7I_3 = 0$$

Example 8.12 Determine the current magnitude and direction through each resistor of Figure 8.7, using the three equations just determined from the application of Kirchhoff's voltage law. Equations (8.26) through (8.30) specify the currents through each resistor according to the choice of loops and assumed direction in Figure 8.7. Using determinants, solve Eqs. (8.31) through (8.33) for the currents I_1, I_2, and I_3.

$$I_1 = \frac{\begin{vmatrix} 10 & 0 & -2 \\ 10 & 12 & 4 \\ 0 & 4 & 7 \end{vmatrix}}{\begin{vmatrix} 5 & 0 & -2 \\ 0 & 12 & 4 \\ -2 & 4 & 7 \end{vmatrix}}$$

$$= \frac{(840 - 80) - (160)}{(420) - (48 + 80)} = \frac{600}{292} = 2.05 \text{ mA}$$

From Eq. (8.31),

$$2I_3 = 5I_1 - 10 = 5(2.05) - 10 = 0.25 \text{ mA}$$

$$I_3 = \frac{0.25}{2} \text{ mA} = 0.125 \text{ mA}$$

From Eq. (8.32),

$$12I_2 = 10 - 4I_3 = 10 - 4(0.125) = 9.5 \text{ mA}$$

$$I_2 = \frac{9.5}{12} \text{ mA} = 0.79 \text{ mA}$$

The current directions were assumed correctly, as all answers were positive. Figure 8.9 depicts the current distribution and directions for the "bridge" circuit of Figure 8.7.

Example 8.13 Solve for the currents in Figure 8.10. Current flow is assumed to be in the directions indicated. Note that, in keeping with Kirchhoff's voltage law, each loop equation is equated to the net voltage source within that loop. If any loop current had been chosen in the direction of electron current flow, the net voltage source within that loop would then have been designated with a negative sign preceding its absolute magnitude.

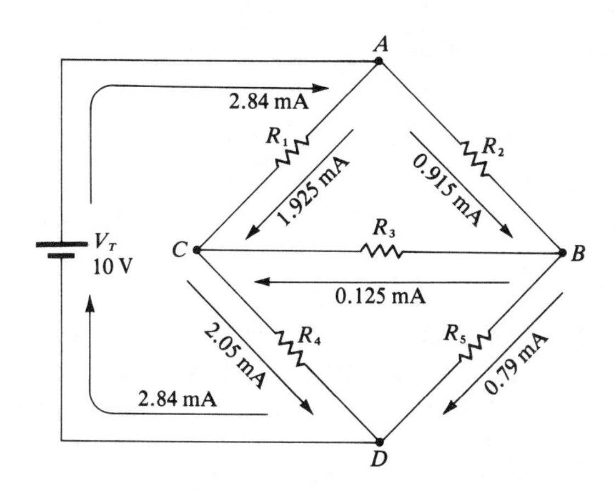

Figure 8.9 Current Distribution for Figure 8.7

From loop 1,

$$9.3I_1 - 4.7I_2 + 2.4I_3 = 10$$

From loop 2,

$$-4.7I_1 + 14.8I_2 + 6.8I_3 = 12$$

From loop 3,

$$2.4I_1 + 6.8I_2 + 13.4I_3 = 20$$

$$I_1 = \frac{\begin{vmatrix} 10 & -4.7 & 2.4 \\ 12 & 14.8 & 6.8 \\ 20 & 6.8 & 13.4 \end{vmatrix}}{\begin{vmatrix} 9.3 & -4.7 & 2.4 \\ -4.7 & 14.8 & 6.8 \\ 2.4 & 6.8 & 13.4 \end{vmatrix}}$$

$$= \frac{(1983.2 + 195.8 - 639.2) - (710.4 - 755.8 + 462.4)}{(1844.4 - 76.7 - 76.7) - (85.3 + 296 + 430)}$$

$$= \frac{1122.8}{879.7} = 1.28 \text{ mA}$$

$$I_2 = \frac{\begin{vmatrix} 9.3 & 10 & 2.4 \\ -4.7 & 12 & 6.8 \\ 2.4 & 20 & 13.4 \end{vmatrix}}{D}$$

$$= \frac{(1495.4 - 225.6 + 163.2) - (69.1 + 1264.8 - 629.8)}{879.7}$$

$$= \frac{728.9}{879.7} = 0.83 \text{ mA}$$

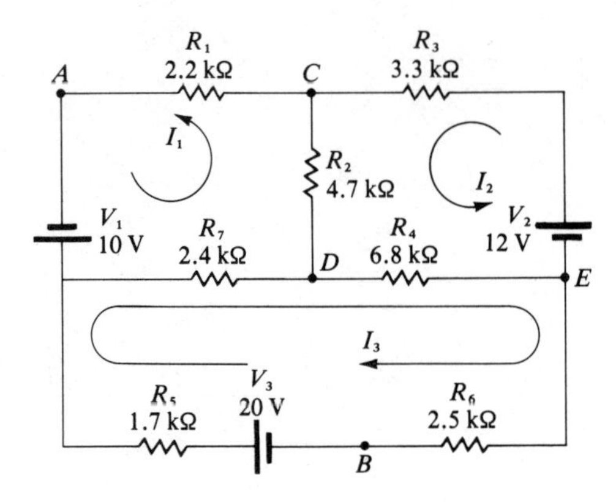

Figure 8.10 Reference Examples 8.13 and 8.14

From loop 2,

$$6.8I_3 = 12 + 4.7I_1 - 14.8I_2$$

$$6.8I_3 = 12 + 6.02 - 12.28 = 5.74$$

$$I_3 = \frac{5.74}{6.8} = 0.84 \text{ mA}$$

Example 8.14 Using the current magnitudes and directions determined in Example 8.13, find V_{AB} in Figure 8.10. Observe subscript notation. One path between points A and B could be as follows:

$$\begin{aligned} V_{AB} &= V_1 + V_{R_5} + V_3 \\ &= 10 \text{ V} + (1.7)(0.84) \text{ V} - 20 \text{ V} \\ &= -8.57 \text{ volts} \end{aligned}$$

Example 8.15 Using the current magnitudes and directions determined in Example 8.13, find V_{AB} in Figure 8.10 using the following path: $V_{AC} + V_{CD} + V_{DE} + V_{EB}$. Verify that the answer is the same for V_{AB} as determined in Example 8.14. Figure 8.11 illustrates the desired path, accounting for the specified current direction and voltage polarities.

$$V_{AC} = I_1 R_1 = (1.28)(2.2) \text{ V} = 2.82 \text{ volts}$$

$$V_{CD} = (I_1 - I_2)R_2 = (0.45)(4.7) \text{ V} = 2.12 \text{ volts}$$

$$V_{DE} = -(I_2 + I_3)R_4 = -(1.67)(6.8) \text{ V} = -11.36 \text{ volts}$$

$$V_{EB} = -I_3 R_6 = -(0.84)(2.5) \text{ V} = -2.1 \text{ volts}$$

$$\therefore \quad V_{AB} = (2.82 + 2.12 - 11.36 - 2.1) \text{ V} = -8.52 \text{ volts}$$

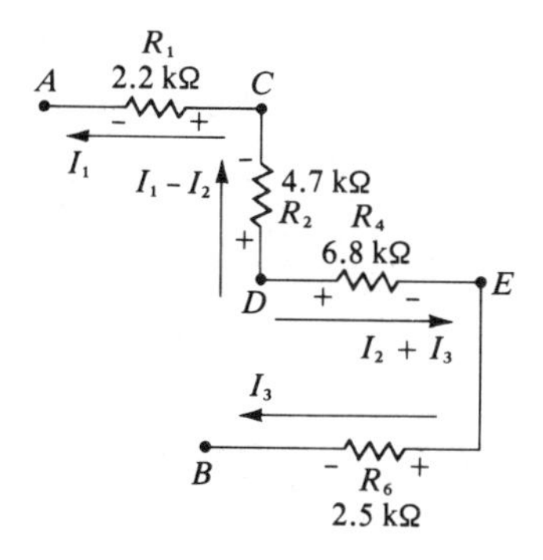

Figure 8.11 V_{AB} along Prescribed Path (Reference Example 8.14)

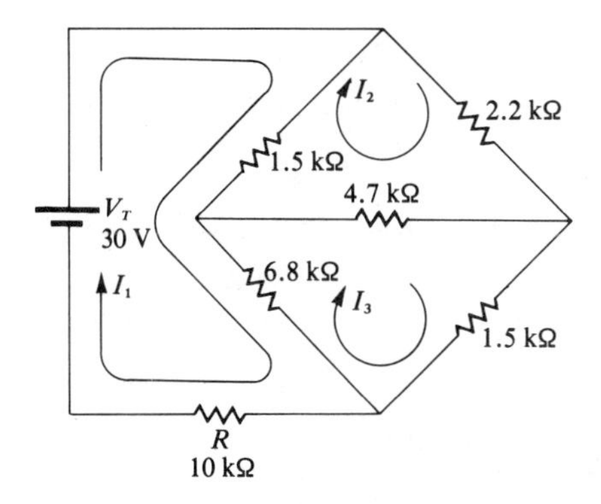

Figure 8.12 Reference Example 8.16

This is the same as was determined in Example 8.14. The minor difference in voltage is due to the roundoff of numbers.

Example 8.16 Determine the current through R in Figure 8.12. Again note that in dropping the kΩ term, current is automatically in mA. Choosing the current directions shown, the three loop equations are as follows:

$$18.3I_1 - 1.5I_2 - 6.8I_3 = 30$$
$$-1.5I_1 + 8.4I_2 - 4.7I_3 = 0$$
$$-6.8I_1 - 4.7I_2 + 13I_3 = 0$$

Since I_1 is the only current through R, only one determinant need be solved.

$$I_R = I_1 = \frac{\begin{vmatrix} 30 & -1.5 & -6.8 \\ 0 & 8.4 & -4.7 \\ 0 & -4.7 & 13 \end{vmatrix} \equiv N}{\begin{vmatrix} 18.3 & -1.5 & -6.8 \\ -1.5 & 8.4 & -4.7 \\ -6.8 & -4.7 & 13 \end{vmatrix} \equiv D}$$

Column 1 of the numerator has two zeros, hence it is ideally suited for solution by cofactors.

$$\begin{aligned} N &= 30 \begin{vmatrix} 8.4 & -4.7 \\ -4.7 & 13 \end{vmatrix} = 30(109.2 - 22.09) \\ &= 2613.3 \\ D &= (1998.4 - 47.94 - 47.94) - (388.42 + 29.25 + 404.25) \\ &= 1080.6 \end{aligned}$$

$$I_1 = \frac{N}{D} = \frac{2613.3}{1080.6} = 2.42 \text{ mA}$$

Example 8.17 Solve the following fourth order determinant:

$$x = \begin{vmatrix} 2 & 6 & 2 & -1 \\ 1 & 0 & -7 & 0 \\ -5 & -1 & 1 & 4 \\ 3 & 4 & 9 & 5 \end{vmatrix}$$

Make cofactors from row 2.

$$x = (\text{sign})\ 1 \begin{vmatrix} 6 & 2 & -1 \\ -1 & 1 & 4 \\ 4 & 9 & 5 \end{vmatrix} + 0 + (\text{sign}) - 7 \begin{vmatrix} 2 & 6 & -1 \\ -5 & -1 & 4 \\ 3 & 4 & 5 \end{vmatrix} + 0$$

$$\begin{aligned} &= (-)1 \begin{vmatrix} 6 & 2 & -1 \\ -1 & 1 & 4 \\ 4 & 9 & 5 \end{vmatrix} (-) - 7 \begin{vmatrix} 2 & 6 & -1 \\ -5 & -1 & 4 \\ 3 & 4 & 5 \end{vmatrix} \\ &= -[(30 + 9 + 32) - (-4 - 10 + 216)] \\ &\quad + 7[(-10 + 20 + 72) - (3 - 150 + 32)] \\ &= -[(71 - 202)] + 7[82 - (-115)] \\ &= 131 + 1379 = 1510 \end{aligned}$$

8.3 THEORY OF THE MATRIX

A form of mathematics called *determinants* was introduced in the previous two sections to facilitate the solution of circuits in other than a simple configuration. It was found that a determinant is really nothing more than an array of numbers. A set of rules, concerning the manipulation of these numbers, produces a solution of each of the unknowns, one at a time. This section will continue the study of yet another, more advanced form of number manipulation. It is known as solution by *matrix*. This methodology has two distinct advantages.

1. Unlike the determinant, a matrix provides for the solution of all unknown quantities simultaneously. This means that one manipulation of the number array gives the answer to all unknown quantities of the simultaneous equations.
2. As will be demonstrated in the following section, matrix arrays are ideally suited for computer solution because a few simple programming instructions can produce all the required solutions simultaneously.

Figure 8.13 illustrates the general configuration of a matrix array. Again keep in mind that each of the subscripted a's merely represents a coefficient as determined by an original set of simultaneous equations. It is important to understand the ordering of the subscripts. The first subscript refers to the row in which a particular a is located, and the second subscript indicates the column in which the a is located. Together, the two-digit subscript defines a unique position in the number array for each of the coefficients.

$$\begin{bmatrix} a_{11} & a_{12} & a_{13} & \cdots & a_{1j} \\ a_{21} & a_{22} & a_{23} & \cdots & a_{2j} \\ a_{31} & a_{32} & a_{33} & \cdots & a_{3j} \\ \vdots & \vdots & \vdots & & \\ a_{i1} & a_{i2} & a_{i3} & \cdots & a_{ij} \end{bmatrix}$$

Figure 8.13 General Configuration of a Matrix

The theory that follows will set forth the rules on how each coefficient is manipulated so that the final answer will be a simultaneous solution of all variables. Suppose an equation is written as

$$AB = C \tag{8.34}$$

and it is desired to solve for B. If both sides of the equation are multiplied by A^{-1}, then the following results:

$$\begin{aligned}(A^{-1})(AB) &= (A^{-1})(C)\\ A^0B &= A^{-1}C\\ B &= A^{-1}C\end{aligned} \tag{8.35}$$

The term on the right-hand side (A^{-1}) is the entire key to the matrix. All the rules which follow will deal with the meaning of this term, which is referred to as the *inverse* matrix.

To illustrate matrix solution, the following three equations, having three unknowns, will be solved:

$$2A - B + C = 3 \tag{8.36}$$

$$-A - B + 2C = 3 \tag{8.37}$$

$$2A + B - 3C = -5 \tag{8.38}$$

Write the initial matrix in the following fashion:

$$\underbrace{\begin{bmatrix} 2 & -1 & 1 \\ -1 & -1 & 2 \\ 2 & 1 & -3 \end{bmatrix}}_{X} \underbrace{\begin{bmatrix} A \\ B \\ C \end{bmatrix}}_{Y} = \underbrace{\begin{bmatrix} 3 \\ 3 \\ -5 \end{bmatrix}}_{Z} \tag{8.39}$$

Notice that within the matrix designated X, the coefficients of the unknown variables are written (reading from left to right). This assumes, of course, that the equations have been aligned. The matrix designated Y contains the three variables. The matrix designated Z contains the coefficients to which each equation is equated. Multiplying both sides of Eq. (8.39) by the inverse of matrix X produces the following:

$$\underbrace{\begin{bmatrix} A \\ B \\ C \end{bmatrix}}_{Y} = \underbrace{\begin{bmatrix} 2 & -1 & 1 \\ -1 & -1 & 2 \\ 2 & 1 & -3 \end{bmatrix}^{-1}}_{X^{-1}} \underbrace{\begin{bmatrix} 3 \\ 3 \\ -5 \end{bmatrix}}_{Z} \tag{8.40}$$

The first rule in solving X^{-1} is to replace each term by its cofactor. For example, the cofactor of a_{11} is

$$\begin{vmatrix} -1 & 2 \\ 1 & -3 \end{vmatrix} = (-1)(-3) - (1)(2) = 1$$

If all terms up to and including a_{33} are solved for their respective cofactors in like fashion, this results in the following new matrix when each cofactor is substituted for its respective a term:

$$\begin{bmatrix} 1 & -1 & 1 \\ 2 & -8 & 4 \\ -1 & 5 & -3 \end{bmatrix}$$

The second rule in the solution of X^{-1} is to precede each of the a_{ij} cofactors by the sign that a determinant cofactor would have according to its location in the array. This sign designation is redrawn again as Figure 8.14. Following the application of rule 2, X^{-1} now takes the following form:

$$\begin{bmatrix} +(1) & -(-1) & +(1) \\ -(2) & +(-8) & -(4) \\ +(-1) & -(5) & +(-3) \end{bmatrix} = \begin{bmatrix} 1 & 1 & 1 \\ -2 & -8 & -4 \\ -1 & -5 & -3 \end{bmatrix}$$

The third rule in solving the inverse matrix is to transpose each row for each column. Row 1, which is now 1, 1, 1, becomes column 1 of the same coefficients. In a similar fashion, rewrite Row 2 as column 2 and rewrite Row 3 as column 3. At this point, X^{-1} now appears as

$$\begin{matrix} \text{Formerly} & \text{Formerly} & \text{Formerly} \\ \text{Row 1} & \text{Row 2} & \text{Row 3} \\ \downarrow & \downarrow & \downarrow \end{matrix}$$
$$\begin{bmatrix} 1 & -2 & -1 \\ 1 & -8 & -5 \\ 1 & -4 & -3 \end{bmatrix} \tag{8.41}$$

$$\begin{bmatrix} +(a_{11}) & -(a_{12}) & +(a_{13}) \\ -(a_{21}) & +(a_{22}) & -(a_{23}) \\ +(a_{31}) & -(a_{32}) & +(a_{33}) \end{bmatrix}$$

Figure 8.14 Cofactor Sign Designation

The fourth and final step is to multiply the matrix of Eq. (8.41) by $1/D$. D is the solution of the determinant of the original matrix X:

$$D = (6 - 1 - 4) - (-2 + 4 - 3) = 2$$

The final solution of X^{-1} is:

$$X^{-1} = \frac{1}{2}\begin{bmatrix} 1 & -2 & -1 \\ 1 & -8 & -5 \\ 1 & -4 & -3 \end{bmatrix} \equiv S \tag{8.42}$$

Substituting Eq. (8.42) back into Eq. (8.40), the matrix array of the three unknown variables is

$$\underbrace{\begin{bmatrix} A \\ B \\ C \end{bmatrix}}_{Y} = \underbrace{\left(\frac{1}{2}\right)\begin{bmatrix} 1 & -2 & -1 \\ 1 & -8 & -5 \\ 1 & -4 & -3 \end{bmatrix}}_{S} \underbrace{\begin{bmatrix} 3 \\ 3 \\ -5 \end{bmatrix}}_{Z}$$

The evaluation of the first unknown (A) is as follows:

> $A = 1/D$ times the sum of the products formed by multiplying each term of row 1 in matrix S by each corresponding term of matrix Z.

Figure 8.15 illustrates the mechanics involved.

$$A = \frac{1}{2}[(1)(3) + (-2)(3) + (-1)(-5)]$$

$$A = 1$$

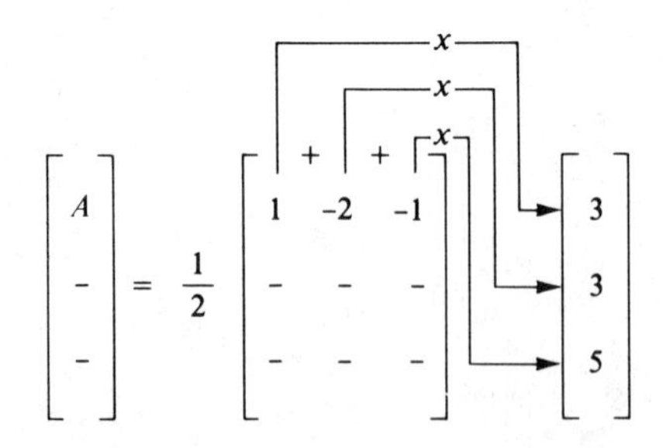

Figure 8.15 Determination of First Matrix Term

In a similar fashion,

$$B = \frac{1}{2}[(1)(3) + (-8)(3) + (-5)(-5)]$$

$$B = 2$$

$$C = \frac{1}{2}[(1)(3) + (-4)(3) + (-3)(-5)]$$

$$C = 3$$

Again, let it be stated that the above process is merely the mechanical application of a set of rules to an array of numbers. The final resultant is the simultaneous solution of all unknowns. The following section will apply these rules to several examples of electronic circuits. The subsequent section will give the reader considerable insight into how modern computers can be employed to aid in the solution of electrical problems which would otherwise consume extensive periods of calculation, coupled with the inherent probability of human error.

8.4 EXAMPLES OF MATRIX SOLUTIONS

The following examples will help to clarify how to write and solve a matrix for simultaneous solutions of unknown quantities.

Example 8.18 Determine the inverse of the following matrix:

$$X = \begin{bmatrix} 3 & -1 & 7 \\ -2 & 6 & -3 \\ 4 & -5 & 1 \end{bmatrix}$$

Rule 1. Replace each a_{ij} term by its cofactor.

$$\begin{bmatrix} -9 & 10 & -14 \\ 34 & -25 & -11 \\ -39 & 5 & 16 \end{bmatrix}$$

Rule 2. Precede each new a_{ij} term by the sign configuration of Figure 8.14.

$$\begin{bmatrix} -9 & -10 & -14 \\ -34 & -25 & 11 \\ -39 & -5 & 16 \end{bmatrix}$$

Rule 3. Transpose the rows and columns.

$$\begin{bmatrix} -9 & -34 & -39 \\ -10 & -25 & -5 \\ -14 & 11 & 16 \end{bmatrix}$$

Rule 4. Multiply the new matrix by $1/D$.

$$D = [(18 + 70 + 12) - (168 + 2 + 45)] = -115$$

$$\therefore \quad [X]^{-1} = \frac{1}{-115} \begin{bmatrix} -9 & -34 & -39 \\ -10 & -25 & -5 \\ -14 & 11 & 16 \end{bmatrix}$$

Example 8.19 Solve the following simultaneous equations using matrix solutions:

$$\begin{aligned} 3b - c &= -(4 + 2a) \\ -3a - 13 &= 2(b - c) \\ a - 3b - 3 &= -3c \end{aligned}$$

Simplify by expanding and aligning like unknowns.

$$\begin{aligned} 2a + 3b - c &= -4 \\ -3a - 2b + 2c &= 13 \\ a - 3b + 3c &= 3 \end{aligned}$$

$$\begin{bmatrix} a \\ b \\ c \end{bmatrix} = \begin{bmatrix} 2 & 3 & -1 \\ -3 & -2 & 2 \\ 1 & -3 & 3 \end{bmatrix}^{-1} \begin{bmatrix} -4 \\ 13 \\ 3 \end{bmatrix}$$

Rule 1:

$$\begin{bmatrix} a \\ b \\ c \end{bmatrix} = \begin{bmatrix} 0 & -11 & 11 \\ 6 & 7 & -9 \\ 4 & 1 & 5 \end{bmatrix} \begin{bmatrix} -4 \\ 13 \\ 3 \end{bmatrix}$$

Rule 2:

$$\begin{bmatrix} a \\ b \\ c \end{bmatrix} = \begin{bmatrix} 0 & 11 & 11 \\ -6 & 7 & 9 \\ 4 & -1 & 5 \end{bmatrix} \begin{bmatrix} -4 \\ 13 \\ 3 \end{bmatrix}$$

Rule 3:

$$\begin{bmatrix} a \\ b \\ c \end{bmatrix} = \begin{bmatrix} 0 & -6 & 4 \\ 11 & 7 & -1 \\ 11 & 9 & 5 \end{bmatrix} \begin{bmatrix} -4 \\ 13 \\ 3 \end{bmatrix}$$

Rule 4:

$$\begin{bmatrix} a \\ b \\ c \end{bmatrix} = \frac{1}{22}\begin{bmatrix} 0 & -6 & 4 \\ 11 & 7 & -1 \\ 11 & 9 & 5 \end{bmatrix}\begin{bmatrix} -4 \\ 13 \\ 3 \end{bmatrix}$$

Solution:

$$\begin{bmatrix} a \\ b \\ c \end{bmatrix} = \frac{1}{22}\begin{bmatrix} (-78 & +12) & \\ (-44 & +91 & -3) \\ (-44 & +117 & +15) \end{bmatrix} = \begin{bmatrix} -3 \\ 2 \\ 4 \end{bmatrix}$$

$$\therefore \quad a = -3, \qquad b = 2, \qquad c = 4$$

Example 8.20 Solve for the currents through each resistor of Figure 8.16 using matrix analysis.

$$7.18I_1 - I_2 + 2.2I_3 = 10$$
$$-I_1 + 5.8I_2 + 1.5I_3 = 15$$
$$2.2I_1 + 1.5I_2 + 8.4I_3 = 20$$

$$\begin{bmatrix} I_1 \\ I_2 \\ I_3 \end{bmatrix} = \begin{bmatrix} 7.18 & -1 & 2.2 \\ -1 & 5.8 & 1.5 \\ 2.2 & 1.5 & 8.4 \end{bmatrix}^{-1}\begin{bmatrix} 10 \\ 15 \\ 20 \end{bmatrix}$$

Rules 1 and 2 combined:

$$\begin{bmatrix} I_1 \\ I_2 \\ I_3 \end{bmatrix} = \begin{bmatrix} 46.5 & 11.7 & -14.3 \\ 11.7 & 55.5 & -13.0 \\ -14.3 & -13.0 & 40.6 \end{bmatrix}\begin{bmatrix} 10 \\ 15 \\ 20 \end{bmatrix}$$

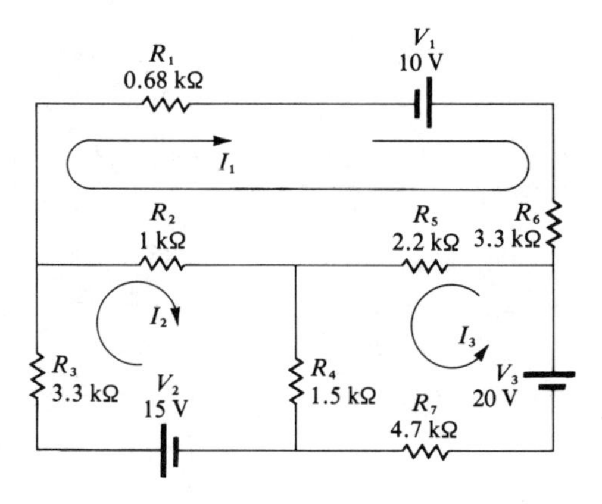

Figure 8.16 Reference Example 8.20

Rules 3 and 4 combined:

$$\begin{bmatrix} I_1 \\ I_2 \\ I_3 \end{bmatrix} = \frac{1}{290.5} \begin{bmatrix} 46.5 & 11.7 & -14.3 \\ 11.7 & 55.5 & -13 \\ -14.3 & -13 & 40.6 \end{bmatrix} \begin{bmatrix} 10 \\ 15 \\ 20 \end{bmatrix}$$

Solutions:

$$\begin{bmatrix} I_1 \\ I_2 \\ I_3 \end{bmatrix} = \frac{5}{290.5} \begin{bmatrix} 46.5 & 11.7 & -14.3 \\ 11.7 & 55.5 & -13 \\ -14.3 & -13 & 40.6 \end{bmatrix} \begin{bmatrix} 2 \\ 3 \\ 4 \end{bmatrix} = \begin{bmatrix} 1.22 \\ 2.37 \\ 1.63 \end{bmatrix} \text{mA}$$

$$\therefore \quad I_1 = 1.22 \text{ mA}, \qquad I_2 = 2.37 \text{ mA}, \qquad I_3 = 1.63 \text{ mA}$$

The assumptions made for the current directions were correct, as all currents resulted in positive values.

$$I_{R_1} = I_1 = 1.22 \text{ mA}$$
$$I_{R_2} = I_2 - I_1 = (2.37 - 1.22) \text{ mA} = 1.15 \text{ mA}$$
$$I_{R_3} = I_2 = 2.37 \text{ mA}$$
$$I_{R_4} = I_2 + I_3 = (2.37 + 1.63) \text{ mA} = 4 \text{ mA}$$
$$I_{R_5} = I_1 + I_3 = (1.22 + 1.63) \text{ mA} = 2.85 \text{ mA}$$
$$I_{R_6} = I_1 = 1.22 \text{ mA}$$
$$I_{R_7} = I_3 = 1.63 \text{ mA}$$

The currents and voltage polarities of Figure 8.16 are summarized in Figure 8.17.

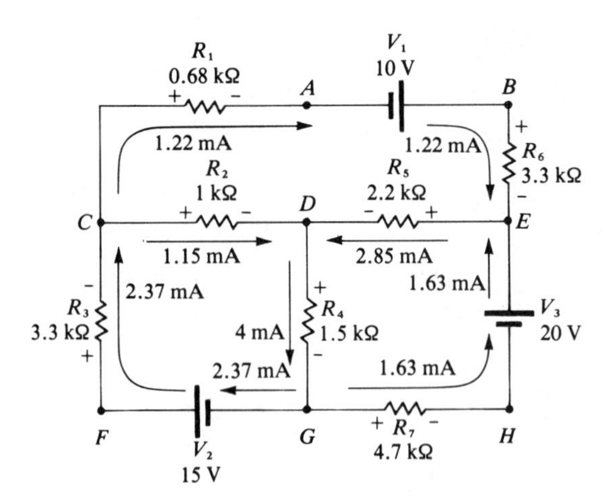

Figure 8.17 Current and Voltage Polarity Distribution of Figure 8.16

Example 8.21 Using the currents found in Example 8.20, find V_{AG} in Figure 8.17 by the following two paths:

Path 1: $V_{AC} + V_{CD} + V_{DG}$

Path 2: $V_{AB} + V_{BE} + V_{EH} + V_{HG}$

Verify that both paths (or any other path between these two points) result in the same answer.

Path 1:
$$V_{AG} = (1.22 \text{ mA})(0.68 \text{ k}\Omega) - (1.15 \text{ mA})(1 \text{ k}\Omega) - (4 \text{ mA})(1.5 \text{ k}\Omega)$$
$$= 0.83 \text{ V} - 1.15 \text{ V} - 6 \text{ V} = -6.32 \text{ volts}$$

Path 2:
$$V_{AG} = 10 - (1.22 \text{ mA})(3.3 \text{ k}\Omega) - 20 + (1.63 \text{ mA})(4.7 \text{ k}\Omega)$$
$$= 10 \text{ V} - 4.0 \text{ V} - 20 \text{ V} + 7.66 \text{ V} = -6.34 \text{ volts}$$

Example 8.22 The current meter in Figure 8.18 is 100 ohms resistance. Using matrix analysis, determine the magnitude and direction of current flow through the meter and all resistors. The currents are assumed in the directions shown.

$$1.33I_1 - 0.33I_2 - 1.0I_3 = 3$$
$$-0.33I_1 + 1.11I_2 - 0.1I_3 = 0$$
$$-1.0I_1 - 0.1I_2 + 1.57I_3 = 0$$

$$\begin{bmatrix} I_1 \\ I_2 \\ I_3 \end{bmatrix} = \begin{bmatrix} 1.33 & -0.33 & -1.0 \\ -0.33 & 1.11 & -0.1 \\ -1.0 & -0.1 & 1.57 \end{bmatrix}^{-1} \begin{bmatrix} 3 \\ 0 \\ 0 \end{bmatrix}$$

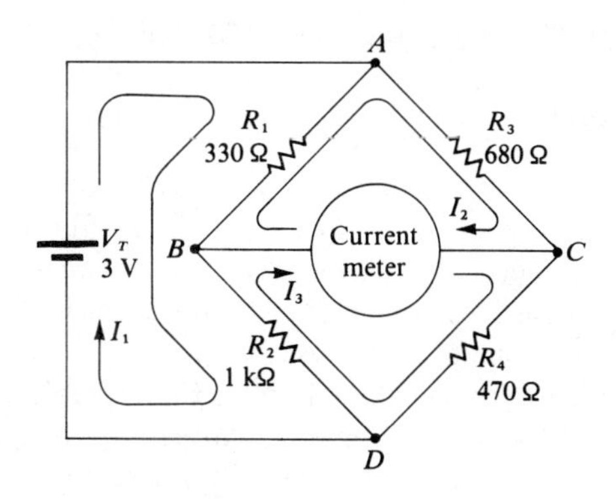

Figure 8.18 Reference Example 8.22

Rules 1 and 2 combined:

$$\begin{bmatrix} I_1 \\ I_2 \\ I_3 \end{bmatrix} = \begin{bmatrix} 1.733 & 0.618 & 1.143 \\ 0.618 & 1.088 & 0.463 \\ 1.143 & 0.463 & 1.367 \end{bmatrix} \begin{bmatrix} 3 \\ 0 \\ 0 \end{bmatrix}$$

Rules 3 and 4 combined:

$$\begin{bmatrix} I_1 \\ I_2 \\ I_3 \end{bmatrix} = \frac{1}{0.958} \begin{bmatrix} 1.733 & 0.618 & 1.143 \\ 0.618 & 1.088 & 0.463 \\ 1.143 & 0.463 & 1.367 \end{bmatrix} \begin{bmatrix} 3 \\ 0 \\ 0 \end{bmatrix}$$

Solutions:

$$I_1 = \frac{(1.733)(3)}{0.958} = 5.43 \text{ mA}$$

$$I_2 = \frac{(0.618)(3)}{0.958} = 1.935 \text{ mA}$$

$$I_3 = \frac{(1.143)(3)}{0.958} = 3.579 \text{ mA}$$

From Figure 8.18,

$$\begin{aligned} I_{R_1} &= I_1 - I_2 = (5.43 - 1.935) \text{ mA} = 3.495 \text{ mA} \\ I_{R_2} &= I_1 - I_3 = (5.43 - 3.579) \text{ mA} = 1.851 \text{ mA} \\ I_{\text{meter}} &= I_3 - I_2 = (3.579 - 1.935) \text{ mA} = 1.644 \text{ mA} \\ I_{R_3} &= I_2 = 1.935 \text{ mA} \\ I_{R_4} &= I_3 = 3.579 \text{ mA} \end{aligned}$$

Figure 8.19 depicts the current distribution of Figure 8.18.

8.5 MATRIX COMPUTER PROGRAMMING (OPTIONAL)

Matrix Statements

In addition to the simple variables that were described in earlier chapters, BASIC allows the use of subscripted variables. Subscripted variables provide the programmer with additional computing capabilities for dealing with lists, tables, matrices, or any set of related variables. In BASIC, variables are allowed one or two subscripts (or more, in many versions).

The name of a subscripted variable is any acceptable BASIC variable name followed by one or two integer expressions in parentheses. For example, a list might be described as $A(I)$, where I goes from 1 to 5 as follows. (All matrices are created with a zero element that can be used but is often unspecified.)

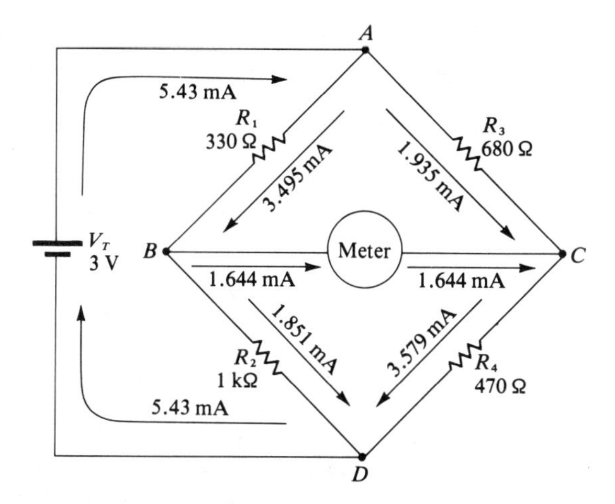

Figure 8.19 Current Distribution of Figure 8.18

A(Ø), A(1), A(2), A(3), A(4), A(5)

This allows the programmer to reference each of six elements in the list, which can be considered a one-dimensional algebraic matrix as follows:

A(Ø)
A(1)
A(2)
A(3)
A(4)
A(5)

A two-dimensional matrix, $B(I,J)$, can be defined in a similar manner and graphically illustrated as follows:

	Column Ø ↓	*Column 1* ↓	*Column 2* ↓		*Column J* ↓
Row Ø →	$B(\emptyset,\emptyset)$	$B(\emptyset,1)$	$B(\emptyset,2)$	· · ·	$B(\emptyset,J)$
Row 1 →	$B(1,\emptyset)$	$B(1,1)$	$B(1,2)$	· · ·	$B(1,J)$
Row 2 →	$B(2,\emptyset)$	$B(2,1)$	$B(2,2)$	· · ·	$B(2,J)$
	⋮	⋮	⋮		⋮
Row I →	$B(I,\emptyset)$	$B(I,1)$	$B(I,2)$	· · ·	$B(I,J)$

Subscripts used with subscripted variables throughout a program can be explicitly stated or be any legal expression.

It is possible to use the same variable name as both a subscripted and an unsubscripted variable. Both A and $A(I)$ are valid variables and can be used in the same program. However, BASIC does not accept the same variable name as both a singly and a doubly subscripted variable name in the same program (that is, $A(I)$ and $A(I,0)$ would refer to the same data item).

DIM Statement A dimension (DIM) statement is used to define the maximum number of elements in a matrix. The DIM statement is of the form:

line number DIM 〈*variable* (n)〉, 〈*variable*(*n,m*)〉, . . .

where the variables specified are indicated with their maximum subscript value(s). For example:

```
10 DIM X(5), Y(4,2), A(10,10)
12 DIM I4(100)
```

Only integer values (such as 5 or 5070) can be used in DIM statements to define the size of a matrix. Any number of matrices can be defined in a single DIM statement as long as their representations are separated by commas. If a subscripted variable is used without appearing in a DIM statement, most versions of BASIC (except, for example, Atari 400/800) will assume it to be dimensioned to length 10—to have 11 elements, 0 through 10. All matrices, however, should be correctly dimensioned in a program. DIM statements are usually grouped together among the first lines of a program.

The first element of every matrix is automatically assumed to have a subscript of zero. Dimensioning $A(6,10)$ sets up room for a matrix with 7 rows and 11 columns. This zero element is illustrated in the following program:

```
LIST
10 REM – MATRIX CHECK PROGRAM
20 DIM A(6,10)
30 FOR I=0 TO 6
40 LET A(I,0) = I
50 FOR J=0 TO 10
60 LET A(0,J) = J
70 PRINT A(I,J);
80 NEXT J: PRINT: NEXT I
90 END

RUN
 0  1  2  3  4  5  6  7  8  9  10
 1  0  0  0  0  0  0  0  0  0  0
 2  0  0  0  0  0  0  0  0  0  0
 3  0  0  0  0  0  0  0  0  0  0
 4  0  0  0  0  0  0  0  0  0  0
 5  0  0  0  0  0  0  0  0  0  0
 6  0  0  0  0  0  0  0  0  0  0
```

Notice in this example that the variables not assigned values by the program were printed out with value zero: They were automatically *initialized* to zero. Most versions of BASIC (except, for example, Atari 400/800) initialize matrix variables automatically in this way.

If the user wishes to conserve memory space, he or she can use the extra variables set up within the matrix. A user could, for example, say DIM

A(5,9) to obtain a 6 × 10 matrix, which would then be referenced beginning with the A(0,0) element.

The size and number of matrices that can be defined depend upon the amount of user storage space available.

A DIM statement can be placed anywhere in a multiple statement line. A DIM statement can appear anywhere in the program and need not appear prior to the first reference to an array. Generally, however, DIM statements are among the first statements of a program so that they may be easily found if any alterations are later required.

Special Matrix Statements Versions of BASIC used on large-scale computer systems have special statements that allow them to perform operations on matrices much as though they were single variables. Thus, matrices can be read, printed, added, subtracted, multiplied, inverted, and so forth. These matrix manipulations are particularly valuable for solving simultaneous equations since, as we have seen earlier in this chapter, the solutions to simultaneous equations can be conveniently expressed in matrix form. (See also Example 8.23.)

Most microcomputer versions of BASIC do not have these capabilities. Appendix D shows how to use conventional BASIC statements to solve two or three equations simultaneously. The special matrix statements described in the rest of this chapter are applicable to BASIC-PLUS.

MAT READ Statement A BASIC-PLUS program can define the size of a matrix in one of two ways: *explicitly*, by including the matrix in a dimension statement, or *implicitly*, where the matrix does not appear in any dimension statement. Implicitly dimensioned matrices are assumed to have 10 elements in each dimension referenced (size 10 for a one-dimensional matrix and size 10 × 10 for a two-dimensional matrix, with each dimension also having a zero row and column). Implicitly dimensioning the matrix $A(I,J)$, for example, has the same effect as explicitly including the following statement:

```
10 DIM A(10,10)
```

Dimensioning a matrix (explicitly or implicitly) establishes two quantities for the system: the default number of elements in each row and column, and the maximum number of elements in the matrix. Through use of the MAT commands, the program can alter the number of elements in each row and the number of columns in the matrix as long as the total number of elements does not exceed the number defined when the matrix was dimensioned. Changing the number of elements in either or both dimensions is termed *redimensioning the matrix.*

The MAT READ statement is used to read the value of each element of a matrix from DATA statements. The format of the statement is as follows:

line number MAT READ ⟨*list of matrices*⟩

Each element in the list of matrices indicates the maximum amount of the matrix to be read (which cannot be greater than the dimensioned size of the matrix). The individual elements are separated by commas. If the matrix name is used without a subscript, the entire matrix is read. For example:

```
10 DIM A(20,20)
20 MAT READ A
```

The above lines read a 20 × 20 matrix of data. Data is read row by row; that is, the second subscript varies most rapidly. If line 20 had read:

```
20 MAT READ A(5,15)
```

a 5 × 15 matrix would be read and the matrix A would be redimensioned.

When a matrix is redimensioned, the user program should take care not to reference elements outside the currently dimensioned range of the matrix. For example, if the range of matrix A is 5 × 7, referencing $A(3,8)$ is improper and, although no error is generated, generally results in some element elsewhere in the matrix being destroyed.

The MAT operations do not set the zero elements $A(\emptyset)$, or $B(\emptyset,n)$ and $B(n,\emptyset)$ of the specified matrix to conform with the requested operation.

MAT PRINT Statement The MAT PRINT statement prints each element of a one- or two-dimensional matrix. The statement is of the form:

line number MAT PRINT ⟨*matrix name*⟩

If the matrix name consists of an unsubscripted matrix name, the entire matrix is printed. If the matrix name is subscripted, then the subscript indicates the maximum size of the matrix to be printed (but does *not* redimension the matrix). Only one matrix can be output by a single MAT PRINT statement.

If the matrix name is followed by a semicolon (;), the data values are printed in a packed fashion. If the matrix name is followed by a comma (,), the data values are printed across the line with one value per print zone. If neither character follows the matrix name (the null case), each element is printed on a separate line.

```
10 DIM A(10,10),B(20,20)

120 MAT PRINT A;          !PRINT 10*10 MATRIX,PACKED FORMAT
130 MAT PRINT B(N,M)      !PRINT N*M MATRIX, 5 ELEMENTS
                          !PER LINE
```

One-dimensional arrays can be printed in either row or column format. For example,

```
MAT PRINT V
```

where V is a singly dimensioned array, prints the array V as a column matrix, and

```
MAT PRINT V,
```

prints the array V as a row matrix, five values per line.

```
MAT PRINT V;
```

prints the array V as a row matrix, closely packed. For example:

```
LIST
1Ø DIM A(7),X(5)
2Ø MAT READ A,X
3Ø MAT PRINT A;:PRINT:MAT PRINT X
4Ø DATA 21,22,23,24,35,36,37,51,52,53,54,55
5Ø END

RUN
 21  22  23  24  35  36  37

 51
 52
 53
 54
 55
```

MAT INPUT Statement The MAT INPUT statement is used to input the value of each element of a predimensioned matrix. The statement is of the form:

line number MAT INPUT ⟨*list of matrices*⟩

Input is read from the keyboard, as with a normal INPUT statement, and a (?) character is printed when the program is ready to accept the input. The LINE FEED key can be used to continue typing data on succeeding lines. The RETURN or ESCAPE key is used to enter the data to the system. MAT INPUT does not affect row zero or column zero of the matrix.

Matrix Initialization Statements A matrix initialization statement allows the user to create initial values for the elements of a matrix. The statement is of the form:

$$\textit{line number}\ \text{MAT}\ \langle \textit{name} \rangle = \langle \textit{value} \rangle \begin{Bmatrix} (DIM1,\ DIM2) \\ (DIM1) \end{Bmatrix}$$

The name specified is the name of a predimensioned matrix, and the optional DIM1 and DIM2 specifications indicate the size of the matrix to be initialized. When specified, DIM1 and DIM2 cause the matrix to be redimensioned. The value can be one of the following:

Value	*Meaning*
ZER	Sets all elements of the matrix to 0 (this is true of all matrices when they are first created). (Function does not set row 0 or column 0.)
CON	Sets all elements of the matrix to 1. (Function does not set row 0 or column 0.)
IDN	Sets up an identity matrix (all elements are 0 except for those on the diagonal, $A(I,I)$, which are 1). (Function does not set row 0 or column 0.)

If no dimensions are indicated (DIM1 and DIM2 are not specified) in a matrix initialization statement, the existing dimensions of the matrix are assumed to be unchanged. For example:

```
10 DIM A(10,10),B(15),C(20,20)
20 MAT A=ZER            !SETS ALL ELEMENTS OF A=0
30 MAT B=CON(10)        !SETS FIRST 10 ELEMENTS OF B=1
40 MAT C=IDN(10,10)
```

It should be noted that these instructions do not set row zero or column zero.

Matrix Operations

The operations of addition, subtraction, and multiplication can be performed on matrices using the common BASIC mathematical symbols.

Each of the matrix operation statements is begun with the word MAT and followed by the expression to be evaluated. Each matrix involved must be predefined in a DIM statement. The subscripts of the matrices need not be indicated on the statement. The matrices indicated for any operation must be conformable to that operation. A subset of one matrix cannot be indicated as part of an operation.

```
110 DIM A(50), B(25), C(50)
120 MAT C=A+B

RUNNH
MATRIX DIMENSION ERROR AT LINE 120

READY
```

In order for line 120 to execute properly, line 110 should read:

```
110 DIM A(50),B(50),C(50)
```

Multiplication of conformable matrices is indicated as follows:

```
10 DIM D(10,5),C(5,10),R(10,10)
200 MAT R = D*C
```

By *conformable matrices* is meant that the number of columns in matrix D is equal to the number of rows in matrix C. The dimensions of the matrix R must be large enough to contain the number of columns in D and the number of rows in C.

Matrix Functions

Among the various matrix functions performed by BASIC, one of the most powerful and useful to the solution of electronic problems is the inverse (INV) function. It will be recalled from Section 8.3 that computing the inverse of a matrix was the key to the simultaneous solutions of all unknown variables within a given set of equations. It involved replacing each a_{ij} term by its cofactor, prefixing the appropriate polarity assigned each a_{ij} matrix position, transposing between rows and columns, and finally multiplying each matrix term by the reciprocal of the value of the original matrix determinant. All these complicated functions are calculated by the one simple INV statement.

When coupled with the previously discussed MAT statements, the programming of electronic loop equations becomes very simple. The INV statement takes the form

```
10   DIM A(4,4), B(4,4)
20   MAT B = INV (A)
```

In the above example, matrix B will be computed as the inverse of matrix A. Both matrices must be square, but this is an automatic consequence of loop equations, since there are as many equations as unknown variables.

Matrix inversion, like the other BASIC-PLUS matrix operations, does not operate on the elements of the row 0 and column 0 of the matrix; however, inversion destroys the previous contents of these elements. The operation MAT A = INV(A) is legal.

8.6 MATRIX PROGRAMMING EXAMPLES (OPTIONAL)

The following examples will help to illustrate how to program a computer to solve simultaneous equations.

Example 8.23 Write a program to solve for the following three unknowns: a, b, and c. Print out the answers.

$$2a - 3.1b + 6.8c = 12$$
$$-1.5a + 2.6b - 4.1c = -6.2$$
$$8.1a - 2.4b - 1.68c = 2.75$$

```
LIST
10      DIM A(3,3),B(3),C(3)
20      MAT INPUT A: MAT INPUT B
30      MAT A= INV(A)
40      MAT C=A*B
50      PRINT "A="C(1)
55      PRINT "B="C(2)
60      PRINT "C="C(3)
70      END

RUN
? 2,-3.1,6.8,-1.5,2.6,-4.1,8.1,-2.4,-1.68
? 12,-6.2,2.75
A= 1.39331
B= 1.97718
C= 2.25627
```

Example 8.24 Write a general program to solve simultaneous equations with up to 10 unknowns. Test the general program by solving the following two sets of simultaneous equations for all the unknowns:

Set 1:

$$1.57x - 3.21y + 6.95z = -21.2$$
$$-8.1x + 13.6y - 10.12z = 15.1$$
$$3.25x - 1.7y + 2.84z = -11.92$$

Set 2:

$$1.81s - 2.613t - 1.89u + 6.14v = 1.115$$
$$-2.31s + 9.25t - 7.811v = -15.2$$
$$1.825t + 4.96u - 2.85v = 4.55$$
$$3.59s - 5.62u + 1.42v = -7.61$$

```
LIST
10      INPUT"NUMBER OF UNKNOWNS ARE";N
20      DIM A(10,10),B(10),C(10)
25      MAT A=ZER(N,N): MAT B=ZER(N): MAT C=ZER(N)
30      PRINT"COEFFICIENTS OF MATRIX 'A' ARE" MAT INPUT A
40      PRINT"COEFFICIENTS OF MATRIX 'B' ARE" MAT INPUT B
50      MAT A=INV(A): MAT C=A*B
60      PRINT"IN THEIR RESPECTIVE ORDER, THE VALUES OF"
65      PRINT"THE UNKNOWNS ARE"
70      FOR K = 1 TO N: PRINT C(K) NEXT K
75      PRINT
80      END
```

Solution/set 1

```
RUN
NUMBER OF UNKNOWNS ARE? 3
COEFFICIENTS OF MATRIX 'A' ARE? 1.57,-3.21,6.95,-8.1,13.6,
-10.12,3.25,-1.7,2.84
COEFFICIENTS OF MATRIX 'B' ARE? -21.2,15.1,-11.92
IN THEIR RESPECTIVE ORDER, THE VALUES OF
THE UNKNOWNS ARE-1.67392  -2.85705  -3.99181
```

Solution/set 2

```
RUN
NUMBER OF UNKNOWNS ARE? 4
COEFFICIENTS OF MATRIX 'A' ARE? 1.81,-2.613,-1.89,6.14,
-2.31,9.25,0,-7.811,0,1.825,4.96,-2.85,3.59,0,-5.62,1.42
COEFFICIENTS OF MATRIX 'B' ARE? 1.115,-15.2,4.55,-7.61
IN THEIR RESPECTIVE ORDER, THE VALUES OF
THE UNKNOWNS ARE .227038  -1.73516  1.45465  -.175998
```

Example 8.25 Write a program to determine the magnitudes of I_1, I_2, and I_3 in Figure 8.20. Print out the results, identifying each current in mA values.

For the figure shown, the loop equations are as follows:

$$11.5I_1 - 4.7I_2 - 6.8I_3 = 22.5$$
$$-4.7I_1 + 41.7I_2 - 22I_3 = 0$$
$$-6.8I_1 - 22I_2 + 38.8I_3 = 0$$

```
LIST
10      DIM A(3,3), B(3), C(3)
20      PRINT"COEFFICIENTS OF MATRIX 'A' ARE"  MAT INPUT A
30      PRINT"COEFFICIENTS OF MATRIX 'B' ARE"  MAT INPUT B
40      MAT A= INV(A): MAT C= A*B
50      FOR K= 1 TO 3
60      PRINT"I"K"="C(K)"mA"
70      NEXT K
80      END

RUN
COEFFICIENTS OF MATRIX 'A' ARE? 11.5,-4.7,-6.8,-4.7,41.7,-22,
-6.8,-22,38.8
COEFFICIENTS OF MATRIX 'B' ARE? 22.5,0,0
I 1 = 2.88328 mA
I 2 = .844062 mA
I 3 = .983908 mA
```

Example 8.26 For any value of V in Figure 8.21, R can vary from 1 kΩ to 10 kΩ in 1 kΩ increments. Write a program to print out (in tabular form) the value of V chosen, the value of R in kΩ, and the value of the current through R in mA.

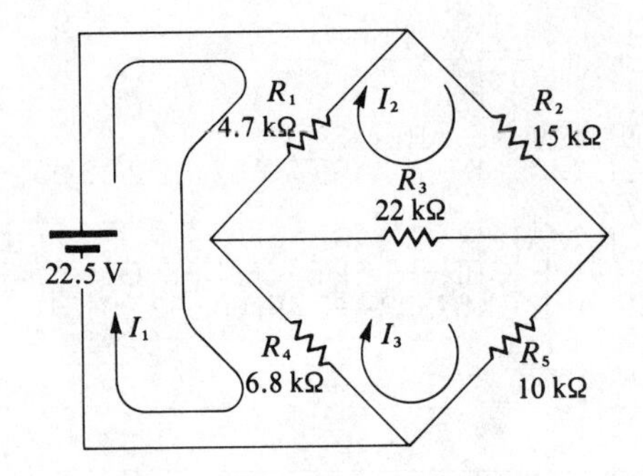

Figure 8.20 Reference Example 8.25

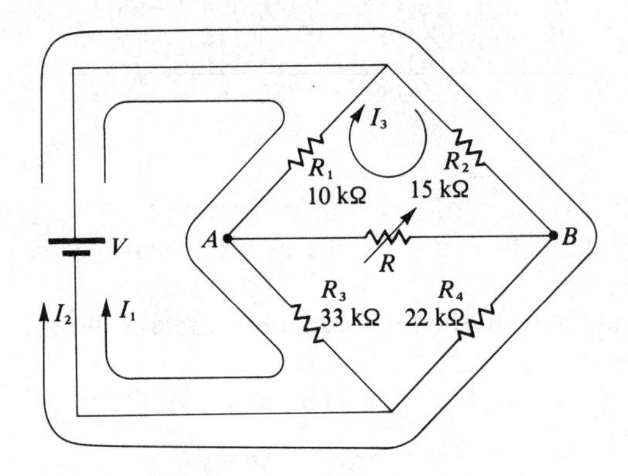

Figure 8.21 Reference Example 8.26

For the directions chosen, the loop equations are as follows:

$$43I_1 + 0I_2 - 10I_3 = V$$

$$0I_1 + 37I_2 + 15I_3 = V$$

$$-10I_1 + 15I_2 + (25 + R)I_3 = 0$$

```
LIST
10 DIM A(3,3), B(3), C(3)
20 INPUT"VALUE OF VOLTAGE SOURCE IS" V
30 B(1),B(2)=V: B(3)=0
35 PRINT"    V","R in K ohms","I(R) in mA"
40 FOR R= 1 TO 10
45 MAT READ A(3,3): LET A(3,3)=25+R
50 MAT A =INV(A): MAT C= A*B
60       PRINT V,R,C(3)
70       RESTORE
80       NEXT R
85       DATA 43,0,-10,0,37,15,-10,15,0
90       END
```

```
RUN
VALUE OF VOLTAGE SOURCE IS? 32.75
  V               R in K ohms     I(R) in MA
 32.75              1             -.321755
 32.75              2             -.30445
 32.75              3             -.288912
 32.75              4             -.274882
 32.75              5             -.262153
 32.75              6             -.250549
 32.75              7             -.23993
 32.75              8             -.230174
 32.75              9             -.221181
 32.75              10            -.212863
```

Note that the negative value of I (R) merely implies that the actual direction of current flow is from Point A to Point B.

SUMMARY

Application of determinants, like matrices, involves a mechanical manipulation of number arrays. The end result of this manipulation is to produce the solutions to the variables of simultaneous equations. Determinants solve for one solution at a time. If only one variable solution is desired, such as the current through one specific resistor, then determinants are easier to apply than a matrix. If all (or most) variable solutions are desired, then matrix solution represents the easier application, especially if computer assistance is available.

Matrix function and operational statements provide for ready application of computers to solve extensive arrays of simultaneous equations. These equations are usually the resultant of the application of Kirchhoff's voltage laws to the closed loops of complex circuits. The primary BASIC statements for matrix application are the INV statement to determine the inverse matrix and the matrix multiplication operator to form the product of two matrices.

PROBLEMS

Reference Sections 8.1 and 8.2

1. Solve the following determinants:

a. $\begin{vmatrix} 3 & 6 \\ -2 & 4 \end{vmatrix}$

b. $\begin{vmatrix} -2 & 5 \\ -4 & 12 \end{vmatrix}$

c. $\begin{vmatrix} -3 & 5 & 4 \\ 2 & -5 & -3 \\ 1 & -10 & 8 \end{vmatrix}$

d. $\begin{vmatrix} 3 & 6 & -9 \\ 2 & 5 & 18 \\ -4 & 1 & -21 \end{vmatrix}$

e. $\begin{vmatrix} 4 & 2 & 0 \\ 0 & -3 & -5 \\ -6 & 1 & -7 \end{vmatrix}$

f. $\begin{vmatrix} 4 & 2 & 0 & 2 \\ -1 & -6 & 9 & -3 \\ 0 & 8 & -5 & 0 \\ 3 & -4 & -1 & 7 \end{vmatrix}$

2. Solve the following equations, using determinants:

$$\begin{aligned} 6a - 5c &= -14 - 4b \\ -3b - 4c &= 11 + 2a \\ a - 2b + 4c &= -20 \end{aligned}$$

3. Write an expression for the solution of the following determinant, using cofactors of row 2:

$$x = \begin{vmatrix} a & d & g \\ b & e & h \\ c & f & i \end{vmatrix}$$

4. Solve the following equations, using determinants:

$$\begin{aligned} -w + 3x - z &= 3 \\ 2x - 3y + 4z &= 29 \\ w + x + y + z &= 2 \\ 2w + 3y - 4z &= -27 \end{aligned}$$

5. Using determinants, solve for the current through each resistor in the circuit of Figure 8.22.

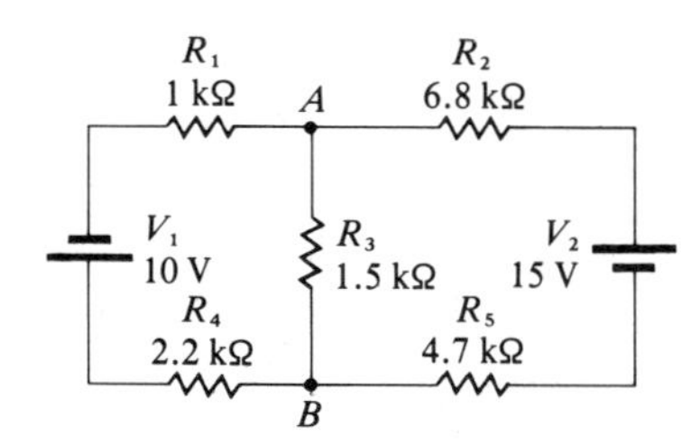

Figure 8.22 Reference Problem 5

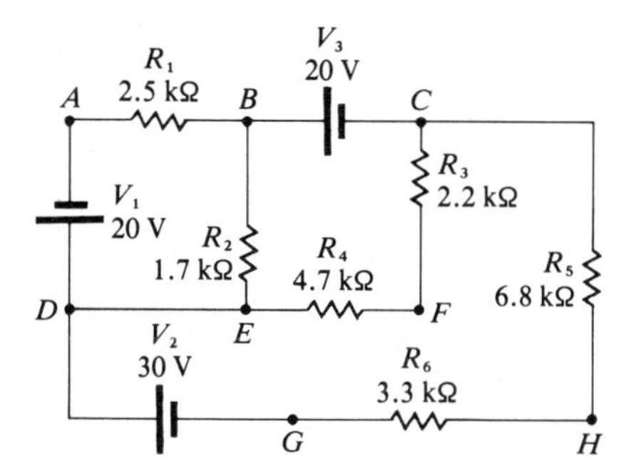

Figure 8.23 Reference Problems 6, 7, and 8

6. Using determinants, solve for the current in each resistor of Figure 8.23.
7. Find V_{BG} in Figure 8.23 by the following path:

$$V_{BE} + V_{EF} + V_{FC} + V_{CH} + V_{HG}$$

8. Find V_{BG} in Figure 8.23 by the following path:

$$V_{BA} + V_{AD} + V_{DG}$$

   Verify that this is the same answer as in Problem 7.
9. Using determinants, solve for the current through each resistor in Figure 8.24.
10. Find V_B with reference to ground in Figure 8.24. Find V_F with reference to ground. What is V_{BF}?
11. Determine V_{BF} in Figure 8.24 by the following path:

$$V_{BC} + V_{CE} + V_{ED} + V_{DF}$$

   Verify that this is the same answer determined for V_{BF} in Problem 10.

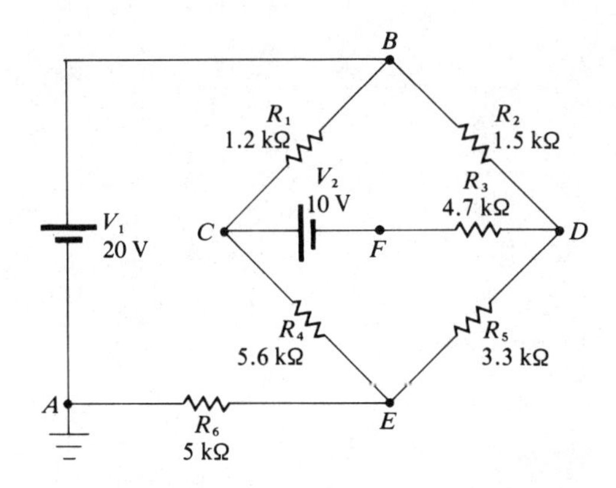

Figure 8.24 Reference Problems 9, 10, and 11

Reference Sections 8.3 and 8.4

12. Find the inverse matrix of the following:

a. $\begin{bmatrix} 2 & -4 & -1 \\ -1 & -2 & 7 \\ 3 & 6 & 5 \end{bmatrix}$

b. $\begin{bmatrix} 0 & 2 & -1 \\ -2 & -3 & 6 \\ 1 & 5 & -4 \end{bmatrix}$

c. $\begin{bmatrix} 1 & -2 & -7 \\ -3 & 4 & 0 \\ 5 & -6 & 8 \end{bmatrix}$

d. $\begin{bmatrix} 1 & -1 & 1 \\ -1 & 1 & -1 \\ 1 & -1 & 1 \end{bmatrix}$

13. Solve the equations of Problem 2 using matrix analysis. Verify that the answers are the same as were determined in Problem 2.
14. Repeat Problem 6, but solve for the currents using matrix solution.
15. Repeat Problem 9, but solve for the currents using matrix solution.

Reference Sections 8.5 and 8.6 (Optional Computer Problems)

16. In the bridge circuit of Figure 8.25, print out the current in mA through each resistor if V is 28 volts.
17. Print out a table of V and the current through each resistor of Figure 8.25 if V varies from 10 to 17.5 volts in 0.5 volt increments.

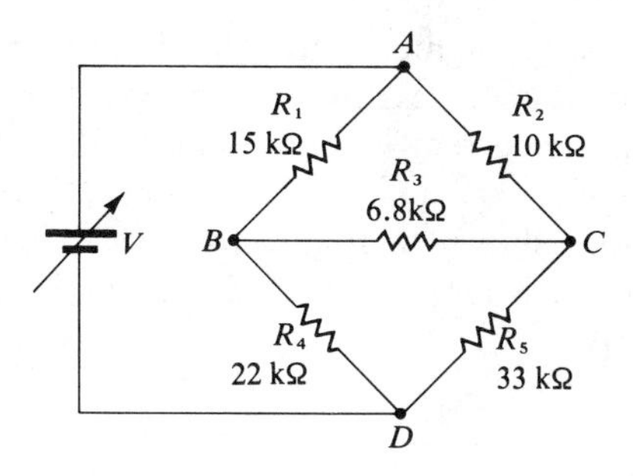

Figure 8.25 Reference Problems 16 through 20

18. Assume a value of 30 volts for V in Figure 8.25. Initially, conventional current flow is from point C to point B for the resistive values shown. If R_5 is decremented in 1 kΩ increments, determine the lowest whole kΩ value just before conventional current flow changes direction and flows from point B to point C. Print out this value of R_5.
19. Reference Figure 8.25. Assume V varies from 1 to 10 volts in 1 volt increments, and R_2 varies from 1 to 10 kΩ in 1 kΩ increments. Print out a table of these 100 combinations, listing V, R_2 in kΩ, and the current through each resistor in mA. *Note*: Use a descriptive heading for the units and arrange the table in pure numeric format.
20. Reference Figure 8.25. Assume V is 52.75 volts. Initially, conventional current flow is from point C to point B for the resistive values shown. If R_5 is decremented, determine the value of R_5 (within 1 ohm) when $I(R_3)$ drops to 0.135 mA in a direction from point C to point B.

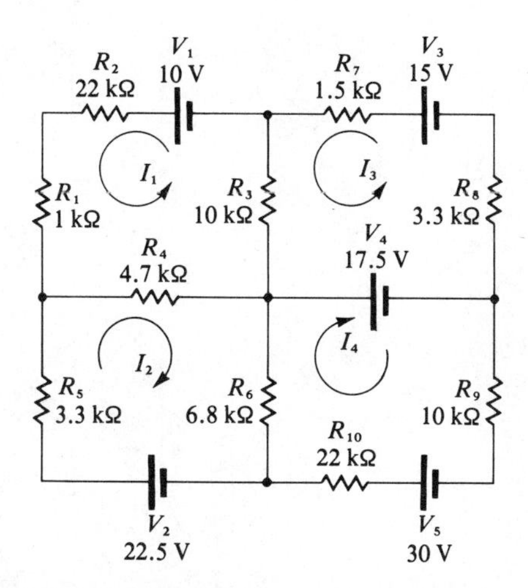

Figure 8.26 Reference Problem 21

Hint: Use successively tighter loops; that is, decrement in 1 kΩ units until the current value is less than 0.135 mA, then increment in 100 ohm units until the current is more than 0.135 mA, then decrement in 10 ohm units, etc.

21. For the directions shown in Figure 8.26, write a program to print out the value of each loop current, identifying each current and its magnitude in mA.

CHAPTER 9

NETWORK THEOREMS

9.1 INTRODUCTION

In earlier chapters, network problems were solved by writing equations based on Kirchhoff's laws. This resulted in a series of equations, which in general, produced as many unknowns as there were loops in the total network. Using standard mathematical solutions for simultaneous equations, the student could solve for any or all of the unknowns. No attempts were made to simplify the networks prior to their solution.

This chapter introduces the application of a series of so-called network theorems for solving various problems. The specific theorem applied usually depends on the electronic network being analyzed, and on the particular solutions being sought.

Generally speaking, network theorems imply the application of a set of rules (determined by the theorem being utilized) to a circuit, the end result of which is to simplify the circuit and thus make its solution easier. This is not unlike the concept of applying a set of rules to the solution of a determinant or matrix as was studied in Chapter 8.

With continued application of these theorems, the specific theorem most applicable to the solution of a problem will become more apparent.

9.2 MILLMAN'S THEOREM

In the preceding chapters, the simple voltage divider was studied. Recalling methods employed earlier, which were essentially an amplification of the principle that voltage is dropped in direct proportion to resistance, permits us to calculate the voltage at point A (with respect to ground) in Figure 9.1.

Application of the voltage divider principle is independent of the current, therefore either voltage source may be chosen as the reference point.

Choosing -20 V as reference,

$$V_A = -20 + \frac{50}{50 + 90}[50 - (-20)]$$

$$= -20 + \frac{5}{14}(70) = 5 \text{ volts}$$

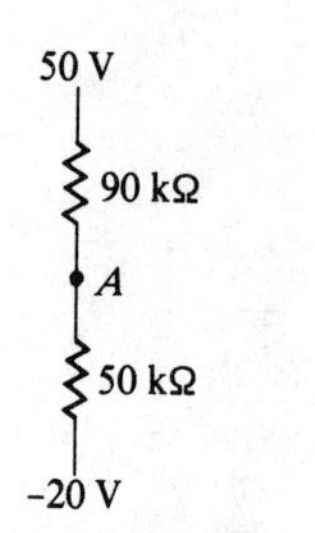

Figure 9.1 Simple Voltage Divider

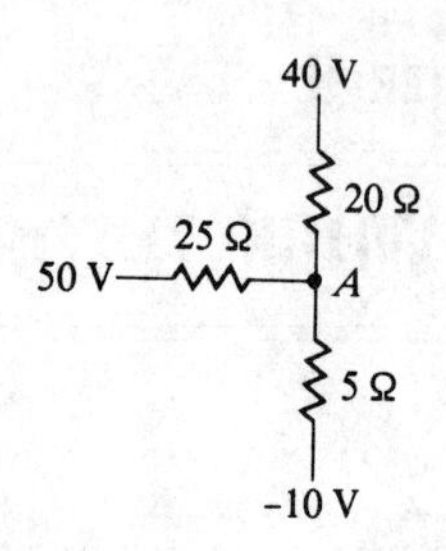

Figure 9.2 Complex Voltage Divider

Choosing $+50$ V as reference,

$$V_A = 50 - \frac{90}{50 + 90}[50 - (-20)]$$

$$= 50 - \frac{9}{14}(70) = 5 \text{ volts}$$

These solutions are not difficult, because current can only flow through one path between the two sources of voltage. But suppose there are several paths for current, such as in Figure 9.2. It is in these types of applications that Millman's theorem is extremely useful.

> "The voltage at the common point where several branches meet is the ratio of the sum of the currents through each branch, calculated as though that were the only branch present, to the sum of the conductances of each branch, again calculated as though that were the only branch present."

Expressed mathematically,

$$\boxed{V_A = \frac{\sum I}{\sum G}} \tag{9.1}$$

Example 9.1 Determine the voltage at point A in Figure 9.2.

If the 20 ohm resistor were the only branch present, and point A were temporarily considered at zero volts potential, then

$$I_{20} = \frac{40 \text{ V}}{20\ \Omega} = 2 \text{ amperes}$$

$$G_{20} = \frac{1}{20\ \Omega} = 0.05 \text{ siemens}$$

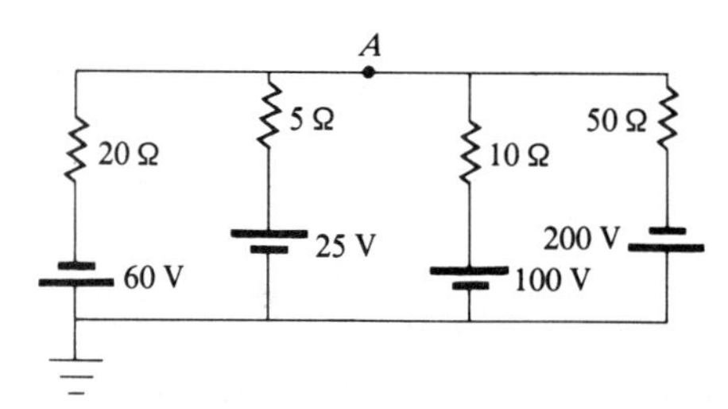

Figure 9.3 Reference Example 9.2

In a similar fashion,

$$I_{25} = \frac{50}{25} = 2 \text{ amperes}$$

$$G_{25} = \frac{1}{25} = 0.04 \text{ siemens}$$

and

$$I_5 = \frac{-10}{5} = -2 \text{ amperes}$$

$$G_5 = \frac{1}{5} = 0.2 \text{ siemens}$$

Applying Eq. (9.1),

$$V_A = \frac{\sum I}{\sum G} = \frac{(2 + 2 - 2) \text{ ampere}}{(0.05 + 0.04 + 0.2) \text{ siemens}}$$

$$= \frac{2}{0.29} \approx 6.9 \text{ volts}$$

Example 9.2 Find the voltage at point A in Figure 9.3.

$$V_A = \frac{\sum I}{\sum G} = \frac{\frac{-60}{20} + \frac{25}{5} + \frac{100}{10} + \frac{-200}{50}}{\frac{1}{20} + \frac{1}{5} + \frac{1}{10} + \frac{1}{50}}$$

$$= \frac{-3 + 5 + 10 - 4}{0.05 + 0.2 + 0.1 + 0.02} = 21.6 \text{ volts}$$

Example 9.3 What should be the voltage V_X in Figure 9.4 so that the voltage at point A equals $+5$ V with respect to ground?

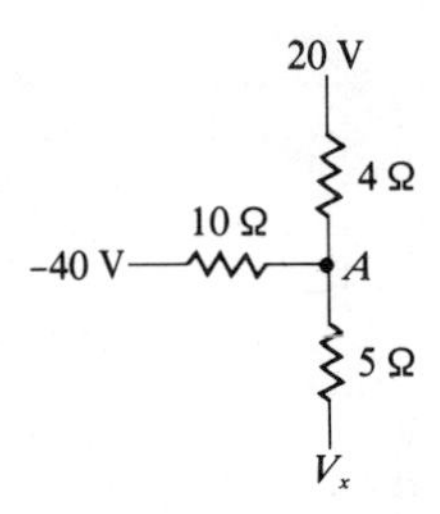

Figure 9.4 Reference Example 9.3

$$5 = V_A = \frac{\frac{20}{4} + \frac{-40}{10} + \frac{V_X}{5}}{\frac{1}{4} + \frac{1}{10} + \frac{1}{5}}$$

Clearing the fraction results in

$$5 = \frac{100 - 80 + 4V_X}{5 + 2 + 4}$$

$$55 = 20 + 4V_X$$

$$4V_X = 35$$

$$V_X = 8.75 \text{ volts}$$

The preceding examples employed unusually small resistors, conveniently chosen so as to stress the principles and not become involved in the mathematics. When encountering resistors in the kilohm or megohm range, many students at first tend to write all the zeros associated with these numbers. If it is convenient to convert all resistors into the same unit, then calculations involving Millman's theorem can be simplified by deleting the zeros associated with the unit designation. This is valid, as the same power of 10 will be present in both the numerator and the denominator of the resultant fraction, and therefore cancel.

Example 9.4 Determine the voltage at point A in Figure 9.5. Note that the 470 ohm resistor and the 680 ohm resistor are converted to kilohms so that all resistors may be expressed in like units.

$$V_A = \frac{\frac{12.5}{2.2} + \frac{0}{3.3} + \frac{-3.5}{0.68} + \frac{-7.25}{0.47}}{\frac{1}{2.2} + \frac{1}{3.3} + \frac{1}{0.68} + \frac{1}{0.47}}$$

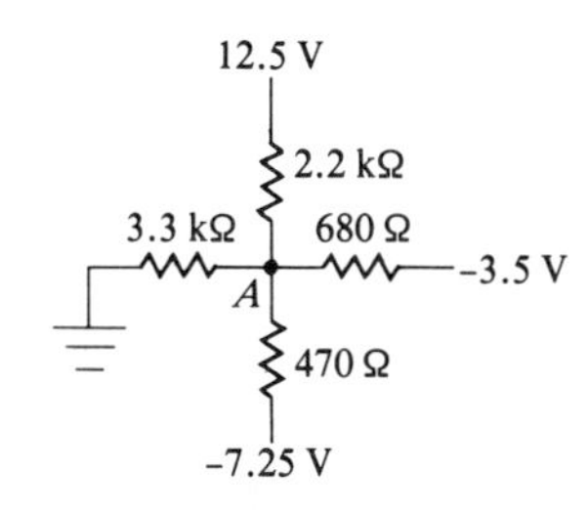

Figure 9.5 Reference Example 9.4

$$= \frac{5.68 + 0 - 5.15 - 15.43}{0.45 + 0.30 + 1.47 + 2.13}$$

$$= \frac{-14.9}{4.35} = -3.43 \text{ volts}$$

Example 9.5 In the circuit of Figure 9.6(a), it is desired to find V_0 for the input shown. The source V varies between the two dc voltage levels of $+5$ volts and $+10$ volts.

When the input is at $+5$ volts,

$$V_0 = \frac{\frac{5}{3.3} + \frac{-5}{2.2} + \frac{0}{1.5}}{\frac{1}{3.3} + \frac{1}{2.2} + \frac{1}{1.5}}$$

$$= \frac{1.52 - 2.27}{0.30 + 0.45 + 0.67} = \frac{-0.75}{1.42} = -0.53 \text{ volt}$$

When the input is at $+10$ volts,

$$V_0 = \frac{\frac{10}{3.3} + \frac{-5}{2.2}}{1.42} = \frac{3.03 - 2.27}{1.42} = 0.53 \text{ volt}$$

Figure 9.6(b) illustrates V_0.

9.3 THEVENIN'S THEOREM

It is often necessary to know the current through only one element in a circuit. If the circuit has a simple series-parallel configuration, Ohm's law is usually sufficient to determine the answer. If, however, the circuit is a complex arrangement, such as the Wheatstone bridge, higher forms of

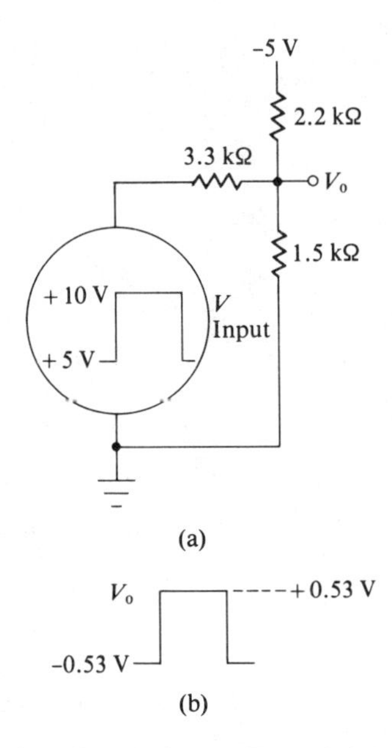

Figure 9.6 Reference Example 9.5

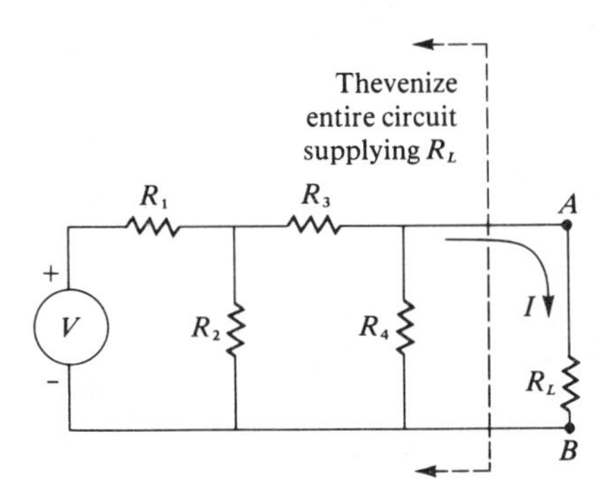

Figure 9.7 Complex Circuit Supplying R_L

mathematical manipulation, such as determinants or matrix analysis, must be used to solve the problem. If this latter approach is not advisable, a method of "circuit reduction" is first necessary to reduce a complex circuit into a more simple series configuration. This would then permit the application of Ohm's law, rather than the use of determinants or matrix analysis. A theorem has been developed to accomplish this task—to reduce all elements of a complex circuit other than the element in question to a single source voltage and a single source resistance supplying power to the resistance(s) in question. This theorem is called Thevenin's theorem. As an illustration of simplifying a complex circuit, consider Figure 9.7. Thevenin's theorem states:

> If R_L were removed from between points A and B, and a voltmeter connected between these points, then the indicated voltage is the value of the Thevenin generator. Figure 9.8(a) illustrates this measurement.
>
> If the original source is replaced with its own internal resistance, and an ohmmeter connected between points A and B (R_L is still removed), then the indicated resistance is the internal resistance of the Thevenin generator. Figure 9.8(b) illustrates this measurement.

Replace the original circuit (everything except R_L) with this new generator and new resistance. Connect R_L back between points A and B. The equivalent circuit is shown in Figure 9.9. The current through R_L is the same in both figures, but notice how much simpler this latter circuit is than was the original circuit of Figure 9.7.

Looking back from R_L into either circuit, both circuits are indistinguishable as to their effects on R_L. It is usually assumed, when replacing a source with its internal resistance (for taking the ohmmeter readings), that voltage sources are replaced with a short circuit and current sources are replaced with an open circuit. If either case is not valid, then the source's actual internal resistance should be substituted for the source before taking the ohmmeter measurement.

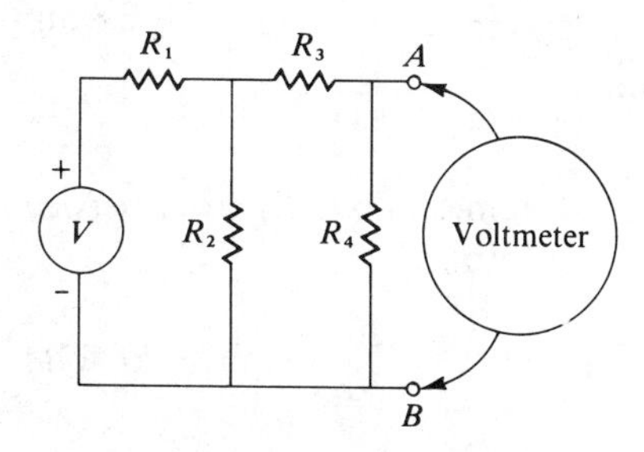

Figure 9.8(a) Determining V_{th}

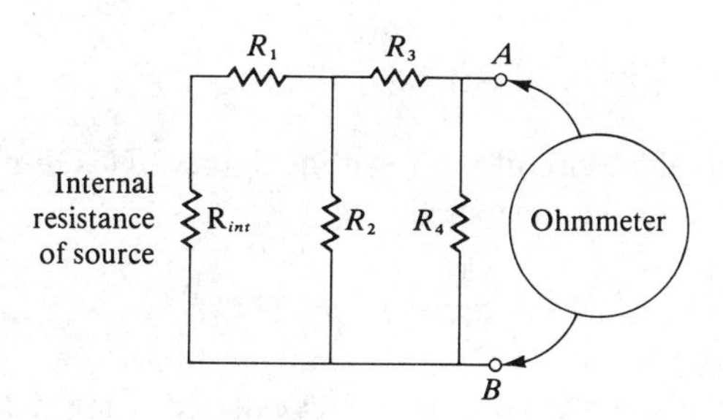

Figure 9.8(b) Determining R_{th}

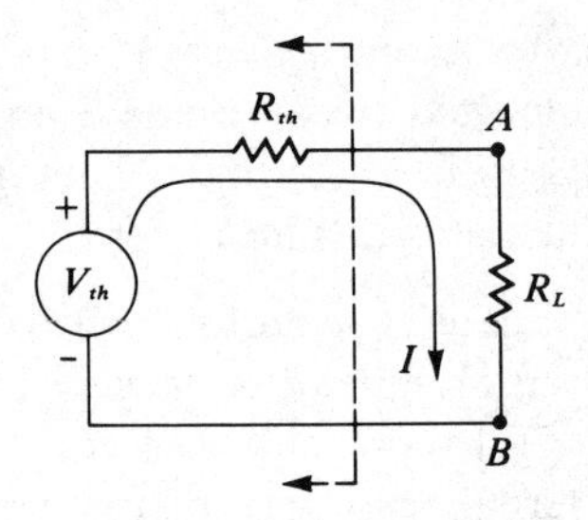

Figure 9.9 Figure 9.7 Simplified by Thevenin Theorem

Example 9.6 Determine the current through R_L in Figure 9.10(a), first by loop equations and then by using Thevenin's theorem.

First write the loop equations.

$$50I_1 - 20I = 50$$
$$-20I_1 + 33I = 0$$

Solve for I the current through R_L.

$$I = \frac{\begin{vmatrix} 50 & 50 \\ -20 & 0 \end{vmatrix}}{\begin{vmatrix} 50 & -20 \\ -20 & 33 \end{vmatrix}} = \frac{1000}{1250} = 0.8 \text{ ampere}$$

Using Thevenin's theorem, remove R_L from between points AB. Measure V_{AB}, as in Figure 9.10(b).

$$V_{\text{Th}} = V_{AB} = \frac{20}{50}(50 \text{ V}) = 20 \text{ volts}$$

After replacing the 50 V source with a short circuit, measure R_{Th} as in Figure 9.10(c).

$$R_{\text{Th}} = 8 + \frac{20 \times 30}{20 + 30} = 20 \text{ ohms}$$

The Thevenin equivalent circuit is shown in Figure 9.11. Ohm's law may now be applied to find the current through R_L

$$I_{RL} = \frac{20 \text{ V}}{25\ \Omega} = 0.8 \text{ ampere}$$

Note that this is the same answer as was calculated using determinants.

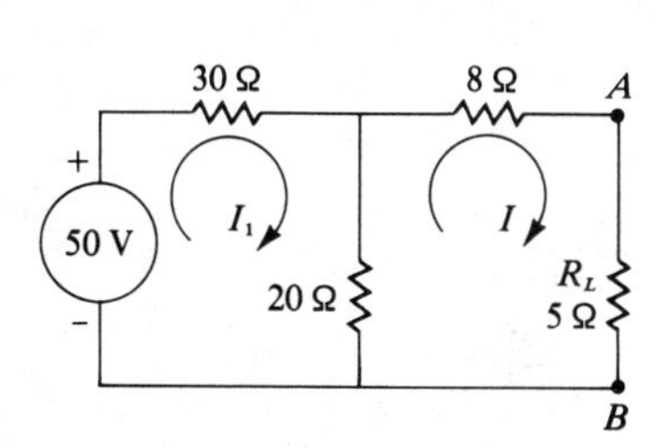

Figure 9.10(a) Reference Example 9.6

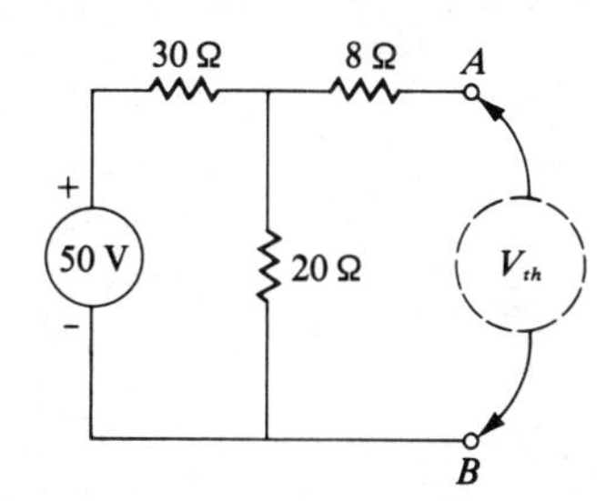

Figure 9.10(b) V_{th} of Example 9.6

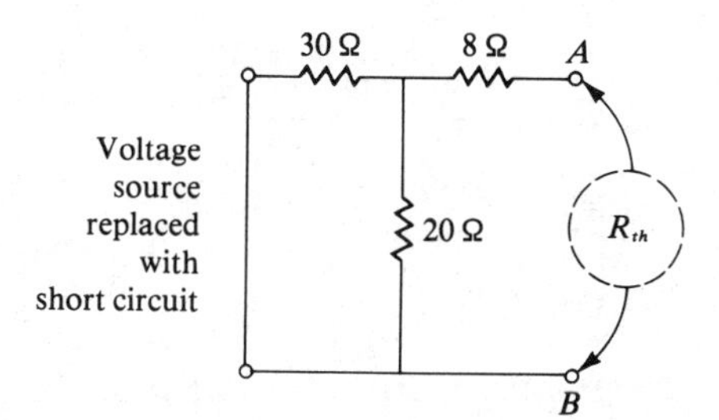

Figure 9.10(c) R_{th} of Example 9.6

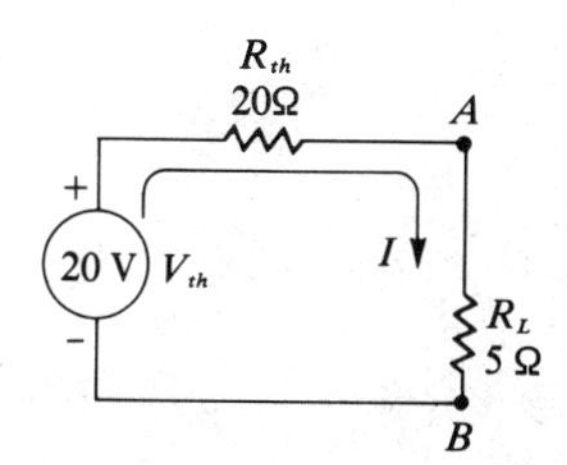

Figure 9.11 Figure 9.10(a) Simplified by Thevenin Theorem

Example 9.7 Determine the current through R_L in Figure 9.12 by using Thevenin's theorem.

Removing R_L from between points AB reduces the circuit to the simple series-parallel arrangement of Figure 9.13. Consider point D (the reference point) at ground potential.

$$V_A = \frac{30}{50}(100 \text{ V}) = 60 \text{ volts}$$

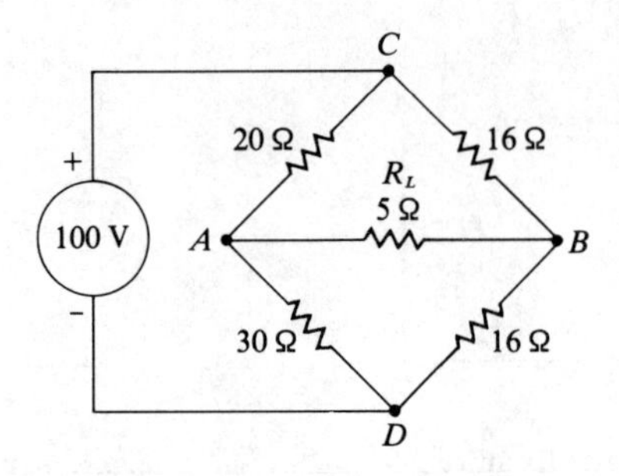

Figure 9.12 Reference Example 9.7

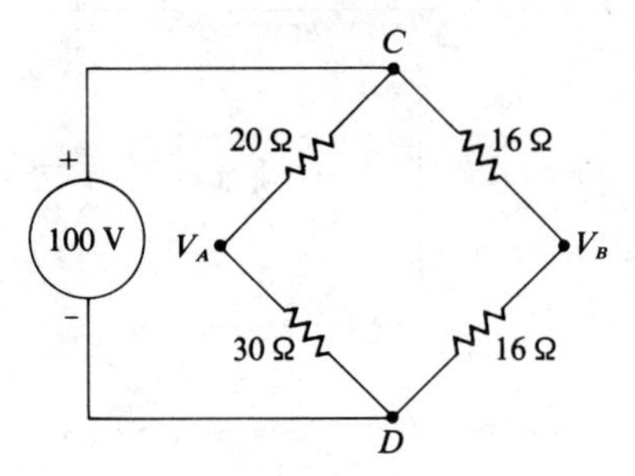

Figure 9.13 Determining Thevenin Voltage of Figure 9.12

Similarly,

$$V_B = \frac{16}{32}(100 \text{ V}) = 50 \text{ volts}$$

$$\therefore \quad V_{\text{Th}} = V_{BA} = 60 - 50 = 10 \text{ volts}$$

Apply the second step of Thevenin's theorem. Short points C and D together; that is, replace the voltage generator by its internal resistance (considered here as a short) and measure the resistance between points A and B. This is illustrated in Figure 9.14.

$$R_{\text{Th}} = (20 \| 30) + (16 \| 16)$$
$$= 12 + 8 = 20 \text{ ohms}$$

Replacing all the circuit (except R_L) with the Thevenin generator and Thevenin resistance, and then replacing R_L back between points AB, results in the circuit of Figure 9.15. I_{R_L} is now easily calculated:

$$I_{R_L} = \frac{V_{\text{Th}}}{R_{\text{Th}} + R_L} = \frac{10}{25} = 0.4 \text{ ampere}$$

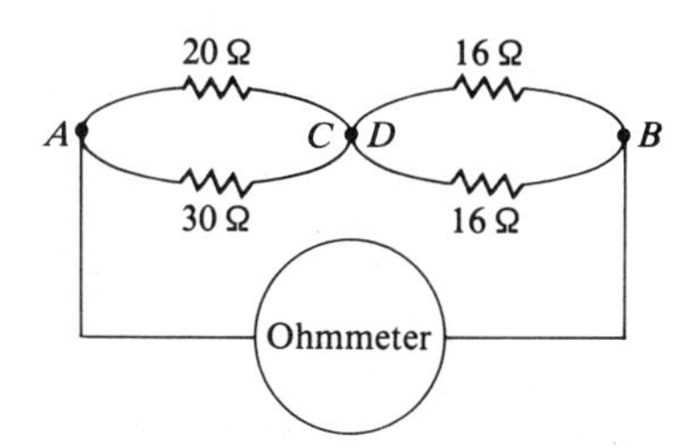

Figure 9.14 Determining Thevenin Resistance of Figure 9.12

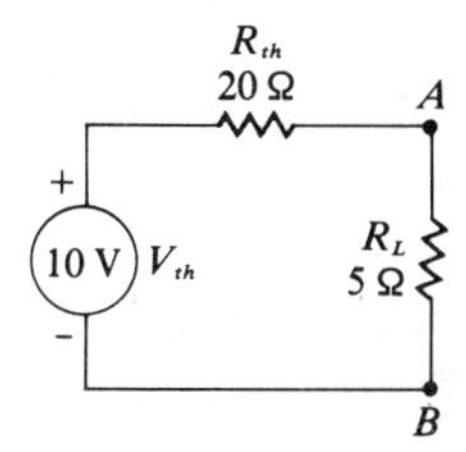

Figure 9.15 Figure 9.12 Simplified by Thevenin Theorem

Note the simplicity of solving this problem by Thevenin's theorem rather than by determinants.

The values and configurations of the preceding examples were again chosen so as to illustrate the mechanics of application. The following example is more practically oriented toward the solution of modern circuits.

Example 9.8 Figure 9.16 illustrates a simple dc biasing arrangement for a transistor. Determine the value of I_B.

Current flows from +10 volts through the 47 kΩ resistor to point A. From there, current splits up as I_B and I_2. This example makes use of both Millman's theorem and Thevenin's theorem. If the transistor circuit between points A and B is removed, and then considered the "load resistor" for a Thevenin's circuit, Figure 9.17 results.

$$V_A = V_{\text{Th}} = \underbrace{\frac{\dfrac{10}{47} + \dfrac{0}{10}}{\dfrac{1}{47} + \dfrac{1}{10}}}_{\text{Millman's theorem}} = 1.75 \text{ volts}$$

$$R_{\text{Th}} = 10 \text{ k}\Omega \| 47 \text{ k}\Omega = \frac{10 \times 47}{10 + 47} \text{ k}\Omega = 8.25 \text{ k}\Omega$$

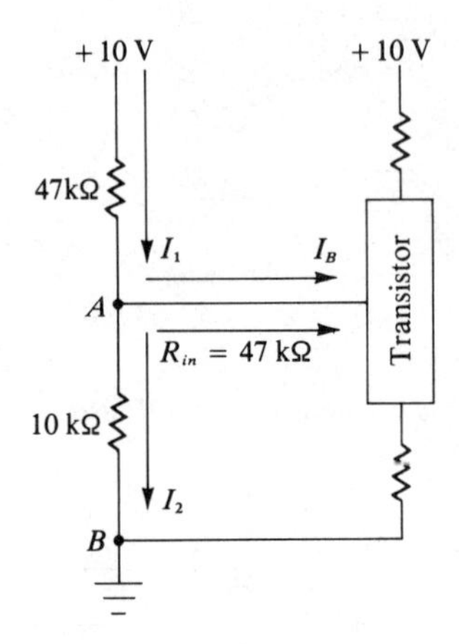

Figure 9.16 Reference Example 9.8

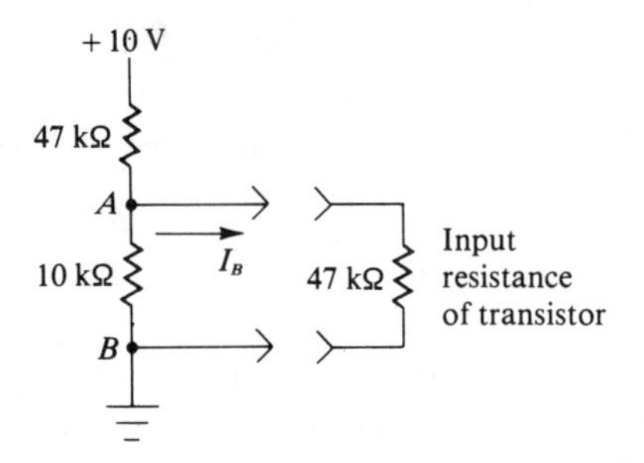

Figure 9.17 Separate Load (Transistor) from Input Circuitry

Combining these calculations results in the circuit of Figure 9.18. Solve for I_B.

$$I_B = \frac{1.75 \text{ V}}{8.25 \text{ k}\Omega + 47 \text{ k}\Omega} = \frac{1.75}{55.25} \text{ mA} = 31.67 \ \mu\text{A}$$

9.4 NORTON'S THEOREM

In Section 9.3, you learned how to reduce all elements of a complex circuit (except the element(s) in question) into a single voltage source and its associated series resistance, to permit the use of Ohm's law when solving for a given unknown, instead of the more complex mathematics of determinants or matrix analysis. The method that explains this reduction process is known as Thevenin's theorem. However, sometimes it is desirable to reduce these same complex arrangements into a single current source with

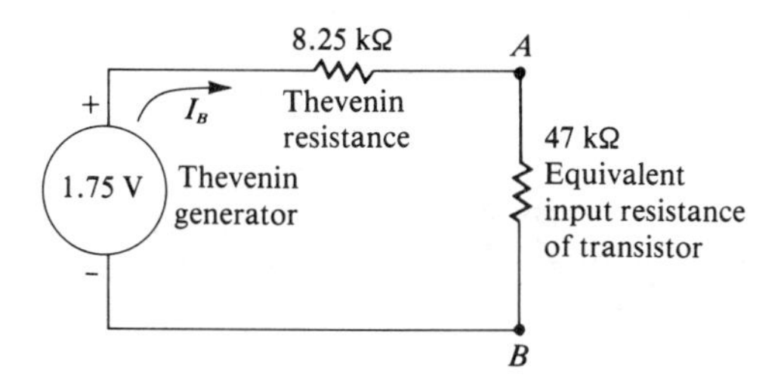

Figure 9.18 Thevenin Equivalent of Example 9.8

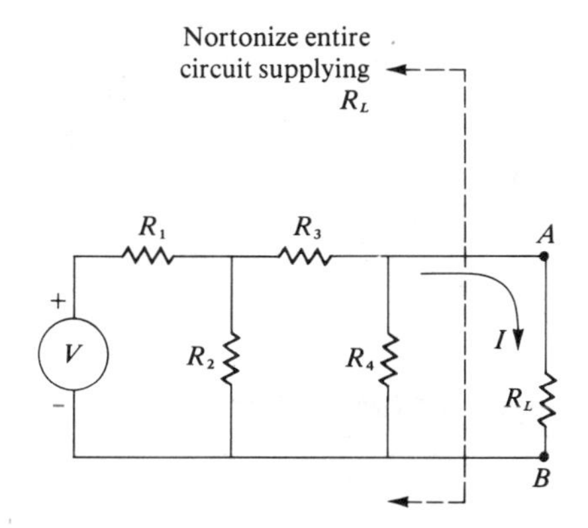

Figure 9.19 Complex Circuit Supplying R_L

its associated parallel resistance. When connected with the element in question, the inverse current ratio law may then be utilized to determine the desired unknown quantities. This theorem is known as Norton's theorem. With reference to Figure 9.19, Norton's theorem states:

> If R_L were removed from between points A and B, and a current meter connected between these points, then the indicated current is the value of the Norton generator. Figure 9.20(a) illustrates this measurement.
>
> If the original source is replaced with its own internal resistance, and an ohmmeter connected between points A and B (R_L is still removed), then the indicated resistance is the resistance of the Norton generator. Figure 9.20(b) illustrates this measurement.

Notice that the measurement of R_{Nor} is identical to the measurement of R_{Th} of Figure 9.8(b). This is to be expected, since the conversions between voltage and current source resistances differ only in their series or parallel

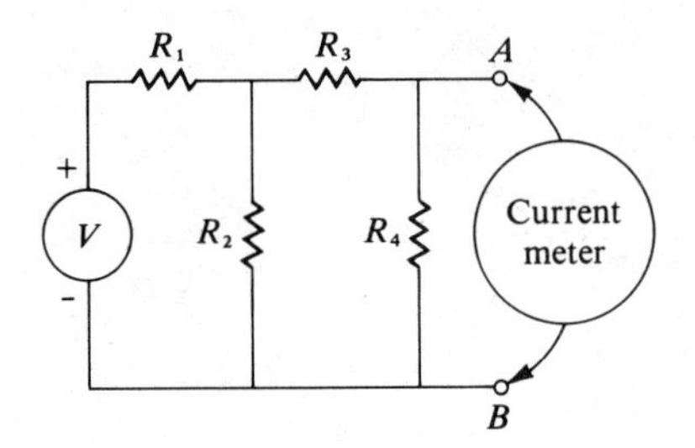

Figure 9.20(a) Determining I_{nor}

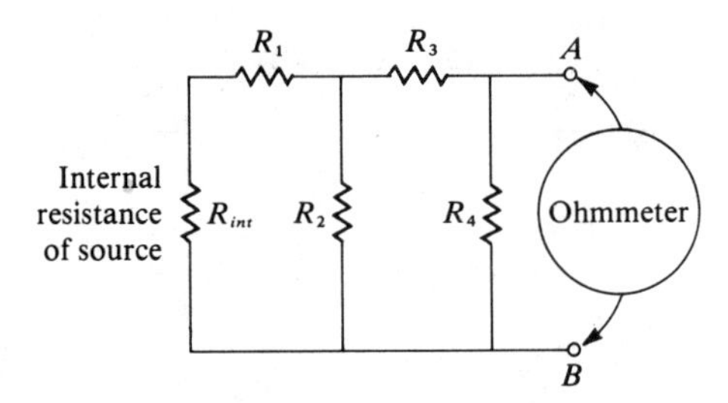

Figure 9.20(b) Determining R_{nor}

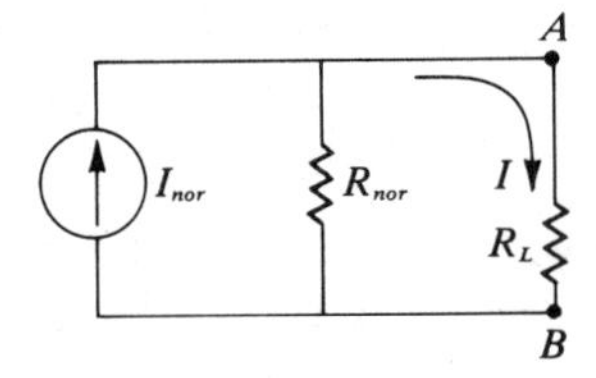

Figure 9.21 Complex Circuit of Figure 9.19 Simplified by Norton's Theorem

placement (in relation to their respective source), and not in their magnitude. As was true in the case for the ohmmeter measurement of R_{Th}, the original source(s) supplying R_L is replaced with its internal resistance(s) when making the ohmmeter measurement for R_{Nor}. Following completion of the two Norton measurements illustrated in Figure 9.20, the original complex circuit of Figure 9.19 can be redrawn as Figure 9.21. Notice that the calculation of I_{R_L} is now reduced to a simple application of the inverse current ratio law.

Example 9.9 Determine the current through R_L of Figure 9.10(a) using Norton's theorem. Compare this answer with I_{R_L} of Example 9.6.

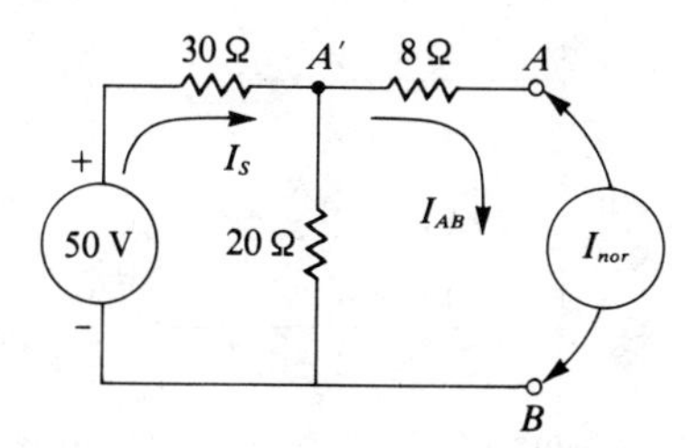

Figure 9.22(a) Determining I_{nor}

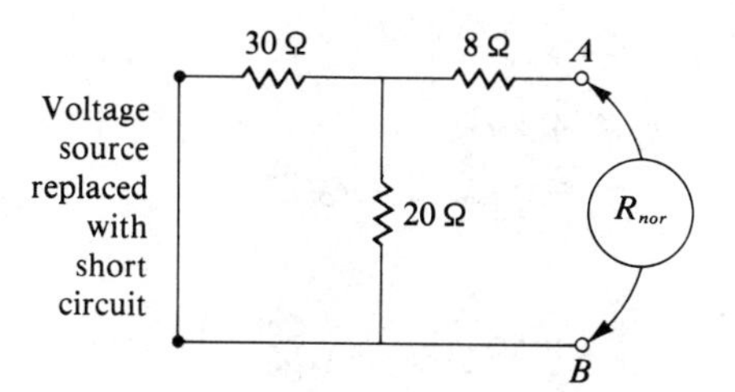

Figure 9.22(b) Determining R_{nor}

Using Norton's theorem, remove R_L from between points AB. Measure I_{AB} as in Figure 9.22(a).

$$20 \text{ ohms} \parallel 8 \text{ ohms} = \frac{20 \times 8}{20 + 8} = \frac{40}{7} \text{ ohms}$$

$$V_{A'B} = \frac{\frac{40}{7}}{30 + \frac{40}{7}} (50 \text{ V}) = 8 \text{ volts}$$

$$I_{AB} = I_{\text{Nor}} = \frac{V_{A'B}}{8} = \frac{8 \text{ V}}{8 \ \Omega} = 1 \text{ ampere}$$

From Figure 9.22(b),

$$R_{\text{Nor}} = 8 + \frac{20 \times 30}{20 + 30} = 20 \text{ ohms}$$

The equivalent circuit is redrawn as Figure 9.23. Using the inverse current ratio law,

$$I_{R_L} = \frac{20}{20 + 5} (1 \text{ ampere}) = 0.8 \text{ ampere}$$

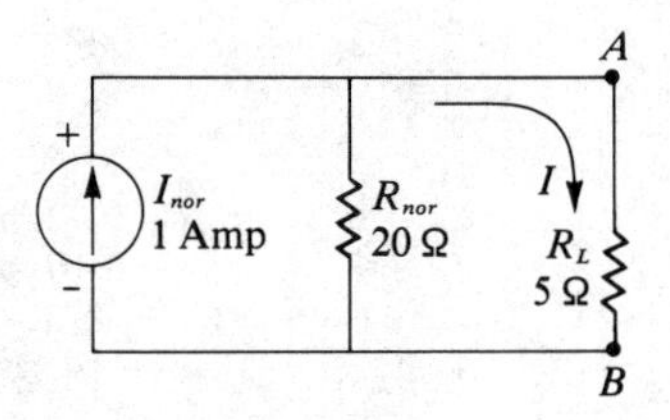

Figure 9.23 Figure 9.10(a) Simplified by Norton's Theorem

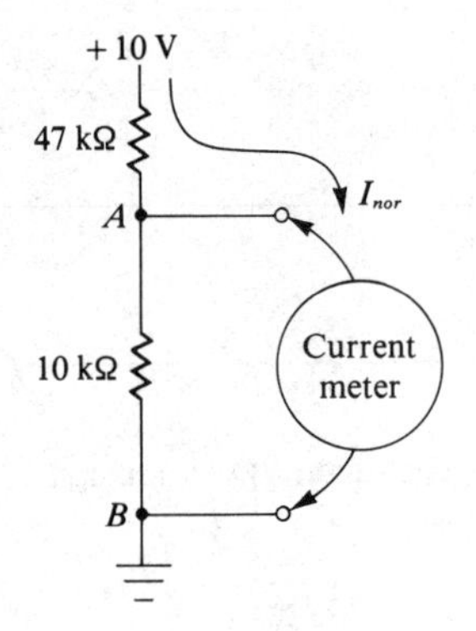

Figure 9.24 Determining I_{nor} of Figure 9.16

Notice that this is the same answer as determined by Thevenin's theorem in Example 9.6.

Example 9.10 Using Norton's theorem, determine I_B of Figure 9.16. Compare this to the answer of Example 9.8.

Removing the transistor and substituting a current generator between points AB results in Figure 9.24. The current meter shorts the 10 kΩ resistor, therefore

$$I_{\text{Nor}} = \frac{10 \text{ V}}{47 \text{ k}\Omega} = \frac{10}{47} \text{ mA}$$

In a manner identical to Example 9.8,

$$R_{\text{Nor}} = \frac{10 \text{ k}\Omega \times 47 \text{ k}\Omega}{10 \text{ k}\Omega + 47 \text{ k}\Omega} = 8.25 \text{ k}\Omega$$

The equivalent circuit is redrawn as Figure 9.25. Using the inverse current ratio law,

$$I_B = \frac{8.25}{8.25 + 47}\left(\frac{10}{47} \text{ mA}\right) = 31.77 \ \mu\text{A}$$

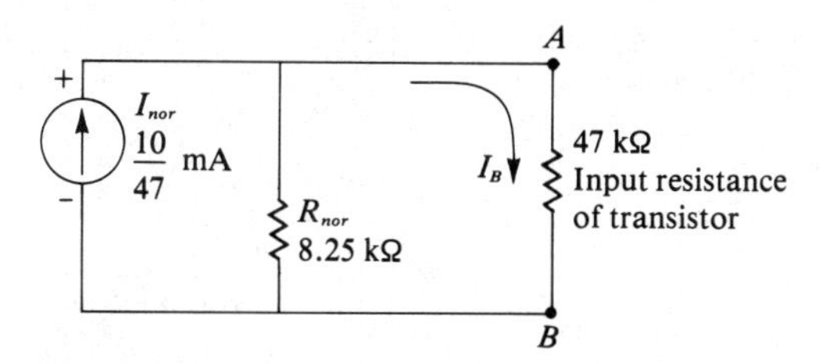

Figure 9.25 Figure 9.16 Simplified by Norton's Theorem

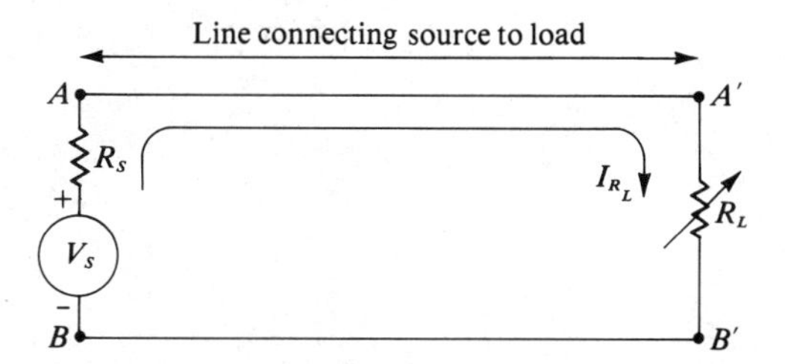

Figure 9.26 Matching R_L for Maximum Power

This compares to the answer previously determined. Minor differences are due solely to the roundoff of numbers during calculations.

9.5 MAXIMUM POWER TRANSFER THEOREM

In Figure 9.26, V_S is a voltage source connected to a power delivery line at points *AB*. R_S is the internal resistance of the source. At the other end of the line (connected to points *A'B'*) is a variable load. A question now arises: "What should be the value of R_L so that the maximum amount of power is delivered to it?" This question will first be solved mathematically in general terms, and then answered specifically by computer computations. From Figure 9.26,

$$I_{R_L} = \frac{V_S}{R_S + R_L}$$

$$P_{R_L} = I_{R_L}^2 \times R_L = \frac{V_S^2 R_L}{(R_S + R_L)^2} \qquad (9.2)$$

At this point it is appropriate to discuss the general shape of a graph relating the power delivered to a load resistor vs the value of the load resistance. At the one extreme, if R_L were zero ohms, there would be no voltage across it. From the equation

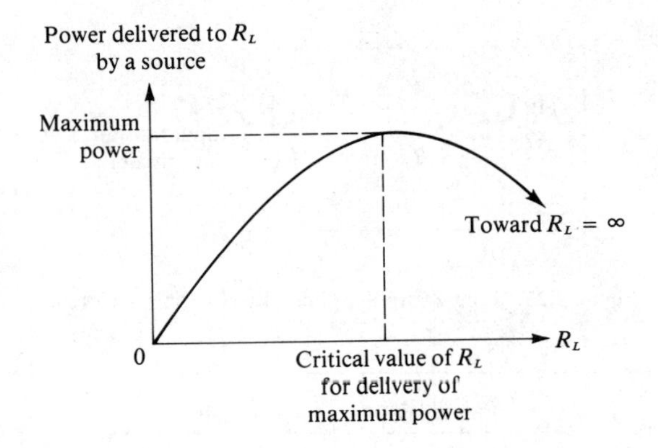

Figure 9.27 Graph of P_{R_L} vs R_L

$$P_{R_L} = \frac{V_{R_L}^2}{R_L}$$

it can be seen that P_{R_L} must equal zero watts if V_{R_L} is equal to zero volts. At the other extreme, if R_L were infinite in value, then I_{R_L} would be zero. From the equation

$$P_{R_L} = I_{R_L}^2 \times R_L$$

it can again be seen that P_{R_L} is zero watts if I_{R_L} is zero amperes. The following is then concluded:

> The power in R_L is zero watts when R_L is zero ohms, rises to a peak as R_L increases to some critical value, then decreases again toward zero watts as R_L continues to increase toward infinite ohms resistance.

Figure 9.27 illustrates a generalized graph of this concept.

Approaching the problem of Figure 9.26 via computer solution, consider V_S to equal 20 volts, R_S to equal 10 kΩ, and R_L to vary from zero to 20 kΩ. The following program will calculate the power delivered to the load resistance for various values of R_L, as it is incremented from 0 to 20 kΩ in 0.5 kΩ increments.

```
LIST
10      PRINT"RL IN OHMS    PWR IN MW": PRINT
20      LET V=20: R1=10E3
30      FOR R=0 TO 20E3 STEP .5E3
40      P=(R*(V/(R1+R))**2)*1E3
45      PRINT R,P
50      NEXT R
55      END
```

```
RUN
RL IN OHMS    PWR IN MW

 0             0
 500           1.81406
 1000          3.30579
 1500          4.53686
 2000          5.55556
 2500          6.4
 3000          7.10059
 3500          7.68176
 4000          8.16327
 4500          8.56124
 5000          8.88889
 5500          9.15713
 6000          9.375
 6500          9.55005
 7000          9.68858
 7500          9.79592
 8000          9.87654
 8500          9.93426
 9000          9.9723
 9500          9.99342
 10000         10
 10500         9.99405
 11000         9.97732
 11500         9.95133
 12000         9.91736
 12500         9.87654
 13000         9.82987
 13500         9.77818
 14000         9.72222
 14500         9.66264
 15000         9.6
 15500         9.53479
 16000         9.46746
 16500         9.39836
 17000         9.32785
 17500         9.2562
 18000         9.18367
 18500         9.1105
 19000         9.03686
 19500         8.96294
 20000         8.88889
```

Note that computer solutions substantiate that maximum power was delivered to R_L when it was 10 kΩ in magnitude, or the same value as R_S.

$$\therefore \quad \boxed{R_L \text{ (maximum power)} = R_S} \tag{9.3}$$

The magnitude of this maximum power is calculated as follows:

$$P_{R_L}\text{ (max)} = I_{R_L}^2 R_L = \frac{V_S^2}{4R_L^2}(R_L) = \frac{V_S^2}{4R_L} \tag{9.4}$$

As shown in preceding sections, regardless of how complex the network containing R_L, determining that particular value of R_L that will receive the maximum power is easily accomplished. First, apply Thevenin's theorem to determine R_{Th}. Secondly, in accordance with the Maximum Power transfer theorem, equate R_L to this value. Note again how the application of two or more different network theorems can produce the desired answers.

Example 9.11 With reference to Figure 9.28, write a program in BASIC to determine the value of R that will receive the maximum power for any input values of R_1, R_2, R_3, and R_4 and for any input value of V. Print out the optimum value of R and the value of the maximum power delivered to R.

An arbitrary choice of resistor values and voltage value was made in running the program.

```
LIST
10      INPUT"THE VALUES OF R1,R2,R3 AND R4 ARE";R1,R2,R3,R4
20      INPUT"THE VALUE OF V IS";V
30      LET V1=V*R3/(R1+R3): V2=V*R4/(R2+R4): V3=V1-V2
40      LET R5=R1*R3/(R1+R3): R6=R2*R4/(R2+R4): R7=R5+R6
50      LET P=(V3**2)/(4*R7)
60      PRINT"VALUE OF R FOR MAXIMUM POWER IS"R7"OHMS"
70      PRINT "MAXIMUM POWER DELIVERED TO R IS"P"WATTS"
80      END

RUN
THE VALUES OF R1,R2,R3 AND R4 ARE? 2.2E3,10E3,4.7E3,6.8E3
THE VALUE OF V IS? 24.75
VALUE OF R FOR MAXIMUM POWER IS 5546.17 OHMS
MAXIMUM POWER DELIVERED TO R IS .210943E-2 WATTS
```

Note that statement 30 calculated the Thevenin generator (with R removed) and statement 40 calculated the Thevenin resistance (R_7). R was

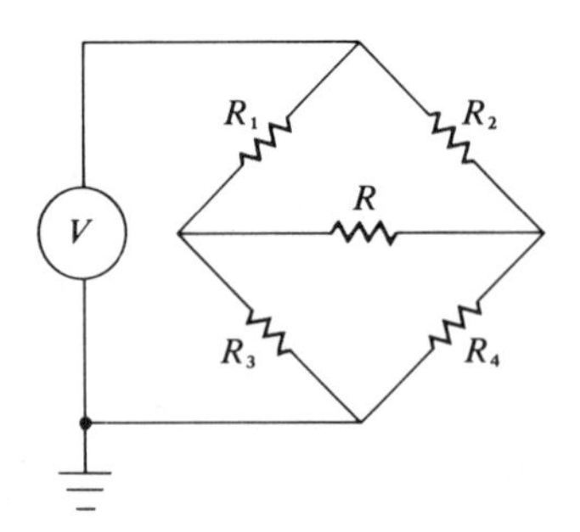

Figure 9.28 Reference Example 9.10

then equated to R_7. Equation (9.4) was used in statement 50 to calculate the maximum power (P).

9.6 SUPERPOSITION THEOREM

Up to this point, networks have been considered with only one source, be it voltage or current. The Superposition theorem deals with networks having two or more sources, even if they are mixtures of both voltage and current sources. It states:

> The total current drawn by a load resistance is equal to the sum of the currents delivered by each source, as if that were the only source present, and all other sources were replaced by their equivalent resistance.

This is really just another way of saying that the whole is nothing more than the sum of its parts. The following examples will illustrate the theorem.

Example 9.12 Determine the current through R_L in Figure 9.29. Use loop equations first, and then compare that answer with the resultant answer determined by using the Superposition theorem.

From Figure 9.29, the following loop equations may be written.

$$40I_1 + 20I_2 = 32$$

$$20I_1 + 50I_2 = 20$$

$$I_1 = \frac{\begin{vmatrix} 32 & 20 \\ 20 & 50 \end{vmatrix}}{\begin{vmatrix} 40 & 20 \\ 20 & 50 \end{vmatrix}} = \frac{1600 - 400}{2000 - 400} = \frac{3}{4} = 0.75 \text{ ampere}$$

$$I_2 = \frac{\begin{vmatrix} 40 & 32 \\ 20 & 20 \end{vmatrix}}{1600} = \frac{800 - 640}{1600} = 0.1 \text{ ampere}$$

Both I_1 and I_2 are positive and in the same direction through R_L, therefore

$$I_{R_L} = I_1 + I_2 = 0.75\text{ A} + 0.1\text{ A} = 0.85 \text{ ampere}$$

The first step in applying the principle of superposition is to replace the 20 V source with a short circuit and then determine only that current due to the 32 V source. From Figure 9.30,

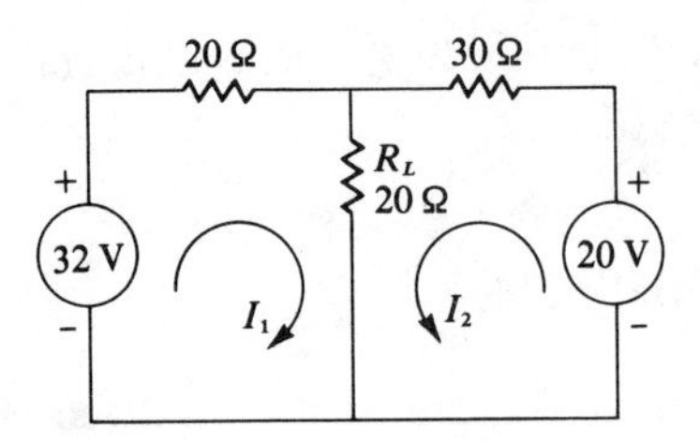

Figure 9.29 Reference Example 9.12

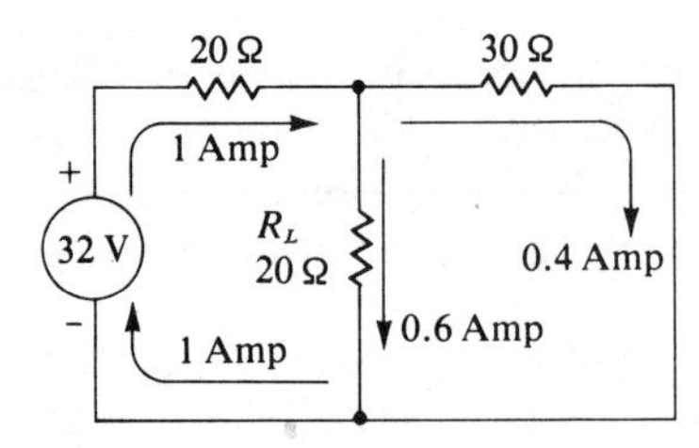

Figure 9.30 Current Distribution Due to 32 V Source

$$R_{T(32)} = 20 + \frac{(30)(20)}{30 + 20} = 32 \text{ ohms}$$

$$I_{T(32)} = \frac{32 \text{ V}}{32 \ \Omega} = 1 \text{ ampere}$$

$$I_{R_L(32)} = \frac{30}{50}(1 \text{ A}) = 0.6 \text{ ampere}$$

Next, the 32 V source is replaced with a short circuit and only the current due to the 20 V source is calculated. From Figure 9.31,

$$R_{T(20)} = 30 + \frac{(20)(20)}{20 + 20} = 40 \text{ ohms}$$

$$I_{T(20)} = \frac{20 \text{ V}}{40 \ \Omega} = 0.5 \text{ ampere}$$

$$I_{R_L(20)} = \frac{20}{40}(0.5 \text{ A}) = 0.25 \text{ ampere}$$

Note that the currents from each source are in the same direction through R_L, therefore they are summed together.

$$I = I_{R_L(32)} + I_{R_L(20)} = 0.6 + 0.25 = 0.85 \text{ ampere}$$

This answer agrees with the previous answer using determinants.

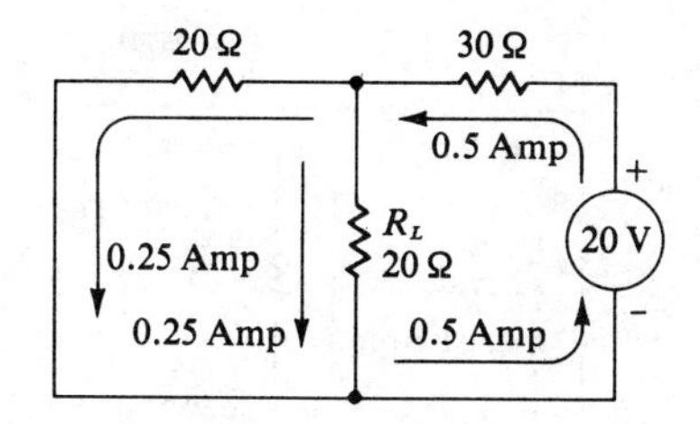

Figure 9.31 Current Distribution Due to 20 V Source

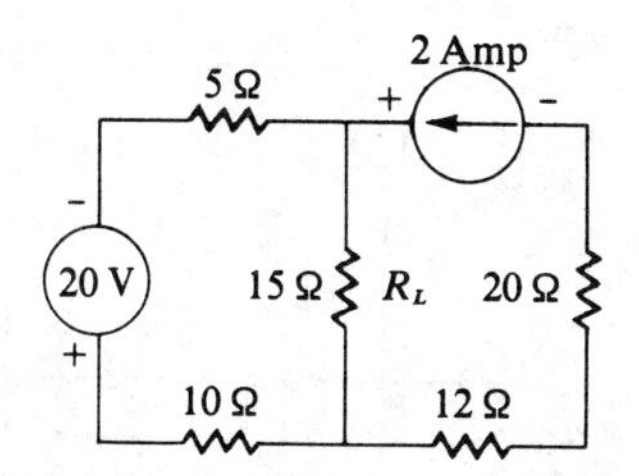

Figure 9.32 Reference Example 9.13

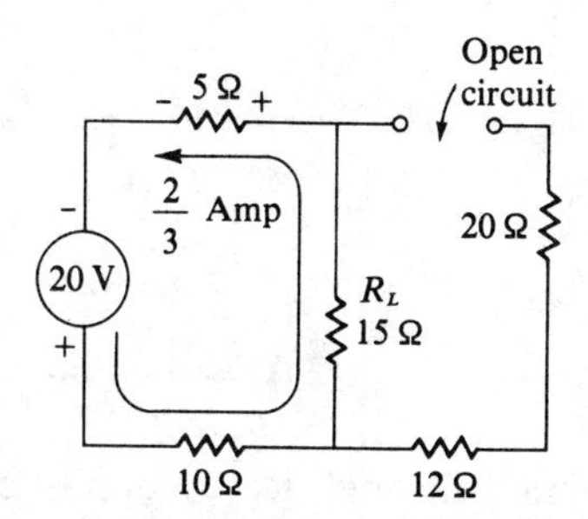

Figure 9.33 Current Distribution Due to 20 V Source

Example 9.13 Determine the current through R_L in Figure 9.32 by using the Superposition theorem.

First consider the current due to the 20 V source. The current generator is replaced with an open circuit, as its resistance is considered infinite. Figure 9.33 results. Notice the direction of this current.

$$I_{R_L(20\text{V})} = \frac{20\text{ V}}{30\ \Omega} = 0.667 \text{ ampere}$$

Next, consider the current due to the current generator. Replacing the voltage source with a short circuit results in Figure 9.34. The current source splits equally between the two paths. Again, notice its direction.

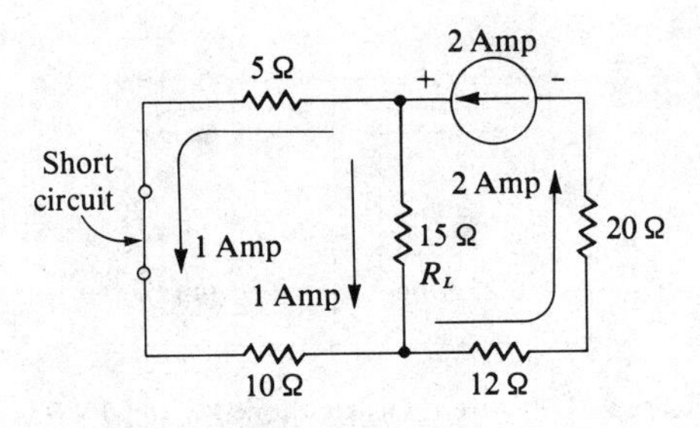

Figure 9.34 Current Distribution Due to 2 A Source

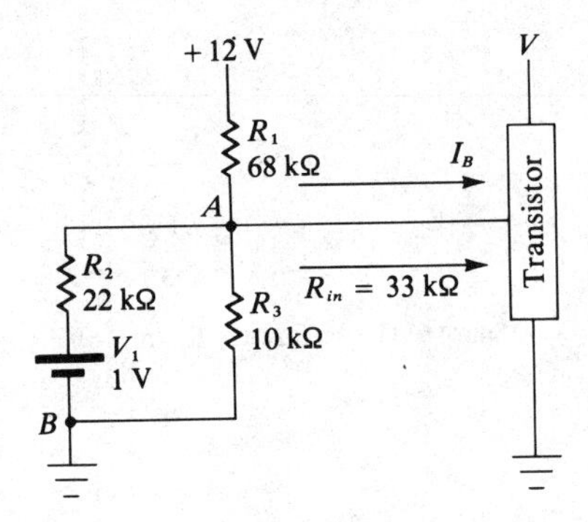

Figure 9.35 Reference Example 9.14

$$I_{R_L(2A)} = \frac{15}{30}(2\text{ A}) = 1.0\text{ ampere}$$

As the current direction through R_L are opposite for both sources, the currents are subtractive.

$$\begin{aligned} I_{R_L} &= I_{R_L(2A)} - I_{R_L(20V)} \\ &= 1.0 - 0.667 \\ &= 0.333\text{ amperes (downward)} \end{aligned}$$

Example 9.14 Figure 9.35 represents a dc transistor amplifier. Determine the input current (I_B), which is a function of both the 12 V source and the 1 V source.

Consider the 12 V source first. After replacing the 1 V source with a short circuit, the resulting equivalent circuit is illustrated in Figure 9.36.

$$R_{AB} = \frac{(22)(10)(33)}{(22)(10) + (22)(33) + (10)(33)}\text{ k}\Omega = \frac{7260}{1276}\text{ k}\Omega = 5.69\text{ k}\Omega$$

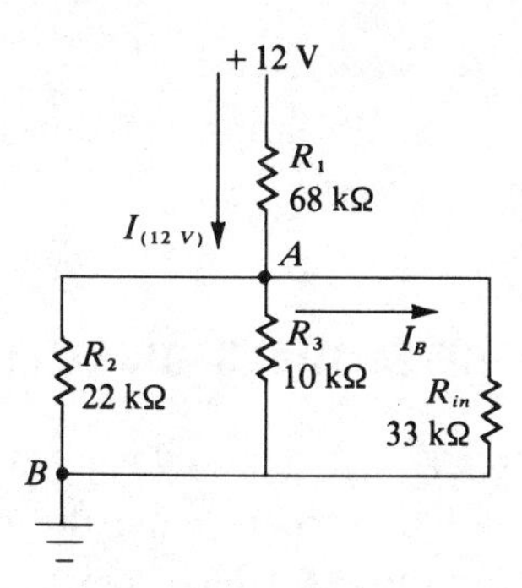

Figure 9.36 Equivalent Circuit Seen by 12 V Source

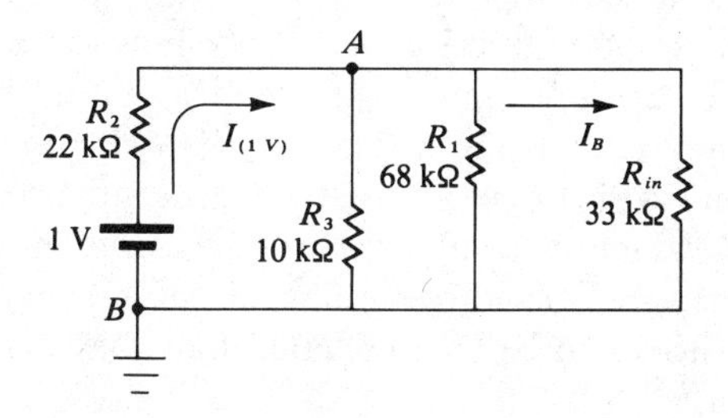

Figure 9.37 Equivalent Circuit Seen by 1 V Source

$$V_A = \frac{5.69}{5.69 + 68}(12\text{ V}) = 0.927\text{ volt}$$

$$I_{B(12\text{V})} = \frac{V_A}{R_{\text{in}}} = \frac{0.927\text{ V}}{33\text{ k}\Omega} = 28.1\ \mu\text{A}$$

Next, consider the 1 V source. After replacing the 12 V source with a short circuit, the resultant equivalent circuit is illustrated in Figure 9.37.

$$R_3 \| R_1 \| R_{\text{in}} = \frac{(10)(68)(33)}{(10)(68) + (10)(33) + (68)(33)}\text{ k}\Omega = 6.896\text{ k}\Omega$$

$$I_{(1\text{V})} = \frac{1\text{ V}}{22\text{ k}\Omega + 6.896\text{ k}\Omega} = 0.035\text{ mA}$$

$$V_A = 1 - (0.035)(22) = 0.23\text{ volt}$$

$$I_{B(1\text{V})} = \frac{V_A}{R_{\text{in}}} = \frac{0.23\text{ V}}{33\text{ k}\Omega} = 6.97\ \mu\text{A}$$

I_B equals the sum of the currents developed by each voltage source, therefore

$$
\begin{aligned}
I_B &= I_{B(12\text{V})} + I_{B(1\text{V})} \\
&= 28.1\ \mu\text{A} + 6.97\ \mu\text{A} \\
&= 35.07\ \mu\text{A}
\end{aligned}
$$

9.7 DELTA-WYE AND WYE-DELTA TRANSFORMATIONS

Observe Figure 9.38. The reader will recognize it as the classical bridge circuit. Suppose it is desired to determine the total generator current (I_g). In a circuit of this nature, in which it is not possible to define explicitly those elements which are in series or those which are in parallel, one must resort to a series of loop equations, such as were studied in Chapter 8. This is often a tedious approach, and one easily subject to error. It is the purpose of this section to describe how to reduce a complex network, such as Figure 9.38, into a simple series-parallel network, and then merely to apply Ohm's law to find the desired solution. If the section between points *ABD* or points *ABC* (delta configurations) were reduced to an equivalent *Y* configuration, the resultant series-parallel network could easily be solved, assuming of course, that the values of the substitute resistors can be so calculated that they are equivalent to the original circuit.

To approach the methodology of such conversions, consider the Δ and *Y* configurations of Figure 9.39. For the two circuits to be equivalent, the resistances *AC*, *AB*, and *BD* of the Δ configuration should respectively equal the resistances *AC*, *AB*, and *BD* of the *Y* configuration.

$$R_{\Delta AC} = \frac{R_A(R_B + R_C)}{R_A + R_B + R_C} = R_{YAD} = R_1 + R_3 \tag{9.5}$$

$$R_{\Delta AB} = \frac{R_B(R_A + R_C)}{R_A + R_B + R_C} = R_{YAB} = R_1 + R_2 \tag{9.6}$$

$$R_{\Delta BD} = \frac{R_C(R_A + R_B)}{R_A + R_B + R_C} = R_{YBD} = R_2 + R_3 \tag{9.7}$$

Adding Eq. (9.5) to (9.6) and subtracting Eq. (9.7) produces the following interesting result:

$$\frac{R_AR_B + R_AR_C + R_AR_B + R_BR_C - R_AR_C - R_BR_C}{R_A + R_B + R_C}$$

$$= R_1 + R_3 + R_1 + R_2 - R_2 - R_3$$

$$R_1 = \frac{R_AR_B}{R_A + R_B + R_C} \tag{9.8}$$

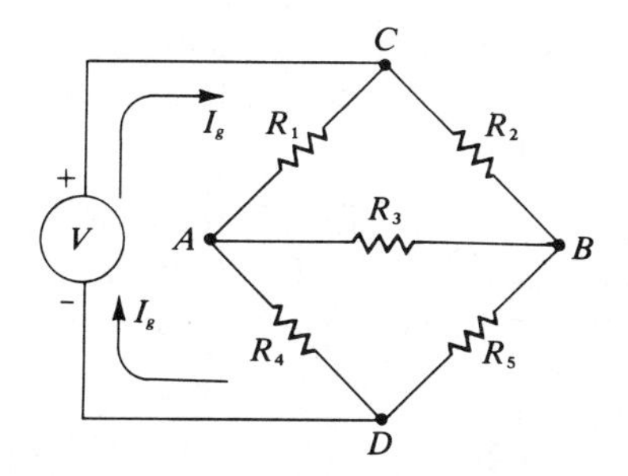

Figure 9.38 Classical Bridge Configuration

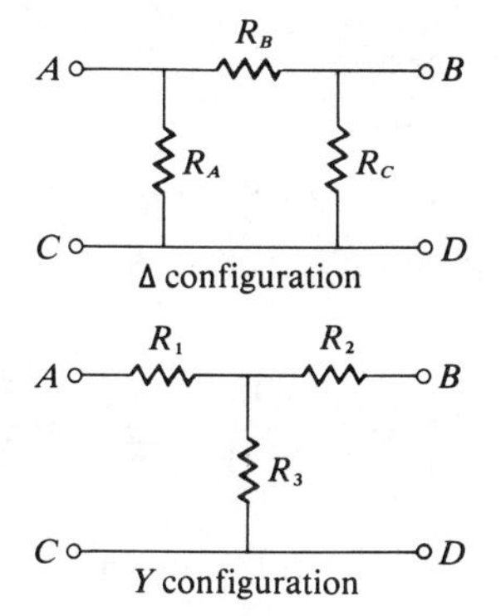

Figure 9.39 Isolated Δ and Y Configurations

In like manner,

$$R_2 = \frac{R_B R_C}{R_A + R_B + R_C} \tag{9.9}$$

$$R_3 = \frac{R_A R_C}{R_A + R_B + R_C} \tag{9.10}$$

Defining $\Delta R = R_A + R_B + R_C$, then

$$\boxed{R_1 = \frac{R_A R_B}{\Delta R}} \tag{9.11}$$

$$\boxed{R_2 = \frac{R_B R_C}{\Delta R}} \tag{9.12}$$

$$\boxed{R_3 = \frac{R_A R_C}{\Delta R}} \tag{9.13}$$

If Figure 9.40 is used to relate the resistors of a Δ configuration to the resistors of a Y configuration, then the following verbal description describes these transformations:

> Each resistor of the Y which is replacing the Δ is equal to the product of the Δ resistors on each side of the Y resistor, divided by ΔR, which is the sum of the three delta resistors.

To convert a Y configuration into a Δ configuration, the following quotients are used:

$$\frac{\text{Equation (9.8)}}{\text{Equation (9.9)}} = \frac{R_1}{R_2} = \frac{R_A}{R_C}$$

In a similar manner,

$$\frac{\text{Equation (9.8)}}{\text{Equation (9.10)}} = \frac{R_1}{R_3} = \frac{R_B}{R_C}$$

and

$$\frac{\text{Equation (9.9)}}{\text{Equation (9.10)}} = \frac{R_2}{R_3} = \frac{R_B}{R_A}$$

Dividing both numerator and denominator of Eq. (9.8) by R_B results in

$$R_1 = \frac{R_A}{\dfrac{R_A}{R_B} + 1 + \dfrac{R_C}{R_B}}$$

Solving for R_A,

$$R_A = R_1\left(\frac{R_3}{R_2} + 1 + \frac{R_3}{R_1}\right)$$

$$= \frac{R_1R_2 + R_1R_3 + R_2R_3}{R_2} \qquad (9.14)$$

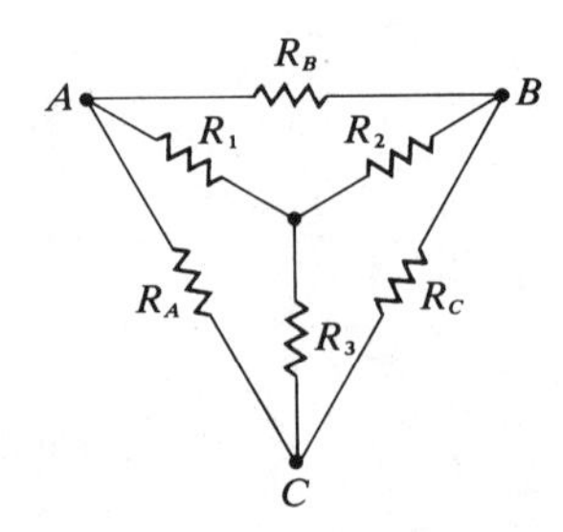

Figure 9.40 Δ-Y Resistors Connected Between Common Points

If we define $YR = R_1R_2 + R_1R_3 + R_2R_3$, then (Eq. (9.14) may be written in the following simplified form:

$$\boxed{R_A = \frac{YR}{R_2}} \tag{9.15}$$

In a similar manner,

$$\boxed{R_B = \frac{YR}{R_3}} \tag{9.16}$$

and

$$\boxed{R_C = \frac{YR}{R_1}} \tag{9.17}$$

If Figure 9.40 is used to relate the elements of a Y configuration to a Δ configuration, then the following verbal description describes these transformations:

> Each Δ resistor is equal to YR divided by the opposite Y resistor, where YR equals the sum of all the products of two of the Y resistors.

Example 9.15 Convert the Δ configuration of Figure 9.41 to a Y configuration.

$$\Delta R = 2 + 3 + 5 = 10 \text{ ohms}$$

Using Figure 9.40 as a reference,

$$R_1 = \frac{(2)(5)}{10} = 1 \text{ ohm}$$

$$R_2 = \frac{(5)(3)}{10} = 1.5 \text{ ohms}$$

$$R_3 = \frac{(2)(3)}{10} = 0.6 \text{ ohm}$$

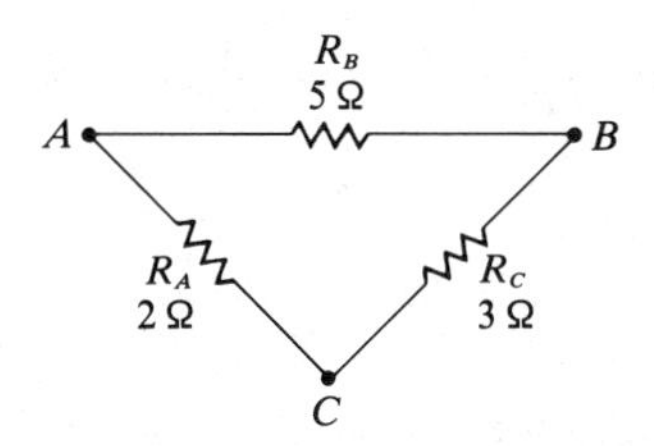

Figure 9.41 Reference Example 9.15

Example 9.16 Convert the Y configuration of Figure 9.42 to a Δ configuration.

$$\begin{aligned} YR &= (5)(10) + (10)(20) + (20)(5) \\ &= 50 + 200 + 100 \\ &= 350 \end{aligned}$$

Using Figure 9.40 as a reference,

$$R_A = \frac{350}{5} = 70 \text{ ohms}$$

$$R_B = \frac{350}{20} = 17.5 \text{ ohms}$$

$$R_C = \frac{350}{10} = 35 \text{ ohms}$$

Example 9.17 Determine the total generator current of Figure 9.43 by using transformations.

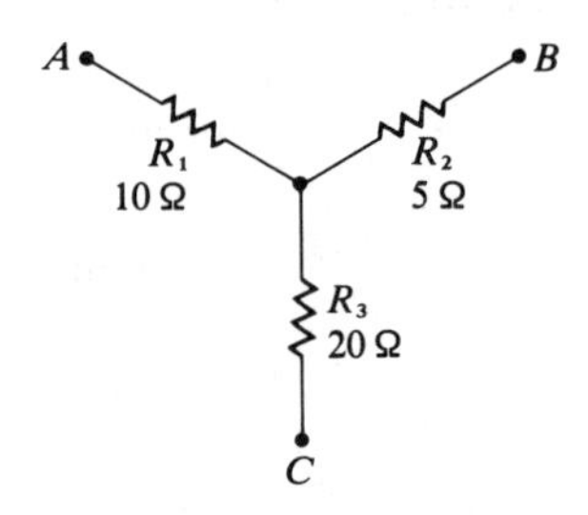

Figure 9.42 Reference Example 9.16

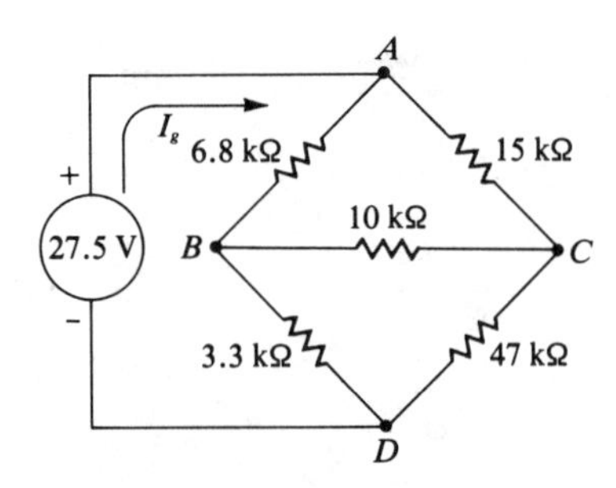

Figure 9.43 Reference Example 9.17

Delta *ABC* of Figure 9.43, when converted in a wye configuration, results in Figure 9.44.

$$\Delta R = (6.8 + 15 + 10)\ \text{k}\Omega = 31.8\ \text{k}\Omega$$

$$R_1 = \frac{(6.8)(15)}{31.8} = 3.21\ \text{k}\Omega$$

$$R_2 = \frac{(6.8)(10)}{31.8} = 2.14\ \text{k}\Omega$$

$$R_3 = \frac{(10)(15)}{31.8} = 4.72\ \text{k}\Omega$$

$$R_2 + R_4 = (2.14 + 3.3)\ \text{k}\Omega = 5.44\ \text{k}\Omega \equiv R_6$$

$$R_3 + R_5 = (4.72 + 47)\ \text{k}\Omega = 51.72\ \text{k}\Omega \equiv R_7$$

$$R_6 \| R_7 = \frac{(5.44)(51.72)}{57.16} = 4.92\ \text{k}\Omega \equiv R_8$$

$$R_1 + R_8 = (3.21 + 4.92)\ \text{k}\Omega = 8.13\ \text{k}\Omega \equiv R_9$$

$$I_g = \frac{V}{R_9} = \frac{27.5\ V}{8.13\ \text{k}\Omega} = 3.38\ \text{mA}$$

9.8 LADDER THEOREM

Circuitry in communication electronics (such as television circuits) or in digital electronics (such as computer delay lines) exhibits a geometric structure resembling a ladder. For example, Figure 9.45 represents part

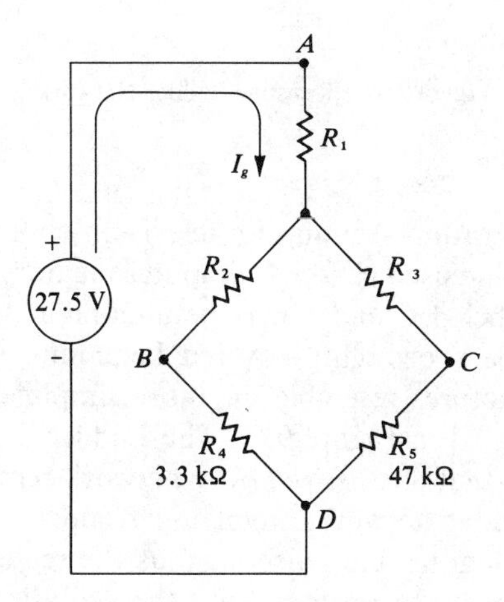

Figure 9.44 ΔABC of Figure 9.43 Transformed into a Y Configuration

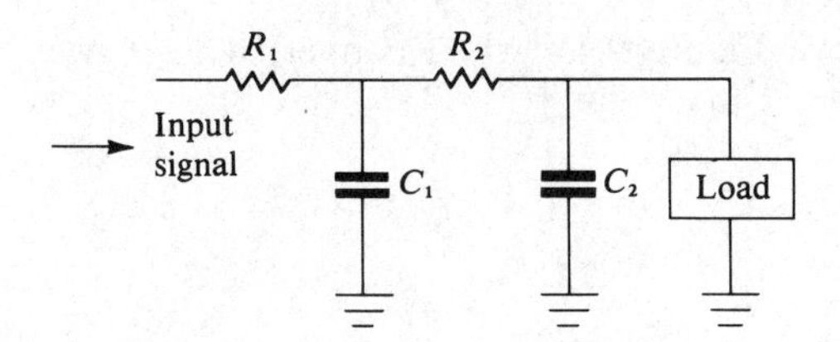

Figure 9.45 Typical T-V Circuitry

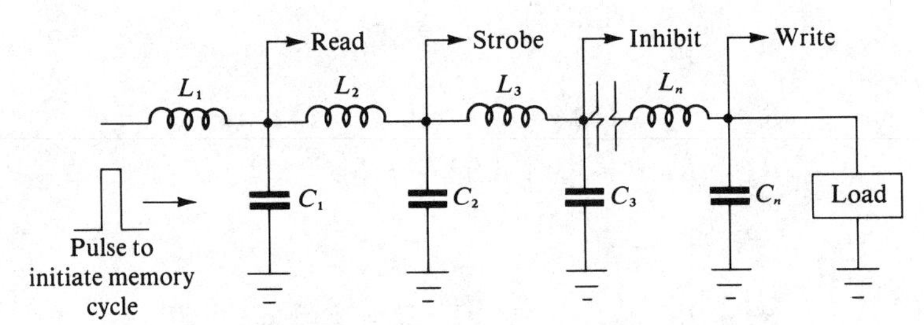

Figure 9.46 Typical Computer Delay Line

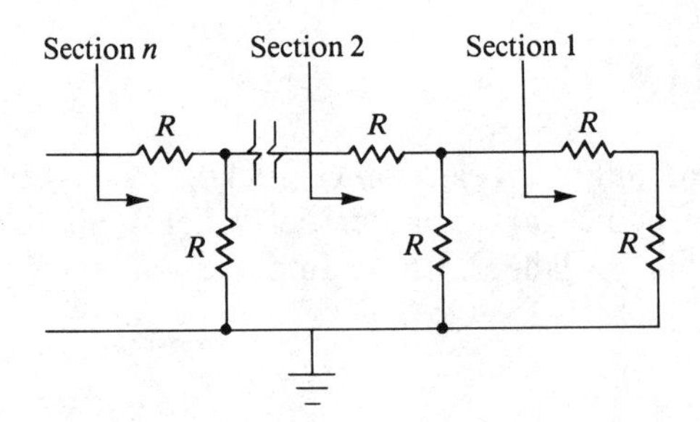

Figure 9.47 Resistive Ladder Network

of the vertical integration section of a television circuit, and Figure 9.46 represents a delay line used in some computer memory circuits. Although the components labeled C and L may be unfamiliar to the reader (these components will be thoroughly covered beginning with the following chapter), these structures resemble the more simplified network of pure resistors, as illustrated in Figure 9.47. The Ladder theorem will analyze the total resistance that a *resistive ladder* network represents to a source. The analysis of ladder networks involving L and C components would proceed along analogous lines, although, as the reader will learn, other factors such as phase relationships enter the analysis. A formula will be presented relating R (total) to both the individual values of resistance and the total number of sections.

Consider Figure 9.48. It has four sections, and in general, the value of each resistor is R ohms. The approach is to label the current through each resistor, the voltage across each resistor, and the voltage at the junction point between each resistor. Computation of all labeled unknowns proceeds as follows.

Assume $V_o = 1$ volt. This may seem a strange approach, since in all probability, the odds are highly against V_o being exactly 1 volt in any given circuit. This really doesn't matter. What is desired is to find the ratio of V_{in} and I_{in} that would produce this 1 volt. Since voltage, current, and resistance are proportionate in dc resistive circuits, any difference between the actual value of V_o and the assumed 1 volt would be accompanied by a corresponding change in all currents and voltages. This ratio would remain a constant, hence R_{in} (as seen by any section) can be evaluated by starting with this simple assumption. From Figure 9.48,

$$V_o = 1 \text{ volt}$$

$$I_o = \frac{V_o}{R} = \frac{1}{R}$$

$$\left.\begin{aligned} I_1 &= I_0 = \frac{1}{R} \\ V_1 &= (I_1)(R) = \left(\frac{1}{R}\right)(R) = 1 \\ V_2 &= V_1 + V_o = 2 \end{aligned}\right\} R_{\text{in }1} = \frac{V_2}{I_1} = \frac{2}{1/R} = \frac{2}{1} R$$

$$I_2 = \frac{V_2}{R} = \frac{2}{R}$$

$$\left.\begin{aligned} I_3 &= I_2 + I_1 = \frac{2}{R} + \frac{1}{R} = \frac{3}{R} \\ V_3 &= (I_3)(R) = \left(\frac{3}{R}\right)(R) = 3 \\ V_4 &= V_3 + V_2 = 3 + 2 = 5 \end{aligned}\right\} R_{\text{in }2} = \frac{V_4}{I_3} = \frac{5}{3/R} = \frac{5}{3} R$$

$$I_4 = \frac{V_4}{R} = \frac{5}{R}$$

$$\left.\begin{aligned} I_5 &= I_4 + I_3 = \frac{5}{R} + \frac{3}{R} = \frac{8}{R} \\ V_5 &= (I_5)(R) = \left(\frac{8}{R}\right)(R) = 8 \\ V_6 &= V_5 + V_4 = 8 + 5 = 13 \end{aligned}\right\} R_{\text{in }3} = \frac{V_6}{I_5} = \frac{13}{8/R} = \frac{13}{8} R$$

$$I_6 = \frac{V_6}{R} = \frac{13}{R}$$

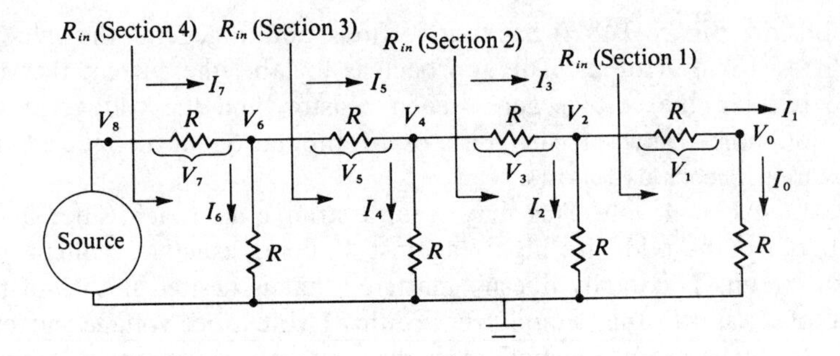

Figure 9.48 Resistive Ladder Network Labeled for Analysis

$$\left.\begin{aligned} I_7 &= I_5 + I_6 = \frac{8}{R} + \frac{13}{R} = \frac{21}{R} \\ V_7 &= (I_7)(R) = \left(\frac{21}{R}\right)(R) = 21 \\ V_8 &= V_7 + V_6 = 21 + 13 = 34 \end{aligned}\right\} R_{\text{in}\,4} = \frac{V_8}{I_7} = \frac{34}{21/R} = \frac{34}{21} R$$

The reader may have already observed an interesting and somewhat unique geometric progression to the total input resistance that would be seen by a source, if looking into any given number of sections of a ladder-type resistive network. Extracting the multiplying factor of R, the coefficients for any given number of sections can be extrapolated as follows. Write down the first two whole integers, namely,

1, 2 (pair 1)

Now continue to expand these integers by successively adding digits together in pairs:

$$1 + 2 = 3$$
$$2 + 3 = 5$$

therefore, the next pair is

3, 5 (pair 2)

Continuing,

$$3 + 5 = 8$$
$$5 + 8 = 13$$

therefore, the next pair is

8, 13 (pair 3)

Continuing,

$$8 + 13 = 21$$

$$13 + 21 = 34$$

therefore, the next pair is

21, 34 (pair 4)

If the first digit of each pair is considered the denominator of a fraction, and the second digit of each pair is considered the numerator, the following fractions would result for the four pairs:

$$\frac{2}{1} \quad \frac{5}{3} \quad \frac{13}{8} \quad \frac{34}{21}$$

Note that the four fractions just happen to be the multiplying factor of R for each of the four sections of the resistive ladder network.

It should not be surprising that such a unique geometric progression exists, since a ladder network possesses symmetry. In the following section, the uniqueness of any symmetrical circuit will be further expanded upon.

Example 9.18 Suppose the resistive network of Figure 9.48 had five sections. If each resistor were 47 kΩ, what is the total input resistance of the ladder network?

Section 4 resulted in a number pair of

21, 34 (pair 4)

Continuing the pairing,

$$21 + 34 = 55$$

$$34 + 55 = 89$$

therefore, R (section 5) is

$$\frac{89}{55} \times (R) = \frac{89}{55}(47\ \text{k}\Omega)$$

$$= 76.05\ \text{k}\Omega$$

Classical mathematics has a formula for such a progression. The sequence of numbers

1, 1, 2, 3, 5, 8, 13, 21, etc.,

in which each succeeding number is the sum of the two previous numbers, is called the *Fibonocci sequence*. The value for the nth fractional coefficient is given as

$$N = \frac{(1+\sqrt{5})^{n+2} - (1-\sqrt{5})^{n+2}}{2[(1+\sqrt{5})^{n+1} - (1-\sqrt{5})^{n+1}]} \tag{9.18}$$

This formula, however, unless calculated by computer, becomes somewhat encumbering for any value of n greater than 3. It is usually easier to calculate successive pair additions to obtain the desired results.

9.9 THEOREMS OF SYMMETRICAL CIRCUITS (BARTLETT'S BISECTIONAL THEOREM)

In the total field of electronics, circuits that exhibit symmetry are very common. Symmetry can occur in simple filters, and extend to complex differential amplifiers. A series of theorems have been developed to simplify the analysis of symmetrical circuits. The following two theorems relate to the simplification:

A. A branch with no voltage across it (or no current through it) may be replaced with either a short circuit or an open circuit and not affect the rest of the network.

B. Two points at the same potential may be replaced with a short circuit and not affect the rest of the network.

To illustrate Theorem A, consider the *balanced bridge* of Figure 9.49. A balanced bridge occurs when the ratio of resistors results in $V_{AB} = 0$ volts; that is, when there is no current through resistor R. It is desired to determine I_1. The loop equations are as follows:

$$\begin{aligned} 55I_1 - 22I_2 - 33I_3 &= 20 \\ -22I_1 + 38.8I_2 - 6.8I_3 &= 0 \\ -33I_1 - 6.8I_2 + 54.8I_3 &= 0 \end{aligned}$$

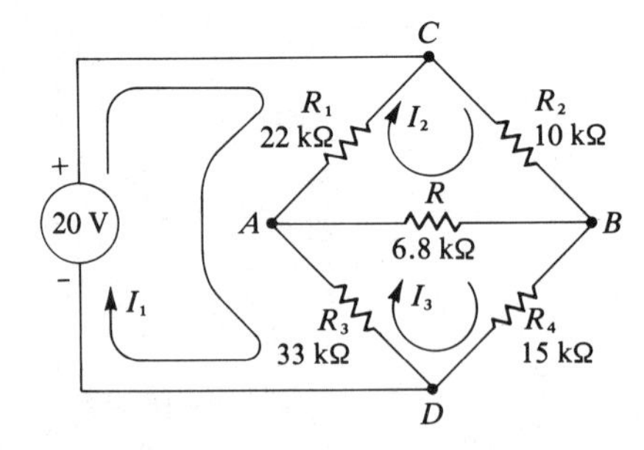

Figure 9.49 Balanced Bridge Circuit to Illustrate Symmetry Principles

When these equations are applied to a computer matrix program, the following current values result:

```
LIST
10      DIM A(3,3), B(3), C(3)
20      MAT INPUT A: MAT INPUT B
30      MAT A =INV(A): MAT C=A*B
40      FOR K=1 TO 3
45      PRINT"CURRENT I"K"="C(K)"MA"
50      NEXT K
60      END

RUN
? 55,-22,-33,-22,38.8,-6.8,-33,-6.8,54.8
? 20,0,0
CURRENT I 1 = 1.16364 MA
CURRENT I 2 = .8 MA
CURRENT I 3 = .8 MA
```

It is seen that I_2 equals I_3, and as they are in opposite directions, there is no current through R. The bridge is balanced; that is, the voltage at point A equals the voltage at point B.

Applying a short circuit between points A and B results in Figure 9.50.

$$R_1 \| R_2 = \frac{22 \times 10}{22 + 10}\ \text{k}\Omega = 6.875\ \text{k}\Omega \equiv R_5$$

$$R_3 \| R_4 = \frac{33 \times 15}{33 + 15}\ \text{k}\Omega = 10.3125\ \text{k}\Omega \equiv R_6$$

$$R_5 + R_6 = 6.875\ \text{k}\Omega + 10.3125\ \text{k}\Omega = 17.1875\ \text{k}\Omega$$

$$I_1 = \frac{20}{17.1875\ \text{k}\Omega} = 1.1636\ \text{mA}$$

If the bridge is opened, Figure 9.51 results.

$$R_1 + R_3 = 22\ \text{k}\Omega + 33\ \text{k}\Omega = 55\ \text{k}\Omega \equiv R_5$$

$$R_2 + R_4 = 10\ \text{k}\Omega + 15\ \text{k}\Omega = 25\ \text{k}\Omega \equiv R_6$$

$$R_5 \| R_6 = \frac{55 \times 25}{55 + 25}\ \text{k}\Omega = 17.1875\ \text{k}\Omega$$

$$I_1 = \frac{20}{17.1875\ \text{k}\Omega} = 1.1636\ \text{mA}$$

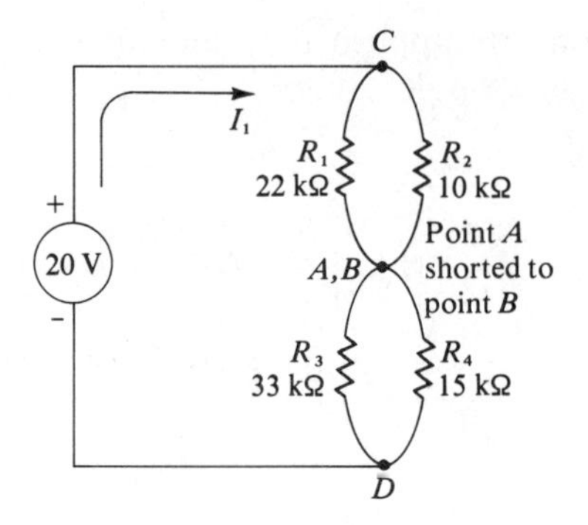

Figure 9.50 Symmetry Theorems Using a Short Circuit between Points AB

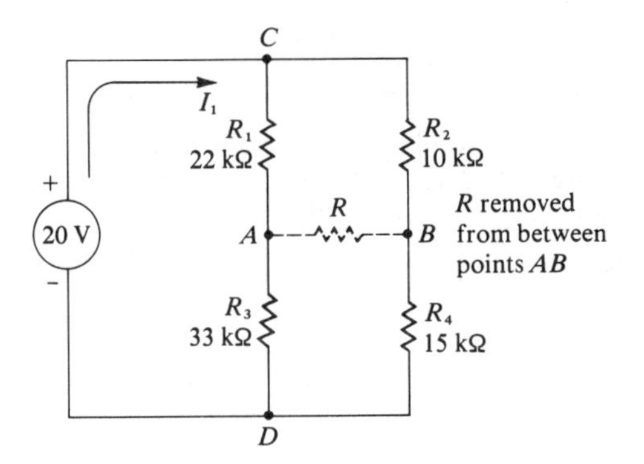

Figure 9.51 Symmetry Theorems Using an Open Circuit between Points AB

Note that the answer for I_1 is the same in all three calculations, but certainly, Figures 9.50 or 9.51 are simpler to analyze than the original circuit of Figure 9.49.

To illustrate Theorem B, consider the somewhat complex resistive cube of Figure 9.52. Assume each resistor has a value of 2 ohms. It is desired to find the total resistance as seen by source V, that is, the total resistance between points A and B. One method is to apply a series of successive delta-wye transformations. This method was left as an exercise (Problem 21). After extensive calculations, it resulted in an answer for R_T of 1.67 ohms.

A much simpler approach, however, is to recognize that by symmetry, point C is at the same potential as point D, and point E is at the same potential as point F. Shorting C to D and E to F results in Figure 9.53, which quickly reduces to Figure 9.54 after combining series and parallel resistors. Finally, using but a single $\Delta - Y$ transform, such as converting $\Delta[A, (E,F), (C,D)]$ to a Y, Figure 9.55 results.

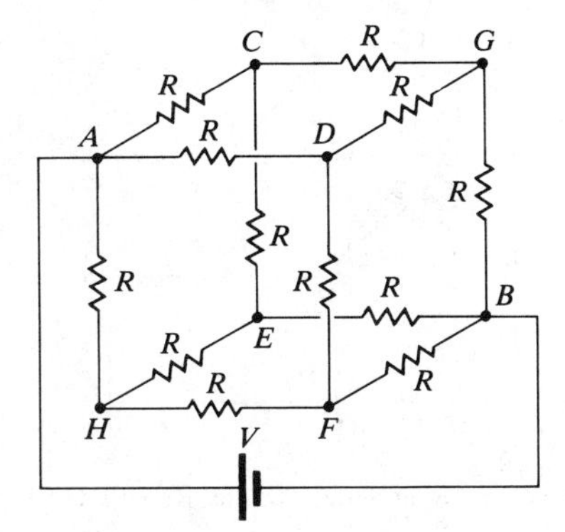

Figure 9.52 Resistive Cube

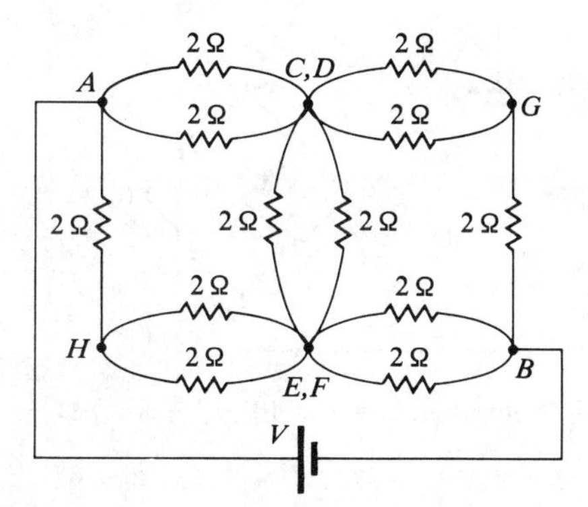

Figure 9.53 Shorting Points of Equal Potential of Figure 9.52

$$R_T = \frac{3}{5} + \frac{(1\tfrac{3}{5})(3\tfrac{1}{5})}{1\tfrac{3}{5} + 3\tfrac{1}{5}} = \frac{5}{3} = 1.67 \text{ ohms}$$

A special application to symmetrical circuits occurs with Bartlett's bisectional theorem. Paraphrasing the theorem, it states that:

> If a circuit is symmetrical, it may be bisected with an infinite resistance, and the conditions applying to any single section are the same as they would apply to the whole circuit.

If the circuit is not already constructed as two equal symmetrical parts, then the "trick" is to manipulate components so as to produce a symmetrical configuration.

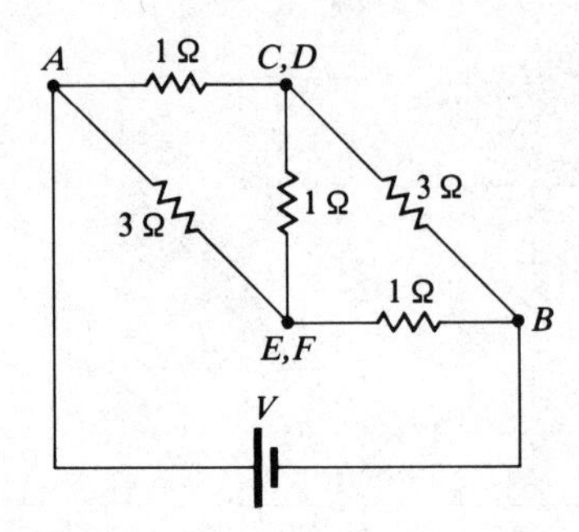

Figure 9.54 Combining Series-Parallel Resistors of Figure 9.53

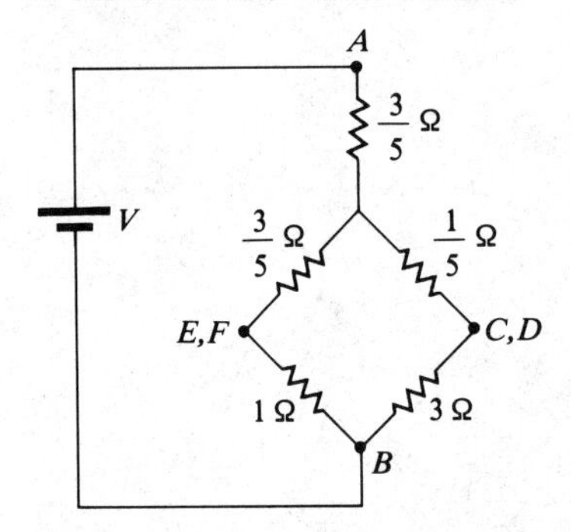

Figure 9.55 Converting Δ(A, CD, EF) of Figure 9.54 into A "Y"

Example 9.19 Determine the current I_1 of Figure 9.56, first by loop equations and then by applying Bartlett's theorem.

Write a determinant for I_1.

$$35I_1 + 10I_2 = 10$$
$$10I_1 + 35I_2 = 10$$

$$I_1 = \frac{\begin{vmatrix} 10 & 10 \\ 10 & 35 \end{vmatrix}}{\begin{vmatrix} 35 & 10 \\ 10 & 35 \end{vmatrix}} = \frac{250}{1125} = \frac{2}{9}\text{ mA} = 0.222\text{ mA}$$

If Figure 9.56 is redrawn as Figure 9.57 and then bisected, the calculation of I_1 (still the current between points AB) is reduced to an Ohm's law equation rather than a determinant. Notice that the two 20 kΩ resistors in parallel replace the original 10 kΩ resistor.

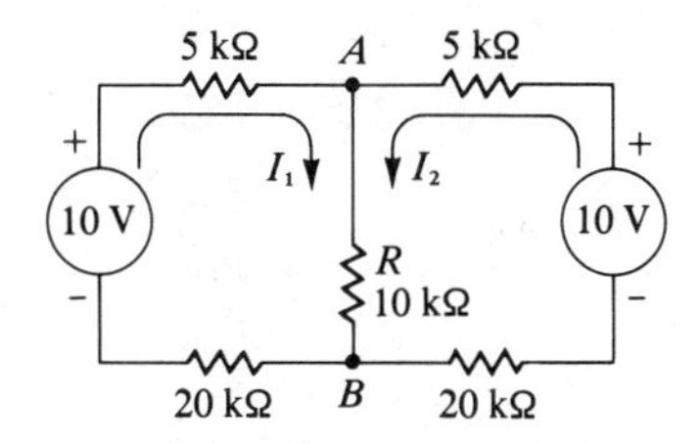

Figure 9.56 Reference Example 9.19

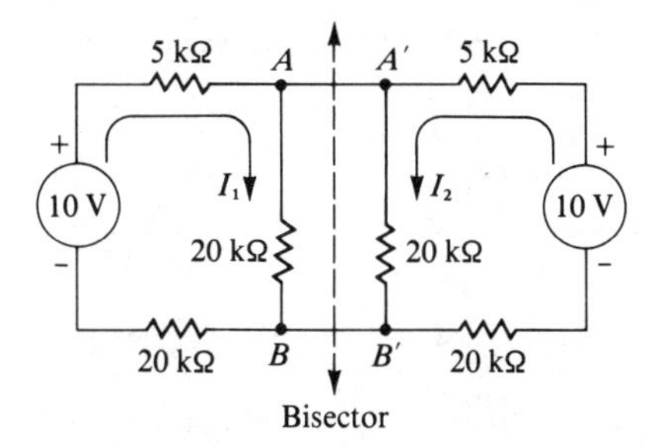

Figure 9.57 Figure 9.56 as Two Equal Symmetrical Parts

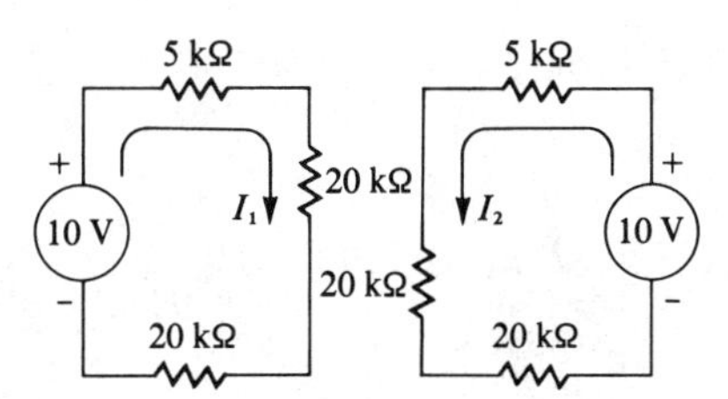

Figure 9.58 Separating Bisected Sections of Figure 9.57

From Figure 9.58 (after separating the original circuit into two equal symmetrical parts),

$$I_1 = \frac{10 \text{ V}}{45 \text{ k}\Omega} = \frac{2}{9} \text{ mA} = 0.222 \text{ mA}$$

which agrees with the answer used in determinants. As stated by the theorem, what applies to one-half of a dual symmetrical circuit also applies to the entire original circuit.

Example 9.20 Figure 9.59 represents a symmetrical circuit. Solve for all three loop currents by a computer matrix program and also by applying Bartlett's theorem.

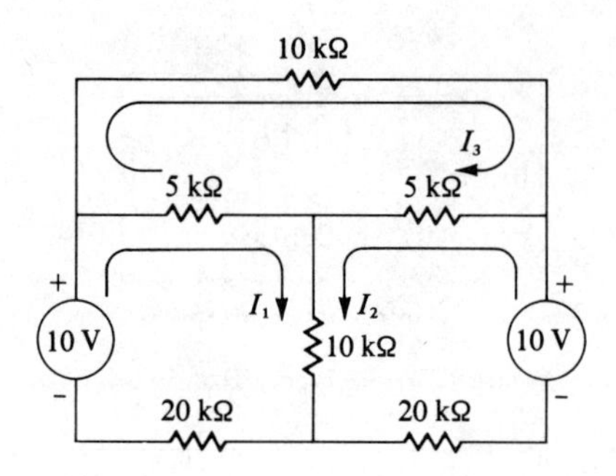

Figure 9.59 Reference Example 9.20

The three loop equations are:

$$35I_1 + 10I_2 - 5I_3 = 10$$

$$10I_1 + 35I_2 + 5I_3 = 10$$

$$-5I_1 + 5I_2 + 20I_3 = 0$$

Solution by computer programming produces the following results:

```
LIST
10      DIM A(3,3), B(3), C(3)
20      MAT INPUT A: MAT INPUT B
30      MAT A =INV(A): MAT C=A*B
40      FOR K=1 TO 3
45      PRINT"CURRENT I"K"="C(K)"MA"
50      NEXT K
60      END

RUN
? 35,10,-5,10,35,5,-5,5,20
? 10,10,0
CURRENT I 1 = .222222 MA
CURRENT I 2 = .222222 MA
CURRENT I 3 = 0 MA
```

If Figure 9.59 is bisected along its lines of symmetry, Figures 9.60 and 9.61 result. It is clear that I_3 must be zero since the bisector has opened its current flow path. Ohm's law may now be applied to I_1 and I_2.

$$I_1 = I_2 = \frac{10 \text{ V}}{45 \text{ k}\Omega} = \frac{2}{9} \text{ mA} = 0.222 \text{ mA}$$

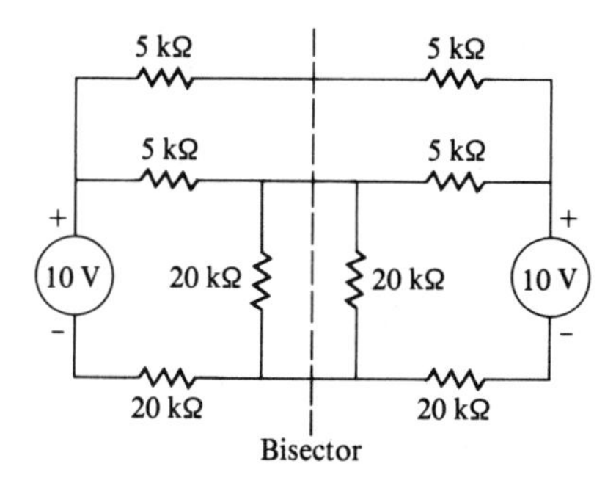

Figure 9.60 Redrawing Figure 9.59 with a Bisector

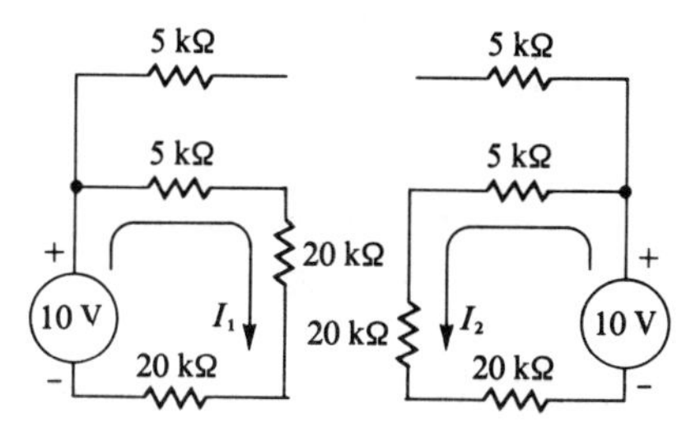

Figure 9.61 Figure 9.60 as Two Symmetrical Sections

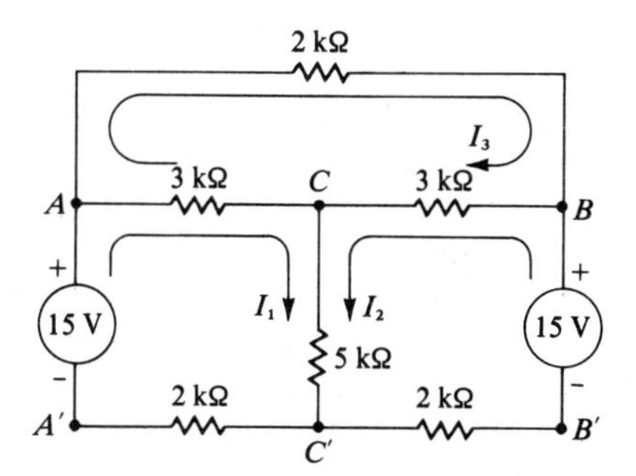

Figure 9.62 Reference Example 9.21

Note how much simpler it is to calculate these currents after applying Bartlett's theorem.

Example 9.21 In this example, points at equal potential will be shorted together. Solve for the total current through the 5 kΩ resistor of Figure 9.62 by first using a computer matrix program and then by shorting together points of equal potential, namely, points A to B and points A' to B'.

The loop equations of Figure 9.62 are:

$$10I_1 + 5I_2 - 3I_3 = 15$$

$$5I_1 + 10I_2 + 3I_3 = 15$$

$$-3I_1 + 3I_2 + 8I_3 = 0$$

```
LIST
10      DIM A(3,3), B(3), C(3)
20      MAT INPUT A: MAT INPUT B
30      MAT A =INV(A): MAT C=A*B
40      FOR K=1 TO 3
45      PRINT"CURRENT I"K" ="C(K)"MA"
50      NEXT K
60      END

RUN
? 10,5,-3,5,10,3,-3,3,8
? 15,15,0
CURRENT I 1 = 1 MA
CURRENT I 2 = 1 MA
CURRENT I 3 = 0 MA
```

The current through the 5 kΩ resistor is the sum of I_1 and I_2.

$$I_{(5\ \text{k}\Omega)} = 1\ \text{mA} + 1\ \text{mA} = 2\ \text{mA}$$

Because of symmetry, the potential at point A must equal the potential at point B, and the potential at point A' must equal that at point B'. If these points of equal potential are now shorted together, Figure 9.63 results. The second generator of 15 volts was added in dashed lines because it may be dropped from the circuit without affecting the analysis,

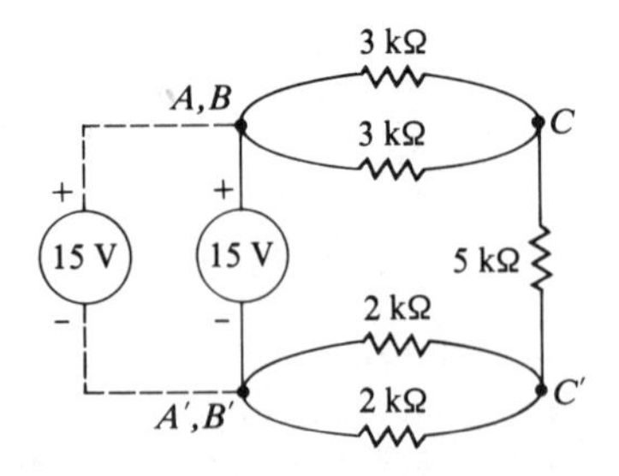

Figure 9.63 Figure 9.62 Simplified

in that equal potentials in parallel are not additive as they would be in series. In summary, the potential difference between points AB and points $A'B'$ is still only 15 volts. From Figure 9.63,

$$I_{(5\ \text{k}\Omega)} = \frac{15}{\left[\frac{(3)(3)}{6} + 5 + \frac{(2)(2)}{4}\right] \text{k}\Omega} = \frac{15}{7.5}\ \text{mA} = 2\ \text{mA}$$

Example 9.22 Figure 9.64 utilizes two current generators. Determine the current through the 5 kΩ resistor by first applying the Superposition theorem and then by shorting together points of equal potential.

Replacing the right-hand generator by an open circuit results in Figure 9.65. Regardless of how the 2 mA divides between points AC and points ABC, the initial current must still return to the generator through the 5 kΩ resistor, as this is the only available current path. The same result would occur if the left-hand generator were replaced by an open circuit, and only the right-hand generator were considered. Therefore,

$$I_{(5\ \text{k}\Omega)} = 2(2\ \text{mA}) = 4\ \text{mA}$$

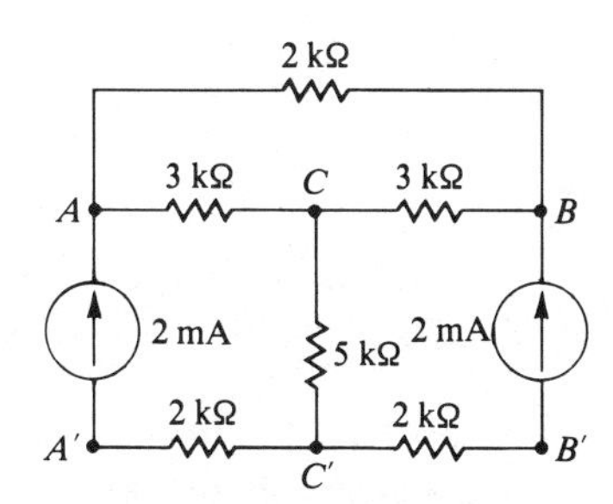

Figure 9.64 Reference Example 9.22

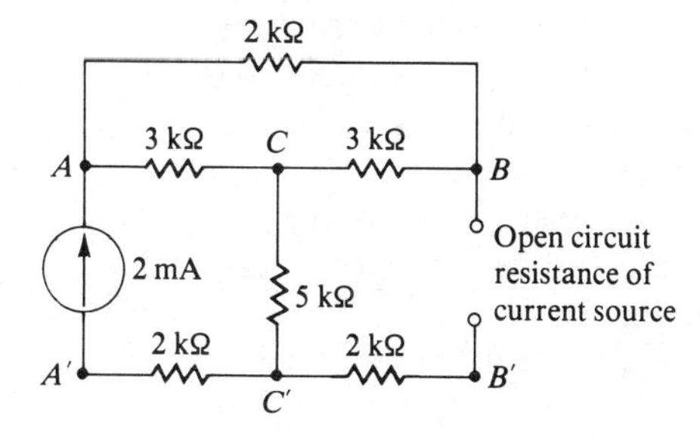

Figure 9.65 Applying Superposition to the Left Current Generator

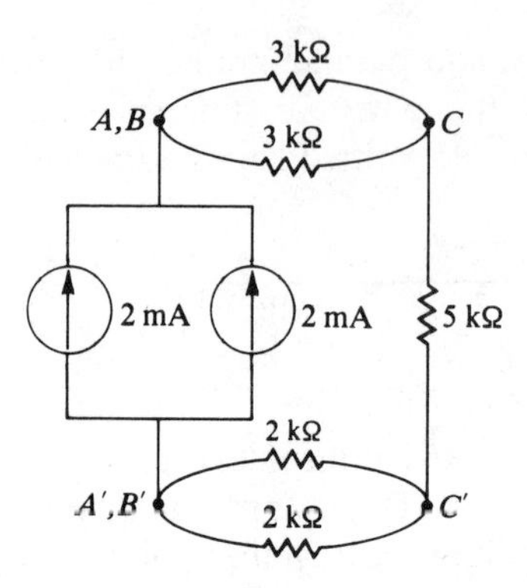

Figure 9.66 Shorting Together Points of Equal Potential in Figure 9.64

Since points *AB* and *A′B′* in Figure 9.64 are (by symmetry) at equal potentials, they may be shorted together as in Figure 9.66. Note that in this example two current generators in parallel are additive and are equivalent to a single 4 mA generator. There is still only one path for current between points *CC′*. All 4 mA, therefore, must flow through the 5 kΩ resistor, which agrees with the answer using the Superposition theorem.

SUMMARY

A number of network theorems have been developed to aid in circuit analysis. Each may be utilized by applying a set of rules germaine to that particular theorem. The theorem employed depends upon the circuit and the unknown variables for which solutions are sought.

Millman's theorem calculates the voltage potential at a common point in a multibranch network. Thevenin's and Norton's theorems facilitate reduction of a complex circuit supplying a load resistor into a single source, with its associated single internal resistance. Solution of the current through a load is then easily calculated by Ohm's law or Kirchhoff's current law. The Maximum Power theorem is the easiest to apply. It states that maximum power is transferred to a load when its resistive value equals the resistance of the source.

Application of the Superposition theorem is based on the premise that the total current through a load is equal to the sum of all the currents supplied by each source in the circuit. Delta-wye transformations permit restructuring *bridge-type* networks into simpler series-parallel arrangements. The Ladder theorem and Bartlett's theorem rely on symmetry for application. The former calculates total resistance looking into any number of resistive ladder-type sections; the latter facilitates circuit calculations by opening or short-circuiting points of equal potential.

Frequently, circuit analysis can be achieved by applying combinations of the network theorems.

PROBLEMS

Reference Section 9.2

1. Determine the voltage at V_A of Figure 9.67 by applying Kirchhoff's proportionate voltage law. Verify this answer by applying Millman's theorem.
2. Determine the value of V_{BA} in Figure 9.67.
3. Determine the value of V_A in Figure 9.68.
4. What is the magnitude of current through R_1, R_2, R_3, and R_4 in Figure 9.68?
5. Determine the value of V in Figure 9.69 so that the voltage of point A is -1 volt.

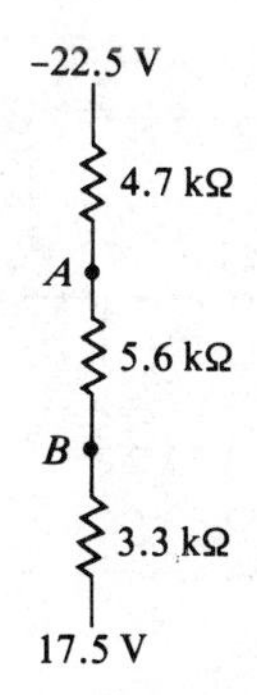

Figure 9.67 Reference Problems 1 and 2

-3.5 V
R_1 10 kΩ
R_4 3.3 kΩ
12 V
R_2 6.8 kΩ
A
R_3 4.7 kΩ
-8.5 V

Figure 9.68 Reference Problems 3 and 4

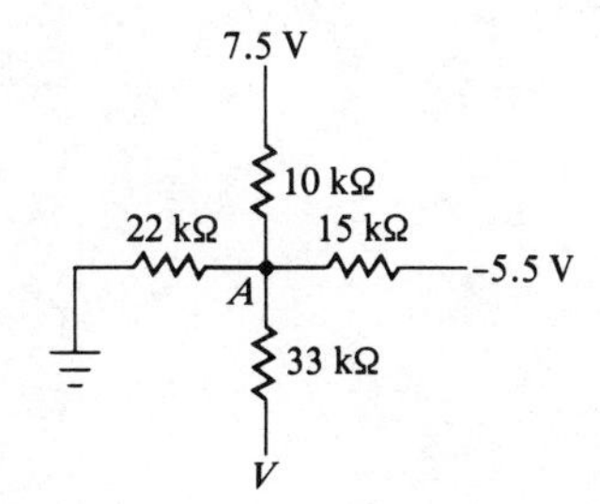

Figure 9.69 Reference Problem 5

Reference Section 9.3

6. Determine the Thevenin generator and Thevenin resistance for R_L in Figure 9.70.
7. How much power is developed across the 3.3 kΩ resistor in Figure 9.70? Determine the answer by using both loop equations and also by Thevenizing the circuit for R_1.
8. Using Thevenin's theorem, determine the current through R in Figure 9.71 if it has a value of 6.8 kΩ.
9. What should be the value of R in Figure 9.71 if the current through it is 0.375 mA?
10. If R in Figure 9.71 were 1.5 kΩ, Thevenize the circuit for R_1.

Reference Section 9.4

11. Calculate a Norton equivalent circuit for R_1 in Figure 9.70.
12. Calculate a Norton equivalent circuit for R in Figure 9.71. *Hint:* Thevenize and then convert to a current source.
13. Develop a Norton equivalent circuit for R_1 in Figure 9.71 if $R = 6.8$ kΩ. What is the current through R_1?

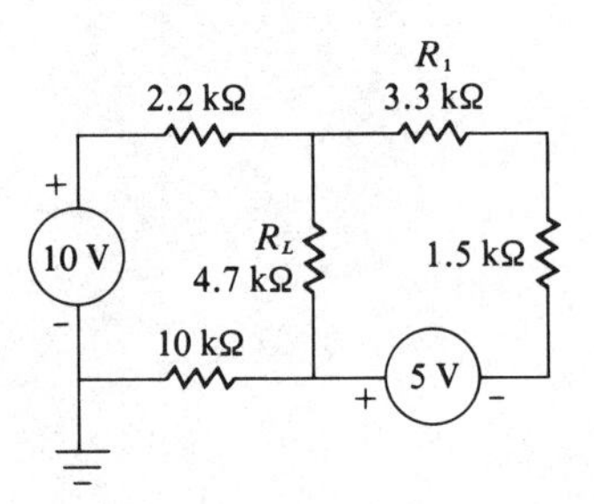

Figure 9.70 Reference Problems 6, 7, and 11

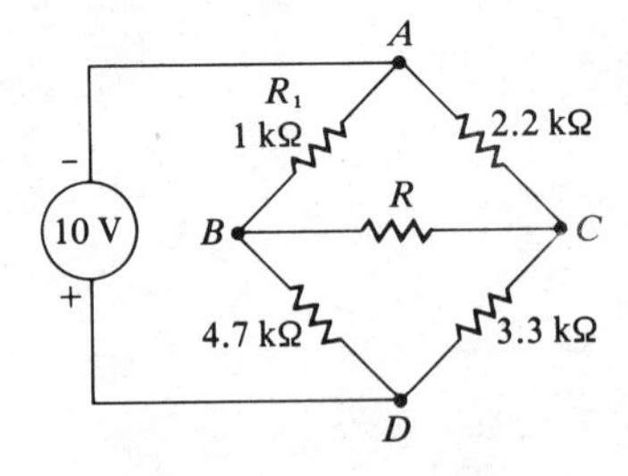

Figure 9.71 Reference Problems 8, 9, 10, 12, and 13

Reference Section 9.5

14. What should be the value of R in Figure 9.72 if it is to receive maximum power? What is the value of this maximum power?
15. What value of R in Figure 9.73 will receive maximum power? What is the value of this maximum power?

Reference Section 9.6

16. Using the Superposition theorem, determine the current through the 5.6 kΩ resistor of Figure 9.74.
17. Determine the current through the 3.3 kΩ resistor in Figure 9.75.
18. Determine V_{AB} in Figure 9.76.

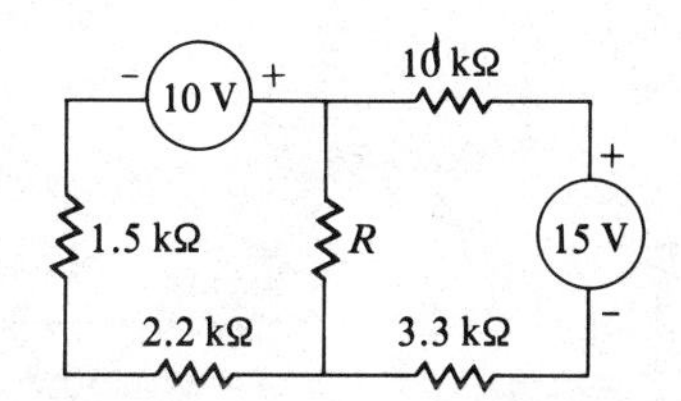

Figure 9.72 Reference Problem 14

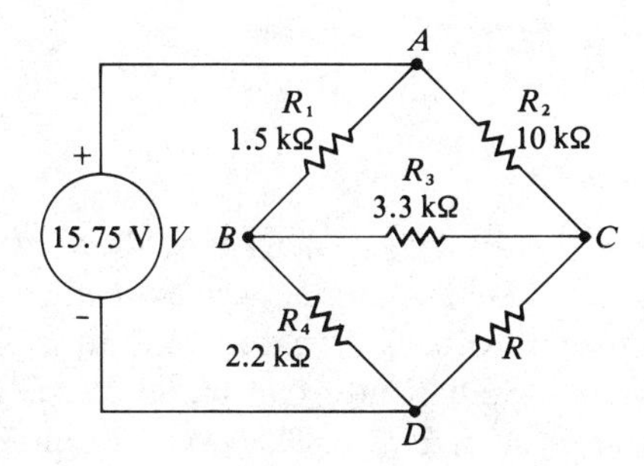

Figure 9.73 Reference Problems 15, 19, 28, and 29

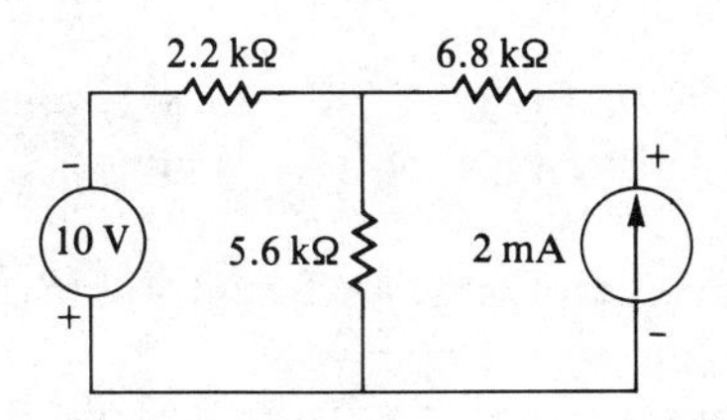

Figure 9.74 Reference Problem 16

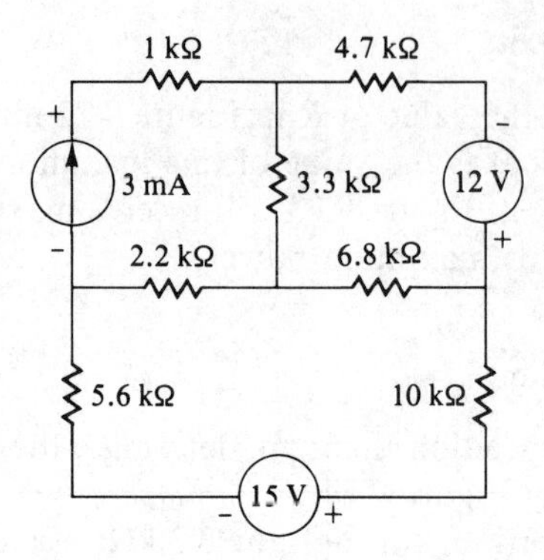

Figure 9.75 Reference Problem 17

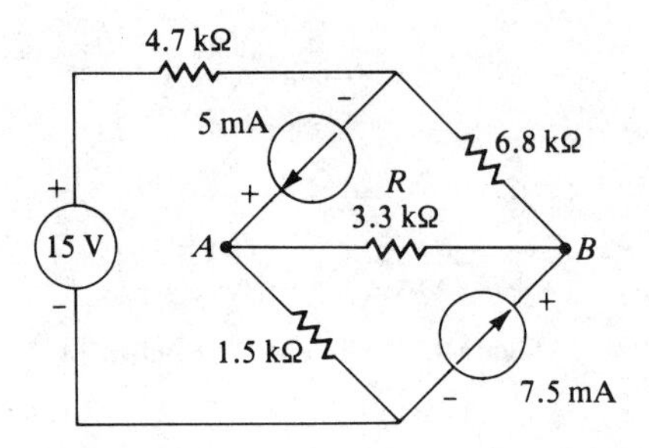

Figure 9.76 Reference Problem 18

Reference Section 9.7

19. If R in Figure 9.73 were 4.7 kΩ, determine the value of the wye equivalent of delta ABC. Determine the wye equivalent of delta BCD.
20. Construct two back-to-back wye networks to replace the circuit of Figure 9.77. (*Hint:* Split R into two parallel resistors.)
21. Assume each resistor of Figure 9.52 were 2 ohms. Using delta-wye transformations, show that R_{AB} equals 1.67 ohms.

Reference Section 9.8

22. If the value of R in Figure 9.78 is 22 kΩ, what is R_{in} for the third section of the ladder-type resistive network? What is R_{in} for the seventh section?
23. Determine the multiplying coefficients of R_{in} for each section of Figure 9.79.
24. Using the methods described in the text, determine the input resistance of the "bridged-T network" of Figure 9.80 if R equals 47 kΩ.

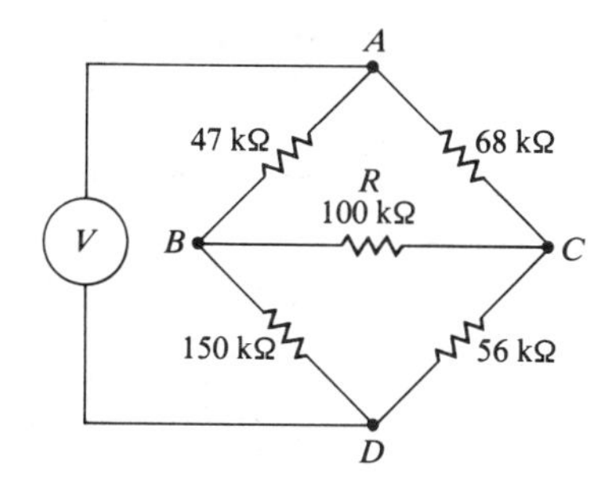

Figure 9.77 Reference Problem 20

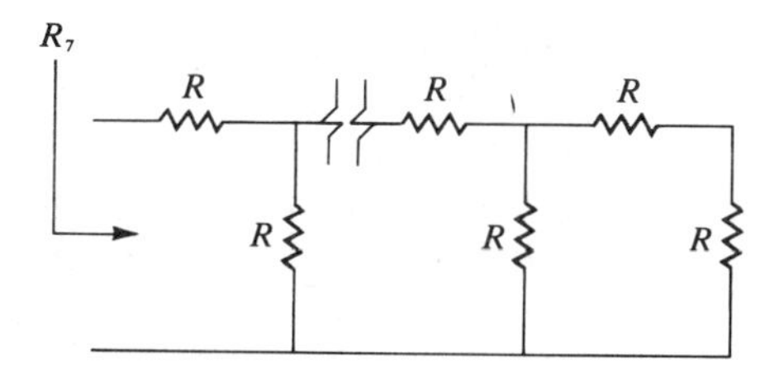

Figure 9.78 Reference Problem 22

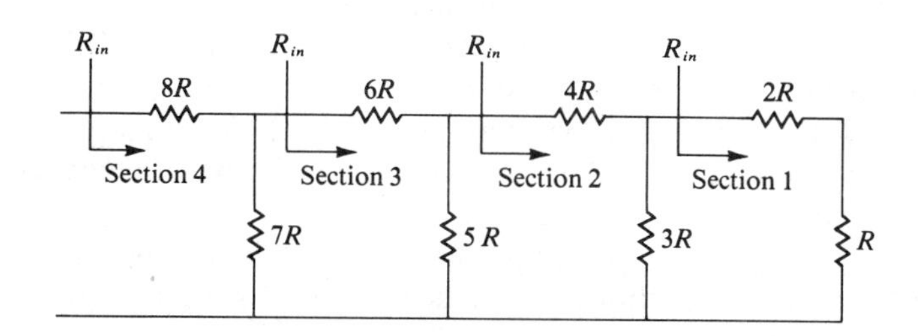

Figure 9.79 Reference Problem 23

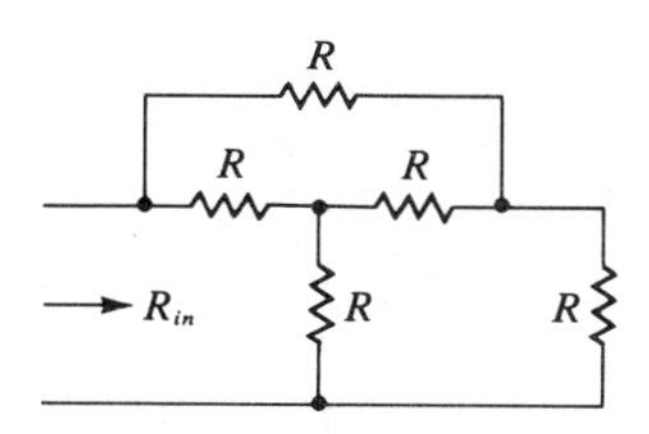

Figure 9.80 Reference Problem 24

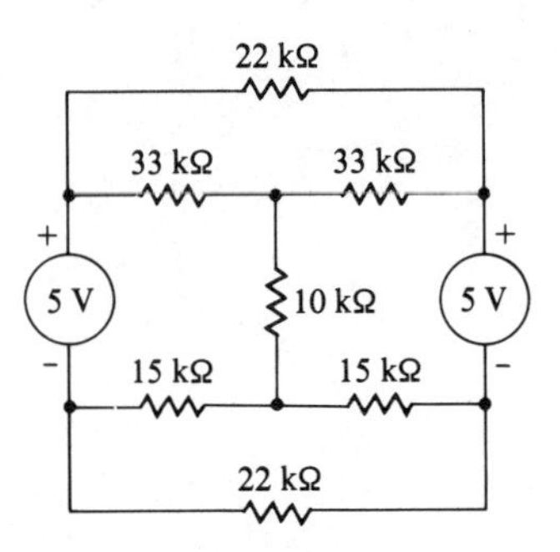

Figure 9.81 Reference Problem 27

Reference Section 9.9

25. Using symmetry theorems, determine the total source current if V of Figure 9.52 were connected between points H and A. Assume V to be 10 volts and each resistor to be 22 kΩ.
26. Repeat Problem 25 for V connected between points H and B.
27. Determine the current through the 10 kΩ resistor of Figure 9.81, first by matrix solution and then by using Bartlett's bisectional theorem.

Optional Computer Problems

28. Write a program in BASIC to determine the power in R of Figure 9.73, if R varies from 1 to 10 kΩ in 0.5 kΩ increments. Print out the value of R and the power developed across it.
29. Reference Figure 9.73, write a program in BASIC that would determine the value of V which would develop 35 mW of power across R if it had a value of 4.7 kΩ.
30. Using BASIC, write a program to calculate and print out the value of the first 25 sections of a ladder-type resistive network for any input value of R.

PART III

The basic theory governing many electronic components and electron devices requires an understanding of magnetism, electromagnetism, and electrostatics. The operation of a large variety of coils, inductors, relays, and switches is based on magnetic fields, cores, and electromagnetic circuits.

In addition to its common application as the electron gun assembly of a television picture tube, electric charge (and its associated electric fields) is the foundation of the operation of capacitors, electrostatic motors, and modern computer memories.

For this reason a reasonably thorough treatment of magnetism, inductance, electrostatics, and capacitance is given in the following chapters.

CHAPTER **10**

MAGNETISM

10.1 BASIC PROPERTIES OF MAGNETISM

The principles of magnetism and electromagnetism offer the student who is new to electrical science some of the greatest challenges for discovery. These principles, being the least understood, are consequently the most fascinating.

History tells us that the ancients were fascinated with a stone that was discovered on the shores of the Mediterranean Sea. They originally believed that this *lodestone* contained mystical powers and forces—forces so strong they could move objects and exert physical pressure on other objects. Today we understand the mystery of the lodestone—an iron oxide called *magnetite*—although we do not fully understand the phenomenon that produces it. The recent discoveries in space research of the unit magnetic pole have excited scientific communities all over the world.

In Chapter 2 we learned that electrons orbited the nucleus of an atom in a predetermined pattern. These orbital patterns are specifically described in terms of energy levels. Each electron has what we might call an *energy address*. This address is related to a distance from the nucleus of the atom. Each electron has a spin direction which further identifies the energy address. Refer to Figure 10.1. Note that this atom of magnesium has two electrons in the first orbit each of which has a rotational pattern as shown (the direction is arbitrary). The remaining outer electrons also have rotational patterns, although the total atom may or may not exhibit magnetic properties.

It is believed that the motion of the moving electron gives rise to the magnetic field that surrounds it. For example, in Figure 10.1(b), note that the motion of the electron is indicated as left to right. The magnetic field that surrounds it has the direction shown. The direction of these magnetic fields can be determined by the right-hand rule: If the thumb points in the direction of conventional current, the remaining fingers point in the direction of the magnetic field.

When we consider a molecule made up of several atoms, the net magnetic field surrounding this molecule is so small as to be almost insignificant. However, if some external magnetic force is applied, the axes and spin directions of some of the orbiting electrons are aligned. The result is a magnetic field surrounding the molecule which has a greater density and intensity than the original.

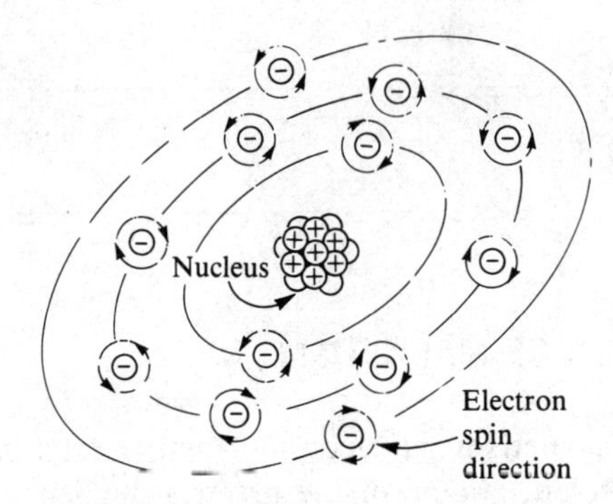

Figure 10.1(a) Pictorial View of Magnesium Atom

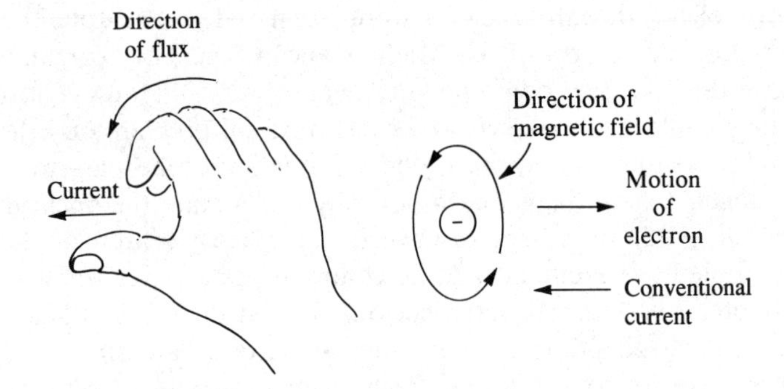

Figure 10.1(b) The Right-Hand Rule

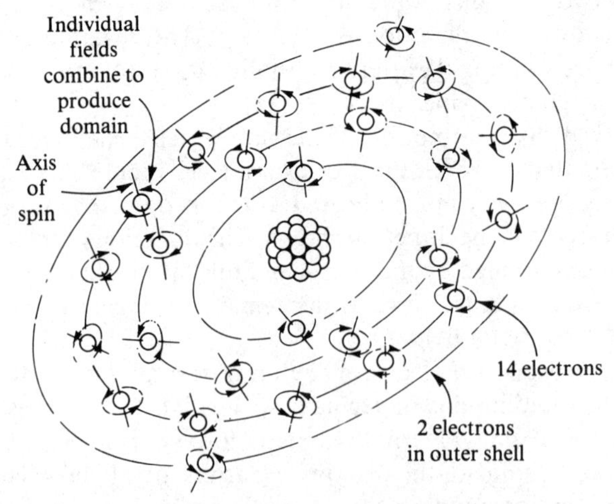

Figure 10.1(c) Iron Atom

Imagine the excitement of those ancient men when they passed the lodestone near other materials and found the new materials to possess the mystical powers of the original!

Magnetic Materials

Today we classify three basic types of magnetic behavior: diamagnetic, paramagnetic, and ferromagnetic. Some of the diamagnetic materials, such as mercury, copper, or gold, are repelled by a magnetic field; for example, an atom of copper placed in a magnetic field will be slightly repelled by that field. Extremely exacting measuring instruments must be used to detect the repelling force. On the other hand, paramagnetic materials are influenced by the forces of the external magnetic flux. The electron spins are altered when subjected to the influence of the external magnetic flux, and the result is a slight attraction. Aluminum is an example of a paramagnetic material. Both of these types are generally considered as nonmagnetic since the influence of a magnetic field on electron spins is very slight.

Ferromagnetic substances, such as iron or nickel, are strongly influenced by external magnetic fields. As an example, iron has an atomic number of 26. The structure, similar to Figure 10.1(c), has two electrons in its outermost shell. In its second outermost shell, it has 14 electrons. The alignment of the axes and spin rotations of these 14 electrons gives rise to the magnetic field surrounding the atom. In essence the individual fields have an additive effect. The composite produces a magnetic field about the molecule that is called a magnetic *domain*.

When they are close enough to each other, the magnetic domains become larger. A bar magnet is an example of the cumulative effect of magnetic domains. The maximum magnetization, known as *saturation*, occurs when all the possible magnetic domains are aligned.

10.2 CHARACTERISTICS OF MAGNETIC FIELDS

We have been discussing magnetic forces and magnetic fields without actually describing their characteristics. No one has ever seen a magnetic line of force, hence at best we speculate about its properties through theories. A magnetic field may be described as that region around a magnetic substance which influences a suspended compass needle. The accuracy of identifying the boundaries of a magnetic field is limited by the quality and sensitivity of our measuring instruments. Eventually, as the magnetic flux diminishes, the mass of the compass needle will cease to be influenced.

Properties of Magnetic Fields

The physicist Michael Faraday first described the field of force that surrounds a magnet by identifying the field by specific lines of force. The line of force theory is sustained partly because of the iron filing experiments used to show the presence of a magnetic field. If a bar magnet is placed underneath a piece of white paper and iron filings are sprinkled over the paper, the filings will arrange themselves in a pattern similar to Figure 10.2(a). Though these appear to be distinct lines of force, they may in fact only represent a region of relative intensity. Faraday's explanation does allow an easy access to understanding magnetism.

Earlier we illustrated through Figure 10.1 that magnetic lines of force have the characteristic of *direction.* The assignment of this direction is purely arbitrary. The scientific community has established that the magnetic line of force moves in a direction from north to south outside a permanent magnet and from south to north within the magnet. Figure 10.2 illustrates some of the characteristics of magnetic lines of force.

The second characteristic of magnetic lines of force is *continuity.* This continuous loop is known as a *maxwell.* It is believed that a magnetic line of force does not exist in segments. Any interruption in the closed or continuous loop of a magnetic line of force causes the entire line to be nonexistent. In Figure 10.2(b) note that the closed loop is continuous even though it is distorted by a second magnet.

The third characteristic is that of *lateral repulsion.* This property causes magnetic lines of force to develop a fairly uniform distribution within a given medium. For example, in Figure 10.2(b) note that six maxwells are shown. Their direction outside the magnet is from north to south. Note, however, that within the magnet the spacing between each line is fairly uniform. This uniformity does not imply that there is equal spacing everywhere the field exists. Rather, the property of mutual repulsion tends to cause symmetry.

The fourth property is that of *elasticity.* Each line of force will occupy as small a space within its closed loop as is possible. This property causes a tension along the length of the line of force. In Figure 10.2(b) the outermost loops are larger than the innermost loops. Because of the properties of lateral repulsion, the density of the domains, and the tendency of magnetic fields to maximize their number, the elasticity of the inner lines is greater.

The fifth property of magnetic forces deals with *attraction and repulsion.* Simply stated, *unlike magnetic poles attract each other and like magnetic poles repel.* Figure 10.2(c) illustrates the force of repulsion. Note how the magnetic fields are distorted. Figure 10.2(d) shows that a force of attraction results from the combining of magnetic domains and their propensity to set up the field of force in the minimum available area. The closer the

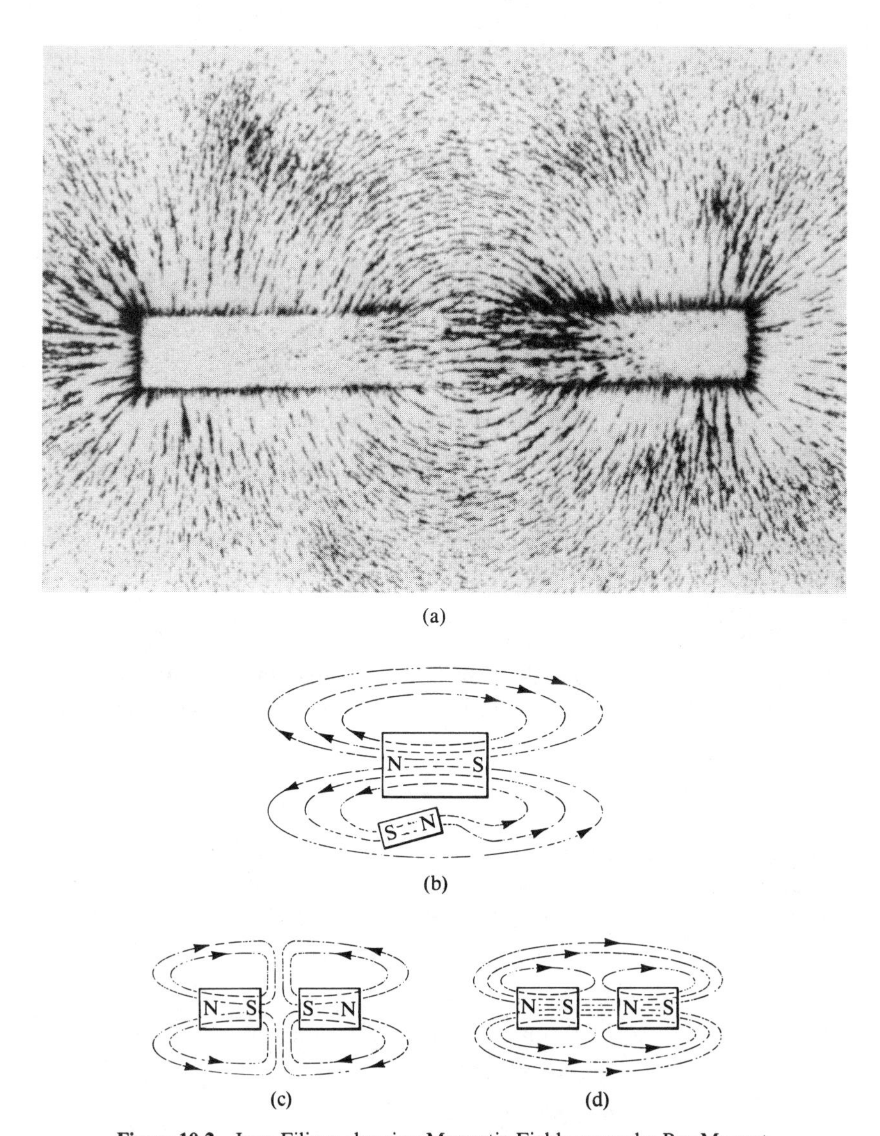

Figure 10.2 Iron Filings showing Magnetic Fields around a Bar Magnet

two unlike poles are brought together, the greater the force exerted between each magnet. This force, the same one that fascinated the ancient wise men, was later explained by Coulomb:

$$F = \frac{M_1 M_2}{\mu d^2} \tag{10.1}$$

where F is the force in dynes, $M_1 M_2$ the magnetic pole strength, μ the permeability (to be discussed below), and d the distance between the poles. Note the distance d in Eq. (10.1) is squared; hence the rapidly diminishing force as the magnetic poles are spread farther apart.

Magnetic Flux

One line of magnetic flux has been identified as a maxwell. Unfortunately the maxwell is an extremely small measure of magnetic flux. The symbol ϕ is used to designate magnetic flux. In the SI system, the unit of magnetic flux is the weber, (wb).

$$1 \text{ weber} = 10^8 \text{ maxwells}$$

The small magnet of Figure 10.2(b) represents a device with 6 lines of force, hence 6 maxwells. The total flux in the SI system is

$$\phi = 6/10^8 \text{ webers}$$

Flux Density

Flux density B is the number of flux lines passing through a given cross-sectional area (A). It is measured in webers per square meter, or Tesla (T).

$$B = \phi/A \qquad (10.2)$$

Example 10.1 A rectangular bar magnet has the following dimensions: 5 mm wide, 4 mm deep, and 100 mm long. If 1000 lines of flux are emanating from the north pole face, what is the flux density at the face of the magnet?

$$B = \phi/A$$

$$A = 4 \text{ mm} \times 5 \text{ mm} = 20 \text{ (mm)}^2$$

$$= (20)(10^{-6}) \text{ m}^2$$

$$\phi = 1000/10^8 = 10^{-5} \text{ wb}$$

$$B = 10^{-5}/(20)(10^{-6}) = 0.5 \text{ wb/m}^2 = 0.5 \text{ T}$$

Permeability

In establishing the characteristics of magnetic flux, little was said about the medium in which the magnetic flux operated. *Permeability* describes the ability of a given medium to support or concentrate magnetic flux.

In Figure 10.3(a), 2 maxwells pass through a given area of free space. If the area is 1 square millimeter, the flux density B is 0.02 wb/m^2. If that same area is replaced by a ferromagnetic material, the magnetic flux will concentrate to a greater density. Figure 10.3(b) illustrates this con-

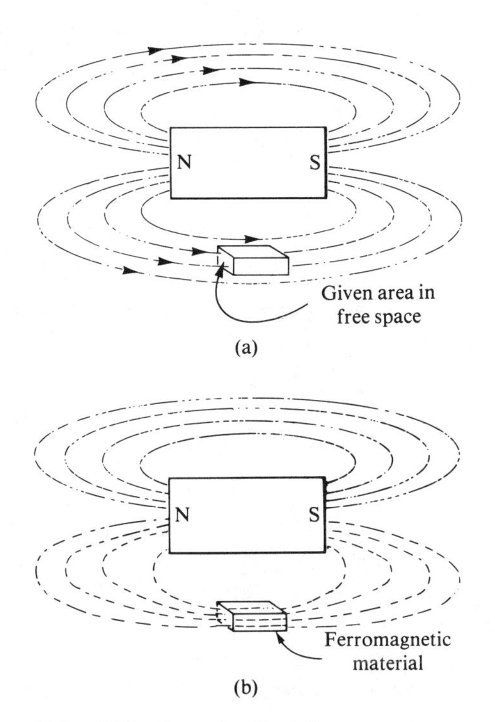

Figure 10.3 Example of Relative Permeability

centration. The new material has 4 maxwells in the same area and has a flux density of 0.04 wb/m^2. By definition, its *relative* permeability is .004/.002 = 2.

$$\mu_r = \text{relative permeability}$$
$$= B_x/B_A \tag{10.3a}$$

where B_x is the flux density in a given material and B_A is the flux density in free space.

Permeability of various substances can be likened to the electrical conductivity materials. Some materials are more conductive than others depending on their molecular structure. The permeability of free space is

$$\mu_o = 4\pi \times 10^{-7} \text{ weber/ampere-meter}$$

The permeability of free space is used as a reference in determining the relative permeabilities of other materials. Most nonmagnetic materials have permeabilities very near that of free space. For example, bismuth, glass, copper, wood, and aluminum all have relative permeabilities very nearly equal to 1.

On the other hand, the permeability of nickel is $2\pi \times 10^{-5}$, so its relative permeability is $2\pi \times 10^{-5}/4\pi \times 10^{-7} = 50$.

$$\mu_r = \frac{\mu_x}{\mu_o} \tag{10.3b}$$

where μ_x = permeability of a given material
$\mu_o = 4\pi \times 10^{-7}$
μ_r = relative permeability

Example 10.2 A given brand of permalloy has a permeability of 0.012π webers/ampere-meter. What is its relative permeability?

From Eq. (10.3),

$$\mu_r = \frac{0.012\pi}{4\pi \times 10^{-7}} = 30{,}000$$

Reluctance

In describing permeability we likened it to the electrical conductivity of a material. In an analogous manner magnetic materials have a counterpart to resistance. It is called *reluctance*. Reluctance, $\mathcal{R}$, describes that property of a medium to oppose the establishment of magnetic flux. Reluctance is not just the reciprocal of permeability since the length of the medium and the cross-sectional area also influence the establishment of the magnetic circuit

$$\boxed{\mathcal{R} = \frac{\ell}{\mu A}} \tag{10.4}$$

where ℓ is length in meters, A is area in square meters, μ is permeability, and $\mathcal{R}$ is reluctance in rels. (The units of rels are ampere-turns/weber.)

Example 10.3 What is the reluctance of a cylindrical tube of iron whose relative permeability is 5000? Assume the tube has a radius of 0.002 m and a length of 15 mm.

$$\mu = \mu_r\mu_o = (5000)(4\pi)(10^{-7}) = 2\pi \times 10^{-3}$$

$$A = \pi r^2 = \pi(0.002)^2 = 4\pi \times 10^{-6}$$

$$\mathcal{R} = \frac{(15)(10^{-3})}{(2\pi)(10^{-3})(4\pi)(10^{-6})}$$

$$= 0.19 \times 10^6 \text{ rels}$$

10.3 ELECTROMAGNETISM

In preceding sections, we described the characteristics of magnetism as it pertained to permanent magnets. A ferromagnetic material such as iron or cobalt, once magnetized, exhibits a specific flux density at the north and south poles. The domains remain established until the magnet is demagnetized. Many of the molecules can be disordered through either severe physical impact or increased temperature, both of which break down the magnetic domains. Either method will vary the flux density at the poles. If a varying density field is to be used in circuits and devices, permanent magnets pose severe limitations.

In 1820, Hans Christian Oersted discovered that a magnetic field can be established by placing an electric current through a coil of wire. Once the current is turned off, the magnetic field surrounding the magnet disappears. The varying current resulted in a varying density magnetic field surrounding the wire. We illustrated in Figure 10.1(b) that an electron in motion possesses a magnetic field around it. The intensity of the magnetic field was dependent on the velocity of the electron. In passing an electric current through a wire, the electrons move through the wire at a fairly constant rate. The intensity of the magnetic field that surrounds the electrons, and hence establishes magnetic domains around the wire, is dependent upon the quantity of electrons.

Figure 10.4 illustrates a wire in which a flow of electrons is passed from the negative to the positive terminal of the source. The compass needles surrounding the wire indicate that the magnetic fields form circular loops extending out from the center of the wire. If the compass is moved around the wire, the needle traces a path which forms the complete magnetic line of force.

The intensity of the magnetic field surrounding the wire is given as H. This intensity varies directly with the current passing through the wire and inversely with the distance d from the center.

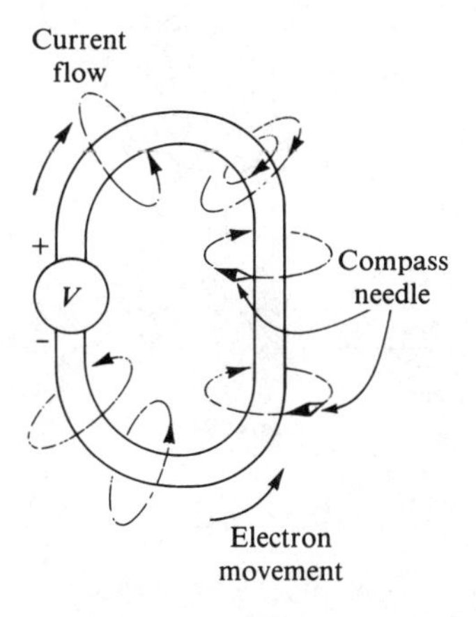

Figure 10.4 Magnetic Field around Current Carrying Conductor

$$H = \frac{I}{2\pi d} \tag{10.5}$$

where I is in amperes and d is in meters.

If the wire forms a complete loop as shown in Figure 10.5, the magnetic field surrounding the loop of wire enters one side of the loop and exits the other. If the right-hand rule is applied to this single loop, the thumb points in the direction of the current through the loop. The magnetic flux enters the loop from the viewer's side of the page. The north pole, therefore, is on the other side of the loop.

When several loops are connected together as shown in Figure 10.6(a), the magnetic domains become cumulative and an electromagnet results

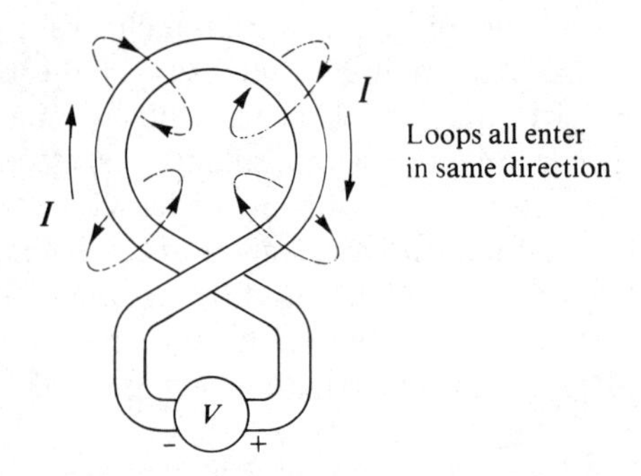

Figure 10.5 Magnetic Field around a Single Loop

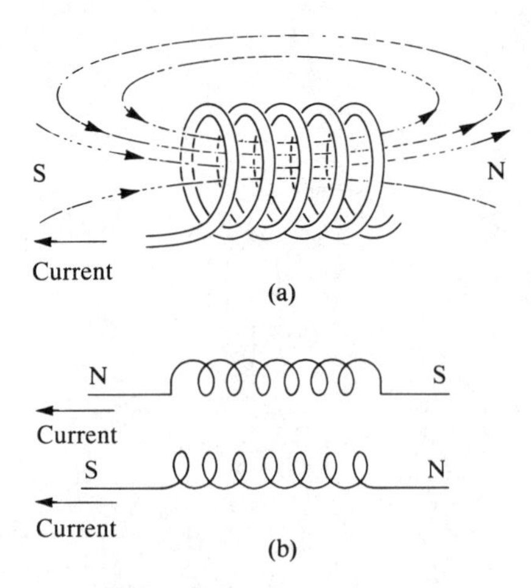

Figure 10.6 Electromagnets

with a specific north and south pole. Figure 10.6(b) shows two coils. At first glance they look the same, with current entering from the right and leaving at the left. Yet the magnetic poles are not the same. Figure 10.7 is an expanded view showing the directions of the winding on a hollow cylinder. Note that the wires in part (a) thread downward in a clockwise direction. In part (b) the wire threads downward but in a counterclockwise direction. The figures indicate the appropriate poles for each case. It is important that the downward threading be observed from the end into which electrons flow.

10.4 MAGNETIZING FORCE

In Figure 10.6(a) we established an electromagnet using loops of wire and passing a current through it. The properties of the magnetic field surrounding the electromagnet are the same as those surrounding a permanent magnet. Since Oersted established that flux intensity is dependent upon the current, it follows, then, that the total flux ϕ and the flux density B will also be influenced by the current.

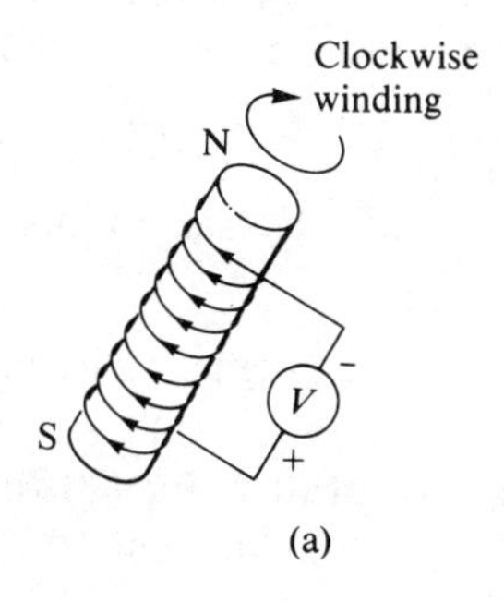

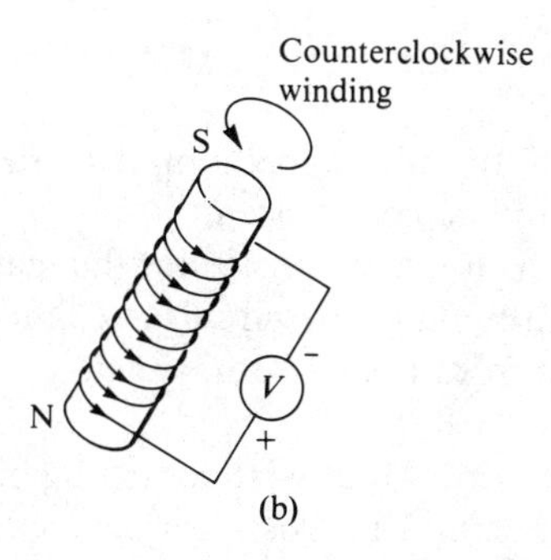

Figure 10.7 Winding Direction and Pole Identification

We can assume that the flux density of either pole of an electromagnet is influenced by the current. The flux density is also dependent upon the geometry of the coil. If we wind 10 turns of wire on a pencil, then slip out the pencil, we will have a coil able to produce a given flux density at either pole end. If the same 10 turns are stretched out to twice the length of the original (diameter unchanged), the density will decrease.

Magnetomotive Force

A coil of wire similar to that of Figure 10.6(a) will produce a magnetomotive force (*mmf*) identified by the symbol $\mathcal{F}$. The units of *mmf* are ampere-turns (At).

$$\boxed{\mathcal{F} = NI} \tag{10.6}$$

where N is the number of turns (or loops) and I is current in amperes.

It is sometimes convenient to show how magnetic circuits compare with electrical circuits. Since the electric current in a circuit does the work (when moved about by voltage), we can compare the flux ϕ of a magnetic circuit to current. The force moving the current is voltage, hence in a magnetic circuit the magnetomotive force $\mathcal{F}$ is what moves the flux. Similarly, the opposition to current is resistance. In magnetic circuits the reluctance $\mathcal{R}$ describes the opposition that a medium offers to the establishment of flux.

$$I = \frac{V}{R} \quad \text{for electric circuits}$$

$$\phi = \frac{\mathcal{F}}{\mathcal{R}} \quad \text{for magnetic circuits} \tag{10.7a}$$

For magnetic circuits if we substitute in Eq. (10.7a) the equivalent values for $\mathcal{F}$ and $\mathcal{R}$ from Eqs. (10.4) and (10.6), then

$$\phi = \frac{\mu ANI}{\ell} \text{ webers} \tag{10.7b}$$

where μ is the permeability of the medium, A is the cross-sectional area, and ℓ is the length of the magnetic circuit.

The question arises as to what constitutes the magnetic circuit. In Figure 10.8 the length of the magnetic circuit is ℓ, and not how far out the magnetic field extends from the electromagnet.

Example 10.4 An air core coil of wire consisting of 45 turns is wound to produce a length of 12 mm. The cross-sectional area is 2×10^{-5} square meters. What total flux is produced when 500 mA of current flows? What is the flux density within the coil?

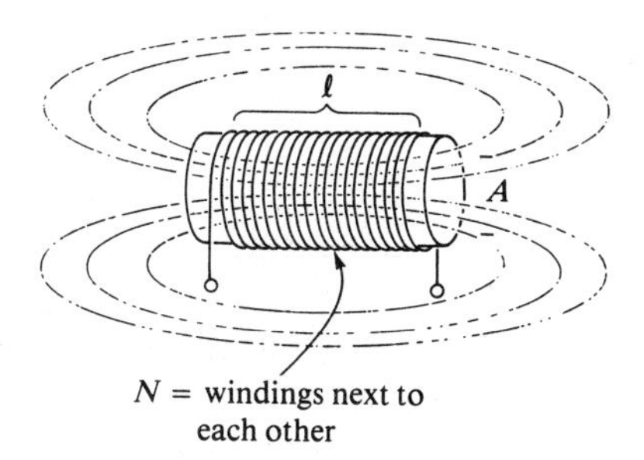

Figure 10.8 Magnetic Field Intensity

From Eq. (10.7b),

$$\phi = \frac{(4\pi)(10^{-7})(2)(10^{-5})(45)(0.5)}{(12)(10^{-3})}$$

$$= 4.71 \times 10^{-8} \text{ weber}$$

$$B = \phi/A$$

$$= \frac{(4.71)(10^{-8})}{(2)(10^{-5})} = 2.355 \times 10^{-3} \text{ wb/m}^2$$

Magnetic Field Intensity

Magnetic field intensity identifies the magnetomotive force over a given magnetic length. For example, in Figure 10.8 the field intensity within the center core area of the coil is determined by the $\mathcal{F}$ and the length occupied by the windings.

$$H = \frac{\mathcal{F}}{\ell} \text{ ampere-turns/meter (At/m)} \qquad (10.8)$$

Often this magnetic field intensity is referred to as a magnetic gradient in the circuit. This magnetic field intensity is also influenced by the medium in which the flux exists. Given

$$H = \mathcal{F}/\ell$$

from Eqs. (10.4) and (10.7) we have

$$\mathcal{F} = \phi\mathcal{R} \qquad \mathcal{R} = \frac{\ell}{\mu A}$$

Substituting in Eq. (10.8),

$$H = \frac{\phi\ell}{\ell\mu A}$$

Since ϕ/A equals B,

$$\boxed{H = \frac{B}{\mu}} \tag{10.9}$$

Example 10.5 A cylinder of iron whose relative permeability is 5000 is to have 1000 turns of wire wound on its length of 10 mm. How much current is required to produce a flux density of 0.02 webers/m^2?

Solving Eqs. (10.2) and (10.7b) for I,

$$I = B\ell/\mu N = B\ell/\mu_r\mu_o N$$

$$= \frac{(0.02)(0.01)}{(5000)(4\pi)(10^{-7})(10^3)} = 31.83\ \mu\text{A}$$

10.5 MAGNETIZATION CURVES

The permeability of most magnetic materials is not linearly related to B, as Eq. (10.9) might indicate. If that equation is written

$$\boxed{\mu = \frac{B}{H}} \tag{10.10}$$

it would appear that the permeability is a constant ratio between the flux density and the intensity. This, in fact, is not the case. If a doughnut-shaped section of mu metal similar to Figure 10.9 has several windings wrapped around the torus and a current is passed through the windings, the field intensity in the toroid increases as the current increases. With the increasing intensity the flux density also increases. As illustrated in Figure 10.11, however, increasing the current beyond a certain point does not

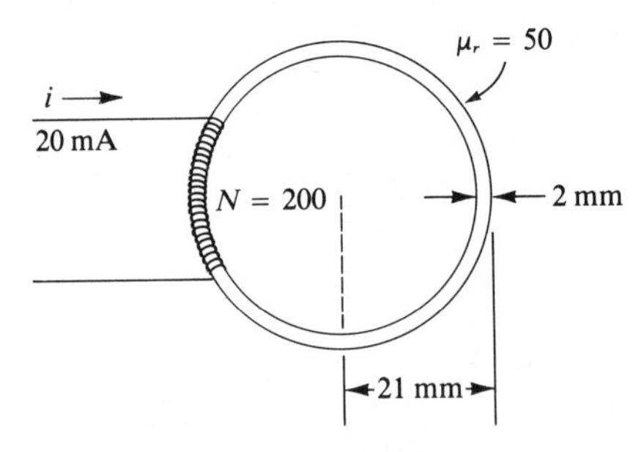

Figure 10.9 Figure for Example 10.6

increase the flux density above 0.6 wb/m^2. This phenomenon deals with the establishment of magnetic domains. The maximum number of magnetic domains has been established when the density of mu metal reaches 0.6 wb/m^2. Equation (10.10) does not reflect this *saturation* condition.

Most materials, such as cast iron, steel, cobalt, and many alloys, have saturation points of maximum flux density. Figure 10.10 shows a typical magnetization curve. Note that the linear region extends from the origin to point A. In this region the ratio of flux density to flux intensity remains constant and is equal to the permeability of the material. Since the region is linear, the ratio can also be determined by the ratio of the B to H increments from the origin to point A on the curve.

The region from point A to point B is identified as the "knee" of the characteristic. The knee region is different for each particular ferromagnetic material. The region from point B to point C is the saturation region. Some materials move abruptly from one region to another, while others slide gradually from one to the other. Each of these materials has particular advantages for certain applications. These applications will be explored later in the chapter.

Figure 10.11 compares the magnetization curves of cast steel and mu metal. Note that the intersection of the two curves represents the same flux density and intensity although the mu metal is already completely saturated while the cast steel is still in its linear region.

The curves shown in Figures 10.10 and 10.11 illustrate how the permeability varies with changes in field intensity. These curves begin at the origin of the B/H characteristic and show the behavior of permeability with increasing density and intensity. For most magnetic materials, this initial curve does not accurately depict the actual pattern of flux variation. The figures show how the magnetization pattern changes if the material starts from an initially demagnetized state.

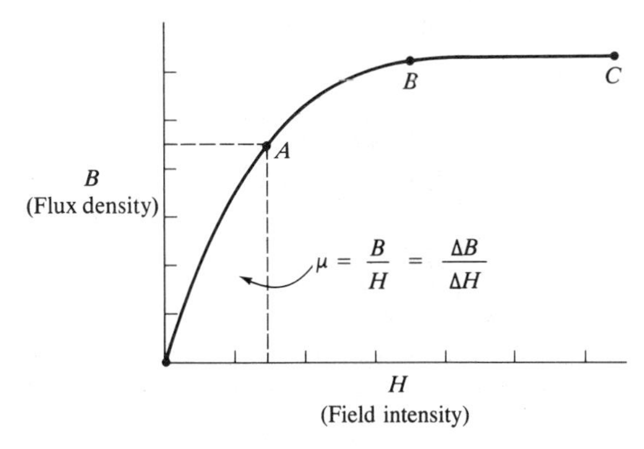

Figure 10.10 Magnetization Curve

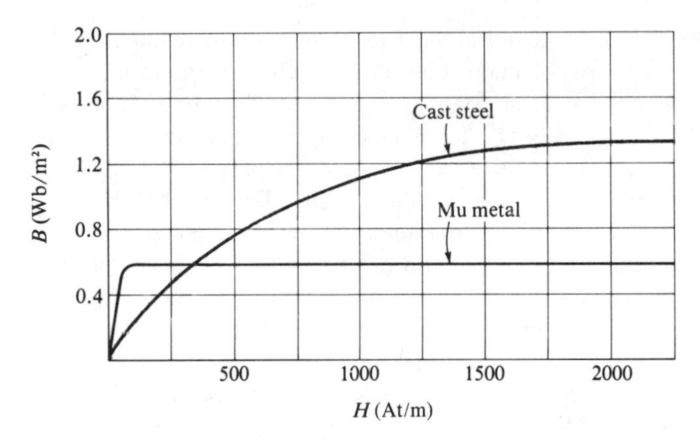

Figure 10.11 Typical B/H Characteristics for Cast Steel and Mu Metal

Consider the toroid illustrated in Figure 10.12(a). The variable resistor allows a measure of control over the current I and therefore the corresponding variation of flux ϕ through the winding and core. As the current increases, the flux density within the toroid increases up to a saturation point; then, as the current is decreased, the flux density decreases but does not follow the same pattern of variation as it did in the upward excursion.

In Figure 10.12(b) note the solid line extending diagonally from the origin to point A. The small arrow indicates the direction of change. Point A represents a field intensity of H_A and a flux density of B_A. These two conditions were brought about as a result of a current level that we will identify as I_A. As the current increases, the field intensity increases to point B with a corresponding increase in the flux density. These two points are indicated as H_B and B_B. A further increase of current increases the intensity but the density has stabilized at a level of B_B. The magnetic lines have become as closely packed as the toroid material will sustain. The intensity of individual lines increases, but the number of those lines within a given area does not.

Decreasing the current down to a level that corresponds to a field intensity of H_A does not decrease the flux density. Indeed the flux density remains at its B_B point even though we have decreased the current through the winding. If we finally decrease the current in the circuit to zero, we know that the field intensity has decreased to zero but the flux density remains at a value E. The toroid has become magnetized as a result of the effects of the electromagnetic windings. This remaining magnetization is called *residual magnetism*. In order to demagnetize the toroid, it would be necessary to reverse the polarity of the power supply and increase the current in the opposite direction. A given reverse current would be required to produce a field intensity of H_G.

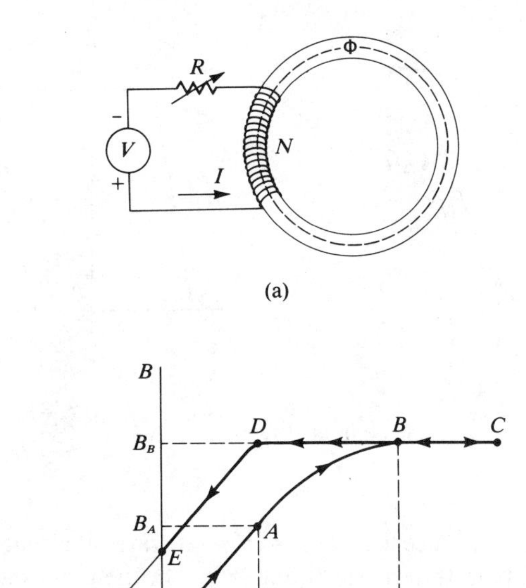

Figure 10.12 Magnetization Characteristics for Unmagnetized Toroid

The field intensity from zero to H_G is identified as the *coercive force*, H_c. The H_c differs for each material. Any further increase in the reverse current would cause the flux in the toroid to change directions and magnetize in the opposite direction. If the magnetization were continued through saturation and the current reduced again to zero, the result would be a remnant flux density again equal to B_E except with the opposite polarization. Figure 10.13 shows the complete B/H characteristics for two ferrite materials. Note in Figure 10.13(b) that the remnant flux density remains at its saturated value B_r after the H has been decreased to zero.

Hysteresis

As the current passing through the toroid of Figure 10.12(a) was increased sufficiently to cause saturation, then reversed to cause saturation in the opposite direction, then reversed once again, the B/H characteristic of Figure 10.13(a) resulted. The characteristic of "lagging behind" of the flux density is called *hysteresis*, B_r. Since the hysteresis is a measure of a specific flux density, it is given in webers per square meter. As an example, if a low carbon content iron is saturated, the remnant flux or

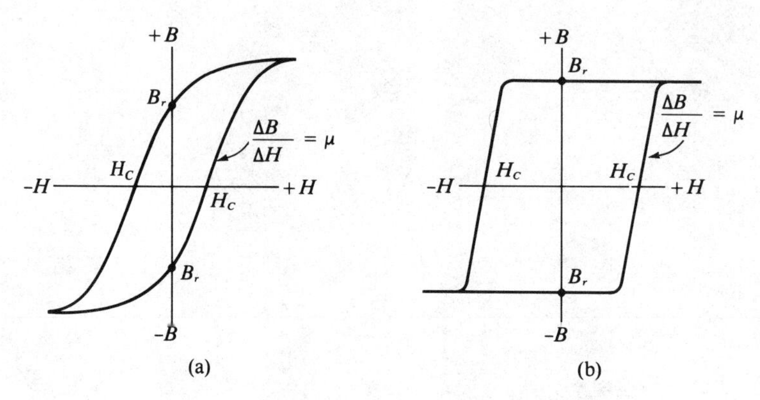

Figure 10.13 Hysteresis Loops

B_r will be 0.6 wb/m^2. Since low carbon iron saturates at about 2.15 wb/m^2, the remaining flux in the iron is about 28%. On the other hand, an alloy of nickel, iron, and copper will saturate at 0.6 wb/m^2 and will have a hysteresis of 0.25 wb/m^2. About 42% of the flux remains.

Coercive Force

The coercive force H_C was defined earlier and illustrated in Figure 10.13. Soft materials such as mu metal, permalloy, and hyperm have low coercive forces of about 2 to 4 ampere-turns per meter. On the other hand, hard materials such as carbon steel, tungsten, and materials containing aluminum and platinum have coercive forces ranging to many thousands of ampere-turns per meter. Once magnetized it takes considerable energy to demagnetize.

10.6 *B/H* CHARACTERISTICS

A considerable amount of information is available from the *B/H* characteristic. Its general shape can imply certain applications. For example, a characteristic similar to Figure 10.13(b) would be useful for magnetic core memories. On the other hand, a characteristic with practically no "window" (hence insignificant B_r and H_c) would be useful for relays and magnetic control devices.

The quantitative values of a curve are also useful in design work. In Figure 10.14 note that several loops are superimposed upon one another. The loop bounded by points *C* and *D* identifies the *B/H* characteristics if the iron material is not saturated. In this case the flux intensity *H* is only increased to about 125 At/m in each direction. The loop bounded by *A* and *B* indicates a greater use of the characteristic.

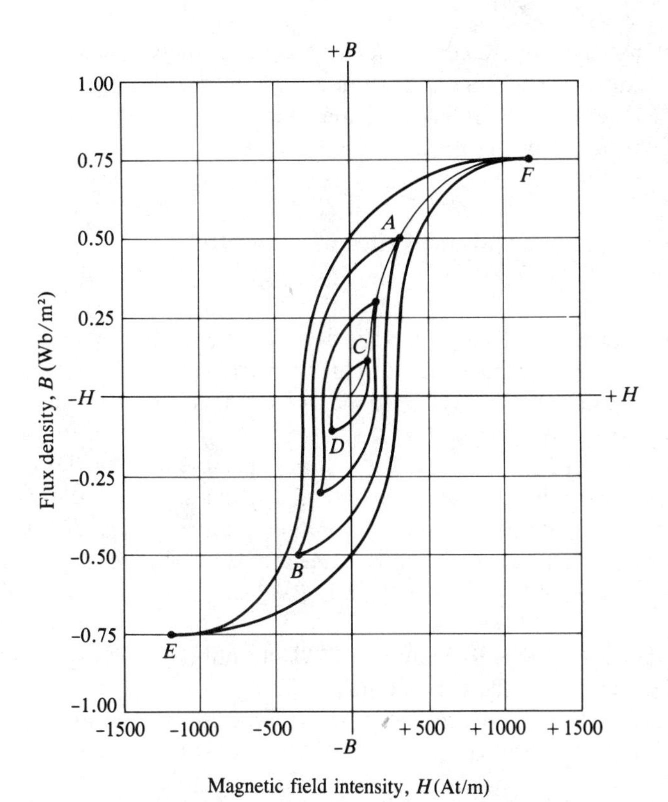

Figure 10.14 Multiple Hysteresis Loops

Example 10.7 Refer to the example of the magnetic material of Figure 10.14. What is the coercive force and the hysteresis? Assume the H is increased to ± 375 At/m.

Since H will vary between points A and B on the loop, the actual values can be determined. From Figure 10.14,

$$H_c = \pm 250 \text{ At/m}$$
$$B_r = 0.375 \text{ wb/m}^2$$

Average Permeability

If a magnetic core is not saturated, the range of operation is somewhere within the smaller hysteresis loops of Figure 10.14. This figure illustrates a case where the flux intensity was not increased into the saturation region. Many devices intentionally do not operate in the saturation region. In fact, a saturated core may reflect disrupting characteristics to the circuit driving the current through the windings.

In many cases where core saturation is not achieved, it is often necessary to know the average permeability. From the average permeability we may determine the relative permeability.

The permeability was given in Eq. 10.10 as

$$\mu = B/H$$

The average value of μ is illustrated in Figure 10.13.

$$\mu_{\text{ave}} = \Delta B/\Delta H \tag{10.11}$$

The average value (10.11) equals the actual value (10.10) only in the linear region of the B/H characteristic. This property can be illustrated from a numerical example. In Figure 10.11, assume an H of 250. The resultant B is approximately 0.38 for cast steel. Using Eq. (10.10), $\mu = B/H = 0.38/250 = 1.52 \times 10^{-3}$. If we assume an increment in the linear region, the permeability results. Assume the H changes from 250 to 500 to produce a ΔH of 250. The corresponding change in B is ΔB. Its value changes from 0.38 to 0.76 for an increment of 0.38. Therefore, $\mu = \Delta B/\Delta H = (0.76 - 0.38)/(500 - 250) = 1.52 \times 10^{-3}$.

Example 10.8 Refer to Figure 10.14. Determine the average value and the relative value of the permeability.

Again H will vary from 375 A/m to -375 A/m. (points A and B). According to Eq. (10.11):

$$\mu = \frac{0.5 - (-0.5)}{375 - (-375)} = 13.33 \times 10^{-4} \text{ wb/A-m}$$

From Eq. (10.3),

$$\mu_r = \frac{13.33 \times 10^{-4}}{4\pi \times 10^{-7}} = 1.061 \times 10^{3}$$

(The relative permeability has no units since it is a ratio.)

Hysteresis Loss

Magnetic devices using iron do not operate in circuits employing only direct current as the energizing agent for flux intensity. Indeed, the current may be pulses (abrupt changes) or it may be alternating in each direction at some specific rate. When this occurs, the flux in the iron is constantly changing its orientation. Since this implies that energy transformations are taking place, some energy is converted to heat. The continual realignment of electron spins and magnetic domains causes some heating depending upon the material. The amount of energy converted to heat, and therefore not usable in the system, is described as *hysteresis loss*. Generally the energy in question can be determined from the B/H characteristic.

Hysteresis loss is measured in joules per cycle (cycle referring to one complete alternation of current first in one direction and then the next). The amount of hysteresis loss is dependent upon the volume of the magnetic material. Understandably, the larger the volume of iron, the greater the loss. It is also dependent upon coercive force and the hysteresis.

$$W_L = V_m A_{\text{loop}} \tag{10.12}$$

where V_m represents the volume of the material in cubic meters, A_{loop} is the area of the hysteresis loop in NI wb/m^2, and W_L is the hysteresis loss in joules/cycle. If we refer to the B/H characteristic of Figure 10.14, we could solve for the approximate area within a loop by counting the number of squares (and portions of squares) enclosed by the loop.

Example 10.9 Assume the toroid of Figure 10.9 has an inside diameter of 9 cm and an outside diameter of 10 cm, and the core is a hollow tube. If 600 mA flows through 2200 turns, what are $\mathcal{F}$, $\mathcal{R}$, ϕ, and B?

$$\mathcal{F} = NI = (2200)(0.6) = 1320 \text{ At}$$

$$\text{Area} = \pi r^2 = (3.1416)(0.5)^2(10^{-2})^2 = 0.785 \times 10^{-4} \text{ m}^2$$

$$\ell = \pi d = (3.1416)(9.5)(10^{-2}) = 29.845 \times 10^{-2} \text{ m}$$

$$\mathcal{R} = \ell/\mu A = \frac{(29.845)(10^{-2})}{(4\pi)(10^{-7})(0.785 \times 10^{-4})} = 3.024 \times 10^9 \text{ rels}$$

$$\phi = \mathcal{F}/\mathcal{R} = 4.365 \times 10^{-7} \text{ wb}$$

$$B = \phi/A = 5.56 \times 10^{-3} \text{ wb/m}^2$$

In the preceding example the flux density is quite low since the core of the ring is actually free space. If the core were replaced by a ferrous material, the density of flux would increase considerably, yet the magnetomotive force would not change because the NI would still be 1320 At.

10.7 THE MAGNETIC CIRCUIT

Section 10.4 established the relationship between Ohm's law of electric circuits and the relational application to magnetic circuits. We will now expand upon this concept. The magnetic structure of Figure 10.12(a) shows a magnetomotive force generated through a winding on a loop of iron. The equivalent magnetic circuit is shown in Figure 10.15. Note that the number of turns times the current is symbolized as a generator $\mathcal{F}$. The ϕ symbolizes the total flux in the iron toroid. The $\mathcal{R}$ represents the reluctance of the iron. The value of $\mathcal{R}$ would be determined by the geometry and type of material used in the core.

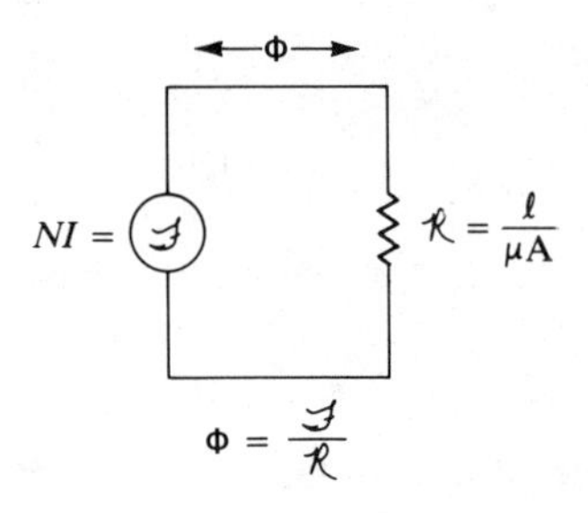

Figure 10.15 Magnetic Circuit

Series Circuits

If the physical device representation were not a toroid but a rectangular-shaped structure similar to Figure 10.16(a), then the equivalent circuit would have to represent the reluctance characteristics of each leg of the rectangle. As in a series electric circuit, the reluctances merely add up to some total reluctance and the circuit would reduce to the one of Figure 10.15.

In Figure 10.16(b) the applied magnetomotive force $\mathscr{F}$ results in magnetomotive force drops around the circuit across each value of reluctance.

$$\mathscr{F}_t = \mathscr{F}_1 + \mathscr{F}_2 + \mathscr{F}_3 + \mathscr{F}_4$$

Since $\mathscr{F} = H\ell$ and $\mathscr{F} = NI$,

$$NI = H_1\ell_1 + H_2\ell_2 + H_3\ell_3 + H_4\ell_4 \tag{10.13}$$

where ℓ is the length of the iron and N, I, and H are as defined earlier. The following two examples illustrate the methods of magnetic circuits.

Example 10.10 Assume the magnetic device of Figure 10.16(a) has the dimensions shown and the magnetic material is transformer iron (Figure 10.17). What current in the winding will set up a total flux of 3.6×10^{-4} weber?

$$A = (2 \times 10^{-2})(2 \times 10^{-2}) = 4 \times 10^{-4} \text{ meter}^2$$
$$B = \phi/A = (3.6 \times 10^{-4})/(4 \times 10^{-4}) = 0.9 \text{ wb/m}^2$$

From Figure 10.17,

$$H = 400 \text{ ampere-turns/meter}$$

The total (mean) length is $(2 \times 16 \text{ cm}) + (2 \times 10 \text{ cm}) = 52$ cm
Since $H_1 = H_2 = H_3 = H_4 = H$, from Eq. (10.13)

$$NI = (400)(52)(10^{-2}) = 208$$
$$\therefore \quad I = 200 \text{ mA}$$

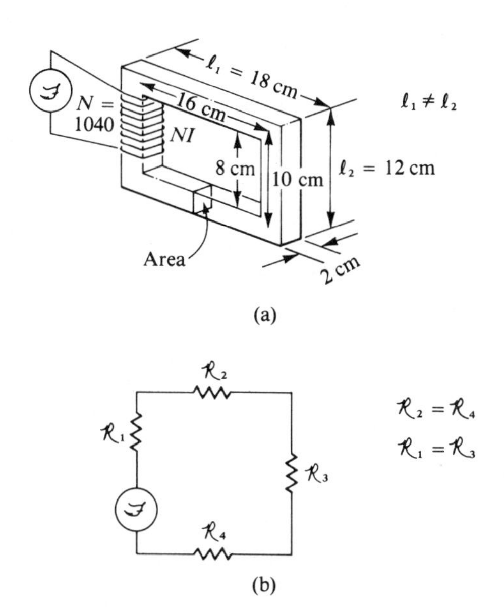

Figure 10.16 Series Magnetic Circuit

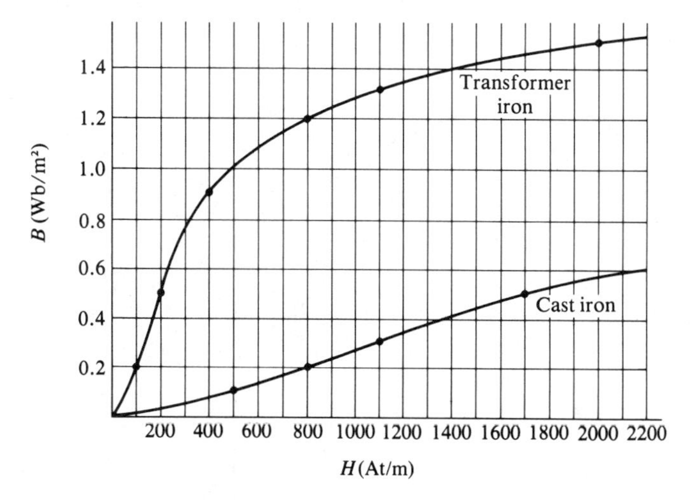

Figure 10.17 *B*/*H* Characteristics

In Example 10.10 we assumed that each iron section of the rectangle consisted of the same magnetic material and that the cross-sectional area was the same. In cases where the dimensions change and the material changes throughout the magnetic circuit, a different approach can be used.

Refer to the iron fixture of Figure 10.18. Although we have shown only three lines of force in the iron, in reality many thousands of lines would be present. The three lines can illustrate our point. In order to sustain three lines of force in the length and cross-sectional area of side *A*, a certain magnetomotive force would be required. If three lines of force are to be established in side *B*, a greater mmf would be required since side *B* is longer than *A*. An even greater magnetomotive force would be required for side *C* because it has a smaller cross-sectional area.

We can conclude that (1) in a series magnetic circuit, the flux will be constant throughout the circuit; and (2) the greater the reluctance, the greater the magnetomotive drop. Now we can establish a general approach to the solution of series magnetic circuits.

1. Identify the areas of each leg of the circuit.
2. Identify the permeability of each leg.
3. Identify the reluctance of each leg.
4. Identify the unknowns such as N, I, and $\mathcal{F}$.

Example 10.11 Refer to the magnetic device of Figure 10.18. Assume that the device is made of transformer iron. The number of turns of the winding is 200. What current will develop a magnetic flux of 200 lines?

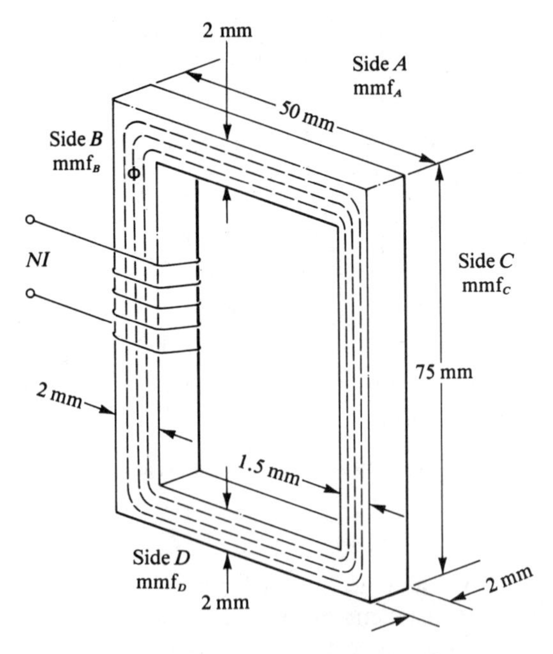

$$NI = \text{mmf}_A + \text{mmf}_B + \text{mmf}_C + \text{mmf}_D$$

Figure 10.18 Series Magnetic Circuit

The cross-sectional area of legs A, B, and D is:

$$A = 2 \text{ mm} \times 2 \text{ mm} = 4 \text{ mm}^2$$

The cross-sectional area of leg C is:

$$A = 2 \text{ mm} \times 1.5 \text{ mm} = 3 \text{ mm}^2$$
$$\phi = 2 \times 10^2/10^8 = 2 \times 10^{-6} \text{ weber}$$

The flux densities of each leg are:

$$B_A = B_B = B_D = \phi/A = \frac{2 \times 10^{-6}}{4 \times 10^{-6}} = 0.5 \text{ wb/m}^2$$

$$B_C = \frac{2 \times 10^{-6}}{(3)(10^{-6})} = 0.667 \text{ wb/m}^2$$

From the curves of Figure 10.17,

$$H_A, H_B, H_D = 200 \text{ At/m}$$
$$H_C = 250 \text{ At/m}$$

At this point, we could solve for I using Eq. (10.13). (Do this as an exercise.) Alternatively, we can proceed as follows:

$$\mu_A, \mu_B, \mu_D = B/H = 0.5/200 = 2.5 \times 10^{-3}$$
$$\mu_C = 0.667/250 = 2.67 \times 10^{-3}$$
$$\ell_A = \ell_D = 48.25 \text{ mm}, \quad \ell_B = \ell_C = 73 \text{ mm}$$

$$\mathcal{R}_A = \mathcal{R}_D = \ell/\mu A = \frac{(48.25)(10^{-3})}{(2.5)(10^{-3})(4)(10^{-6})} = 4.825 \times 10^6$$

$$\mathcal{R}_B = \frac{(73)(10^{-3})}{(2.5)(10^{-3})(4)(10^{-6})} = 7.3 \times 10^6$$

$$\mathcal{R}_C = \frac{(73)(10^{-3})}{(2.67)(10^{-3})(3)(10^{-6})} = 9.125 \times 10^6$$

$$\mathcal{R}_t = 2(4.825 \times 10^6) + (7.3)(10^6) + (9.125 \times 10^6)$$
$$= 26.075 \times 10^6$$

$$I = \frac{\phi\mathcal{R}}{N} = \frac{(2)(10^{-6})(26.075)(10^6)}{200} = 260.75 \text{ mA}$$

One very important series magnetic circuit consists of an iron structure which includes an air gap. The gap itself presents no particular problems if we treat it and its dimensions as an additional magnetic medium within the circuit. Refer to the circuit of Figure 10.19. Note that the doughnut-shaped toroid has a small gap cut into one end. All the continuous flux lines in the iron must pass through the gap. The magnetomotive force required to set up the flux is developed by the current through the windings. However, the reluctance of the air gap is much greater than

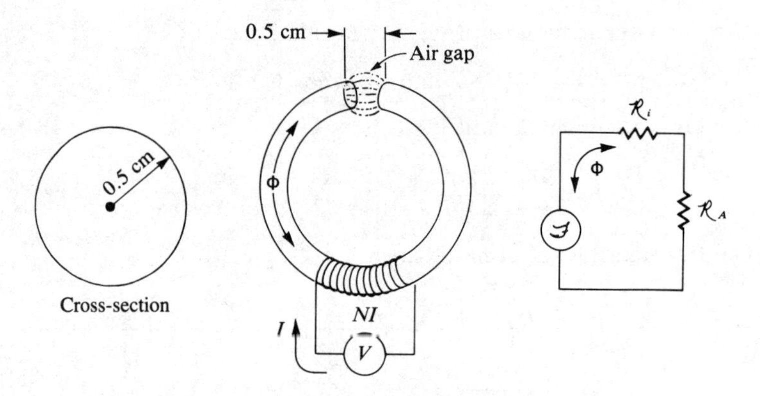

Figure 10.19 Magnetic Circuit Including an Air Gap

the reluctance of the iron; hence most of the magnetomotive force developed by the turns will be dropped across the air gap. This is the case even though the air gap is physically much smaller than the iron. The following example illustrates this case.

Example 10.12 Assume the circular toroidal magnet of Figure 10.19 has the following dimensions: outside diameter = 10 cm, inside diameter = 9 cm, length of the gap = 0.5 cm, and the number of turns = 1000. Assume the magnet is cast iron. What will be the current required to set up a total flux of 5×10^{-5} wb?

$$\text{Cross-sectional area} = A = \pi r^2 = (3.1416)(0.5 \times 10^{-2})^2 = 0.785 \times 10^{-4}$$

$$\ell_i = \text{average length of iron} = (\pi d - 0.5) \text{ cm}$$

$$= 29.84 \text{ cm} - 0.5 \text{ cm} = 29.34 \text{ cm}$$

$$\text{Length of air gap} = \ell_a = 0.5 \text{ cm}$$

From Figure 10.17,

increments arbitrarily chosen for μ

$$\mu = 0.3/1100 = 2.7 \times 10^{-4}$$

$$\mu_A = 4\pi \times 10^{-7}$$

$$\mathcal{R}_a = \frac{\ell}{\mu A} = \frac{(0.5)(10^{-2})}{(4\pi)(10^{-7})(0.785 \times 10^{-4})} = 50.68 \times 10^6$$

$$\mathcal{R}_i = \frac{(29.34)(10^{-2})}{(2.7)(10^{-4})(0.785 \times 10^{-4})} = 1.38 \times 10^7$$

$$NI = \phi \mathcal{R}_t = (5)(10^{-5})(6.448)(10^7) = 3224$$

$$I = 3224/1000 = 3.224 \text{ amperes}$$

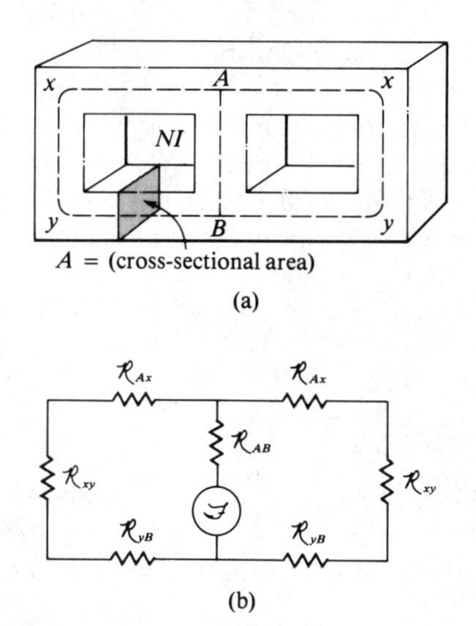

Figure 10.20 Series-Parallel Magnetic Circuit

It should be noted from Example 10.12 that the reluctance of the air gap is 3.7 times that of the iron. In many practical circuits the reluctance of the iron is so small in comparison to that of the air gap that it can be neglected in the calculations.

Series-Parallel Circuits

The basic principles for solving series and parallel magnetic circuits are in many ways the same as for solving electric circuits. If the iron circuit and any gaps or windings are reduced to their appropriate mmf's and $\mathscr{R}$'s, the circuit can be redrawn and unknowns determined.

Refer to Figure 10.20. We assume that a series of windings on the center leg will develop a magnetomotive force which will establish a flux ϕ that takes two paths. If the iron structure in both directions is symmetrical, the flux will divide evenly between the two paths. If the geometries and the materials are different, the flux will distribute according to those characteristics.

The procedure for solving for the current through the winding to produce a given flux is similar to the procedures already discussed. For example, for the structure of Figure 10.20 we might assume two equal flux fields in both branches. We would solve for the areas and the average length of each segment of the hardware. It would be necessary to determine the flux density for each leg. We would then determine the flux intensity and finally solve for the current.

Example 10.13 Assume the circuit of Figure 10.20(a) has the following dimensions. The distance A to B is 4 cm. The distance $Ax + xy + By = 12$ cm. Assume that the transformer iron is used and the cross-sectional area throughout is 0.4×10^{-4} meter^2. What current through the 2000 windings will produce a flux of 0.3×10^{-4} weber in the xy leg?

The equivalent circuit is shown in Figure 10.20(b).

The flux density in each outside branch is

$$B = \phi/A = \frac{(0.3)(10^{-4})}{(0.4)(10^{-4})} = 0.75 \text{ wb/m}^2$$

The flux density in the center leg is

$$B = \phi/A = \frac{(0.6)(10^{-4})}{(0.4)(10^{-4})} = 1.5 \text{ wb/mA}^2$$

A second equivalent circuit can be drawn, as in Figure 10.21, from Figure 10.20(b).

$$H_{AxyB} = 300 \text{ A/m}$$

$$H_{AB} = 2000 \text{ A/m}$$

Solving for one loop,

$$\mathscr{F} = H_{AxyB}\ell_{AxyB} + H_{AB}\ell_{AB}$$

Since $\mathscr{F} = NI$,

$$I = \frac{(300)(12)(10^{-2}) + (2000)(4)(10^{-2})}{(2)(10^3)}$$

$$= 58 \text{ mA}$$

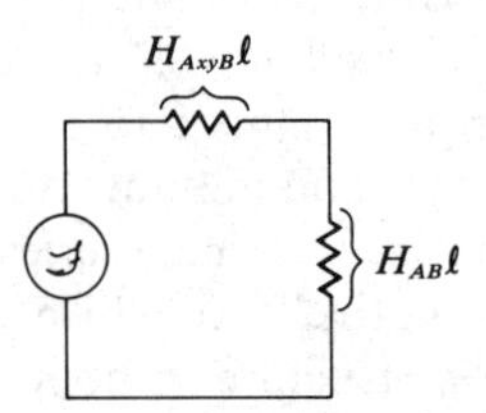

Figure 10.21 Equivalent Circuit of Figure 10.20

10.8 MAGNETIC BUBBLES

Most of the previous paragraphs have dealt with magnetic properties of ferromagnetic substances and electromagnets. Magnetic bubbles in certain substances were discovered in 1960. Not until 1969 did researchers

Bobeck and Perneski recognize the importance of magnetic bubble applications to digital computers.

The term *bubble* is used only to describe how a magnetic domain behaves. There is no actual bubble (air or otherwise) in a given substance. For our purposes a magnetic bubble is a cylindrical domain with a magnetization whose direction is opposite to the area around it.

Refer to Figure 10.22. The square slab of material has a magnetic field passing through it. The field and its direction are labeled as "external magnetic field." Within the material are small irregular shaped "bubbles" whose magnetic domain is in the opposite direction to the external field. The four bubbles shown can be positioned by external control. Their movement can be controlled by varying the flux density of the external magnetic field and a control field.

Magnetic bubble modules are fabricated of several layers of elements. Figure 10.23 shows a basic substrate (slab) of nonmagnetic material

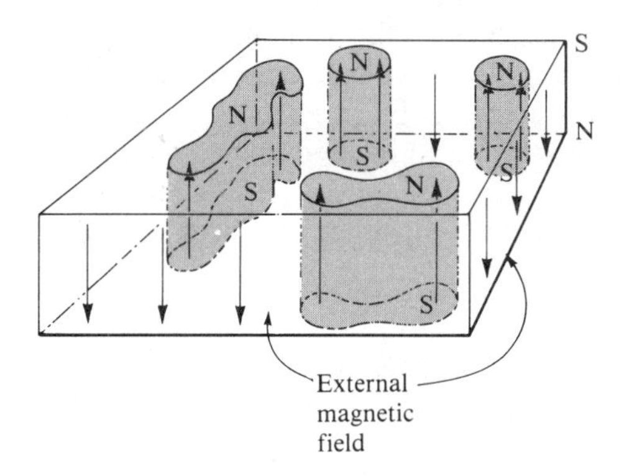

Figure 10.22 Bubble Magnetics

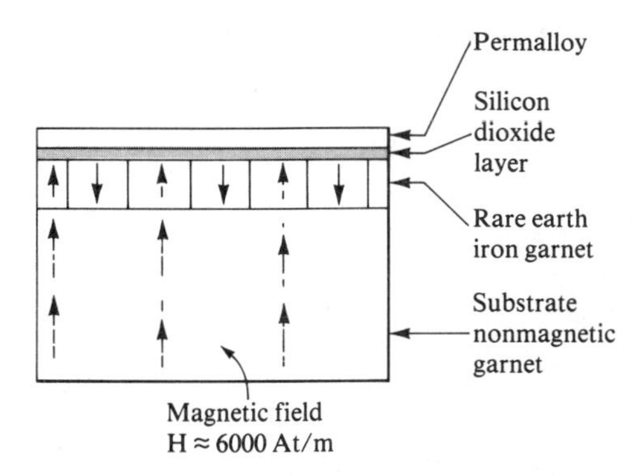

Figure 10.23 Structure of Bubble Slab

(usually about 0.25 mm thick). Chemically grown onto the substrate is the magnetic bubble garnet (about 0.0005 mm thick). Between the permalloy and the bubble garnet is a thin layer of silicon dioxide which is used as an insulator. The bubbles themselves are small; current state of the art shows them to be about 6 microns (0.006 mm).

Note in Figure 10.22 that the bubbles have different shapes. Some of the larger bubbles can change their shape, replicate, combine, split, and elongate depending upon the external control field.

Figure 10.24(a) is a minute representation of an actual data storage element employing magnetic bubble logic. The fixed field is the same field shown in Figure 10.23. Note in row 1 that two bubbles are positioned under three V-shaped patterns in the permalloy. As a group they could represent 101 (or a binary 5). Row 2 has a 110, and row 3 a 010. The bubbles can be propagated in a line, inch-worm style, through control currents in the lines. If a current is passed through control line 3 as in Figure 10.24(b), the bubble under the middle V will move to the end V. In this manner the bubble patterns can be moved along a control line.

Positioning the bubbles on the V-shaped pattern can be accomplished by applying a perpendicular control field to the iron garnet region. Figure 10.25 shows how the bubble domain positions itself relative to the V. The position along the V can be used to identify information states.

Bubble memory devices are now in production and are being delivered. Monsanto is one of the pioneers in this area. Bubble densities are now

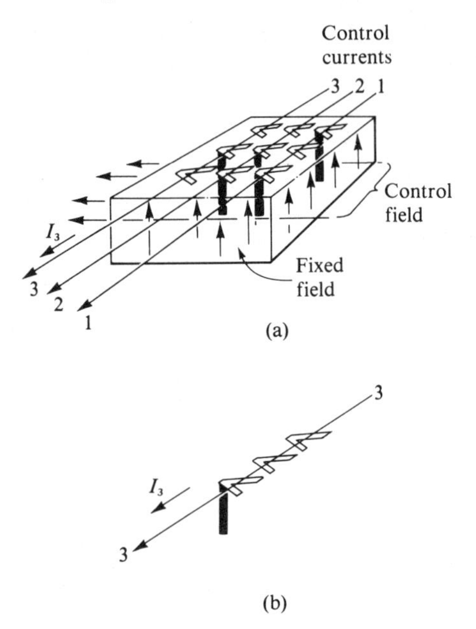

Figure 10.24 Segment of Bubble Magnetics

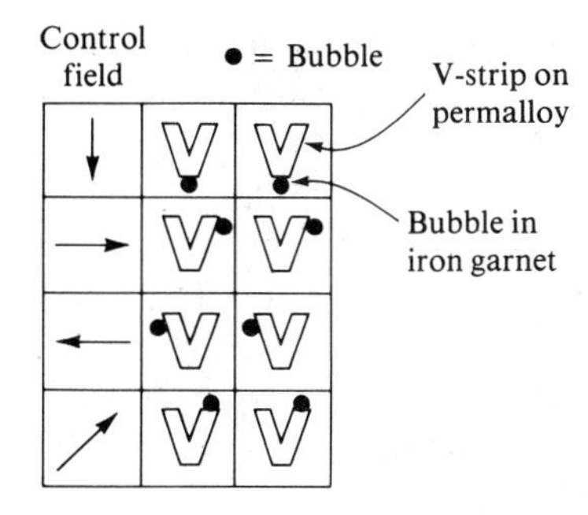

Figure 10.25 Bubble Positioning

as high as 10,000,000 bits per square inch. This density of computer memory is many times greater than other memories.

10.9 APPLICATIONS OF MAGNETIC PRINCIPLES

It would be impossible to explore all the possible applications of magnetism, electromagnetism, and bubbles. The following examples illustrate general principles.

Loudspeakers

The standard permanent magnet-type loudspeaker is actually an electro-magneto-mechanical device. It converts electrical energy to compressions and rarifications of the surrounding air. Figure 10.26 illustrates how this conversion can be accomplished.

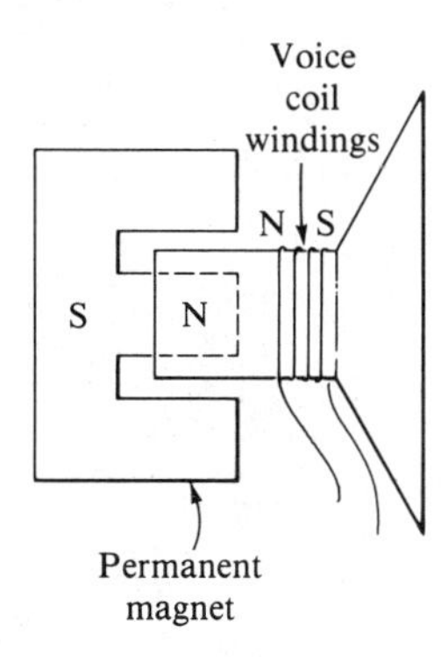

Figure 10.26 Loudspeaker

The large round permanent magnet has an outside ring with one round center pole. A cylinder attached to the cone has a voice coil wound around it. As fluctuations of current pass through the winding, alternating electromagnetic poles appear at the ends of the cylinder. The interaction of the two fields causes the cone to vibrate back and forth, setting up standing waves in the space in front of the cone and behind it. The outer edge of the cone is generally mounted to a frame, while the base and the cylinder upon which the voice coil is wound are allowed to float free over the permanent magnet. A mechanical force in newtons perpendicular to the magnetic pole is generated.

Relays and Switches

Figure 10.27 depicts two more electromagnetic mechanical devices whose objectives are to convert electrical energy into mechanical energy. Figure

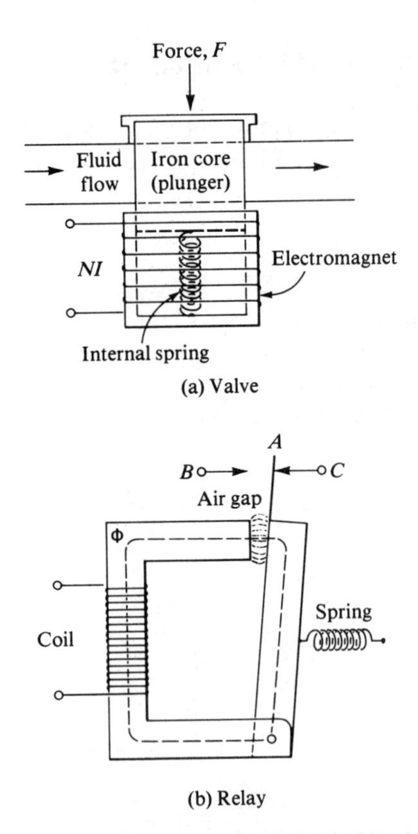

Figure 10.27 Electromagnetic-Mechanical Devices

10.27(a) schematically depicts a water flow valve similar to an electronically controlled fill valve on a washing machine. As current flows through the electromagnet, the iron core plunger is drawn into the core, allowing water to pass through the line. To move the plunger requires a mechanical force. The mass of the plunger must be accelerated.

$$F = ma$$

The energy required to accelerate the mass over a distance is given as

$$w = F\ell \qquad (10.14)$$

where F is the force in newtons and ℓ is the distance in meters. If we assume for now that the system has no loss due to friction, we conclude that the electrical energy in joules must equal the mechanical energy in newton-meters.

This principle can be seen also in the relay circuit. Figure 10.27(b) is a schematic of a three contact relay. When the relay is not energized, contacts A and C are closed. When the relay is energized, contacts A and C open and A and B close. In order to cause the relay to close, however, the force of the spring must be overcome via the magnetic circuit. In this system let us assume the total number of lines of force are to remain constant. The area of the air gap, the permeability of the air, and for now, the spring tension are constants as well. The total length (ℓ) of the magnetic circuit is the only variable. Given that

$$\mathscr{F} = \phi\mathscr{R} \qquad \text{and} \qquad \mathscr{R} = \ell/\mu A$$

then

$$\mathscr{F} = \phi\ell/\mu A$$

Since the magnetic lines of force will seek to occupy the smallest amount of space, $\mathscr{F}$ will vary as ℓ varies. If we calculate $\mathscr{F}$ over the range of change of $\mathscr{F}$, we have the average change

$$\mathscr{F}_{\text{ave}} = \phi\ell/2\mu A$$

Since $\phi/\mu A = H$ and $H\ell = \mathscr{F}$,

$$\mathscr{F}_{\text{ave}} = \mathscr{F}/2 \qquad (10.15)$$

The energy stored in the gap is

$$w = \mathscr{F}_{\text{ave}}\phi$$

Substituting Eqs. (10.14) and (10.15),

$$F = B^2 A/2\mu \qquad (10.16)$$

Equation (10.16) is the force in newtons acting to close the gap between the relay arm and stationary housing.

Shielding

Since we live in a world pervaded by magnetic flux from many sources, this same flux can be a hostile environment for some research and some operations. At present it is impossible to achieve a zero flux environment.

In systems where delicate instrumentation is involved, it is possible to reduce the flux to extremely low levels through shielding. A cylinder or housing of extremely high permeability may be used to "collect" the magnetic lines of force. In reality the existing lines of force take the path of least reluctance and concentrate in the high permeability housing. A material known as alloy 1040 (a combination of nickel, copper, molybdenum, and iron) can have a permeability as high as 110,000 and thus makes an excellent collector of magnetic flux.

Core Memory

The toroidal cores studied in preceding sections are used for the fabrication of core memory modules for digital computers. The cores, however, are much smaller. A typical core used in computer systems has an 18 mils

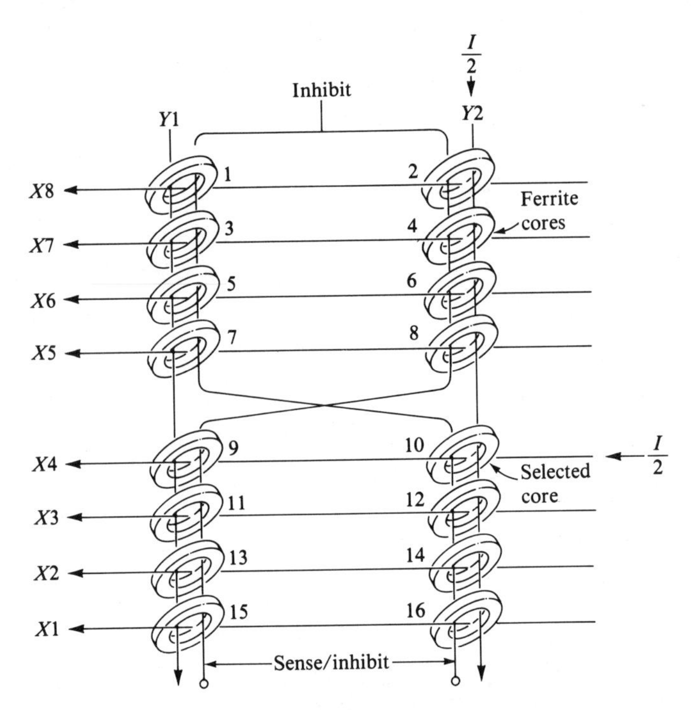

Figure 10.28 Core Memory

outside diameter, 11 mils inside diameter, and is 4.5 mils thick. The inside diameter is smaller than a human hair.

In the case of cores, there is no actual winding around each core loop. Each loop can be magnetized by merely passing a wire through the center. A typical core memory mat has three wires passing through each loop.

Figure 10.28 is a sample of a core memory. The B/H characteristic of each core is shown in Figure 10.13(b). A current passing through either an X winding or a Y winding is not sufficient to magnetize the core to saturation. However, a current passing through both an X and a Y winding will saturate a given core. For example in Figure 10.28, if a current is passed through Y_2 and X_4, only one core (core 10) will be magnetized. Such a magnetization can represent a logic 1. If the currents were reversed, the core would be magnetized a logical 0.

A typical memory mat consists of 8192 ferrite cores arranged in a 128 by 64 array. The third wire shown in Figure 10.28 is used for writing and reading logical data into and out of the core array. We may now compare the bit density of core memories with that of bubble memories. The 8192 bits in a core array can occupy an area of 40 square inches. This represents a density of about 200 bits per square inch. Magnetic bubble density is about 50,000 times greater.

SUMMARY

Magnetic lines are used to denote the force that exists around a permanent magnet or an electromagnet. The magnetic lines have properties of direction, continuity, lateral repulsion, elasticity, and attraction and repulsion. Magnetic flux density is given as $B = \phi/A$. Magnetic materials have permeability, which defines the ability to support magnetic flux. The reluctance of a material defines that material's ability to oppose the establishment of magnetic flux.

Electromagnetism is magnetic flux developed as a result of current flow through wire or a series of loops and windings. The greater the current the greater the flux intensity, H. A magnetization curve defines the B/H characteristics of a given material. The coercive force and the residual magnetism may be obtained from such a set of curves. The average permeability and the hysteresis loss may be determined from the B/H characteristics.

Magnetic circuits may be constructed similar to electric circuits. The magnetic circuit counterpart to Ohm's law is $\mathscr{F} = \phi\mathscr{R}$. Using the magnetic circuit equivalency, series and parallel magnetic circuits may be solved.

Magnetic bubbles are made up of small magnetic domains in a crystalline substrate. These bubbles can be ordered to make up combinations of binary numbers.

PROBLEMS

Reference Section 10.2

1. A magnetic flux of 0.25 weber consists of how many magnetic lines of force?
2. How many webers are represented by a magnetic flux which contains 80 kilolines?
3. If a flux density is given in webers per square meter, what is the flux density of 50 million maxwells passing through an area of 2 square meters?
4. A space of 4 square cm has 20,000 maxwells of flux passing through it. What is the flux density?
5. If a given flux density is 1500 maxwells per square cm, what would be the magnetic flux for an area of 5 square cm? What is the flux density, B?
6. In a given sample of magnetic flux 3 cm by 5 cm, the flux density was measured as 0.25 wb/m^2. How many total lines of force are passing through the area?
7. If a given material will support 5 times more flux than nickel, what is its permeability?
8. If the flux density in a given material is 0.03 wb/m^2, what will the new flux density be if the material is replaced by one which has a permeability of 20?

Reference Sections 10.3 and 10.4

9. What is the field intensity in ampere-turns per meter at a point 50 cm from the center of a wire carrying 20 mA of current?
10. Refer to Problem 9. What is the field intensity at a point twice as far from the center?
11. Refer to Problem 9. At what distance from the center of the wire will the field intensity drop to 10% of the intensity 1 cm from the center?
12. What is the magnetomotive force of a coil of wire with 1000 turns and a current flow of 25 mA through it?
13. If a given solenoid has an mmf rating of 750 and an ammeter indicates 120 mA of current passing through the coil, how many turns of wire are in the coil?
14. What is the reluctance of a substance if the mmf of 800 ampere-turns produces a $\phi = 16{,}000$ maxwells?
15. A coil of wire consists of 200 turns of wire. When 26 mA of current flows through the coil, what is the magnetic force?

16. The flux density of an air space is given as $B = 3000$. If the air space is replaced by an alloy of nickel and iron and the new flux density is 75,000, what is the permeability of the alloy?
17. The field intensity inside a given coil is given as the ratio of mmf/ℓ. A coil 10 cm long consisting of 200 turns carrying 2 mA of current will produce what field intensity?
18. If a coil 5 cm long produces a field intensity of $H = 20$ At/m when 12 mA passes through the coil, how many turns of wire are wound on the coil?
19. If the coil in Problem 18 were stretched out to 20 cm, what would be the new field intensity?

Reference Sections 10.5 and 10.6

20. Refer to the B/H characteristic of Figure 10.14. What is the coercive force and the hysteresis if the H is increased to ± 1200 At/m?
21. Refer to Problem 20. What is the average permeability?
22. Assume a given ferromagnetic material has a volume of 22 cubic centimeters. If the area of the hysteresis loop is 12 ampere-turn webers per square meter, how much energy is lost as hysteresis loss?
23. Refer to Problem 22. If the area of the hysteresis loop is reduced 20%, what is the hysteresis loss?

Reference Section 10.7

24. Assume the magnetic circuit of Figure 10.29 is made up of transformer iron. If the total flux field is 0.4×10^{-4} webers, what is the current through the windings?

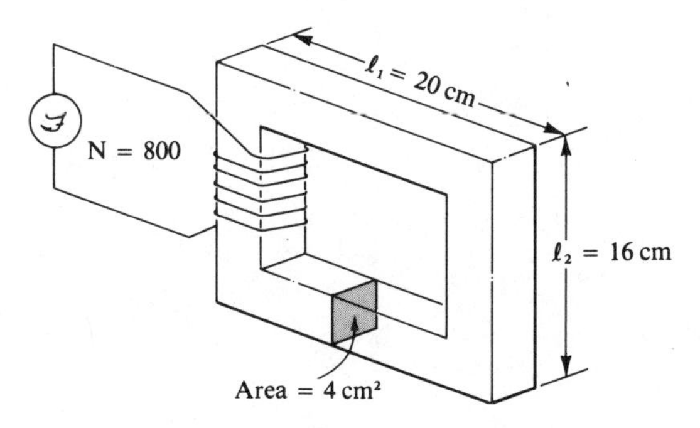

Figure 10.29 Figure for Problem 24

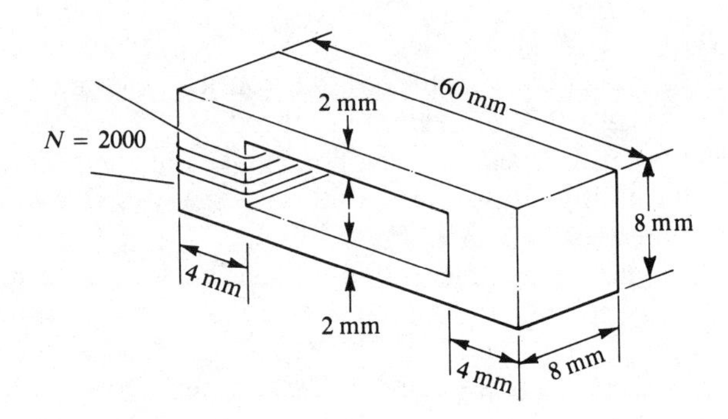

Figure 10.30 Figure for Problem 26

25. If the current in Problem 24 is held constant and the number of turns is doubled, what is the new flux density?
26. Assume the current in the circuit of Figure 10.30 is 200 mA. What is the total flux in the magnetic circuit? If the current is reduced to half, will the number of lines in the magnetic circuit reduce to half? (Assume transformer iron.)
27. Refer to Problem 26. If a 1 mm gap is cut into the 2 mm leg, what total flux would result if 200 mA of current flows through the winding? By how many webers has the flux been reduced?
28. Refer to Problem 27. How much current in the windings would restore the flux to the value of Problem 26?

CHAPTER **11**

INDUCTANCE

11.1 ELECTROMAGNETIC INDUCTION

The process whereby an electromagnetic flux is developed as a result of moving charge carriers through a conductor is reversible. A conductor with no current flowing in it, when moved through a magnetic field, will cause charge carriers to move through the conductor depending upon several circuit conditions. For example, note that the permanent magnet shown in Figure 11.1 illustrates a length of conductor through the center of the magnetic flux. If this conductor is not moved, there will be no movement of the charge carriers in wire. Therefore, the meter will remain on zero. If the wire is moved up and down, through the magnetic flux, the meter needle will deflect positively and negatively depending upon the direction of motion. Consider for the moment that the wire is moving steadily upward perpendicular to the magnetic flux. Electrons in the wire will move toward point A of the meter and away from point B. The needle will deflect to the left, indicating a negative voltage across the terminals of the meter.

The amount of voltage induced will depend upon four factors: the flux density, the length of the conductor within the magnetic flux, the velocity of the wire moving through the flux, and the angle of cutting. If we presume the cutting angle to be 90°, the voltage induced in the conductor is given as:

$$V_i = B\ell s \tag{11.1}$$

where B is the flux density in webers per square meter, ℓ is the length of the conductor in meters, and s is the velocity in meters per second.

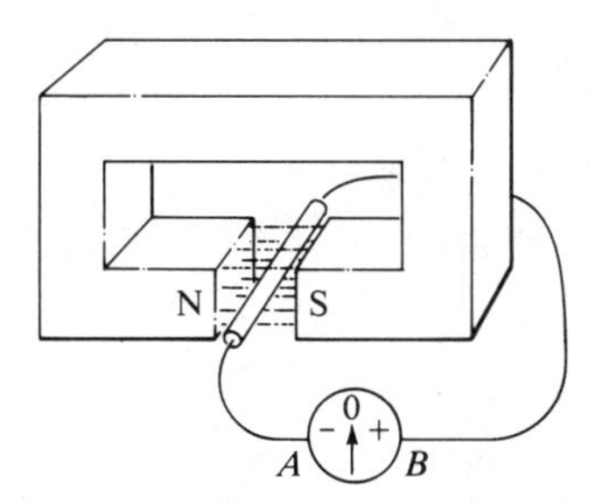

Figure 11.1 Induced Voltage

The direction of current in the wire can be determined by the right-hand rule. Figure 11.2 shows a cutaway view of a wire moving down through a magnetic field. Below it, note the right hand showing the index finger pointing in the direction of the magnetic flux (north to south). The thumb points in the direction of the motion of the wire. The middle finger points toward the reader, indicating the direction of current in the wire. The length of wire within the magnetic flux will have the polarity shown if the wire moves down through the flux. The end of the wire closest to the reader will be positive, since the electrons are moving toward the other end.

The amount of voltage developed can be increased by increasing either the density of the magnetic flux or the length of the wire passing through the field, or by increasing the velocity of the wire. Figure 11.3 illustrates a coiled wire consisting of three loops. Note one dimension of the horseshoe magnet is ℓ. If this magnet is passed back and forth across one side

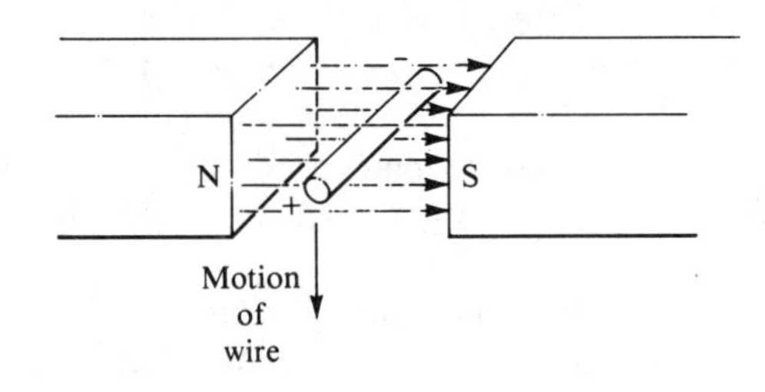

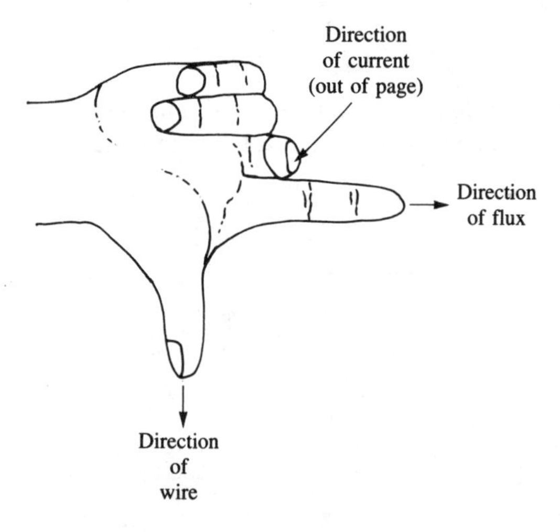

Figure 11.2 Right-Hand Rule for Induction

of the loop, the voltage developed will be $V = B(3\ell)s$. By increasing the number of loops, the induced voltage can be increased for a given flux density, because the effective length of the field is increased.

Example 11.1 Assume a conductor is moved through a flux density of 0.2 wb/m². If the circuit is similar to Figure 11.1 and the gap is 80 mm deep, what is the maximum induced voltage if the maximum velocity is 125 meters per second?

$$V = B\ell s = (0.2)(0.08)(125) = 2 \text{ volts}$$

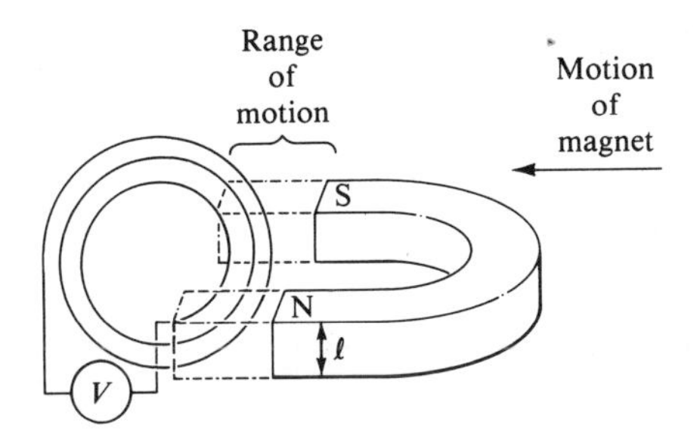

Figure 11.3 Induction in Multistrand Conductor

Mutual Electromagnetic Induction

The principle of electromagnetic induction applies equally well in cases where the changing magnetic field is captured within an iron core. Refer to Figure 11.4(a). Note that the toroid has two sets of windings, N_1 and N_2. When the switch is closed, current flows through winding N_1, establishing a magnetic flux in the toroid. During the time that the field is changing (increasing toward the point of stabilization), the magnetic lines of force are cutting through the winding N_2, establishing a voltage V_2 across N_2. Once the current through N_1 has reached its maximum value (V_1/R), the flux established by N_1 becomes a constant field. The voltage at V_2 drops to zero. When the switch is opened, the collapsing magnetic flux will once again cause a voltage to be induced at V_2, but this time of the opposite polarity.

Figure 11.4(b) illustrates the changing voltage across N_2 when the switch is opened and closed. This transference through the two windings is magnetically interlinked through the *changing* flux. This condition is called *mutual induction.* It is important to note that there is no electrical connection between the circuit involving N_1 and the circuit of N_2. Electrical coupling has been the effect of mutual induction.

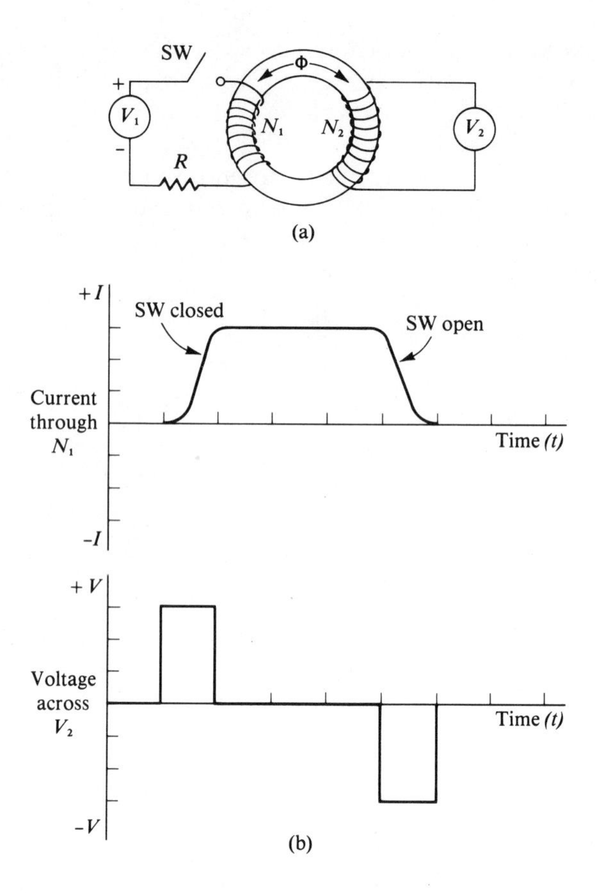

Figure 11.4 Mutual Induction

Faraday's Law

Faraday observed that when a coil of wire was moved through a magnetic field or when a magnetic field was moved across a coil, a voltage was developed at the terminals of the coil. In Figure 11.3 the magnet was moved back and forth through one side of the coil. If the magnetic flux is moved across the entire coil, the voltage induced is proportional to the number of turns and the rate of change of the flux (Faraday's law).

$$V_i = N\frac{\Delta\phi}{\Delta t} \tag{11.2}$$

where V_i is the induced voltage, N is the number of turns, and $\Delta\phi/\Delta t$ is the rate of change of flux with time. As an example, if one million lines of force cut 20 turns of a loop in a quarter of a second, the voltage induced will be 800 mV (10^6 lines = 0.01 weber). If the coil is just sitting there within the flux but not moving, no voltage will be induced because the $\Delta\phi/\Delta t$ is zero.

Example 11.2 Assume a 100 turn coil has a flux of 0.4 wb when the current is 200 mA. When the current was increased, the flux increased to 0.52 wb in 40 ms. What is the average voltage during the time of the flux change? What would the voltage be if the flux then decreases to zero in 200 ms?

From Eq. (11.2),

$$V_i = N\Delta\phi/\Delta t$$

$$\Delta\phi = \phi_2 - \phi_1 = 0.52 - 0.4 = 0.12 \text{ wb}$$

$$\Delta t = 40 \text{ ms}$$

$$V_i = \frac{(100)(0.12)}{(40)(10^{-3})} = 300 \text{ volts}$$

When the current is turned off, the flux collapses to zero.

$$\Delta\phi = 0 - 0.52 = -0.52 \text{ wb}$$

$$\Delta t = 0.2 \text{ second}$$

$$V_i = \frac{(100)(-0.52)}{0.2} = -260 \text{ volts}$$

(The minus sign indicates that the polarity across the coil has reversed during the time the field was collapsing.)

11.2 INDUCTANCE

We have seen that current flowing through a conductor establishes a magnetic field around that conductor. When the current is changing, the magnctic ficld is also changing. Therefore, we have magnetic field lines cutting the conductor itself and, according to Faraday's law, this induces a (new) voltage in the conductor. The property that a conductor has to induce a voltage in itself through the flow of a time-varying current is called *self-inductance*.

The *inductance* (L) of a conductor is a measure of how well it is able to induce a voltage in itself when the current through it is changing. The degree of self-induction depends on the change in the number of flux linkages (λ) that are established when the current changes; thus

$$L = \frac{\Delta\lambda}{\Delta i} \tag{11.3}$$

It is also true that the voltage induced in the conductor depends on the rate of change of the number of flux linkages.

$$V_L = \Delta\lambda/\Delta t \tag{11.4}$$

where V_L is the magnitude of the self-induced voltage. From Eq. (11.3) we have

$$\Delta\lambda = L\Delta i$$

Substituting this result in Eq. (11.4), we find

$$V_L = L\frac{\Delta i}{\Delta t} \tag{11.5}$$

Equation (11.5) states the very important fact that the voltage induced in a conductor is proportional to the rate of change of the current flowing through it. The inductance L, which is constant for a given conductor, is thus a measure of the magnitude of the voltage induced in it by a given rate of change of current.

From Eq. (11.3), we see that the units of inductance L are webers per ampere. This unit of inductance is given the name *henry*, after the physicist Joseph Henry.

A current change of one ampere through one henry in one second will develop one volt across the inductance.

Example 11.3 A given inductance develops 12 volts when a linear rise in current of 100 mA occurs in 40 ms. What is the value of inductance?

$$L = \frac{V_L \Delta t}{\Delta i} = \frac{(12)(40)(10^{-3})}{(100)(10^{-3})} = 4.8 \text{ henries}$$

11.3 PHYSICAL PROPERTIES OF INDUCTANCE

The principal difficulty of establishing the value of inductance using the method of Example 11.3 is the need to make actual laboratory measurements. It would be far more useful and easier if we could predict the value of an inductance from its physical properties and geometric characteristics. Using the properties of magnetics from Chapter 10, we may derive another equation for inductance.

From Eq. (11.3),

$$L = \Delta\lambda/\Delta i$$

We assume the $\Delta\lambda$ is equal to flux linkages. These flux linkages can be defined in algebraic terms.

$$L = \lambda / i \tag{11.6}$$

The standard notation for electronic symbols is to use capital letters for direct current and a lower case letter for instantaneous values. The lower case "i" in Eq. (11.6) indicates a current at a given instant of time. The flux linkages are dependent upon the number of turns in the winding.

$$\lambda = N\phi$$

However, from Eq. (10.2),

$$\phi = BA$$

And from Eq. (10.9),

$$B = \mu H$$

Therefore

$$\phi = \mu H A$$

$$L = N\mu H A / i \tag{11.7}$$

However, $H\ell = Ni$ and $i = H\ell / N$.
Substituting into Eq. (11.7),

$$\boxed{L = N^2 \mu A / \ell} \tag{11.8}$$

where N is the number of turns, μ is the permeability of the core, A is the cross-sectional area in square meters, ℓ is the length of the inductor in meters, and L is the inductance in henries.

The value of the permeability in Eq. (11.8) is the actual permeability of the core material. If only the relative permeability of a given core material is given, it must be multiplied by the permeability of free space.

Example 11.4 Assume 1800 turns of wire are wound in a single layer on a doughnut-shaped toroid whose relative permeability is 500. If the outside diameter is 20 cm and the inside diameter is 18 cm, what is the inductance?

$$A = \pi r^2 = \pi(10^{-2})^2 = 3.1416 \times 10^{-4}\ \text{m}^2$$

$$\ell = \pi d = (3.1416)(19)(10^{-2}) = 59.69 \times 10^{-2}\ \text{meter}$$

$$\mu = \mu_r \mu_o = (500)(4\pi)(10^{-7}) = 6283.19 \times 10^{-7}$$

$$L = \frac{(1800)^2(6283.19)(\pi)(10^{-4})(10^{-7})}{(59.69)(10^{-2})}$$

$$= 1.071\ \text{henries}$$

Equation (11.8) applies to a particular class of inductors. The results are reasonably accurate for a single layer of wire wound on a core. Single-layer air core coils and some ferrous core coils are used extensively for high frequency operation. Power type devices generally use multilayer windings. Calculations for these inductors require special equations. One good reference for these equations is a publication by International Telephone and Telegraph Corporation, *Reference Data for Radio Engineers*.

11.4 INDUCTORS IN SERIES AND PARALLEL

Often it is necessary to connect several inductors in various circuit combinations. Depending upon the application, the magnetic flux of one inductor may or may not interact with the magnetic flux of another. In those cases, where the magnetic flux of each inductor in a circuit does not interact with the flux of any other inductor, some simple rules governing the total inductance of the circuit may be followed. In Section 11.3 we studied the physical characteristics of an inductor and noted that the length and the number of turns both influenced the total inductance. These concepts can be applied to analyzing inductors when connected in series, parallel, and series-parallel combination. Consider a given inductor of length ℓ and cross-sectional area A. The inductance is given as:

$$L = \frac{N^2 \mu A}{\ell}$$

If we divide the number of turns in half and the length of the coil in half, we get:

$$N_1 = N_2 = N/2$$

$$\ell_1 = \ell_2 = \ell/2$$

Assume that the magnetic flux of each of the separate coils does not interact with the other. The total inductance is:

$$L_T = \frac{(N/2)^2 \mu A}{\ell/2} + \frac{(N/2)^2 \mu A}{\ell/2} = \frac{N^2 \mu A}{\ell}$$

Series Inductors

From the preceding discussion we see that when two or more inductors are connected in series with no interaction of the magnetic flux, the total inductance is the sum of the individual inductances.

$$\boxed{L_T = L_1 + L_2 + \cdots + L_N} \tag{11.9}$$

The derivation of Eq. (11.9) is actually based upon Faraday's law. Refer to Figure 11.5. The voltages in a closed loop must equal zero; hence the applied voltage must equal $V_1 + V_2$. Since this is a series circuit connection, the rate of change of current is the same everywhere in the circuit.

$$L_T \frac{\Delta i}{\Delta t} = L_1 \frac{\Delta i}{\Delta t} + L_2 \frac{\Delta i}{\Delta t} + \cdots + L_N \frac{\Delta i}{\Delta t}$$

Factoring and canceling the $\Delta i/\Delta t$ results in Eq. (11.9).

A similar method of analysis can be applied to show that voltages in a series inductive circuit distribute by the ratio of the sizes of the respective inductors. Refer to the circuit of Figure 11.5. Assume the rate of change of current in the circuit to be constant. The voltages will distribute according to the ratio.

$$V_1 = \left(\frac{L_1}{L_1 + L_2}\right) V \tag{11.10}$$

Note the similarity of Eq. (11.10) with the resistive ratio law.

Example 11.5 The total instantaneous voltage across two series-connected inductors is 20 V. If $L_1 = 2$ henries and 5 volts is developed across it, what is the inductance of L_2? (Assume no interactive flux.) What is the total inductance?

$$V_1 = \left(\frac{L_1}{L_1 + L_2}\right) V$$

$$5 = \frac{2(20)}{2 + L_2}$$

$$L_2 = 6 \text{ henries}$$

$$L_T = 2 + 6 = 8 \text{ henries}$$

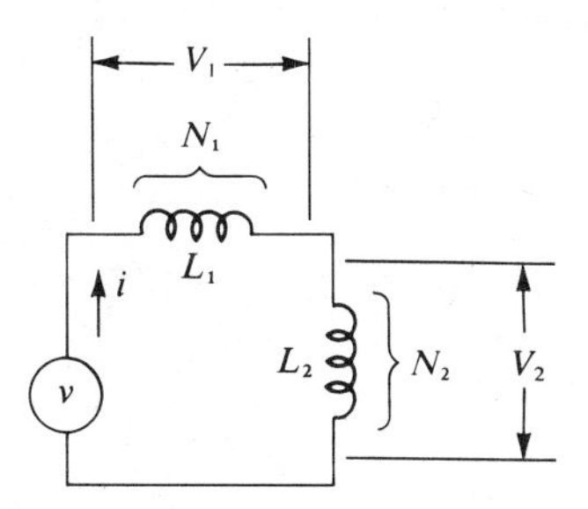

Figure 11.5 Inductors in Series

Parallel Circuit

The total inductance of parallel connected inductors can be determined using the circuit of Figure 11.6. Note that the current through L_1 and L_2 must add up to the total current supplied by the source. If Kirchhoff's current law is applied, the total inductance becomes

$$L_T = \frac{1}{1/L_1 + 1/L_2 + \cdots + 1/L_N} \tag{11.11}$$

where N is the number of parallel inductances. Here again, note the similarity of L_T to the total resistance of parallel circuits.

Series-Parallel Inductance

When no interactive magnetic flux is involved between inductors connected in a series-parallel configuration, the total inductance can be found by applying rules similar to those for resistors in series-parallel. For example, the total inductance of Figure 11.7 is equal to the parallel combination of L_2 and L_3 added to the inductance of L_1. It is important

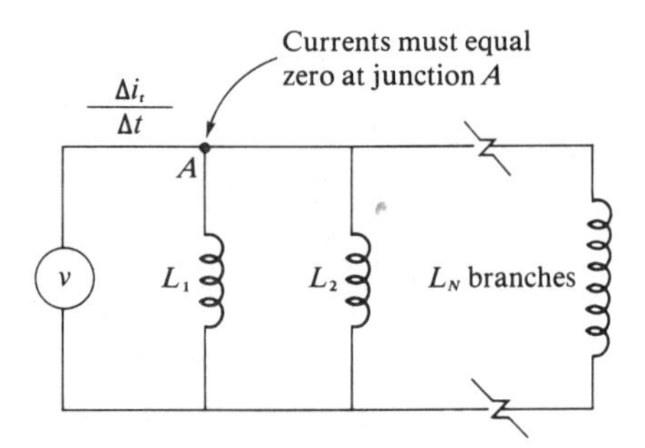

Figure 11.6 Inductors in Parallel

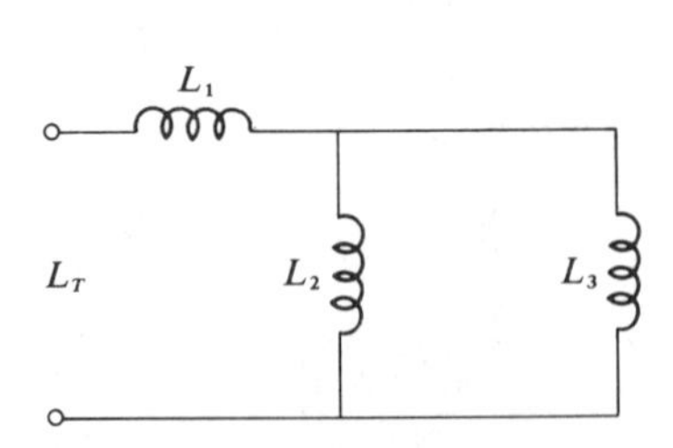

Figure 11.7 Inductors in Series-Parallel

to note that these equations involving total inductance apply to inductors where the magnetic flux is captured within closed iron cores. In the case of open air cores the inductor's magnetic flux emanating from the coils will interact with nearby conductive materials within the boundaries of the magnetic flux. These interactions will influence the total inductance. In the case of air cores, Eqs. (11.10) and (11.11) do not apply.

Example 11.6 Assume the bridge network of Figure 11.8(a) consists of five inductors with no interactive flux. What is the total inductance?

The delta network consisting of three 30 mH inductors must be converted to a *Y* network (Figure 11.8(b)).

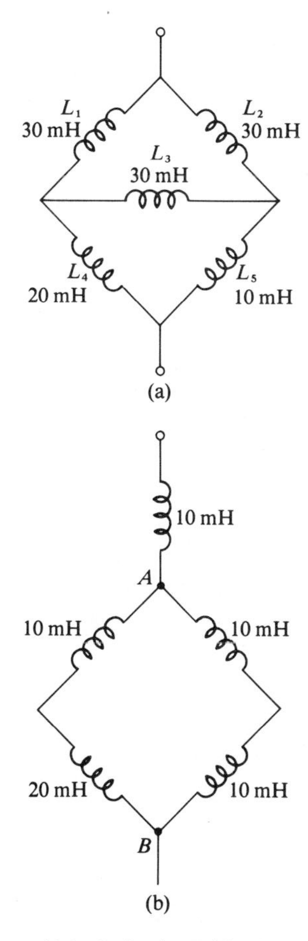

Figure 11.8 Inductive Bridge Network

The remaining circuit may now be solved.

Inductance between A and B:

$$L_{AB} = \frac{(20 \text{ mH})(30 \text{ mH})}{30 \text{ mH} + 20 \text{ mH}}$$

$$= 12 \text{ mH}$$

$$L_T = 22 \text{ mH}$$

11.5 MUTUAL INDUCTANCE

When the magnetic flux of two inductors is interactive, the total circuit is influenced by the two inductors. The amount of influence depends upon the physical relationship of the two coils. In Figure 11.4 the toroid core has two sets of windings, N_1 and N_2. If a current is passed through N_1, magnetic flux is established in the core, causing electrons to move through the winding of N_2. The moving electrons in N_2 correspondingly set up a magnetic flux influencing the circuit of N_1. The amount that these two sets of windings influence each other depends upon the percentage of flux that mutually interacts between them. It also depends upon the directional relationship of the magnetic flux.

Lenz's Law

H. L. Lenz observed that the polarity of a voltage induced in a winding depends on the direction of the magnetic lines of force as they move across the coil. Further, he observed that the direction of the induced voltage which caused a current in the coil would result in a magnetic flux that opposed the original flux producing the current.

The principle of Lenz's law is best seen with an example. Refer to Figure 11.4(a). When the switch is closed, the current in N_1 produces a rising flux in a clockwise direction in the toroid. This rising flux will in turn induce a voltage and hence a current in winding N_2. The current in N_2 will produce an opposing secondary flux (counterclockwise) in the toroid. When the switch is opened, the collapsing flux in N_1 will reverse the polarity across N_2.

Similarly, a self-induced voltage has a polarity that opposes the build-up of the current that produced it.

Coefficient of Coupling

In Figure 11.9 two coils are wound on a common coil form. If a current is passed through the circuit, magnetic flux will be established. Without

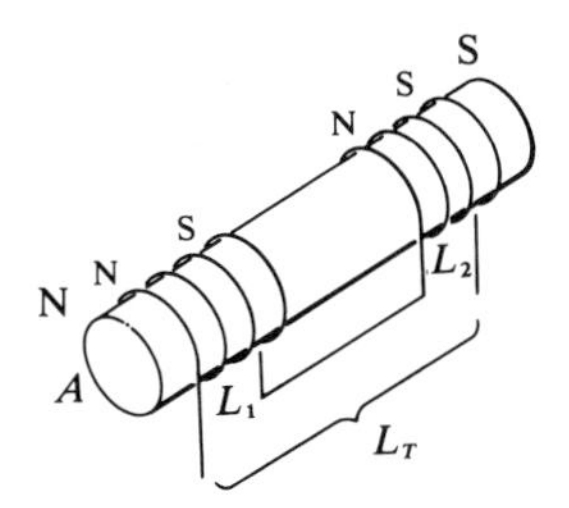

Figure 11.9 Coupling Between Coils

delving into a quantitative analysis, let us assume that 40 percent of the flux generated by L_1 cuts L_2; then, because of the physical and geometric relationship, 40 percent of the flux generated by L_2 will cut L_1. This percentage of mutual interacting flux is described as the *coefficient of coupling*, *k*. It is the ratio of the mutually coupled flux to the total flux produced by one coil.

$$k = \phi_m/\phi_1$$

In the construction of Figure 11.9 only a small amount of the flux interacts, hence the coefficient of coupling is only about 0.4 or 40 percent. Many devices, such as transformers and magnetic amplifiers, are manufactured where the coefficient of coupling approaches its maximum value of 1.

Mutual Inductance

Since mutual flux implies an effective inductance mutually shared, there will be a mutual inductance. Referring again to Lenz's law, we conclude that the voltage induced in winding N_2 of Figure 11.4 is dependent upon the current in N_1.

$$V_2 = M\frac{\Delta i_1}{\Delta t} \qquad (11.12a)$$

and conversely,

$$V_1 = M\frac{\Delta i_2}{\Delta t} \qquad (11.12b)$$

where M is the mutual inductance. However, according to Faraday,

$$V_2 = kN_2\frac{\Delta \phi_1}{\Delta t} \qquad \text{and} \qquad V_1 = kN_1\frac{\Delta \phi_2}{\Delta t}$$

where k is the coefficient of coupling.

If the two Faraday equations are equated (Eqs. 11.12a and 11.12b), the result is

$$M\frac{\Delta i_1}{\Delta t} = kN_2\frac{\Delta \phi_1}{\Delta t} \qquad (11.13)$$

$$M\frac{\Delta i_2}{\Delta t} = kN_1\frac{\Delta \phi_2}{\Delta t} \qquad (11.14)$$

Solving for M,

$$M = kN_2\frac{\Delta \phi_1}{\Delta i_1} \qquad (11.15)$$

$$M = kN_1\frac{\Delta \phi_2}{\Delta i_2} \qquad (11.16)$$

Multiplying Eq. (11.15) by Eq. (11.16),

$$M^2 = k^2\left(N_1\frac{\Delta \phi_1}{\Delta i_1}\right)\left(N_2\frac{\Delta \phi_2}{\Delta i_2}\right) \qquad (11.17)$$

However,

$$L_1 = N_1\frac{\Delta \phi_1}{\Delta i_1} \quad \text{and} \quad L_2 = N_2\frac{\Delta \phi_2}{\Delta i_2}$$

therefore, we may rewrite Eq. (11.17) as

$$M = k\sqrt{L_1 L_2} \qquad (11.18)$$

where only the positive root of L_1L_2 is defined.

Example 11.7 Assume the two inductors of Figure 11.9 have inductances of 2 henries and 8 henries, respectively. If the coefficient of coupling is 10 percent, what is the mutual inductance?

$$\begin{aligned} M &= (0.1)\sqrt{(2)(8)} \\ &= (0.1)\sqrt{16} \\ &= 0.4 \text{ henry} \end{aligned}$$

We may now wonder what effect the mutual inductance has on the total inductance of a circuit containing coils with interactive flux. The amount of mutual inductance shared by the circuit of winding N_1 and winding N_2 is defined as mutual inductance and given as Eq. (11.18).

Refer again to Figure 11.9. Note that winding L_1 is wound counterclockwise on the core and winding L_2 is also wound counterclockwise as viewed from side A. This means that the north poles of the two inductors will be on the same side. The direction of the magnetic fields will be aiding each other, resulting in a larger magnetic field over the combined pair of inductors. These two coils are said to be *series aiding* because the interacting flux of both coils is in the same direction. The total inductance of the circuit not only reflects the individual inductances L_1 and L_2, but the mutual inductance as well.

$$L_T = L_1 + L_2 + 2M \tag{11.19}$$

If the windings of L_2 in Figure 11.9 are changed to a clockwise direction, the magnetic field of L_2 will oppose the field of L_1. The inductors are said to be *series opposing*. The total inductance becomes the sum of the individual inductance minus the mutual inductance.

$$L_T = L_1 + L_2 - 2M \tag{11.20}$$

Example 11.8 Assume two inductors are wound on the same coil form. $L_1 = 20$ mH and $L_2 = 40$ mH. If the coefficient of coupling is 0.8, what is the total inductance if they are connected series aiding? What is the total inductance if they are connected series opposing?

$$M = (0.8)\sqrt{(40)(20)} = 22.63 \text{ mH}$$

$$L_T = 20 + 40 + (2)(22.63) = 105.25 \text{ mH}$$

$$L_T = 20 + 40 - (2)(22.63) = 14.75 \text{ mH}$$

11.6 ENERGY IN AN INDUCTOR

Assume for a moment that we have been able to construct an inductor from wire which has no resistance. Then, when a voltage is applied, current will flow in the circuit. Since the current is not passing through any resistance, none of the energy is converted to heat. All of the energy delivered to the circuit is being stored in the magnetic flux. When the applied voltage is removed, the energy delivered to the inductor is returned to the circuit via the collapsing magnetic flux.

According to Eq. (5.10),

$$W = Pt$$

Stated, using Ohm's law

$$W = ivt \tag{11.21}$$

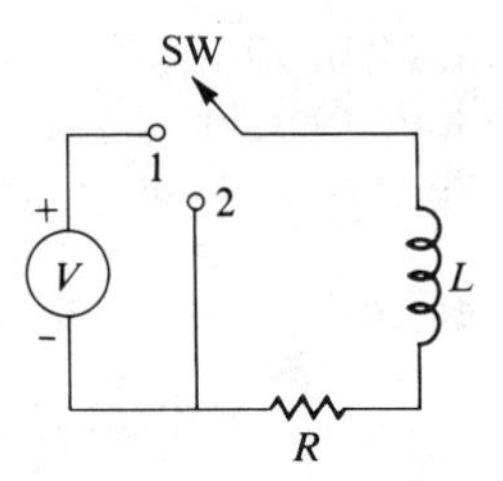

Figure 11.10 Energy in an Inductor

Using advanced mathematics, Eqs. (11.5) and (11.21) reduce to

$$\boxed{W = \tfrac{1}{2}Li^2} \tag{11.22}$$

where L is the inductance in henries, i is the current in amperes, and w is the energy in joules.

Refer to the circuit of Figure 11.10. When switch SW is moved to position 1, the current will rise to a maximum value in the circuit limited by the value of R. It will stabilize at V/R. A constant magnetic field will surround the inductor. This magnetic field will have energy stored in the form of magnetic flux. The energy can be used to couple magnetically other circuitry through mutual inductance or it can be returned to the original circuit. Assume that the switch in Figure 11.10 has been in position 1 long enough to have a constant magnetic field surrounding the inductor. If the switch is moved to position 2 (in zero time), the magnetic flux will collapse and current will continue to flow in the circuit. In essence, the energy stored in the magnetic flux is being returned to the circuit in terms of the current flowing into resistor R. Once the magnetic field has collapsed totally, no additional current will flow; all the energy stored in the magnetic field will have been transformed into heat.

Example 11.9 How much energy is stored in an inductive field of a coil of 12 henries when it is connected in series with a 100 ohm resistor across 50 volts? (Assume negligible resistance of the inductor's windings.)

$$I = V/R = 50/100 = 0.5 \text{ ampere}$$

$$W = \frac{LI^2}{2} = \frac{(12)(0.5)^2}{2} = 1.5 \text{ joules}$$

11.7 TIME CONSTANT

We have observed that Faraday's law establishes that the voltage across an inductance depends upon the rate of change of current through the inductor. If a dc voltage is applied to an inductive circuit, the current in the circuit will rise up to some maximum value. Generally, this maximum value is limited by the amount of resistance in the circuit. As an example, the circuit of Figure 11.10 will conduct a current out of the generator through the switch, the resistor, and the inductor when the switch is in position 1. If we assume that the inductor is made up of wire which has a negligible amount of resistance, then $I_{max} = V/R$. The current may not rise to a value beyond that allowed by resistor R.

We have seen in preceding sections that an inductor, according to Lenz's law, tends to oppose any change in the current flowing through it. Hence, when the switch in Figure 11.10 switches to position 1, the current does not rise instantaneously. The current rises to its maximum value in a manner shown in Figure 11.11(a). Note that just after the switch is closed (t_0), the current starts with a very rapid rise and then tapers gradually to a steady value.

If we apply Kirchhoff's law to the circuit of Figure 11.10, then at any instant of time the voltage across the resistor plus the voltage across the inductor must equal the applied voltage.

$$V = V_r + V_L$$

However, according to Ohm's law, $V_r = IR$; and according to Faraday's law

$$V_L = L\frac{\Delta i}{\Delta t}$$

Therefore, the voltage equation becomes:

$$V = iR + L\frac{\Delta i}{\Delta t} \tag{11.23}$$

Refer again to Figure 11.11(a). Note that at t_0 time no current is flowing in the circuit; therefore there is no voltage drop across the resistor. All the applied voltage must appear across the inductor. Equation (11.23) reduces to

$$V = L\frac{\Delta i}{\Delta t} \tag{11.24}$$

If both sides of this equation are divided by L, then the *initial rate of change of the current* in the circuit is equal to the applied voltage divided by the inductance. In Figure 11.11(a) the dotted line from the origin projects the rate of change and causes an intersection with I_{max} at point A. Point A represents a time identified as the *time constant* of the circuit. Time constant is symbolized by the Greek letter τ (tau). If at that point of time the current is maximum, Eq. (11.24) becomes

$$\frac{V_L}{L} = \frac{I_{max}}{\tau} \tag{11.25}$$

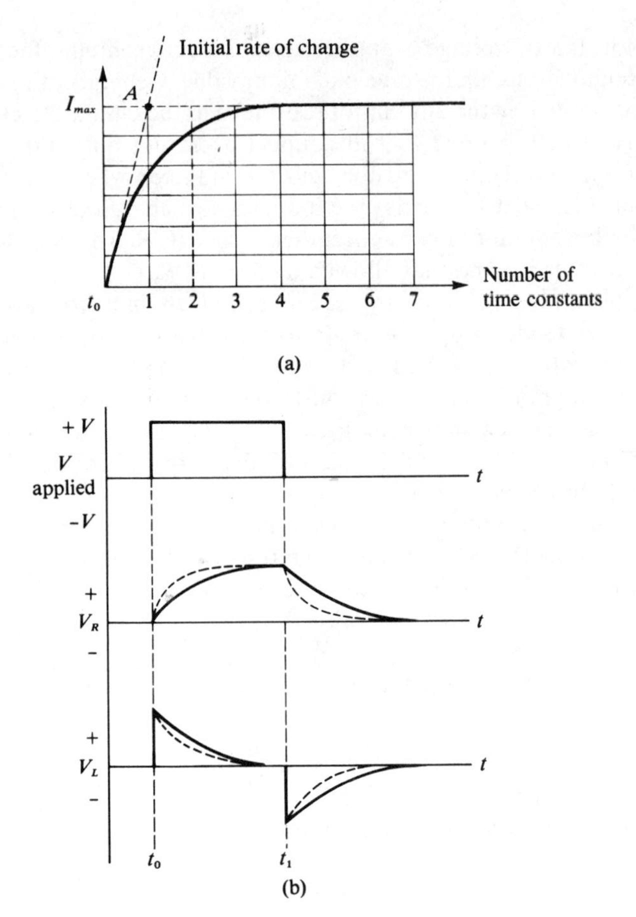

Figure 11.11 Time Constants

During the initial rise from t_0, $V_L = V_{max}$. However, $V_{max} = I_{max}R$. Substituting into Eq. (11.25),

$$\tau = L/R \tag{11.26}$$

where L is the inductance in henries, R the resistance in ohms, and τ the time in seconds.

Refer again to the current characteristic of the circuit (Figure 11.11(a)). Note that after one time constant the actual current in the circuit is not equal to I_{max}. Had the current continued at its initial rate of change, it would have arrived at I_{max} in one time constant. The resistance in the circuit begins to limit the current right after t_0. *At one time constant, the actual current is approximately 63 percent of I_{max}*. In order for the current to reach its maximum value, it would require approximately five time constants. Refer to the original circuit of Figure 11.10, and note that the current in the circuit is rising as shown in Figure 11.11(a). Therefore, the voltage across the resistor is rising according to the same relationship since $V_r = iR$. Kirchhoff's law requires that the sum of the voltages must equal the applied voltage. It follows that the voltage across the inductor is decreasing from its maximum value down to zero. Figure 11.11(b) illustrates the three voltages in the circuit: the applied voltage, the voltage across the resistor, and the voltage across the inductor. Note how these voltages change from the point in time identified as t_0. Increasing the value of the resistance would cause the voltage across the resistor to rise to its maximum value in a shorter length of time, shown as the dotted line.

At time t_1 the switch in Figure 11.10 is moved to position 2 (we will assume instantaneous switching). At that time the applied voltage drops to zero. The magnetic field begins to collapse, establishing a voltage of V across the coil. This voltage is now the applied voltage of the circuit. The net result is that current continues to flow in the circuit. The decaying current causes a decaying voltage across the resistor which follows the pattern of the current. It will take five time constants for the current to decay to zero; hence both voltages across the resistor and the inductor will be zero.

Example 11.10 A 12 V battery is switched into a series circuit consisting of a 2 kΩ resistor and a 4 henry inductor. What is the time constant of the circuit? What is the maximum current? What is the current 2 ms after the battery is switched in? What is the current after 10 ms?

$$\tau = L/R = 4/2\text{k} = 2 \text{ ms}$$
$$I_{max} = 12 \text{ V}/2\text{k} = 6 \text{ mA}$$

Since $i = .63I_{max}$ after one time constant (2 ms), $i = .63(6 \text{ mA}) = 3.78$ mA. Since 10 ms is 5 time constants, $i = I_{max} = 6$ mA.

Universal Time Constant Curve

It has been shown that the current in an L/R circuit will rise to 63 percent of its maximum value in one time constant. It will require approximately five time constants for it to rise to its maximum value. The current in an L/R circuit acts very predictably in that the current rises 63 percent of its remaining value in each time constant. An L/R circuit consists of a 1 kΩ resistor and a 2 mH coil and an applied dc voltage of 10 volts; the maximum value of current is 10 mA. The current will rise to 6.3 mA in the first time constant of 2 microseconds. During the second time constant, the current will rise 63 percent of the remaining 3.7 mA, or 2.33 mA. In the third time constant, the current will rise 63 percent of the remaining current of 1.37 mA. Because of this predictable pattern, L/R circuits may

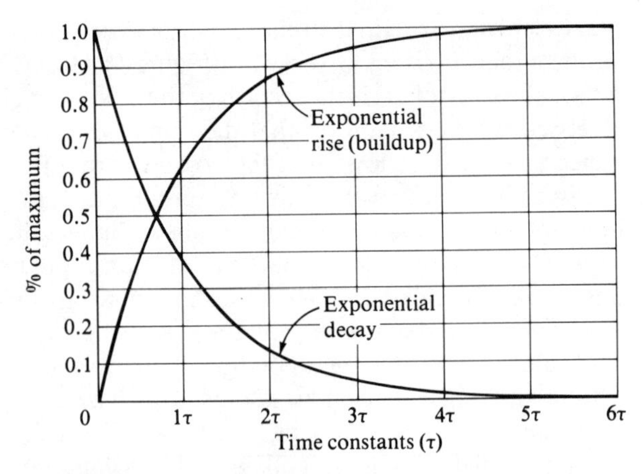

Figure 11.12 Universal Time Constant Characteristics

be solved using a universal time constant characteristic. Figure 11.12 illustrates such a universal characteristic.

Note that in one-half of a time constant, the current has risen to about 40 percent of its maximum value. The universal time constant characteristic of Figure 11.12 shows both a charge and a discharge characteristic. These curves are also referred to as *standard buildup and decay*. If the current in an L/R circuit is decaying, it would follow the predictable pattern shown. For example, in the circuit of Figure 11.10 when the switch is in position 2, the current is decaying from its maximum value. In three time constants the current is only 5 percent of its original value. The universal curves may be used with fairly good accuracy.

Example 11.11 In the circuit of Figure 11.13 assume $R_1 = 40\ \text{k}\Omega$, $R_2 = 10\ \text{k}\Omega$, $L = 5$ H, and $V = 20$ V. If the switch is placed in position 1 for 0.25 ms, then switched to position 2 for 0.25 ms, what is the current through the coil?

With switch in position 1:

$$\tau_1 = \frac{L}{R_2} = \frac{5}{10^4} = 0.5\ \text{ms}$$

$$I_{max} = 20/10^4 = 2\ \text{mA}$$

The 0.25 ms represents one-half of a time constant, hence, from the curve,

$$I_L = 40\%\ \text{of}\ 2\ \text{mA} = 0.8\ \text{mA}$$

With switch in position 2:

$$\tau_2 = \frac{L}{R_1 + R_2} = \frac{5}{(50)(10^3)} = 0.1\ \text{ms}$$

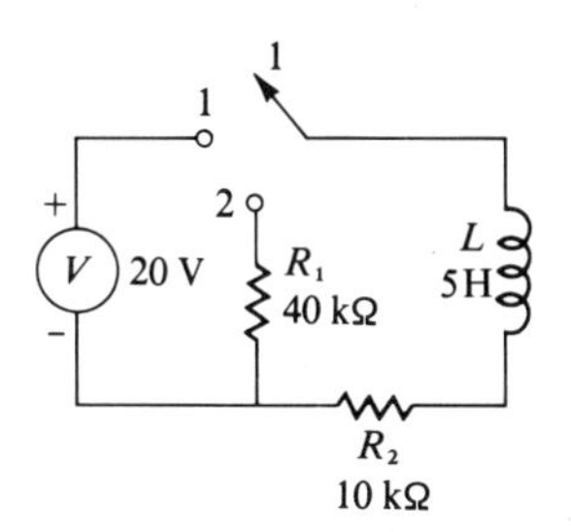

Figure 11.13 Figure for Example 11.11

The 0.25 ms represents two and one-half time constants. Therefore, from the curves the current will decay to

$$I_L = 8\% \text{ of } I_0 = (0.08)(0.8) \text{ mA} = 64\ \mu\text{A}$$

Note how difficult the solution becomes when dealing with small or large percentages of the time constant curve. If we consider Example 11.11 with the switching times changed to 0.12 ms each, the graphical solution yields unreliable results. A more exacting method of calculating circuit conditions is required. Such an approach is covered in Section 11.8.

L/R Networks

The *L/R* circuits shown in Figures 11.10 and 11.13 are simple series circuits. The amount of resistance and inductance in each circuit is readily identified, and therefore the time constants are easily calculated. In some series-parallel circuits such characteristics are not so immediately obvious.

Figure 11.14 illustrates various circuit configurations. These too are fairly elementary. Note in Figure 11.14(a) that the equivalent circuit is used only to determine the time constant of the circuit. In Figure 11.14(b) the two inductors must be considered in parallel and the two resistors in series when calculating the time constant of the circuit.

Example 11.12 In the circuit of Figure 11.14(b), assume $R_1 = 2\text{ k}\Omega$, $R_2 = 4\text{ k}\Omega$, $L_1 = 2\text{ mH}$, and $L_2 = 3\text{ mH}$. If V is 24 volts, what is the current in L_2 after one and a half time constants?

We use the equivalent circuit to determine the time constant.

$$\tau = L/R \qquad L = 1.2\text{ mH},\ R = 6\text{ k}\Omega$$

Note that R is the Thevenin equivalent resistance of the circuit lying to the left of L_1.

$$\tau = 0.2\ \mu\text{s}$$

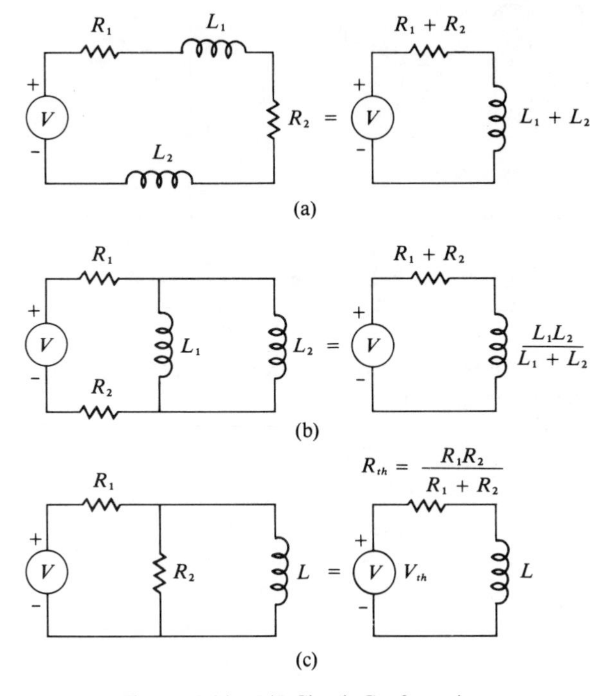

Figure 11.14 *R/L* Circuit Configurations

The maximum current is

$$I = 24/(6\ \text{k}\Omega) = 4\ \text{mA}$$

From the universal curves,

$$i = (0.77)(4) = 3.08\ \text{mA}$$

This current divides between the two inductors by the inverse ratio.

$$i_2 = \frac{(L_1)(3.08)}{L_1 + L_2}$$

$$= 1.232\ \text{mA}$$

The preceding examples illustrate that not only do equivalent circuits show timing conditions, but also that Ohm's and Kirchhoff's laws must be employed for circuit analysis. Often circuits will be such that basic theorems will be required for an analysis as well. Figure 11.14(c) shows how Thevenin's theorem is employed in developing the equivalent circuit.

Example 11.13 Assume the circuit of Figure 11.15(a) has 22 volts dc applied, $R_1 = 3\ \text{k}\Omega$, $R_2 = 4\ \text{k}\Omega$, $R_3 = 2\ \text{k}\Omega$, $R_4 = 1\ \text{k}\Omega$, and $L = 38$ henries. What is the time constant of the circuit? When the switch is

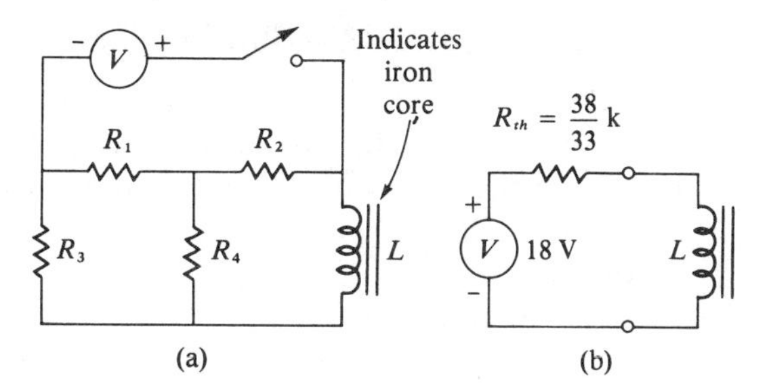

Figure 11.15 Figure for Example 11.13

closed, what is the maximum current through L? After one time constant, what is the voltage across L and across R_3? What is the current through R_4?

Solve for the Thevenin equivalent circuit seen by L. (Replace V with a short and do a wye-delta conversion. See Figure 11.15(b).)

$$\tau = \frac{L}{R_{\text{Th}}} = \frac{38}{(38/33)(10^3)} = 33 \text{ ms}$$

$$I_{\text{max}} = \frac{V_{\text{Th}}}{R_{\text{Th}}} = \frac{18}{(38/33)(10^3)} = 15.63 \text{ mA}$$

The voltage across the inductor after one τ is 37 percent of the applied Thevenin voltage.

$$V_L = (0.37)(18) = 6.67 \text{ volts}$$

After one time constant the current through L is 63 percent of its maximum value.

$$i_L = (0.63)(15.63)(10^{-3}) = 9.85 \text{ mA}$$

The voltage across R_3 can be determined from Kirchhoff's law. Refer to Figure 11.15(a). If we add the voltages around the outside loop,

$$V = V_{R_3} + V_L$$

$$22 = V_{R_3} + 6.67$$

$$V_{R_3} = 15.33 \text{ volts}$$

Since R_3 is 2 kΩ, the current flowing up through R_3 after one τ is

$$i_3 = 7.66 \text{ mA}$$

The current through R_4, therefore, must be

$$\begin{aligned} i_4 &= i_L - i_3 \\ &= 9.85 \text{ mA} - 7.66 \text{ mA} \\ &= 2.19 \text{ mA} \end{aligned}$$

11.8 ALGEBRAIC SOLUTIONS OF INSTANTANEOUS VALUES

In the preceding section we noted the difficulty in solving for circuit currents and voltages using the universal time constant curves. We will now consider the algebraic method of solution for the current as a function of time in an L/R circuit.

Equation (11.23) is restated here as:

$$V = iR + L\frac{\Delta i}{\Delta t}$$

If this equation is rearranged

$$\frac{\Delta i}{i - V/R} = \frac{-R}{L}\Delta t$$

However, from Eq. (11.26),

$$\frac{\Delta i}{i - V/R} = -\frac{1}{\tau}\Delta t$$

Using higher mathematics this equation can be solved for i_t.

$$\boxed{i_t = \frac{V}{R}(1 - e^{-t/\tau})} \qquad (11.27)$$

where V is in applied dc volts, R is resistance in ohms, t is time in seconds, and τ is the time constant.

In Eq. (11.27) the symbol e is used. The value of e is the base of natural logarithms. It is a constant similar in nature to π. Its value is given as $e = 2.71828$. In L/R circuits the value of e is raised to various powers.

It is important to gain a general concept of this exponent, since many needless calculations may be avoided. Study the following short list for e raised to plus and minus values of N.

N	e^N
$N >$ 50	$\to \infty$
10	22,026
5	148.41
2	7.389
0	1
−2	0.1353
−5	6.738×10^{-3}
−10	4.540×10^{-5}
< -50	$< 1.929 \times 10^{-22}$

Small hand calculators often provide a key function for e^N. If the calculator does not have such a key, it is possible to find values of e^N using logs base 10.

$$e^N = \text{antilog}\,(0.4343N)$$

Example 11.14 Refer to the circuit of Figure 11.14(a). $R_1 = 2\ \text{k}\Omega$, $R_2 = 4\ \text{k}\Omega$, $L_1 = L_2 = 2$ henries. What is the voltage across R_1, 0.5 ms after 10 volts is applied?

$$\tau = \frac{L}{R} = \frac{4}{(6)(10^3)} = 0.667\ \text{ms}$$

The current in the circuit at any time t is given by Eq. (11.27). At $t = 0.5$ ms

$$i = \frac{10}{(6)(10^3)}(1 - e^{[-(0.5)(10^{-3})]/[(0.667)(10^{-3})]})$$

$$= (1.67)(10^{-3})(1 - e^{-0.75})$$

Since $e^{-0.75} = 0.47236$, we have

$$i = (1.67)(10^{-3})(0.5276) = 0.88\ \text{mA}$$

The voltage across R is iR.

$$V_{R_1} = (0.88)(10^{-3})(2)(10^3) = 1.76\ \text{volts}$$

Voltage Across *R*

Since Eq. (11.27) defines the instantaneous current in a series L/R circuit, we may identify the general equation for the voltage across the resistance. From Ohm's law, $V = iR$.

Therefore, substituting Eq. (11.27),

$$V_R = (V/R)(1 - e^{-t/\tau})R$$

Canceling R's,

$$\boxed{V_R = V(1 - e^{-t/\tau})} \qquad (11.28)$$

where V is the applied voltage in volts.

If there is more than one series resistance in a circuit, the voltage of Eq. (11.28) represents the total voltage across all the series resistance.

Example 11.15 Refer to Figure 11.16(a). Assume $V = 20$ V, $R_1 = 1\ \text{k}\Omega$, $R_2 = 2\ \text{k}\Omega$, $R_3 = 3\ \text{k}\Omega$, and $L = 3$ henries. Assume switch 2 is open. What is the voltage across R_1 1.5 ms after switch 1 is closed?

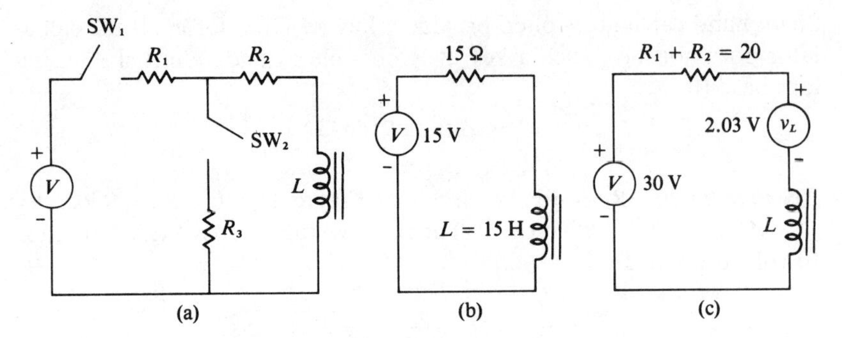

Figure 11.16 Figure for Examples 11.15 and 11.16

With switch 2 open,

$$\tau = \frac{L}{R} = \frac{3}{(3)(10^3)} = 1 \text{ ms}$$

From Eq. (11.28),

$$\begin{aligned} V_R &= 20(1 - e^{[-(1.5)(10^{-3})]/(10^{-3})}) \\ &= 20(1 - 0.223) \\ &= 15.54 \text{ volts} \end{aligned}$$

However, this voltage distributes across R_1 and R_2 by ratios.

$$V_{R_1} = \frac{15.54}{3} = 5.18 \text{ volts}$$

Voltage Across *L*

By Kirchhoff's voltage law, the voltage across the inductor in a series L/R circuit is

$$V_L = V - V_R \tag{11.29}$$

If Eq. (11.28) is multiplied out,

$$V_R = V - Ve^{-t/\tau}$$

Substituting into Eq. (11.29),

$$\boxed{V_L = Ve^{-t/\tau}} \tag{11.30}$$

Example 11.16 Refer to the circuit of Figure 11.16(a). Assume $R_1 = R_2 = R_3 = 10$ ohms, $L = 15$ H, $V = 30$ volts, and switch 2 is closed. If switch 2 is opened 2 seconds after switch 1 is closed, what is the voltage across L, 4 seconds after switch 1 is closed?

The Thevenin equivalent circuit with switches 1 and 2 closed is shown in Figure 11.16(b).

$$\tau = 1 \text{ second}$$
$$V_L = 15(e^{-2/1}) = 2.03 \text{ volts}$$

The 2.03 volts is the instantaneous voltage across L at $t = 2$ seconds.

The equivalent circuit with switch 2 open and switch 1 still closed is shown in Figure 11.16(c). Note that the flux buildup in L has been represented as a separate voltage of 2.03 volts in the equivalent circuit. The voltage across L jumps to 12.7 volts when switch 2 is open. The new voltage across L after 2 more seconds is:

$$V_L = V_i e^{-t/\tau_1}$$

where V_i is the initial voltage of 27.97 volts and τ_1 is the new time constant.

$$\tau_1 = 15/20 = 0.75 \text{ second}$$
$$V_L = 27.97e^{-2/0.75} = 1.96 \text{ volts}$$

L/R circuits with direct current applied must be analyzed using all the available tools of circuit analysis. Not only are network theorems powerful tools, but basic concepts of device theory are useful as well.

For example, in the circuit illustrated in Figure 11.17(a), assume it is desired to determine the current flowing out of V_1 at a particular instant of time. We can make several observations. Let us presume that the two inductors are not independent iron core devices. They are wound on a common core series aiding, and have a coefficient of coupling of 0.6. The presence of the bridge network and the two voltage sources appears to complicate the analysis of the circuit.

If, however, we redraw the circuit, incorporating the mutual inductance in each loop, we get the circuit of Figure 11.17(b).

Since the circuit is symmetrical, we may use Bartlett's theorem. If the circuit is redrawn as in Figure 11.17(c) and a Bartlett bisector is passed through the center, the final circuit of Figure 11.17(d) results. We may now identify the current flowing out of supply V_1 for any instant of time t.

SUMMARY

A conductor moving through a magnetic flux will develop a voltage at the terminals of the conductor. The amount of voltage is directly proportional to the flux density, the length of the conductor, and the velocity of the conductor. If a coil of wire is moved through a magnetic flux, the

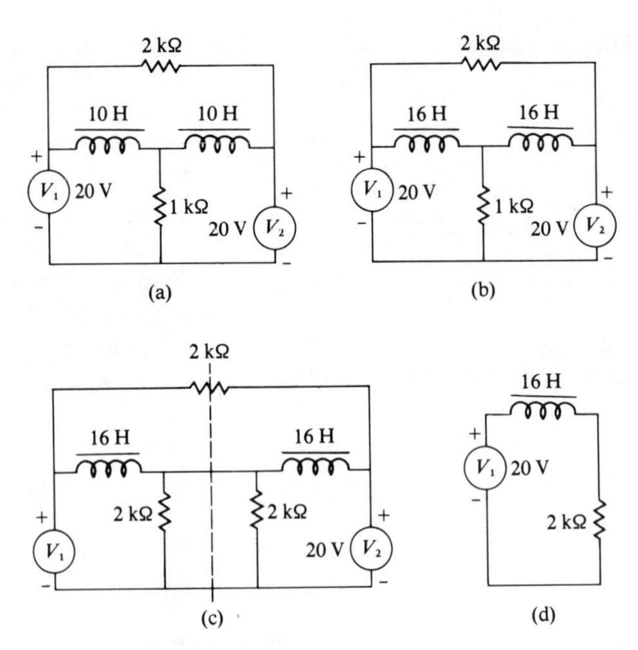

Figure 11.17 Magnetic-Coupled Time Constant Circuit

voltage induced depends upon the rate of cutting of flux and the number of turns. Inductance is defined as a constant which represents the rate of change of flux linkages with respect to the current producing the flux. The voltage developed across a coil depends upon the inductance and the rate of change of current through the coil. The inductance of a coil is directly proportional to the number of turns squared, the permeability, and the cross-sectional area, and is inversely proportional to the length of the coil.

Inductors connected in series and parallel are like resistors in series and parallel.

Two windings which share mutual flux exhibit a mutual inductance depending upon the percentage of interactive flux. The total inductance of two interactive coils depends upon whether the flux of each is aiding or opposing each other.

When a dc is applied across an L/R circuit, the current builds to its maximum value in five time constants. A time constant is given as L/R seconds. The current, hence the voltages, may be determined for any instant during the five time constant interval. The energy stored in the flux will cause the current to continue to flow until the magnetic flux has completely collapsed.

PROBLEMS

Reference Section 11.1

1. Assume a 10 cm length of wire passes through a magnetic field at a right angle to the field. The flux density is 0.12 weber/m^2; the velocity is 5 meters per second. How much voltage is developed at the terminals of the wire?
2. A length of wire is passed through a linear magnetic field at an angle of 90° with the field. 1.96 inches of the wire is in the field whose flux density is 0.48 weber/m^2. If the velocity of the wire is 10 meters per second, what voltage is developed at the terminals of the wire?
3. Refer to Problem 2. If the speed with which the wire passes through the flux doubles, what is the voltage developed at the terminals?
4. In Problem 2, if the flux density were decreased by 20 percent and all other factors remained the same, what voltage would be developed at the terminals?
5. A 12 cm conductor is passed through a linear magnetic flux of 0.8 weber/m^2 at a 90° angle with the flux. At what velocity must the conductor pass through the flux to produce 1.44 volts?

Reference Sections 11.2 and 11.3

6. A coil of fine wire whose total length is 200 meters is wound on an air core 4 cm in diameter. The coil is passed through a linear magnetic flux (0.8 wb/m^2) such that all the loops of the coil cut the flux at 90°. If the velocity is quite slow (0.2 meter/second), what voltage will be developed at the terminals of the coil?
7. If the current through a coil of 20 henries is changing at the rate of 0.2 mA per millisecond, what voltage will be developed across the coil?
8. If the rate of change of the current through the coil of Problem 11.7 were doubled while the magnitude of the current remained the same, what would the new induced voltage be?
9. The current through an unknown coil is changing at the rate of 3 mA per second. If the voltage across the coil is 15 mV, what is the value of the coil in henries?
10. Assume a single layer of adjoining turns of wire is wound on an air core coil form whose area is 0.8 square cm. The coil length of the 15 turns is 3 cm. What is the inductance of the coil?
11. If 40 turns of rigid wire are turned on a pencil, and the pencil is then slipped out, the coil stretches slightly to 8 inches. What is the inductance of the coil? (Assume the diameter of a pencil to be $\frac{1}{4}$ inch.)

12. If the wire in Problem 10 is replaced by a higher gage wire, such that the number of turns is increased to 30 while the length of the coil remains the same, what is the new inductance?
13. Stretching the coil of wire in Problem 11 to 14 inches, will cause how much of a change in the inductance?
14. Refer to Problem 10. If a cylinder of low carbon content iron with a relative permeability of 250 is inserted into the air core, what is the new inductance? (Assume the iron core is 3 cm long.)
15. Which air core coil will have a greater inductance: (a) 200 turns on 1 inch diameter coil form whose coil length is two inches, or (b) 200 turns on a 2 inch diameter coil form whose coil length is 1 inch?
16. When 200 feet of #30 gage wire are wound on an air core coil with each winding adjacent to the next, what is the total inductance if the diameter of the coil form is 1 inch?

Reference Section 11.4

17. A coil has 60 turns of wire wound on an iron core of 5 cm length and a permeability of 4×10^{-4}. If the radius of the core is 1 cm, what is the inductance?
18. Two iron core chokes have an inductance of 8 henries and 12 henries, respectively. What is the total inductance when they are connected (a) in series, (b) in parallel? (Assume no interacting magnetic flux.)
19. Three iron core coils, 12-henry, 20-henry, and 30-henry, are connected in parallel. (a) What is the total inductance? (b) What is the total inductance when they are connected in series?
20. If the value of L_3 in Example 11.6 is doubled, what is the value of the total inductance?
21. Refer to Example 11.6. What value of L_5 will balance the bridge? What is the total inductance for the balanced bridge?
22. If the total inductance of two series coils is 27 mH and L_1 is twice the value of L_2, what is the inductance of L_2?

Reference Section 11.5

23. Two coils, $L_1 = 10$ mH and $L_2 = 40$ mH, have a unity coefficient of coupling. What is the mutual inductance?
24. If two coils are wound on the same iron core, such that one has an inductance of 2 henries and the mutual inductance is 2 henries, what is the inductance of L_2 if $k = 1.0$?
25. If the coefficient of coupling between two coils is 0.8, and the coils are each 20 mH, what is the mutual inductance?
26. If the coefficient of coupling in Problem 24 is reduced to 0.4, what is the new mutual inductance?

27. Two coils are connected series aiding: $L_1 = 120$ mH, $L_2 = 30$ mH, and $k = 0.9$. What is the total inductance of the combination?
28. If the two coils of Problem 25 are connected series opposing, what is the total inductance of the combination?
29. If two coils are connected series aiding, their total inductance is 50 mH. When the two coils are electrically connected series opposing, their total inductance is 10 mH. What is the mutual inductance?
30. When a 10 mH coil and a 3.6 mH coil have a mutual inductance of 3 mH, what is the coefficient of coupling?

Reference Section 11.6

31. How much energy in joules will a 10 mH coil store in its magnetic flux if the current flowing through it is 25 mA?
32. Refer to Problem 31. If the current is doubled, how much energy is stored in the magnetic flux?
33. Assume a coil with 1000 turns stores 2.5 joules of energy when 350 mA flows through it. What is the inductance of the coil?
34. Refer to the circuit of Figure 11.18. How much energy is stored in the flux?
35. Refer to Problem 34. If the resistance is reduced to half its value, what will be the change in stored energy?

Reference Section 11.7

36. A series L/R circuit consists of a coil of 10 mH and a resistance of 1500 ohms. What is the time constant of the circuit? What is the maximum current when 30 volts dc is applied?
37. In the circuit of Problem 36, how long would it take for the current to reach its steady state value? How long will it take for the voltage across the inductor to diminish to zero?
38. A circuit consists of an inductance of 20 henries and a resistance of 100 ohms. What would the voltage be across the resistor 0.05 second after 100 volts is applied to the circuit? What is the voltage across the inductor at that time?

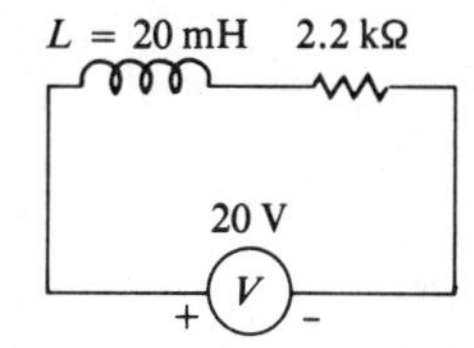

Figure 11.18 Figure for Problem 34

39. In the circuit of Problem 38, what would the current be in the circuit 0.1 second after the voltage was applied?

Reference Section 11.8

40. A relay with a time delay has an inductance of 2 H and a resistance of 200 ohms. If the relay closes at 31.3 mA, and a time delay of 20 ms is desired, find the value of voltage which must be applied.
41. A circuit consists of a 100 mH coil and a 10 kΩ resistor. What will be the voltage across the resistor 20 ms after 20 volts is applied to the circuit?
42. A series L/R circuit consists of a 2 henry inductor. The time constant of the circuit is 2 ms. When 10 volts is applied, 6.3 mA flows after one time constant. What is the resistance?

CHAPTER 12

ELECTROSTATICS

12.1 ELECTRIC CHARGE

Almost everyone has experienced the effects of electrostatics. The sudden spark when touching metal after walking across a nylon rug and when sliding across a seat cover are good examples. Children experience electrostatics by rubbing a blown up balloon on their hair and then sticking the balloon against a wall.

Some of these electrostatics can be demonstrated in a physics laboratory. As an example, if a rubber rod is rubbed with a piece of wool, the rod picks up the lost electrons from the wool. The rod now has a greater quantity of electrons than it did in its neutral state.

We have established that the basic atomic particles, namely electrons and protons, have fields of forces surrounding them. The direction of these fields has been arbitrarily chosen. Figure 12.1 illustrates the isolated electron and proton and their associated fields. The direction of the field in each case, as shown by the arrows, is the direction that an isolated electron would move in if placed in the vicinity of the particle. Basically, the electric charge of an electron is the same as, though opposite to, that of a proton.

An atom that has an equal number of electrons and protons is said to be *electrostatically neutral*; that is, it exhibits no net electrostatic characteristics. For example, the atom of sodium has 11 electrons in orbit around a nucleus containing 11 protons. The valence shell has one electron orbiting at the highest energy address. Chlorine too is electrostatically neutral when 17 electrons are orbiting about the 17 protons of the nucleus. Chlorine has high electron grabbing properties, hence, when the one valence electron from a sodium atom is given up to the chlorine

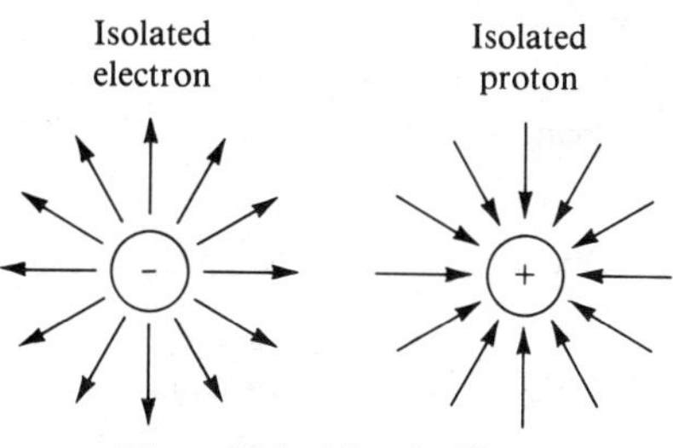

Figure 12.1 Electric Charge

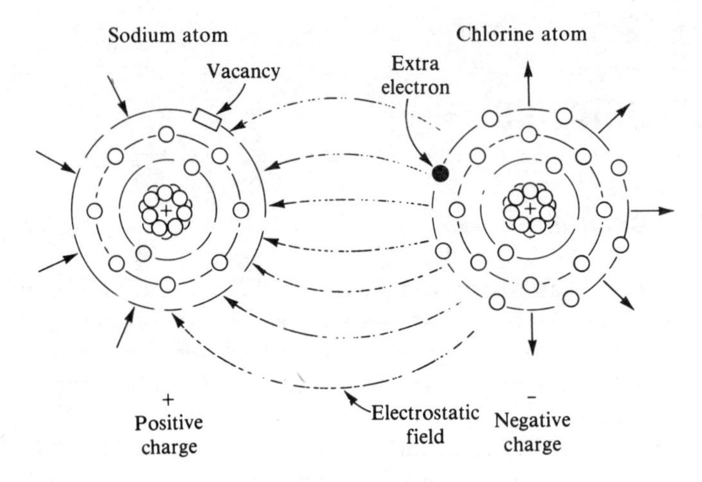

Figure 12.2 Charged Atoms

atom, two ions are produced. The atom of sodium is now a positively charged atom since there are only 10 electrons and 11 protons. The chlorine is negatively charged since it now has 18 electrons and 17 protons. The total chlorine atom has a net electric charge of one electron. Figure 12.2 illustrates these two atoms and the electric field associated with them. The chlorine and sodium ions are bound in an ionic attraction. In the crystal form they become sodium chloride (NaCl), just ordinary table salt.

The physical relationship between the forces of positive and negative charge is best seen using a gold leaf electroscope. The device is shown in Figure 12.3. Assume that the rubber rod of our physics experiment is rubbed with wool, then brought near the collector. Remember that the rod is negatively charged; that is, it has an excess of electrons. When the rod touches the collector, some of the electrons transfer to it. When the rod is removed, the electrons disperse down to the leaves. The leaves then spread apart. The conclusion, then, must be that there exists a force repelling the leaves from each other.

These forces may be restated as:

Like charges repel and unlike charges attract.

The magnitude of the attraction or repulsion depends on the total quantity of charge present.

12.2 COULOMB'S LAW

Coulomb demonstrated the force of attraction using pith balls suspended by silk threads about 5 cm apart. Negative charge was transferred to one ball and positive charge to the other. The force of attraction moved the

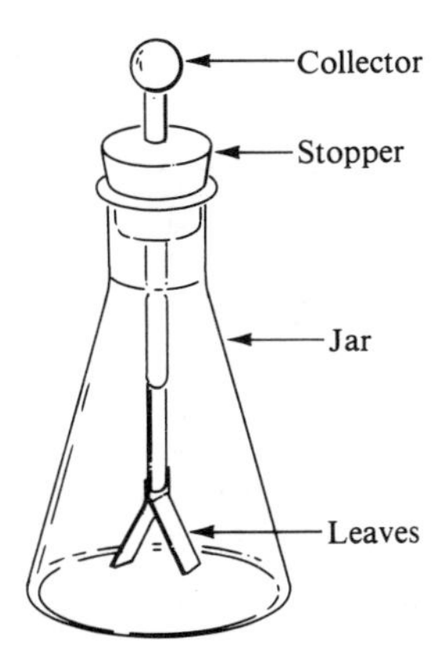

Figure 12.3 Electroscope

pith balls toward each other where they remained in a suspended state. The mathematical relationship was stated in Chapter 2 (Eq. 2.1) and is restated here for convenience.

$$F = \frac{kQ_1Q_2}{r^2} \tag{12.1}$$

where F is the force in newtons, Q_1 and Q_2 are the quantities of the charges in coulombs, r is the distance between charges in meters, and k is a constant depending upon the environment. In free space, $k = 9 \times 10^9$.

Recall that one coulomb equals 6.24×10^{18} electrons. Therefore, the charge of one electron is 1.6026×10^{-19} coulomb. The constant k of Eq. 12.1 is given in units of newton-meters^2 per coulomb^2.

Example 12.1 What force exists between two charges of 20×10^{-9} coulomb and -15×10^{-9} coulomb? Assume the air space between them is 5 cm.

$$\begin{aligned} F &= \left| \frac{(9)(10^9)(20)(10^{-9})(15)(10^{-9})}{(5 \times 10^{-2})^2} \right| \\ &= 1.08 \times 10^{-3} \text{ newtons} \end{aligned}$$

F is a force of attraction, because the charges were of opposite sign. In Example 12.1 it should be noted that the force of attraction is less than four thousandths of an ounce.

The preceding example assumed that the two charges were fixed in space, that is, immovable. Therefore, the force that exists has the potential to do work. In Figure 12.4 the two clusters of charge are fixed. If an electron were suspended in the field, it would move from Q_1 toward Q_2. Since

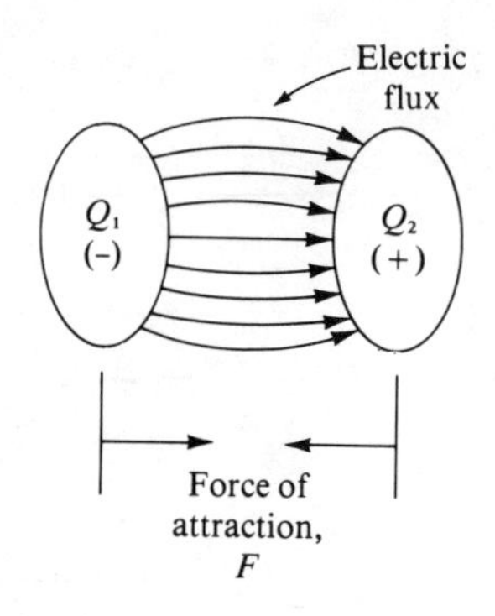

Figure 12.4 The Electric Field

the electron of a given mass has been set in motion, potential energy is being converted to kinetic energy. Therefore, the potential energy represented by the system will be decreased.

Coulomb's law can be expanded to include more than the force between two charges. In fact, many forces at many angles may be considered. For example, in semiconductor materials varying clusters of charge exert influences on charge carriers moving them through the crystal. The net force becomes the sum of the individual forces.

Example 12.2 Assume that three stationary clusters of charge are arranged in a straight line as shown in Figure 12.5(a). $Q_1 = 2 \times 10^{-8}$, $Q_2 = 5 \times 10^{-8}$, $Q_3 = -10 \times 10^{-8}$. What is the resultant force F acting on Q_2? (Assume an air space environment.)

The force upon Q_2 exerted by Q_1 is:

$$F_1 = \frac{(9)(10^9)(2)(10^{-8})(5)(10^{-8})}{2^2(10^{-2})^2}$$

$$= 0.0225 \text{ newton (repel)}$$

The force upon Q_2 exerted by Q_3 is:

$$F_3 = \frac{(9)(10^9)(10)(10^{-8})(5)(10^{-8})}{(10^2)(10^{-2})^2}$$

$$= 0.0045 \text{ newton (attract)}$$

Using the concepts of superposition, the net force is the sum of F_1 and F_3.

$$F = F_1 + F_3$$

$$= 0.027 \text{ newton acting to move } Q_2 \text{ toward } Q_3.$$

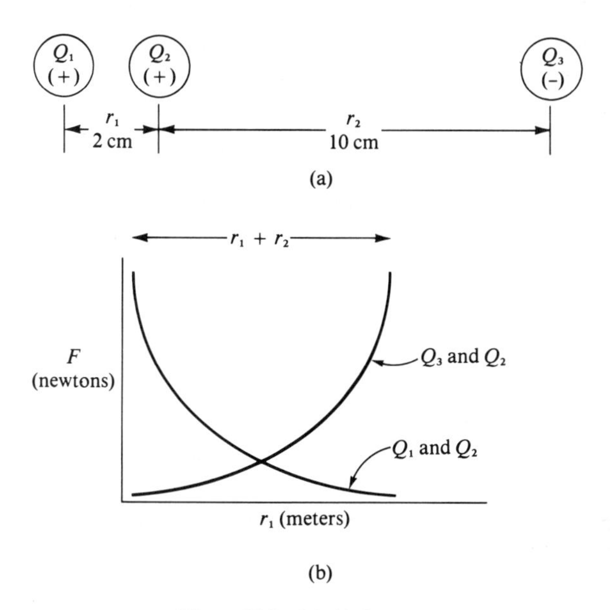

Figure 12.5 Multiple Forces

In Example 12.2 the three charges were in one straight line. When clusters of quantity charge are at different angles, the vector sum of the forces must be considered.

Refer once again to Figure 12.5. If we assume that charges Q_1 and Q_3 are stationary and Q_2 is not, then Q_2 is free to move between the two. As Q_2 moves away from Q_1, the repelling force decreases. As Q_2 moves toward Q_3, the force of attraction increases. At some point between the two, the force of attraction equals the force of repulsion. Figure 12.5(b) shows the force in newtons acting on Q_2 at different values of r_1. The point of intersection of the two curves represents equal attracting and repelling forces. The actual shapes of the two curves will vary depending upon the quantity charges.

Characteristics of Electric Fields

In magnetics, the field of force surrounding a magnet is characterized as lines of force, though there is no real basis for believing there are actual lines. The notion is one of convenience. Similarly, the force that exists between suspended pith balls or gold leaves is characterized as *lines of force*. Here again we have no basis for believing there are lines other than the expediency of explanation.

Figure 12.4 shows two charged bodies, with the force of attraction depending upon the quantity charge of Q_1 and Q_2. Here too we characterize the field of force with lines and attribute specific characteristics to them. Following are the important characteristics of electric field lines.

1. Electric lines (or electrostatic lines) are not continuous; they begin and end at charged surfaces. In Figure 12.4 each line begins at Q_1 and ends at Q_2.
2. Electric lines have direction. It is the same as the direction in which an electron would move if it were suspended in the electric field in a vacuum.
3. Electric lines exhibit the characteristics of lateral repulsion, that is, the lines tend to equalize the spacing between each other.
4. Electric lines always enter or leave a charged surface at right angles to the surface. Notice in Figure 12.4 that each line leaving Q_1 and entering Q_2 creates a 90° angle with each surface.

In Figure 12.4 the total electric field is called the *electric flux*. The symbol for electric flux is Ψ (the Greek letter *psi*, pronounced "sigh"). Since the intensity of the electric flux depends completely on the quantity of charge producing it, we assign Ψ the same units as charge—coulombs.

Flux Density

Before discussing flux density let us examine what takes place when a voltage is placed across a pair of parallel plates. Refer to Figure 12.6. Each plate has an area A and each is parallel to the other, a distance d apart. When the voltage is applied, an excess number of electrons accumulates on the right-hand plate and a deficiency occurs at the left-hand

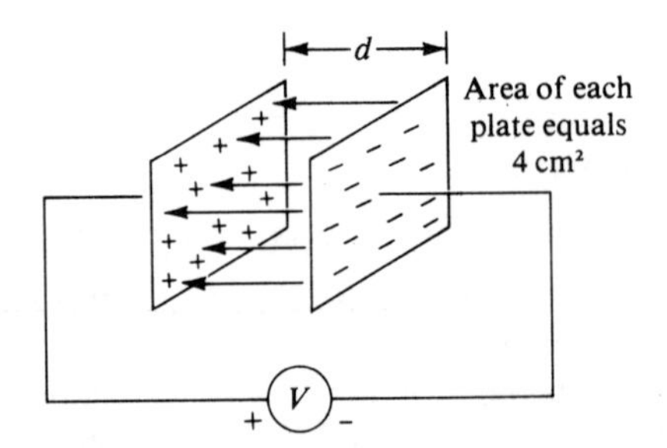

Figure 12.6 Parallel Plates with Electric Charge

plate. Electrons are moved away from the left-hand plate toward the positive terminal of the voltage source. The total charges on both plates are equal but of opposite polarity. The flux density between the plates is the ratio of total flux to the area through which it passes

$$D = \frac{\Psi}{A} \tag{12.2}$$

where Ψ is in coulombs, A is in square meters, and D is flux density in coulombs per square meter.

Example 12.3 Assume the total charge stored on each plate of Figure 12.6 is 5×10^{-6} coulomb. Each plate is 4 cm by 5 cm. What is the flux density?

$$\begin{aligned} D &= \frac{(5)(10^{-6})}{(4)(10^{-2})(5)(10^{-2})} \\ &= 2.5 \times 10^{-3} \text{ coulomb/m}^2 \end{aligned}$$

12.3 ELECTRIC FIELD INTENSITY

Electric field intensity can be thought of as the electric field strength. The strength or intensity is defined as the force exerted upon a unit charge within an electric field. We have already seen the effect of forces acting upon a charge (Figure 12.5 and Example 12.2).

Refer to Figure 12.7. Assume a unit positive charge is located between the two parallel plates. A force is exerted upon the unit charge, tending to move it toward the negative plate. The greater the magnitude of the charge Q, the greater the force exerted upon the unit positive charge. Field intensity is the ratio of this force to the quantity of charge. It is symbolized by the letter $\mathscr{E}$

$$\mathscr{E} = \frac{F}{Q_1} \tag{12.3}$$

where force F is in newtons, Q_1 is in coulombs, and $\mathscr{E}$ is in newtons per coulomb.

Since we are dealing with a unit charge, we can substitute into Eq. (12.1)

$$F = k\frac{Q_1 Q}{r^2} \tag{12.4}$$

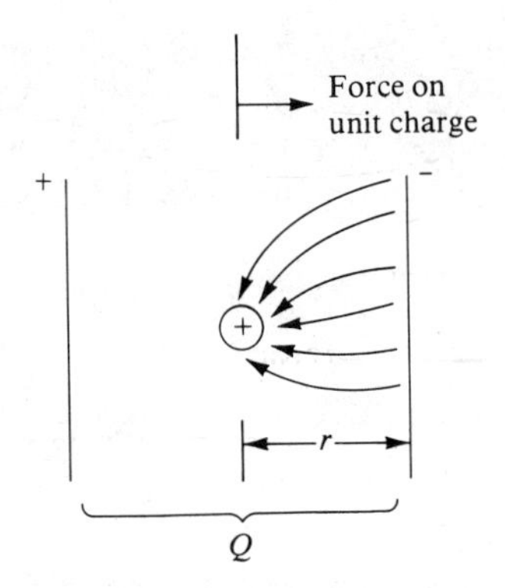

Figure 12.7 Electric Field Intensity

Substituting Eq. (12.4) into (12.3),

$$\mathscr{E} = \frac{kQ_1 Q}{Q_1 r^2} = \frac{kQ}{r^2} \tag{12.5}$$

k in Eq. (12.1) was a constant used for air, and is identified as the proportionality constant of free space.

The property of a material called *permittivity* (ϵ) is a measure of how well an electric field can establish electric flux in it. The greater its permittivity, the greater the concentration of flux it *permits* a given electric field to establish. Permittivity is analogous to its counterpart in a magnetic field, permeability. The units of ϵ are ampere-seconds per volt-meter. The permittivity of free space is

$$\epsilon_o = 1/4\pi k = 8.854 \times 10^{-12} \text{ A-s/V-m}$$

The *relative* permittivity of a material is defined as the ratio of its actual permittivity to the permittivity of free space.

$$\epsilon_r = \epsilon / \epsilon_o \tag{12.6}$$

We may now substitute for k in Eq. (12.5)

$$\mathscr{E} = \frac{Q}{4\pi\epsilon_o r^2} \tag{12.7}$$

where $\mathscr{E}$ is the field intensity, ϵ_r and ϵ_o are the relative and vacuum permittivities, Q is the charge reference, and r is the distance of a unit charge from the reference.

Example 12.4 Assume the two charges of Figure 12.8 are contained in a vacuum and the charges are $\pm 20 \times 10^{-8}$ coulomb. What is the electric field intensity at point a?

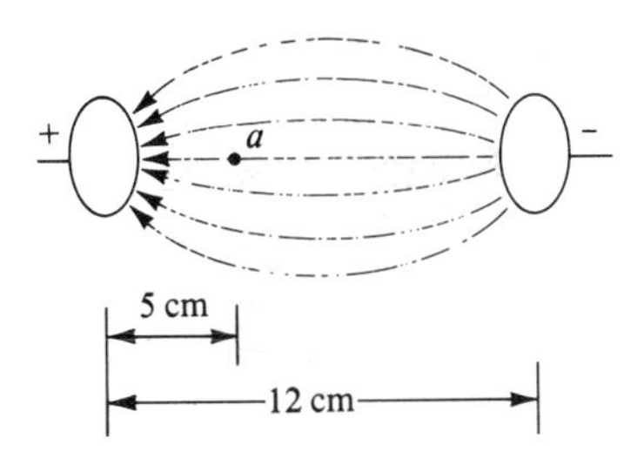

Figure 12.8 Figure for Example 12.4

The effect of the positive charge is given by Eq. (12.5).

$$\mathscr{E}^+ = \frac{(9 \times 10^9)(20 \times 10^{-8})}{(5 \times 10^{-2})^2}$$

$$= 7.2 \times 10^5 \text{ newtons/coulomb}$$

$$\mathscr{E}^- = \frac{(9 \times 10^9)(20 \times 10^{-8})}{(7 \times 10^{-2})^2}$$

$$= 3.67 \times 10^5 \text{ newtons/coulomb}$$

$$\mathscr{E} = \mathscr{E}^+ + \mathscr{E}^- = 1.09 \times 10^6 \text{ newtons/coulomb}$$

12.4 ELECTRIC POTENTIAL DIFFERENCE

The electric field intensity between two parallel plates has been defined as newtons per coulomb. This intensity may be defined in terms convenient to circuit design. Before doing so, let us consider some basic concepts from physics. In Chapter 2 we established voltage as the energy required to move charge through a potential difference.

$$V = \frac{W}{Q} \tag{12.8}$$

where W is energy in joules, Q is charge in coulombs, and V is voltage. Stated in words, *it takes one joule of energy to move one coulomb through a potential difference of one volt.*

We also know that energy is defined as force acting over a distance.

$$W = Fr \tag{12.9}$$

where F is force in newtons and r is distance in meters. To consider the force in an electric field environment, we must substitute Eq. (12.3) into Eq. (12.9).

$$W = \mathscr{E}Qr \tag{12.10}$$

Substituting for W in Eq. (12.8) from Eq. (12.10) and simplifying,

$$\mathscr{E} = \frac{V}{r} \text{ volts/meter} \tag{12.11}$$

Equation (12.11) states that the electric field intensity between a pair of parallel plates is directly proportional to the voltage across the plates and inversely proportional to the distance between the plates. In Figure 12.9 the field intensity is 5000 volts per meter.

Voltage Gradients

In Figure 12.9, the variation of voltage between the plates, called the *voltage gradient*, is linear with respect to distance. If a hypothetical voltmeter were connected between the negative plate and the electric field, we could measure the voltage gradient. If the voltage probe were inserted halfway between the plates, the meter would read 10 volts. One millimeter from the negative plate the meter would read 5 volts. Similarly, 1 millimeter from the positive plate, the meter would read 15 volts.

In order to demonstrate this linear relationship, we must examine the characteristics of the electric field between parallel plates. Equation (12.7) gives the field intensity between a unit positive charge and a reference charge Q. The field intensity between a pair of parallel plates is

$$\mathscr{E} = \frac{Q}{\epsilon A} \tag{12.12}$$

where A is the area of the plates. Setting Eqs. (12.11) and (12.12) equal to each other,

$$\frac{Q}{\epsilon A} = \frac{V}{r}$$

Solving for V,

$$V = \frac{Qr}{\epsilon A} \tag{12.13}$$

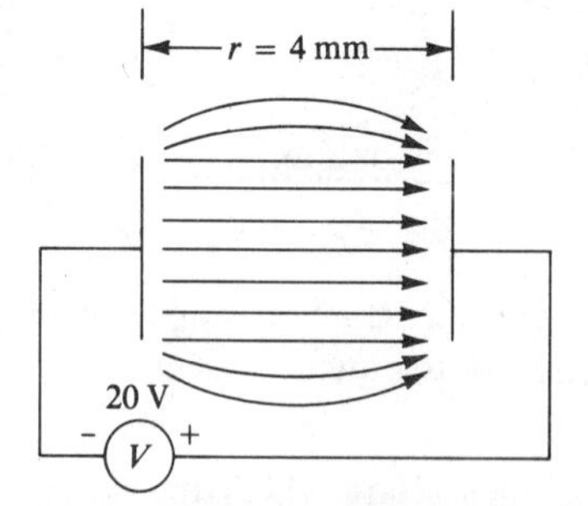

Figure 12.9 Electric Potential Difference

where Q is quantity charge in coulombs, r is the distance between plates, in meters, and A is the surface area of one plate in square meters.

Example 12.5 In Figure 12.10, the dimensions of each plate are 4 cm by 8 cm. The plates are separated by air (the permittivity is essentially the same as free space) and are 2 cm apart. The applied voltage is 20 V. How much charge is stored on each plate? What is the electric field intensity between the plates? What is the voltage (with respect to the negative terminal of the applied voltage) midway between the plates? What is $\mathscr{E}$ midway between the plates?

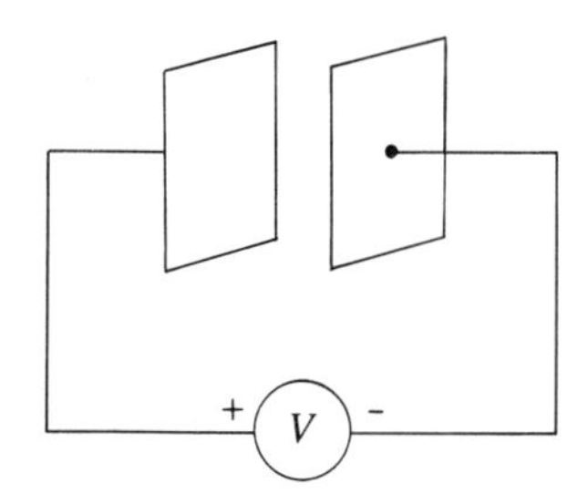

Figure 12.10 Figure for Example 12.5

$$A = (4 \times 10^{-2})(8 \times 10^{-2}) = 32 \times 10^{-4}\text{m}^2$$

By Eq. (12.13),

$$Q = \frac{V\epsilon_0 A}{r} = \frac{20(8.854 \times 10^{-12})(32 \times 10^{-4})}{2 \times 10^{-2}}$$
$$= 28.33 \times 10^{-12} \text{ coulomb}$$

By Eq. (12.11),

$$\mathscr{E} = V/r = \frac{20}{2 \times 10^{-2}} = 1000 \text{ } V/\text{m}$$

Midway between the plates, $r = 1 \text{ cm} = 10^{-2}$ m, so by Eq. (12.11),

$$V = \mathscr{E}r = (1000)(10^{-2}) = 10 \text{ volts}$$

The electric field intensity $\mathscr{E}$ between the plates depends *only* on V and the distance r between the plates. Therefore, $\mathscr{E}$ has the same value midway between the plates (1000 V/m) as it does anywhere else between the plates.

Equipotential Lines

Equipotential lines are lines along which the voltage is the same. These equipotential lines always form 90° angles with the flux lines. Figure

12.11(a) illustrates that equipotential lines exist between two parallel plates. Note how the lines flare out at the top and bottom where a "fringing effect" takes place. Equipotential lines, like flux lines, do not actually exist as lines. Rather, points of equal voltage join to form a representative line.

Figure 12.11(b) illustrates the equipotential lines between two quantity charges. They are unlike charges, and the pattern is somewhat similar to that of the parallel plates. If, however, they are like charges, the pattern differs considerably as can be noted in Figure 12.11(c). In Figure 12.11(d) two cylinders are placed next to each other and a difference of potential placed across them. The flux lines emanate from the smaller cylinder to the larger. The inner sections of the cylinders contain the flux lines bowed out toward the center (shown two-dimensionally in Figure 12.11(e)). The equipotential lines become a series of circular patterns within the flux. This is the equivalent of an electrostatic lens.

12.5 DISPLACEMENT CURRENT

In our study of magnetism we learned that a wire carrying a current generated magnetic flux, and conversely, when flux cut across a wire, a current (or movement of charges) was induced in the wire. In this examination of electrostatics we have learned that a difference in the concentration of quantity charge produces an electric flux. When a material with fairly uniform charge distribution is placed in an electric field, the field will cause a different charge distribution in the material. The charge concentration will, in turn, develop an electric field that will interact with the original electric field. This phenomenon occurs with materials having electric charges that are not tightly bound to their parent atoms. Metals and semiconductors are excellent examples. With materials such as insulators, where valence electrons are not readily given up, there is no redistribution of the electrons. Figure 12.12 illustrates two parallel plates with and without an intervening conductive material. Note how the conductor becomes polarized (polarity distributed across the material). The electric field within the conductor is in a direction opposite to the external field.

To establish the polarity shown in the material, electrons and electron vacancies had to "move." Coulomb has established that a movement of electrons constitutes a current: $i = \Delta Q/\Delta t$. This is called *displacement current*. Such a current is useful in analyzing instantaneous conditions of a circuit.

12.6 DIELECTRIC CONSTANT

In Figure 12.12 we inserted a material with charge carriers that were relatively free to move about. If we replace the material with certain

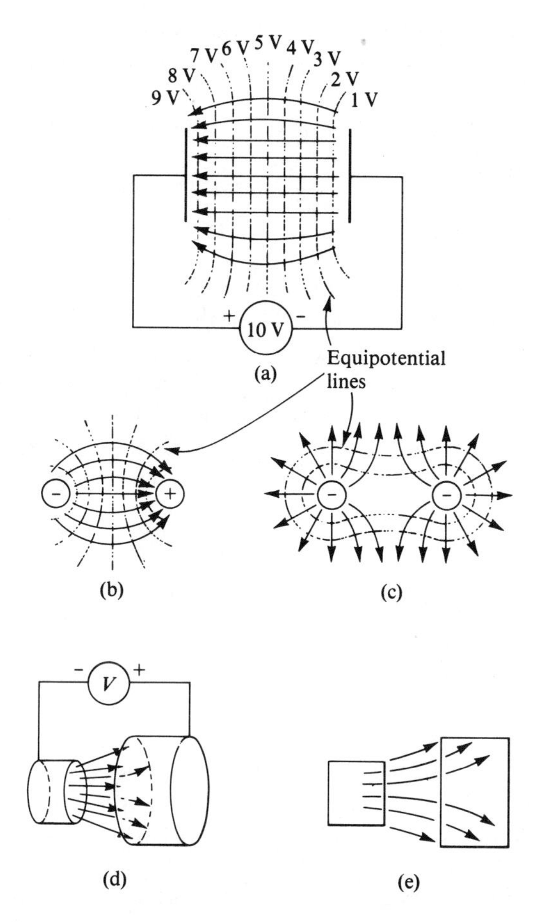

Figure 12.11 Equipotential Lines

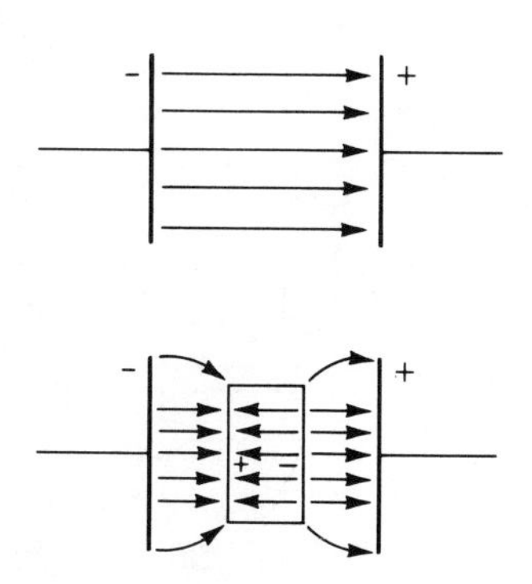

Figure 12.12 Electrostatic Induction

insulating materials, charge concentrations do not occur. The electric field concentration causes the valence electrons to step to higher energy addresses, but does not supply sufficient energy to free the electron from its parent. Essentially, energy is stored in the material's atoms by increasing the potential energy of valence shell electrons. Once these electrons drop back to their equilibrium level, energy is given up. We may think of the atoms within the material as becoming polarized, though no actual polarities exist.

Refer now to Figure 12.13. Assume that a given voltage causes an electric field to be established between the two plates and that the material between the plates is a ceramic such as magnesium silicate. If we compare the flux density in the magnesium silicate with the flux density in a vacuum, we note there is six times greater density than in the vacuum. Thus, the magnesium silicate has a relative permittivity of 6.

Eq. (12.12) represents the field intensity present between two parallel plates:

$$\mathscr{E} = \frac{Q}{\epsilon A}$$

According to Eq. (12.2), however, Q/A equals flux density D, so

$$\mathscr{E} = \frac{D}{\epsilon} \tag{12.14}$$

Substituting Eqs. (12.11) and (12.6) into (12.14), and solving for D,

$$D = \frac{V\epsilon_r\epsilon_o}{r} \tag{12.15}$$

where D is the flux density in coulombs per square meter, V is the voltage applied to a pair of parallel plates, ϵ_r is the relative permittivity, ϵ_o is the permittivity of a vacuum, and r is the distance between the plates in meters.

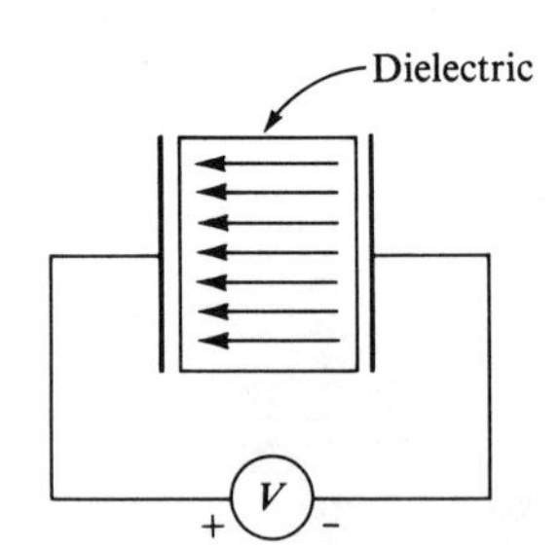

Figure 12.13 Dielectric Constant

TABLE 12.1 Dielectric Constants

Dielectric Material	Dielectric Constant (Relative Permittivity)
Araldite	3.8
Barium-strontium titinate	7500.0
Butadien styrent (rubber)	2.96
Glass	8.1
Magnesium silicate	6.0
Mica	5.0
Paper	2.5
Phenolformaldehyde	8.6
Porcelain	6.5
Silica	3.78
Steatite	5.77
Teflon	2
Air	1.0006
Water	80
Wood	2.42

Many different materials are used between parallel plates to increase the flux density. These materials are called *dielectrics*. The relative permittivity of a material is also called its *dielectric constant*. Table 12.1 lists the dielectric constants of various materials. Note that the ceramic material barium-strontium titinate has a dielectric constant equal to 7500 times that of a vacuum. Some substances (not shown in the table) are in fluid form. Oil, for example, has a dielectric constant of 4.

Example 12.6 Assume a pair of parallel plates, each having a surface area of 10 cm^2, are separated by a material having a dielectric constant of 500. What is the flux density when 5 volts is applied? Assume the plates are spaced 1 mm apart.

From Eq. (12.15),

$$D = \frac{V\epsilon_r\epsilon_o}{r}$$

$$= \frac{(5)(500)(8.854)(10^{-12})}{10^{-3}}$$

$$= 22.135 \times 10^{-6} \text{ coulombs/m}^2$$

Refer once again to Figure 12.13. Note that V is the voltage across the plates. Assume that a specific dielectric is in the space between the plates. If V is increased, eventually it will reach a value which will cause an arc to take place between the surfaces. The arc-over begins when a certain number of electrons break away from their parent atoms in the dielectric. These carriers act as ballistic electrons and bombard other atoms in the dielectric, thereby freeing their valance electrons. The process is cumulative until an arc-over occurs, a condition somewhat similar to an avalanche.

When the arc-over does occur, the dielectric becomes a conductor and current flows between the two plates. The amount of field intensity that a given dielectric can withstand before arc-over occurs is defined as *dielectric strength*. The dielectric strengths of various materials depend upon the basic substance, the manufacturing process, and the thickness of the dielectric. The thicker the dielectric, the greater the voltage required to produce arc-over. Table 12.2 lists various materials and their dielectric strengths. The actual strengths are listed in kilovolts per millimeter, the voltage that causes a 1-millimeter-thick sample to arc over. Often it may be useful to identify the dielectric strength of an insulator in volts per mil, where 1 mil = .001 inches. Multiply the kV/mm value by 25.4 to obtain volts/mil.

Example 12.7 Assume that a pair of parallel plates is to use a dielectric of paraffin. The space between the plates is $\frac{1}{8}$ inch. What is the maximum voltage that can be applied?

Since 39.37 inches equals 1 meter,

$$\tfrac{1}{8} \text{ inch} = 3.175 \text{ mm}$$

From Table 12.2 the voltage breakdown for paraffin is 11.5 kV/mm.

TABLE 12.2 Dielectric Strengths

Material	kV/mm	Material	kV/mm
Air	2.95	Jute (impregnated)	1.2
Asbestos paper	4.2	Lava	3
Asphalt (Byerlyte)	14.0	Marble	6.5
Bakelite, wood molding mixture	17.7	Mica	21
Bakelite, asbestos molding mixture	9.8	Micabond, plate	37.5
Bakelite, Micarta-213	31.4	Micabond, flexible	23.1
Cellophane	51	Oil, insulating	10
Celluloid (clear)	12	Paper	8.7
Cellulose acetate	48.0	Paraffin	11.5
Empire cloth, muslin	48.0	Porcelain	8.0
Fiber, vulcanized, including hard fiber, all colors	3.9	Pressboard (oiled)	29.2
Glass (ordinary)	8	Pressboard (varnished)	15.5
		Rubber (hard)	70
		Slate	1.3
		Wood (maple), paraffined	4.6

For 3.175 mm thickness,

$$V = \left(\frac{11.5\ \text{kV}}{\text{mm}}\right)(3.175\ \text{mm})$$
$$= 36.51\ \text{kV}$$

SUMMARY

Positive and negative electric charges are characterized by electrostatic fields surrounding them. These lines extend outward from the charge perpendicular to the surface of the charge. Two quantity charges exhibit a force of attraction or repulsion between them.

Electric fields have terminal points, direction, and lateral repulsion. Electric flux density defines the number of electric flux lines in a given area.

Permittivity defines the ability of a material to establish electric flux. Electric field intensity is a function of the charge, the permittivity of the material, and the distance between charges.

When two parallel plates have a difference in the quantity charge on them, electric flux lines develop between the plates. Equipotential lines can be described between the plates. These lines define the voltage gradients that exist between the plates.

Various materials have dielectric constants. These constants represent the relative permittivity of the material. The material's ability to support electric flux without breaking down is the dielectric strength of the material.

PROBLEMS

Reference Section 12.2

1. What is the force in newtons between two quantity charges, 5×10^{-8} and -20×10^{-8} coulomb? Assume the space between the charges is air and that the distance is 3 cm.
2. Refer to Problem 1. At what distance will the force decrease to half?
3. If three clusters of charge are placed in a straight line and fixed in position, what is the force acting on Q_1? Assume $Q_1 = 10^{-8}$, $Q_2 = 2 \times 10^{-8}$, $Q_3 = -4 \times 10^{-8}$. Assume the space between Q_1 and Q_2 is 5 cm and the space between Q_2 and Q_3 is 2 cm. (Assume free space.)
4. Refer to Problem 3. What is the force acting upon Q_3? Is it repelling from or attracting to Q_2?
5. Assume two parallel plates have a mutual surface area of 50 cm^2. If the flux density is 0.022 coulomb/m^2, what is the quantity charge stored in the plates?

6. Assume a pair of parallel plates each have a surface area of 40 cm^2. If the charge on the plates is 25×10^{-7} coulomb, what is the flux density? (Assume the plates are 2 cm apart.)

Reference Sections 12.3 and 12.4

7. If the spacing between the plates of Problem 6 is doubled, what is the flux density?
8. Refer to Problem 1. If the two charges are placed in a vacuum, what is the field intensity at the midpoint between the charges? What is the intensity 1 cm from the positive charge? What is it 1 cm from the negative charge?

Reference Section 12.6

9. Assume two parallel plates in a vacuum are spaced 1 cm apart. If the dimensions of each plate are 5 cm by 8 cm, what charge is accumulated on the plates when 5 volts is applied? What is the energy stored? What is the field intensity?
10. Assume the two plates of Problem 9 are moved apart to 3 cm. What is the field intensity? What is the energy stored? What are the voltages 1 cm from each plate?
11. Assume a pair of parallel plates 2 cm by 1 cm are laminated to a sheet of porcelain 0.2 cm thick. What is the flux density when 10 volts is applied? What voltage will cause arc-over through the porcelain?
12. If the porcelain in Problem 11 were replaced with mica, what would be the new flux density and arc-over voltage?
13. Refer to Problem 11. How much energy is stored in the dielectric material in the form of electric flux?
14. Refer to Problem 12. How much energy is stored in the system? How much energy is stored at just below the breakover voltage?
15. Assume a pair of circular plates with a radius of 10 mm and a spacing of 2 cm are to use hard rubber as a dielectric. What is the breakover voltage?
16. The total surface area of each of two parallel plates is 12 mm. The dielectric constant of the material between the plates is 200. What is the flux density if 12 volts is applied?
17. If the dielectric material of Problem 16 was removed, what is the new flux density?
18. If a pair of parallel plates uses cellulose acetate as a dielectric material, what is the maximum voltage which can be applied if the spacing is 0.34 inch?
19. If the material in Problem 18 is replaced with marble, what is the maximum voltage?
20. Which dielectric material should be used between parallel plates to store maximum charge, if the plate dimensions and separation are fixed?

CHAPTER 13

CAPACITANCE

13.1 CAPACITOR

Most of the principles of electrostatics discussed in Chapter 12 apply to the concepts developed in this chapter. Coulomb's law will be explored further, along with the extensions of stored electric field. The physical properties of capacitors and the types of capacitors are discussed, as well as the nature of capacity as a dynamic circuit element.

The word *capacitor* is derived from capacity, which describes an ability to store something. An individual may have a capacity for tolerance or generosity. We refer to capacity as the ability to store electric charge.

One type of capacitor can be formed by a pair of parallel plates separated by a dielectric material. The parallel plates need not be made of rigid material, nor do they need to be fixed in place. Figure 13.1 illustrates several types of capacitors.

In (a) the two parallel plates are separated by an air dielectric. This structure was discussed in Chapter 12. In (b) several parallel plates are interleaved to increase the total surface area. The dielectric may be mica, tantalum, or any high permittivity material. In (c) we note that a capacitor consists of a larger metal surface area of any size interfaced with the upper disk separated by a dielectric. The effective surface area is that of the disk. Figure 13.1(d) illustrates a variable capacitor. If we assume that the back plate is fixed in position and the front plate (an eccentric semicircle) can be rotated, then the effective surface area can be varied between two extremes. In (e) two foil-type plates are separated by a soft dielectric such as paper, then rolled into a small package. Figure 13.1(f) illustrates another variable capacitor where the surface area consists of two cylinders. A dielectric spacer between the two cylinders provides some additional rigidity and capacity. The last illustration (g) will be discussed in more detail later in the chapter. This capacitor is identified as an electrolytic capacitor because of the chemical electrolytes used in its construction.

Although Figure 13.1 illustrates several physical capacitors, many other types are not shown. A bus cable, as found in a computer, consists of many wires within a flat tapelike arrangement. It has a capacity represented by parallel wires. Capacitance exists whenever any two conductors are separated by an insulator (dielectric).

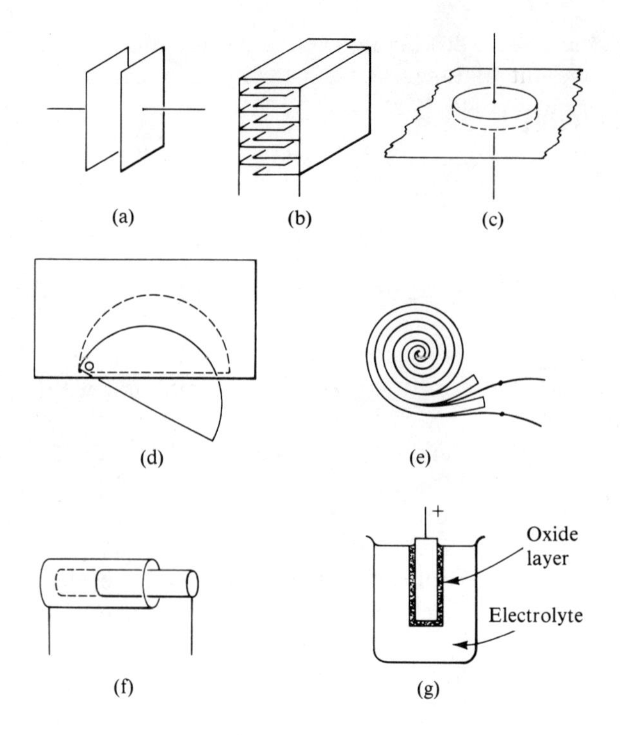

Figure 13.1 Types of Capacitors

If we return to Coulomb's original experiments, we can observe a basic characteristic when a voltage is applied to a pair of parallel plates. Assume for now that when we applied 10 volts dc, the charge that accumulated on the plates was 5×10^{-7} coulomb. When the voltage was doubled, the quantity charge doubled to 10×10^{-7} coulomb. Coulomb's conclusion was that a constant relationship exists between Q and V. This constant is identified by C for capacity.

$$C = Q/V \tag{13.1}$$

where Q is in coulombs, V is in volts, and C is in farads. In the illustration we had 10 volts developing 5×10^{-7} coulomb of charge. In this case the capacity becomes 0.05×10^{-6} farad. Because the unit of capacity, the farad, is quite large and we rarely encounter actual capacities of that size, the units of the microfarads (μF) are more often used to specify the capacity of a capacitor. $1\ \mu\text{F} = 10^{-6}$ F. Our illustration capacity is 0.05 μF. Similarly, the micromicrofarad or picofarad (pF) is 10^{-12} farad.

In Chapter 2 current flowing into a circuit was defined as the movement of charge per unit of time. Mathematically, this can be expressed as $i = \Delta Q/\Delta t$. Equation (13.1) can also be restated as

$$C = \frac{\Delta Q}{\Delta V} \tag{13.2}$$

The increment change in Q, with respect to an incremental change in V, is still the constant C (or capacity). Since:

$$\Delta Q = C\Delta V \tag{13.3}$$

we may divide both sides by Δt, and solve for i.

$$\boxed{i = C\,\frac{\Delta V}{\Delta t}} \tag{13.4}$$

where i is the average current in amperes, C is in farads, ΔV is an increment in volts, and Δt is an increment of time.

Equation (13.4) expresses the very important fact that the current in a capacitor is directly proportional to the rate of change of the voltage across it. The equation explains why very large currents flow into a capacitor when a voltage is suddenly switched across it. Note in Figure 13.2 that a voltage of 20 volts is applied to the capacitor in 2 microseconds. We are assuming the input voltage shown as V_{in} is to step from zero to 20 volts. If

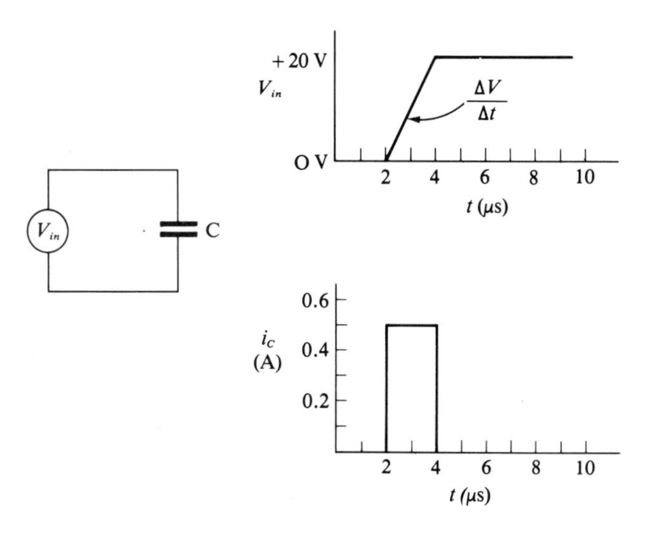

Figure 13.2 Current into a Capacitor

the capacitor is 0.05 μF, the current will be 0.5 ampere during the change in voltage. It is important to note that the large current flowed into the capacitor only *during the change* in input voltage. Once the voltage reached 20 volts, the rate of change of the voltage was zero; hence no current flowed in the circuit. Indeed, the capacitor is fully charged.

Example 13.1 Assume that a 0.01 μF capacitor has a pulse of voltage as shown in Figure 13.3(a). What is the average current flowing into and out of the capacitor?

Using Eq. (13.4) for the interval from 0 to 3 μs,

$$i = \frac{(0.01)(10^{-6})(15)}{(3)(10^{-6})} = 50 \text{ mA}$$

For the interval from 3 to 6 μs,

$$i = \frac{(0.01)(10^{-6})(0)}{(3)(10^{-6})} = 0 \text{ mA}$$

For the interval from 6 to 10 μs,

$$i = \frac{(0.01)(10^{-6})(-15)}{(4)(10^{-6})} = -37.5 \text{ mA}$$

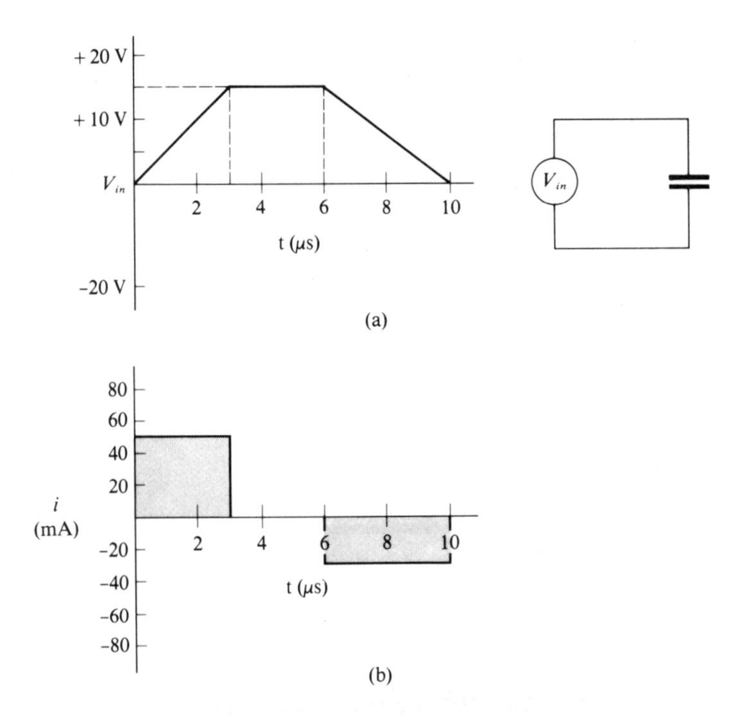

Figure 13.3 Figure for Example 13.1

(The negative sign indicates that a current is flowing out of the capacitor.) Figure 13.3(b) shows a symbolic representation of the current in the circuit.

13.2 PHYSICAL PROPERTIES OF CAPACITANCE

In Chapter 12 we learned that substituting the dielectric for the air gap between the two parallel plates increases the flux density. We can now see how changes in effective surface area, dielectric material, and distance between plates influence the capacity. These factors in turn will influence the amount of stored charge.

Equation (12.13) can be rewritten. Substituting d for r,

$$Q = \frac{\epsilon VA}{d} \tag{13.5}$$

Substituting Eq. 13.5 into Eq. (13.1),

$$\boxed{C = \frac{\epsilon A}{d}} \tag{13.6}$$

where C is the capacity in farads, A is the area in square meters, d the distance between the plates in meters, and ϵ the actual dielectric constant.

If the material between the plates has relative permittivity ϵ_r, the ϵ in Eq. (13.6) can be replaced with

$$\epsilon = \epsilon_r \epsilon_o$$

We see from Eq. (13.6) that increasing the distance between the plates will decrease the capacity. Increasing the area of the plates increases the capacity.

Example 13.2 What is the capacity of the device shown in Figure 13.1(a) if the two plates are spaced 0.2 mm apart and each plate is 2 cm by 2 cm? Assume an air dielectric. What is C if the dielectric is mica?

The area is

$$A = (2)(10^{-2})(2)(10^{-2}) = 4 \times 10^{-4}\ \text{m}^2$$

$$\epsilon = 8.854 \times 10^{-12}$$

Using Eq. (13.6),

$$C = \frac{(8.854)(10^{-12})(4)(10^{-4})}{(0.2)(10^{-3})} = 17.7\ \text{pF}$$

For mica dielectric (Table 12.1),

$$\epsilon_r = 5$$

$$C = \frac{(8.854)(10^{-12})(5)(4)(10^{-4})}{(0.2)(10^{-3})} = 88.5 \text{ pF}$$

This example illustrates the benefit of using a material with a large ϵ_r to separate the plates of a capacitor: Using a material with $\epsilon_r = 5$, we obtained 5 times the capacitance in the same space.

Working Voltage Rating

The working voltage rating of a capacitor is the maximum dc voltage that can be applied across the plates of the device. This rating is influenced by the spacing between the plates, the dielectric material used, and the type of capacitor. Obviously, the closer the plates are to each other, the lower the voltage that can be applied. Table 12.2 lists the dielectric strengths of various materials used in manufacturing capacitors.

Manufacturers list the working voltage as WVDC, which means *working voltage direct current*. These ratings are generally given in round figures such as 25 WVDC, 200 WVDC, 600 WVDC, or sometimes, 1600 WVDC.

Example 13.3 What would be the capacity and the voltage rating of a capacitor similar to Figure 13.1(c)? Assume the thickness of the glass dielectric to be 0.1 mm and the diameter of the disk to be 0.5 cm.

$$A = \pi r^2 = (3.1416)(0.25)^2(10^{-2})^2 = 0.196 \times 10^{-4}$$

$$\epsilon_r = 8.1$$

$$C = \frac{(8.1)(8.85)(10^{-12})(0.196)(10^{-4})}{(0.1)(10^{-3})} = 14 \text{ pF}$$

The voltage breakdown from Table 12.2 for glass is 8 kV/mm

$$V = \frac{(8 \text{ kV})(0.1 \text{ mm})}{\text{mm}} = 800 \text{ volts}$$

Electrolytic Capacitors

The preceding discussion about the geometry of capacitors illustrates that the amount of capacity is limited by the geometry of the plates. Yet many applications of electrical science require very large capacitance. Several hundred microfarads is not uncommon. To construct a capacitor of this size, using the materials and geometries just discussed, would produce an unreasonably large physical structure. True, the distance between plates could be drastically reduced, but that would reduce the working voltage rating for ordinary materials.

The technique of reducing the thickness of the dielectric can be used if an oxide is used as a dielectric. Note in Figure 13.1(g) that a container filled with an electrolyte of aluminum hydroxide has an aluminum electrode placed in it. If the container is a conductor, it may be used as a contact. When a positive polarity is connected to the aluminum rod with respect to the container, an oxide layer forms around the aluminum rod. This oxide layer becomes the dielectric between the electrolyte and the rod.

Care must be taken not to reverse the polarity else the chemistry of the device will not work. Such an electrolytic capacitor must be used only in circuits where polarity can be observed and voltage ratings are comparatively low. The electrolyte may be either a liquid or a paste. These capacitors are generally available in aluminum cans or in tubular paper containers. They are characterized by high leakage currents and chemical deterioration. In most ordinary capacitors the leakage resistance is generally greater than 1000 megohms, yet in electrolytic types it can be as low as 10 megohms, a distinct disadvantage for some circuits.

13.3 CAPACITORS IN SERIES AND PARALLEL

When more than one capacitor is connected in a given circuit configuration it is often necessary to determine the net or total capacity of the system. How this total capacity is influenced can generally be seen physically from Eq. (13.6). There are, however, some fairly simple relationships which can be readily determined using quantity charge.

Series Circuits

When two capacitors are connected in a series circuit configuration as shown in Figure 13.4, the sum of the voltages across each capacitor must equal the applied voltage.

$$V = V_1 + V_2 \tag{13.7}$$

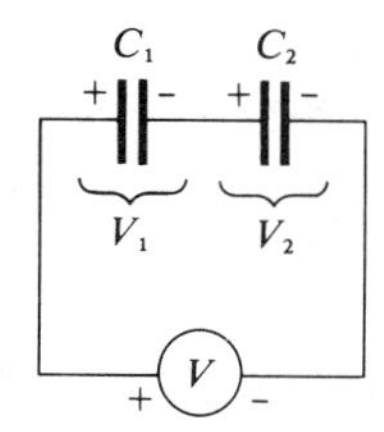

Figure 13.4 Capacitors in Series

Therefore, from Eq. (13.1),

$$\frac{Q_T}{C_T} = \frac{Q_1}{C_1} + \frac{Q_2}{C_2}$$

Since this is a series circuit the amount of Q transferred anywhere in the circuit must be the same: $Q_T = Q_1 = Q_2$.

Factoring and canceling,

$$\frac{Q}{C_T} = Q\left(\frac{1}{C_1} + \frac{1}{C_2}\right)$$

and solving for C_T,

$$\boxed{C_T = \frac{C_1 C_2}{C_1 + C_2}} \tag{13.8}$$

We see, therefore, that the total capacity of two capacitors in series is less than the smallest of the two capacitors. This is similar to resistors in parallel.

By a similar analysis, any number of capacitors connected in series becomes:

$$C_T = \frac{1}{\dfrac{1}{C_1} + \dfrac{1}{C_2} + \cdots + \dfrac{1}{C_N}} \tag{13.9}$$

where N is the number of capacitors.

It is also interesting to note that voltages distribute themselves by an inverse ratio to the relative sizes of the capacitors. Solving Eq. (13.7) for V_2,

$$V_2 = V - V_1$$

But $V_1 = Q/C_1$ and $Q = C_T V$, therefore

$$V_2 = V - \frac{C_T V}{C_1} \tag{13.10}$$

Substituting Eq. (13.8) into Eq. (13.10) results in

$$\boxed{V_2 = \frac{C_1 V}{C_1 + C_2}} \tag{13.11}$$

In a similar manner,

$$\boxed{V_1 = \frac{C_2 V}{C_1 + C_2}} \tag{13.12}$$

We see, therefore, the inverse relationship of the voltage distribution.

Example 13.4 If 0.2 μF and 0.3 μF capacitors are connected in series across 50 volts, what is the total capacity and what is the voltage across each capacitor?

$$C_T = \frac{(0.2)(0.3)}{0.2 + 0.3} = 0.12\ \mu\text{F}$$

If we assume the 0.2 μF to be C_1 and the 0.3 μF to be C_2,

$$V_1 = \frac{(0.3)(50)}{0.3 + 0.2} = 30 \text{ volts}$$

$$V_2 = \frac{(0.2)(50)}{0.3 + 0.2} = 20 \text{ volts}$$

The smaller capacitor (C_1) has the larger voltage across it.

This inverse voltage relationship poses some interesting problems with regard to working voltage ratings. When two or more capacitors are connected in series, care should be taken when selecting the voltage ratings.

Example 13.5 If a 0.1 μF 600 WVDC capacitor is connected in series with a 0.05 μF 200 WVDC capacitor, what is the maximum voltage that may be applied?

Equations (13.11) and (13.12) must be solved for V and then appropriate substitutions made. (Assume C_1 is the 0.1 μF capacitor.)

$$V = \frac{V_1(C_1 + C_2)}{C_2} = \frac{600(0.1 + 0.05)}{0.05} = 1800 \text{ volts}$$

$$V = \frac{V_2(C_1 + C_2)}{C_1} = \frac{200(0.1 + 0.05)}{0.1} = 300 \text{ volts}$$

Clearly the maximum is the lesser of the two voltages, 300 volts. Applying 1800 volts would not exceed the rating of C_1 but would drive V_2 up to 1200 volts: 1000 volts beyond its rating!

Parallel Circuits

The basic circuit is shown in Figure 13.5. Note that the total quantity charge arriving at point A in the circuit is equal to Q_1 plus Q_2. Since this is a parallel circuit, all the voltages are equal: $V = V_1 = V_2$.

$$Q_T = Q_1 + Q_2 \qquad \text{and} \qquad C_T V = C_1 V_1 + C_2 V_2$$

Factoring V and canceling,

$$C_T = C_1 + C_2$$

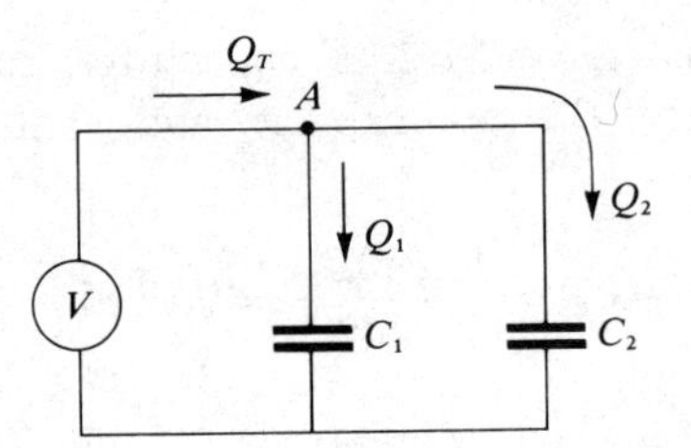

Figure 13.5 Capacitors in Parallel

In a similar manner a general equation for any number of parallel capacitors can be developed.

$$C_T = C_1 + C_2 + \cdots + C_N \tag{13.13}$$

where N is the number of capacitors connected in parallel. Connecting capacitors in parallel is essentially increasing effective plate area, thereby increasing capacity. The working voltage ratings present no special problems because the maximum circuit rating is the smallest capacitor rating.

Example 13.6 Assume three capacitors are connected in parallel across 20 volts dc. If $C_1 = 0.1\ \mu\text{F}$, 200 WVDC, $C_2 = 0.5\ \mu\text{F}$, 600 WVDC, and $C_3 = 0.02\ \mu\text{F}$, 400 WVDC, what are the total capacity and the quantity charge stored in each capacitor?

$$C_T = 0.1 + 0.5 + 0.02 = 0.62\ \mu\text{F}$$
$$Q_1 = C_1 V = (0.1)(10^{-6})(20) = 2 \times 10^{-6}\ \text{coulomb}$$
$$Q_2 = C_2 V = (0.5)(10^{-6})(20) = 10 \times 10^{-6}\ \text{coulomb}$$
$$Q_3 = C_3 V = (0.02)(10^{-6})(20) = 0.4 \times 10^{-6}\ \text{coulomb}$$

Note that the largest capacitor has the greatest charge developed across it.

Series-Parallel Circuits

Series-parallel constructions of capacitive circuits are frequently encountered. No special equations can be developed to aid in circuit analysis. Instead, the basic principle just studied and those of dc circuits can be used. The following example will illustrate this.

Example 13.7 If all the capacitors of Figure 13.6 have the same value C, what is the total capacity?

C_5 and C_6 are in series. Their equivalent capacity is $C/2$.
$C/2$ is in parallel with C_4. This equivalent is $3C/2$.
$3C/2$ is in series with C_3. This equivalent is $3C/5$.
$3C/5$ is in parallel with C_2. This equivalent is $8C/5$.
$8C/5$ is in series with C_1, hence the total capacity is $C_T = 8C/13$.

13.4 ENERGY IN A CAPACITOR

If a capacitor is connected across a dc voltage source and then removed, the charge still remains on the plates of the capacitor. If there were no leakage resistance, the charge would remain indefinitely. If the two leads are touched to each other, a spark will jump between the two wires, equalizing the charge and hence discharging the capacitor. The accumulated charge represents a difference of potential that represents the potential to do work. The energy stored in the dielectric of the capacitor can be transformed into other forms of energy. In the case of our demonstration, the energy stored in the capacitor was transformed into light (photons) and heat (calories).

We may calculate in joules the specific amount of energy stored in the capacitor. Since energy in joules is given as

$$W = Pt$$

and $P = iv$, we can restate Eq. (11.21) as

$$W = ivt$$

The total energy stored by the capacitor can be shown to be:

$$\boxed{W = \tfrac{1}{2}CV^2} \tag{13.14}$$

where W is the energy in joules, C is the capacity in farads, and V is the voltage in volts.

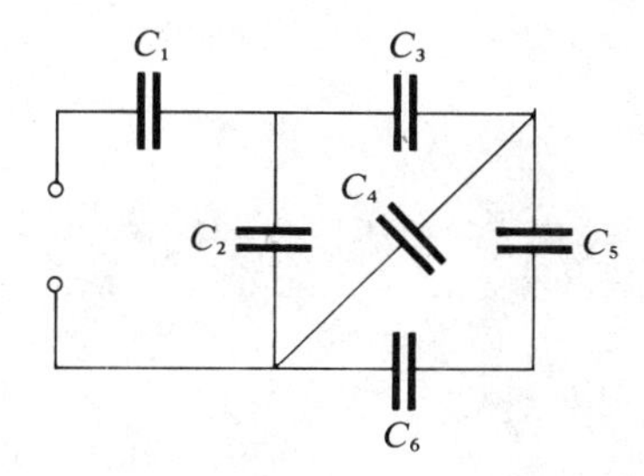

Figure 13.6 Series-Parallel Capacitors

Recall that an inductor stores energy in its magnetic flux. In order to maintain the storage of energy for the inductor, a constant value of current has to flow. In a capacitor, once it becomes fully charged, the charging force can be removed and the energy storage remains static (assuming no leakage resistance).

Example 13.8 If the circuit of Figure 13.4 consists of a 0.01 μF and a 0.04 μF capacitor, how much energy is stored in each capacitor if the applied voltage is 50 volts. What is the total energy stored?

The voltage across the 0.01 μF capacitor is 40 volts and the voltage across the 0.04 μF capacitor is 10 volts.

$$W_1 = \frac{(0.01)(10^{-6})(40)^2}{2} = 8 \times 10^{-6} \text{ joule}$$

$$W_2 = \frac{(0.04)(10^{-6})(10)^2}{2} = 2 \times 10^{-6} \text{ joule}$$

$$W_t = \frac{(0.008)(10^{-6})(50)^2}{2} = 10 \times 10^{-6} \text{ joule}$$

We see from Example 13.8 that the series combination of C_1 and C_2 resulted in the total energy being the sum of the two individual energies. It can be shown that this would be the case whether the circuit was a series, parallel, or series-parallel circuit. In general, the total energy is the sum of the energy stored in each individual element.

$$\boxed{W_t = W_1 + W_2 + \cdots + W_N} \qquad (13.15)$$

It is often useful to relate energy storage to the amount of charge transferred in a given circuit. Solving Eq. (13.1) for V and substituting into Eq. (13.14), the energy equation becomes

$$W = Q^2/2C \qquad (13.16)$$

where Q is in coulombs, W is in joules, and C is in farads.

13.5 TIME CONSTANT

We have seen that current flows into a capacitor or out of a capacitor so long as the voltage across it is changing. Refer again to Figure 13.3 and note that as the input voltage was changing from zero to +15 volts, the average current in the circuit was 50 mA. Theoretically, if the voltage changed from zero to +15 volts in zero time, infinitely large currents would flow in the circuit. Extremely large currents would damage the circuit and possibly the circuit producing the voltage.

In actual circuits there is always some resistance in series with the capacitor. Figure 13.7(a) illustrates a basic circuit. When the switch is placed in position 1, the applied voltage V distributes across R and C; however, the voltage across a capacitor depends upon the accumulation of charge across the plates. Equation (13.1) states that $V = Q/C$. Therefore, at the first instant after the switch is in position 1, there is no charge on C, hence no voltage. All of the applied voltage is across the resistor. The maximum current which can flow at that first instant is limited by the value of R.

$$\boxed{I_m = V/R} \tag{13.17}$$

As charge begins to accumulate on the capacitor, the voltage across C rises and eventually reaches the applied voltage V. When this occurs, there is no voltage across R and the current in the circuit is zero.

Figure 13.7(b) illustrates how the current in the circuit decays from its maximum value to zero. If we assume at t_x that the switch is moved to position 2, the capacitor begins to discharge through R. Here again the maximum current will be limited by R. Again the current will diminish to zero in the same amount of time it takes for the current to go to zero when C was charging.

The voltages in the circuit must at any instant of the charging or discharging of C adhere to Kirchhoff's and Ohm's laws. According to Kirchhoff,

$$V = v_r + v_c \tag{13.18}$$

and according to Ohm's law,

$$v_r = iR \tag{13.19}$$

Since the voltage across a resistor is determined by the current, v_r will follow a pattern exactly like the current. As the voltage across the resistor

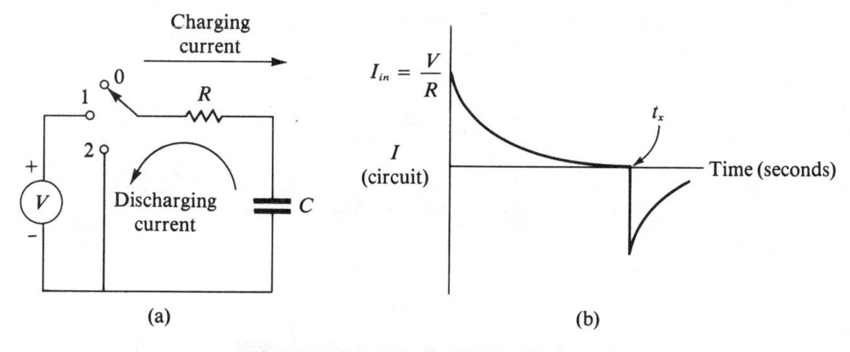

Figure 13.7(a & b) Basic *RC* Circuit

is decaying, the voltage across the capacitor is building up to its maximum value.

Figure 13.7(c) illustrates the relationship of the voltages during the charging of the capacitor. (from time t_0 to t_x). At any instant the sum of v_r and v_c equals the applied voltage V.

At t_x the situation changes. At the first instant after t_x, note that the absolute voltage across C equals the absolute voltage across R. The two then decay toward zero, but at any instant the voltages are equal and opposite to each other.

Let us assume that the resistance of Figure 13.7(a) is 200 kΩ, C is 20 μF, and the applied voltage is 20 V. When the switch is set to position 1, the current begins to decay as shown in Figure 13.7(d). Note that it decays to zero in about 20 seconds. If the current had continued to decay at its initial rate (the dotted line), it would have decayed to zero in 4 seconds. This initial rate of change of the current can be seen quantitatively by substituting into Eq. (13.18) the individual values of v_r from Eq. (13.19):

$$V = iR + v_c$$

However, from Eq. (13.4) we may substitute for i,

$$V = RC\,\frac{\Delta V}{\Delta t} + v_c$$

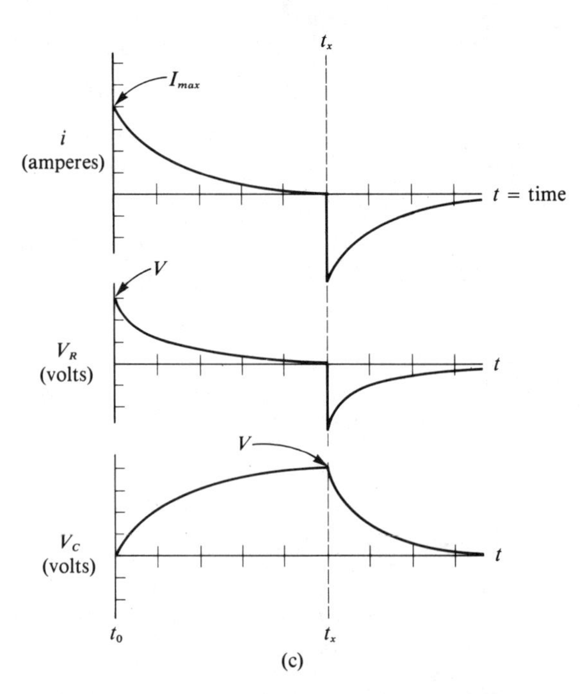

Figure 13.7(c) Voltages and current in an *RC* Circuit

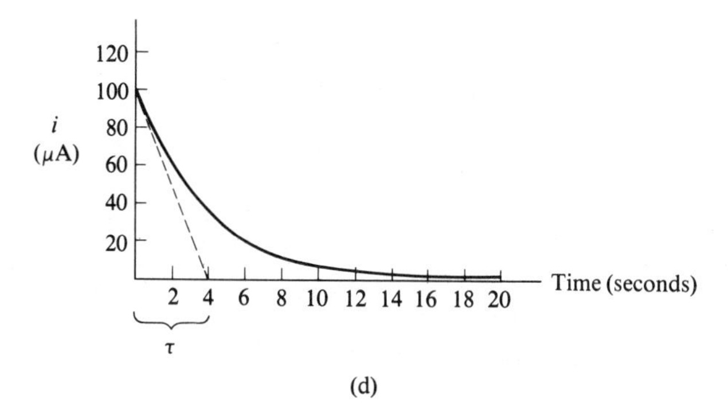

(d)

Figure 13.7(d) *RC* Time Constant

But at that first instant during the initial rate of change the value of v_c is zero.

$$V = RC\,\frac{\Delta V}{\Delta t}$$

However, $V = \Delta V$ over the projected interval of Δt, hence

$$\Delta t = RC = \text{time constant } \tau \qquad (13.20)$$

where τ is in seconds, R in ohms, and C in farads. For the circuit represented by Figure 13.7(d), the time constant is 4 seconds.

$$\tau = RC = (200)(10^3)(20)(10^{-6}) = 4 \text{ seconds}$$

In an *RC* circuit it takes five time constants for the current to decay to (essentially) zero. In Figure 13.7(d) 5τ is equal to 20 seconds.

Example 13.9 Refer to the circuit of Figure 13.8. Assume $V = 40$ volts, $C = 0.02\ \mu\text{F}$, and the time constant is 1 ms. What is the maximum current and how long after the switch is set to position 2 will the current be zero?

$$R = \frac{\tau}{C} = \frac{10^{-3}}{(0.02)(10^{-6})} = 50 \text{ kilohms}$$

$$I_M = \frac{V}{R} = \frac{40}{(50)(10^3)} = 0.8 \text{ mA}$$

$$5\tau = 5 \text{ ms}$$

The timing characteristics of the current decay in an *RC* circuit are similar to the voltage decay characteristics of an *L*/*R* circuit. We note in Figure 13.7(d) that the current decays 63 percent from its initial (or maximum) value in one time constant. In the second time constant the

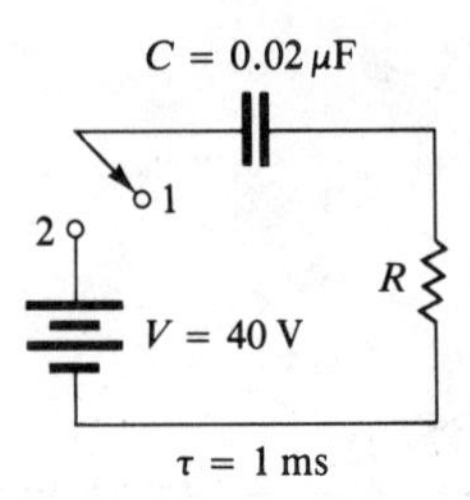

Figure 13.8 Figure for Example 13.9

current decays 63 percent from the value of current it had at the end of the first time constant. During each time constant thereafter, the current decays 63 percent from the value it had at the end of the preceding time constant.

Example 13.10 For the circuit of Figure 13.9(a), assume $R_1 = R_2 = 10$ kilohms, $V = 40$ volts, and $C = 0.05\ \mu$F. The switch is placed in position 1 for 0.5 ms, then switched to position 2. What is the current through R_2, 1 ms after the switch is in position 2?

With switch in position 1,

$$\tau = R_1 C = (10^4)(0.05)(10^{-6}) = 0.5 \text{ ms}$$

$$I_{max} = \frac{V}{R_2} = \frac{40}{10^4} = 4 \text{ mA}$$

Therefore, the current i_1 in R_1 at $t = 0.5$ ms is

$$i_1 = (0.37)(4)(10^{-3}) = 1.48 \text{ mA}$$
$$V_{R_2} = i_1 R_2 = (1.48)(10^{-3})(10^4) = 14.8 \text{ volts}$$

Therefore $v_c = 40 - 14.8 = 25.2$ volts.

When the switch is placed in position 2, notice that the current in the capacitor reverses direction.

$$\tau = (R_1 + R_2)C = (20)(10^3)(0.05)(10^{-6}) = 1 \text{ ms}$$

$$I_{max} = \frac{v_c}{(R_1 + R_2)} = \frac{25.2}{(20)(10^3)} = 1.26 \text{ mA}$$

Therefore, the current i_2 in R_2 1 ms later is

$$i_2 = (0.37)(1.26)(10^{-3}) = 0.466 \text{ mA}$$

Figure 13.9(b) illustrates the voltage across the capacitor and the current through R_2.

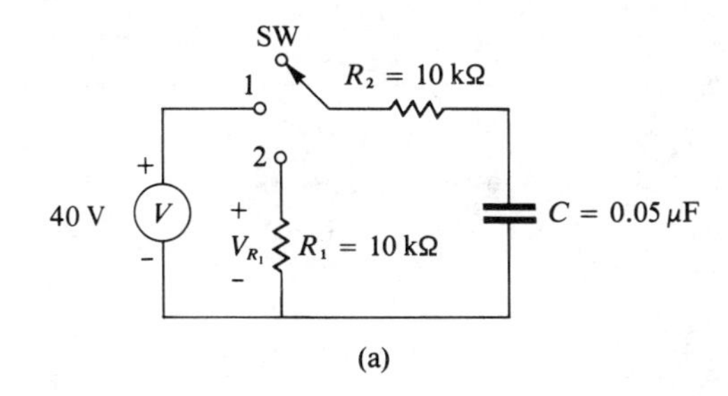

(a)

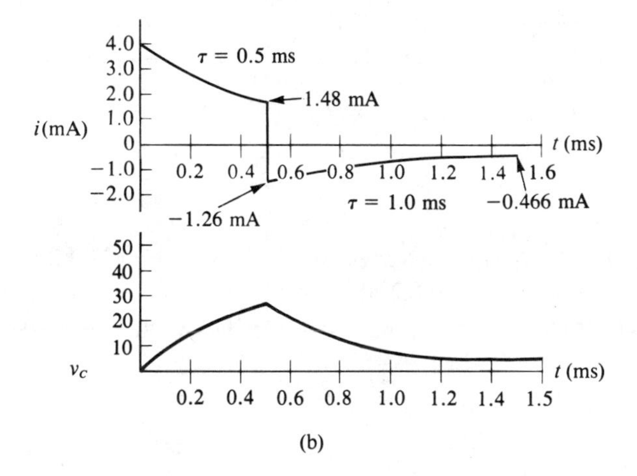

(b)

Figure 13.9 Figure for Example 13.10

Universal Time Constant Curve

The previous discussion, along with Example 13.10, shows that currents and voltages are relatively easy to calculate at exact multiples of the circuit time constant. When we need to know currents and voltages at fractional parts of a time constant or at noninteger multiples, different methods must be employed.

The student of electrical science may wonder, "What's wrong with approximations?" Unfortunately, in current state-of-the-art design using *RC* circuits, fairly exacting measurements are required. For example, the clock in a modern computer sequences instructions, transfers data, and acts as a basic timing element. Such a clock is often constructed of a basic *RC* timing circuit.

The universal time constant curve can be used in a manner similar to that of *L*/*R* circuits. Note once again that a percentage of change from a reference can be determined from the graph of Figure 13.10 and then a calculation completed. The following example will illustrate the case.

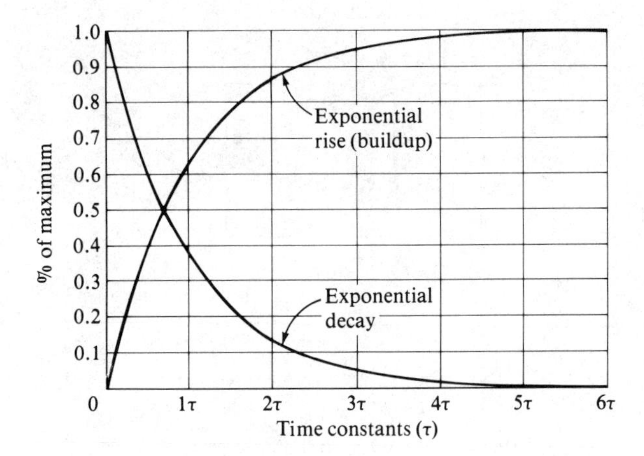

Figure 13.10 Figure for Example 13.11

Example 13.11 Assume a series circuit consists of a 10 kilohm resistor and two capacitors, $C_1 = 0.2\ \mu F$ and $C_2 = 0.8\ \mu F$. What is the current 2 ms after 20 volts dc is applied? What is the voltage across C_1?

$$I_{max} = \frac{V}{R} = \frac{20}{10^4} = 2\text{ mA}$$

$$\tau = \frac{RC_1C_2}{C_1 + C_2} = \frac{(10^4)(0.2)(0.8)(10^{-6})}{0.2 + 0.8} = 1.6\text{ ms}$$

The time is 2 ms/1.6 ms or 1.2τ. From the curve of Figure 13.10, 1.2τ is equivalent to a 70 percent decay or 30 percent of maximum.

$$i = (0.3)(2)(10^{-3}) = 0.6\text{ mA}$$

The voltage across R is 6 volts, hence 14 volts distributes between C_1 and C_2 via an inverse ratio.

$$V_1 = \frac{(0.8)(14)}{0.2 + 0.8} = 11.2\text{ volts}$$

The two preceding examples have illustrated some basic techniques for solving RC circuits. These principles may be applied to almost any circuit configuration. Refer to the circuits of Figure 13.11. In the first two circuits, (a) and (b), it is relatively easy to identify the time constant of the circuit. If $C_1 = C_2$ and $R_1 = R_2$, the time constant of (a) is R_1C_1. In (b) the time constant is $4R_1C_1$. In the circuit of Figure 13.11(c) it is not as easy to find the time constant. If all the resistors are of equal value, the time constant of $0.6\,R_1C$ can be determined using Thevenin's theorem. Solve for the Thevenin equivalent generator "seen" by the capacitor. The Thevenin voltage will be $0.8\ V$ and the resistance will be $0.6\ R$.

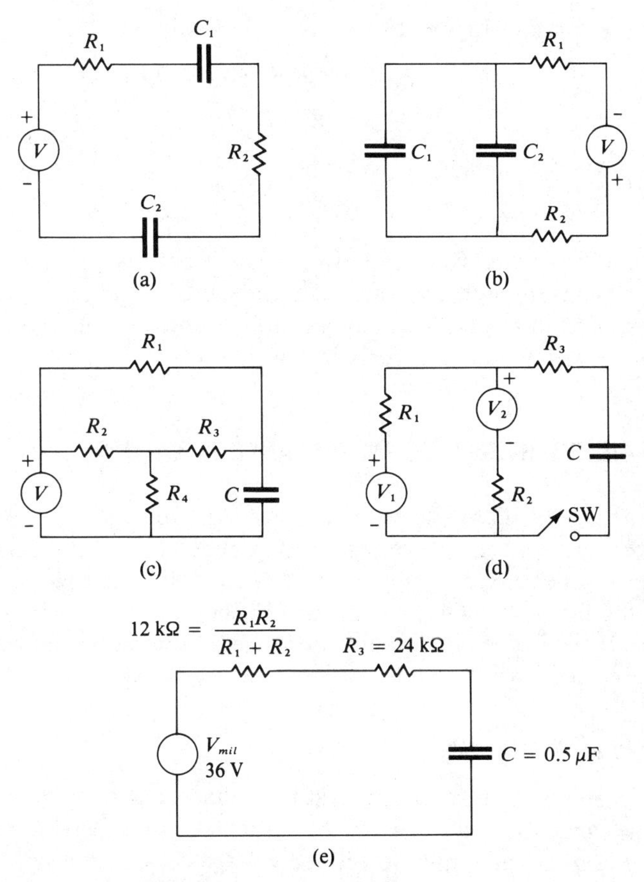

Figure 13.11 *RC* Circuit Configurations

Example 13.12 Assume the capacitor in Figure 13.11(d) is 0.5 μF. If $R_1 = 20$ kΩ, $R_2 = 30$ kΩ, $R_3 = 24$ kΩ, $V_1 = 40$ volts, and $V_2 = 30$ volts, what is the current flowing into the capacitor 40 ms after the switch is closed? What is it 50 ms after the switch is closed?

An equivalent circuit must be developed to obtain τ and I_{max}. A combination of Millman's and Thevenin's theorems can be used. If the combination of R_3 and C is removed from the circuit, what remains can be thevenized. According to Millman's theorem the open circuit voltage of the combination is 36 volts. The Thevenin resistance is 12 kilohms. The equivalent circuit is shown in Figure 13.11(e).

$$I_{max} = \frac{36}{(12 + 24)\text{ k}\Omega} = 1 \text{ mA}$$

$$\tau = (36)(10^3)(0.5)(10^{-6}) = 18 \text{ ms}$$

$$40 \text{ ms}/18 \text{ ms} = 2.22\ \tau$$

From the curve of Figure 13.10,

$$i = 10\% \text{ of } I_{max} = (0.1)(1)(10^{-3}) = 0.1 \text{ mA}$$

$$50 \text{ ms}/18 \text{ ms} = 2.77\ \tau$$

From the curve again,

$$i = 6\% \text{ of } I_{max} = (0.06)(1)(10^{-3}) = 0.06 \text{ mA}$$

Note in the preceding example how difficult it becomes to use the universal time constant curve at small percentages of τ and at several multiples of τ. For this reason a more exacting approach must be used for determining voltages and currents in the circuit.

13.6 ALGEBRAIC SOLUTIONS TO INSTANTANEOUS VALUES

In the analysis of many electrical circuits it is frequently necessary to know the exact current or voltage at a particular instant of time. We have already mentioned timing by an *RC* clock. Other timing circuits, such as the firing of a relay driver, the phasing of control circuits, and the interval time of delay circuits, are some additional examples where time is a critical variable.

Current in an *RC* Circuit

We have seen through Eq. (13.18) that the voltage across the resistor and across the capacitor in a series *RC* circuit must always equal the applied voltage. Let us assume that the applied voltage is a constant value.

$$V = v_r + v_c$$

If we substitute Eq. (13.19) and a version of Eq. (13.4) from advanced mathematics, we may solve for i as a function of t.

$$\boxed{i = \frac{V}{R} e^{-t/\tau}} \tag{13.21}$$

where τ is the time constant of the circuit, i is the current in amperes, V is the applied voltage, and R is the resistance. Recall, however, that V/R is I_{max}. We may generalize Eq. (13.21) to cover a wider range of circuit possibilities by letting V/R equal the initial current value when $t = 0^+$ (an instant of time just after zero). The equation may then be stated as:

$$\boxed{i = I_{(0)} e^{-t/\tau}} \tag{13.22}$$

We have already examined the values of $e^{-t/\tau}$ with variations of t. For clarity we will review some key ideas. If we assume $t = 0$, then the entire exponent of e is equal to zero, and the value e^0 equals 1. The current in the circuit at $t = 0$ is called the *initial current*. Figure 13.2(a) illustrates the value at $t = 0$. When t equals a large value, let us assume 8τ, note that e is raised to a minus 8 and its value is near zero. This means $I_{(0)}$ is multiplied by a value near zero which itself will result in a value near zero. Finally let us assume $t = 2\tau$. The value of e^{-2} is a nonzero value (0.135) multiplied by $I_{(0)}$ and a given current results. We see here the similarity to the universal time constant curve with the exception that we do not have to rely on the inherent errors of graphical solutions.

Example 13.13 Assume the RC circuit of Figure 13.11(a) has the following values: $R_1 = 2\ \text{k}\Omega$, $R_2 = 3\ \text{k}\Omega$, $C_1 = C_2 = 0.2\ \mu\text{F}$. If 15 volts dc is applied, what is the current through R_1 when $t = 230$ microseconds?

The τ of the circuit is $\tau = (R_1 + R_2)\left(\dfrac{C_1}{2}\right)$

$$\tau = (5)(10^3)(0.1)(10^{-6}) = 500 \text{ microseconds}$$

$$I_{(0)} = \frac{V}{R} = \frac{15}{(5)(10^3)} = 3 \text{ mA}$$

$$i = (3)(10^{-3})e^{-230/500} = (3)(10^{-3})e^{-0.46}$$

$$e^{-0.46} = 0.6313$$

$$i = (3)(10^{-3})(0.6313) = 1.89 \text{ mA}$$

Example 13.13 is an excellent illustration of the basic timing concepts in a series RC circuit. The current in the circuit decays toward zero in a pattern as outlined in Figure 13.12. Timing diagrams such as these are an important part of both linear and digital electronics. For this reason we will explore in greater detail the timing considerations of the circuit. For example, we might wish to know what is the change in current in this circuit when $t = 50$ to $t = 75$ microseconds. (This represents a change of 25 microseconds.) What would be the change in current if the same 25 microsecond change took place from 400 to 425 microseconds? Information such as this is available from a timing diagram. Figure 13.12(b) illustrates how such a diagram can be constructed. A modern calculator should be used for the calculation of the instantaneous currents. We can also write a short program in BASIC that would print the values of I. Both will be illustrated here.

Assume Eq. (13.22) is successively solved for i at values of t in increments of 25 microseconds. The result is the following table:

t	i(mA)	t	i(mA)
0	3	275	1.731
25	2.854	300	1.646
50	2.715	325	1.582
75	2.582	350	1.490
100	2.456	375	1.417
125	2.336	400	1.348
150	2.222	425	1.282
175	2.114	450	1.220
200	2.011	475	1.160
225	1.913	500	1.104
250	1.819		

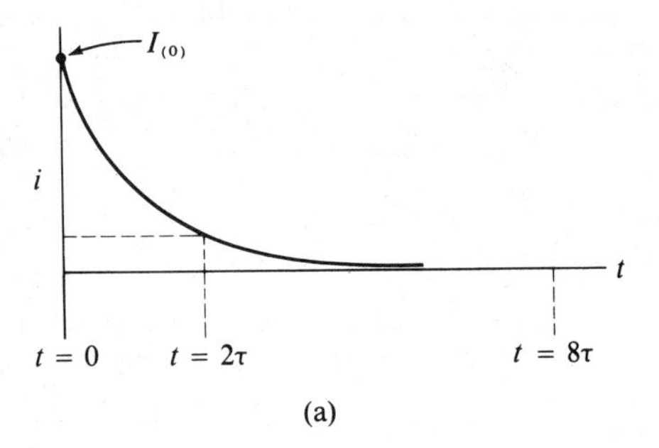

(a)

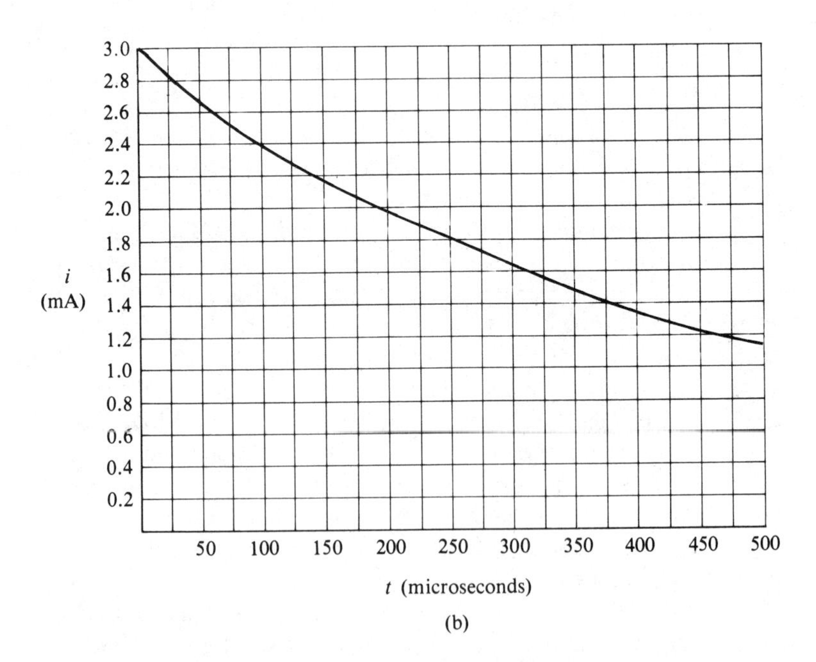

(b)

Figure 13.12 Instantaneous Current in an *RC* Circuit

We note from the timing diagram or from the table that the change in current from 50 to 75 μs is 2.582 – 2.715 or –0.133 mA. The minus sign represents a decrease in current. The change in current from 400 to 425 μs is 1.282 – 1.348 or –0.066 mA. These two increments can be observed graphically in Figure 13.12(b).

A simple program in BASIC can provide a quick table of values.

```
10  FOR T = 0 TO 500 STEP 25
20  I = 3E-3* EXP(-T/500)
30  PRINT T, I
40  NEXT T
50  END
```

Voltage Across *R*

In a series RC circuit, the current at any instant of time is described by Eq. (13.21). According to Ohm's law, the voltage across a resistor is $v = iR$. Substitute Eq. (13.21) for i.

$$\boxed{V_R = Ve^{-t/\tau}} \tag{13.23}$$

where V represents the applied voltage to the circuit. Here again we should note that when $t = 0$, all of the applied voltage is across the resistor. As t increases, the voltage across R decays in the same way the current in the circuit decays. It is also important to note that if there is more than one resistor in the circuit, then V_R is the equivalent resistance. This is best illustrated by Figure 13.11(b). We must first solve the circuit for a simple RC series circuit in order to determine τ, then we may solve for the instantaneous voltage across the equivalent R. This voltage distributes by ratios in the actual circuit.

Voltage Across *C*

The voltage across the capacitor is one which builds up or charges to some applied voltage. When the applied voltage is decreased, the capacitor voltage decays or discharges. It may in some instances discharge from one polarity and then charge to the opposite polarity. We will examine both of these cases with some detail. Equation (13.18) states

$$V = V_R + V_C$$

Substitute Eq. (13.23) into this equation.

$$V = Ve^{-t/\tau} + V_C$$

Solving for V_C,

$$\boxed{V_C = V(1 - e^{-t/\tau})} \tag{13.24}$$

This equation describes how the voltage across the capacitor changes with time during buildup or charging conditions. Note that when $t = 0$, V is multiplied by a value $(1 - e^0)$ or zero. The voltage across the capacitor is zero at that first instant. As t increases to a large value, the V is multiplied by a value $(1 - e^{-\infty})$ or 1, hence the capacitor is charged to the applied voltage in the circuit.

Example 13.14 Refer to the circuit of Figure 13.13. Assume $V = 50$ volts, R is 100 kilohms, and C is 0.01 μF. What is the instantaneous voltage across R, 0.45 ms after the switch is closed (moved to position 1)? What is the voltage across C, 0.9 ms after the switch is closed?

The $\tau = (10^5)(10^{-8}) = 1$ ms.

$$V_R = 50e^{-0.45/1} = (50)(0.6376) = 31.88 \text{ volts}$$

$$V_C = 50(1 - e^{-0.9/1}) = 50(1 - 0.4066) = 29.67 \text{ volts}$$

Example 13.14 illustrates how the voltage on a capacitor builds up or charges as t increases. After 5 ms the voltage on C would be 50 volts. When the switch in Figure 13.13 is moved to position 2, the capacitor begins a decay or discharge sequence. Let us presume that the switch can move to position 2 at any time of the charge sequence or even after the capacitor is fully charged. The general equation that identifies the instantaneous voltage on a capacitor during discharge is

$$\boxed{v_C = V_{(0)}e^{-t/\tau}} \tag{13.25}$$

where $V_{(0)}$ is the initial voltage across the capacitor at the beginning of discharge. In Example 13.14, if we had moved the switch to position 2 after it had been in position 1 for 0.9 ms, the $V_{(0)}$ would have been 29.67 volts. Had we waited for the capacitor to fully charge, the $V_{(0)}$ would have been 50 volts.

We will explore Example 13.14 in more detail in order to bring home some very important aspects of series RC circuits.

We have seen that V_C builds to 29.67 volts in 0.9 ms. Let us assume the switch is placed in position 2 for 0.9 ms, then returned to position 1 for

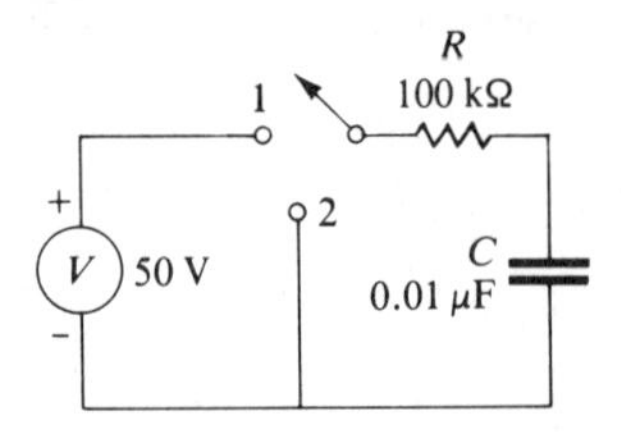

Figure 13.13 Figure for Example 13.14

0.9 ms. What voltage will be across C? It will *not* be 29.67 volts. During the decay phase the voltage did not drop to zero. Instead, using Eq. (13.25) where $V_{(0)}$ is 29.67 volts, it drops to approximately 12.06 volts. When the switch returns to position 1, there is already 12.06 volts on the capacitor. In the charge equation we can calculate the additional buildup on the capacitor above the 12.06 volts.

$$V_C = (50 - 12.06)(1 - e^{-0.9/1}) = 37.94(0.5934)$$
$$= 22.51 \text{ volts (additional buildup)}$$

The actual voltage is 22.51 + 12.06 or a voltage of 34.57 volts.

Several important concepts can be learned by studying the following example very carefully. Note in Figure 13.14(a) that when the switch is in position 1 the current source is causing the capacitor to charge positive with respect to ground. When the switch is moved to position 2 the capacitor will discharge and then charge to the opposite polarity.

Example 13.15 Refer to Figure 13.14(a). Assume the switch is in position 1 for 200 μs and is then switched to position 2 for 200 μs. What is the voltage across C?

The Thevenin equivalent circuit seen by C at switch position 1 is shown in Figure 13.14(b). τ is 156 μs.

$$V_C = 6(1 - e^{-200/156}) = 6(1 - e^{-1.282})$$
$$= 4.34 \text{ volts}$$

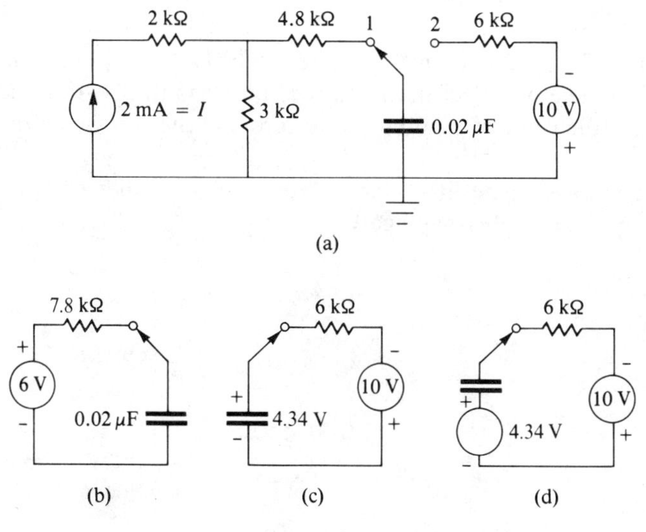

Figure 13.14 Figure for Example 13.15

When the switch is moved to position 2, the equivalent circuit is that of Figure 13.14(c). The time constant is 120 μs; but, now the capacitor is discharging. We may assess the *change* in V_C from the equivalent circuit of Figure 13.14(d). Note that the two voltages add up to 14.34 volts. Therefore, $\Delta V_C = 14.34(1 - e^{-200/120}) = 14.34(0.811) = 11.63$ volts. The actual capacitor voltage after 200 μs is

$$V_C = 4.34 - 11.63 = -7.29 \text{ volts}$$

The case of applied voltages to series *RC* circuits where the additional voltage aids or opposes an existing charge occurs very frequently. Two general equations can be developed from basic concepts.

Assume the circuit of Figure 13.15 represents an *RC* circuit where an existing voltage $V_{(0)}$ already exists on the capacitor. If voltage V aids polarity $V_{(0)}$, it will add to the charge. If V opposes $V_{(0)}$, it will charge in the opposite polarity. The voltage applied is taken with reference to ground.

For the case where V is charging C to the same polarity as $V_{(0)}$, the voltage across C will be the existing voltage plus any additional change due to V and $V_{(0)}$.

$$V_C = +V_{(0)} + [(V - V_{(0)})(1 - e^{-t/\tau})]$$

Multiplying through and canceling,

$$V_C = V + V_{(0)}e^{-t/\tau} - Ve^{-t/\tau}$$

Factoring,

$$\boxed{V_C = V + (V_{(0)} - V)e^{-t/\tau}} \qquad (13.26)$$

Consider Eq. (13.26) when $t = 0$. The value of e raised to the zero power results in 1. The two V's cancel and the voltage is the existing charge. As t increases, the value of e diminishes to zero and the voltage on the capacitor becomes V.

Note that $V_{(0)}$ may be either + or −, depending on whether it aids (+) or opposes (−) the applied voltage V.

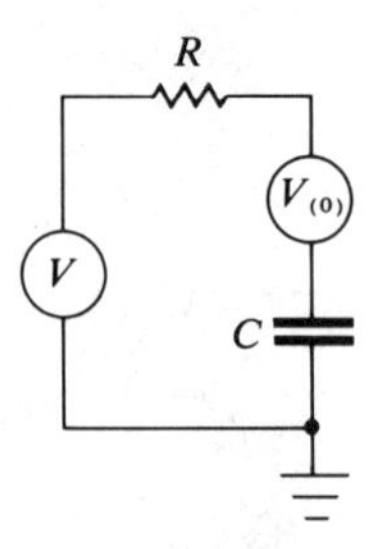

Figure 13.15 Additional Charge Buildup

Example 13.16 Assume an RC circuit is charged to -10 volts. $R = 10$ kΩ, $C = 0.5$ μF. If 10 volts of the opposite polarity is applied to the circuit, how long after the voltages are applied will the voltages across C equal zero?

$$\tau = RC = (10^4)(0.5)(10^{-6}) = 5 \text{ ms}$$

Set Eq. (13.26) equal to zero.

$$0 = 10 + (-10 - 10)e^{-t/5 \text{ ms}}$$
$$0.5 = e^{-t/5 \text{ ms}}$$

Taking the natural log (ln) of both sides, we obtain $-.69 = t/(5 \times 10^{-3})$. Therefore, $t/5$ ms $= 0.69$ and $t = 3.45$ ms.

General Form Time Constant Equations

The solutions to the problems and examples of Sections 13.5 and 13.6 required specific equations. Similar equations were developed for L/R circuits in Chapter 11. It should now be apparent that the current and voltage relationships in series RC or series L/R circuits follow exponential relationships. In a capacitive circuit, current decays exponentially to zero; whereas in an inductive circuit, it rises exponentially to maximum. Any decaying exponential follows the form

$$y = K_1 e^{-x}$$

while any rising function follows the form

$$y = K_1 - K_2 e^{-x}$$

where K_1 and K_2 are constants. These simple relationships are not really altered even if there exist some initial conditions of current and voltage. Equation (13.26) and Example 13.16 illustrate the case. After studying these examples, the student may readily conclude that there are far too many equations to remember and too many circuit conditions to account for.

The approach for any series RC or L/R circuit is simplified using the general form equations of Appendix F. Three examples are provided. After studying Appendix F, the student should rework Examples 13.14, 13.15, and 13.16. Then return to Chapter 11 and rework a few of the later examples in the chapter.

SUMMARY

Coulomb observed that the ratio of quantity charge to voltage for parallel plates is a constant. This constant is identified as C, the capacity.

Physical characteristics and geometry determine the amount of capacity offered by parallel plates and a dielectric. The larger the plate area, the greater the capacity. The closer the plates are to each other, the greater the capacity. The dielectric constant also affects the capacity directly. Finally, the working voltage rating of a capacitor is determined by the spacing between plates and the dielectric strength.

When capacitors are connected in parallel, the total effective capacity is the sum of the individual capacities. When they are connected in series, the total capacity is less than the smallest capacitor. The voltage distribution of series-connected capacitors follows an inverse relationship; that is, the smaller capacitor will develop the greater voltage.

A capacitor stores energy in its dielectric. The amount of energy stored depends directly on the capacitance and the square of the voltage.

Circuits involving capacity and resistance exhibit a specific time constant. One time constant is the time required for a capacitor to charge to 63 percent of its maximum value. The circuit requires five time constants to stabilize to its steady state conditions.

PROBLEMS

Reference Section 13.1

1. If a capacitor stores 20 microcoulombs when 40 volts is applied, what is the capacitance?
2. If the voltage on a 0.02 μF capacitor is increased from 10 volts to 18 volts, how much additional quantity charge is stored in the plates?
3. If 200 volts dc is connected across a 0.22 μF capacitor, how many electrons are transferred into the plates?
4. Which capacitor, a 0.02 μF or a 0.2 μF capacitor, will store more charge when 10 volts is applied? How much more?
5. What is the required rate of change of voltage across a 2000 pF capacitor if 2 mA is to flow into the capacitor?
6. How much current will flow into a 220 pF capacitor when a 5 volt pulse of voltage with a rise time of 0.01 μs is applied?
7. What is the current flowing into and out of a 0.40 μF capacitor when a 5.5 volts pulse is applied? Assume the rise time is 0.5 μs, the duration time is 2 μs, and the decay time is 1.5 μs.
8. Refer to Problem 7. If the amplitude of the pulse is doubled, but the rise, duration, and decay times are unchanged, what is the new current?
9. If the voltage across a capacitor changes from 0 to 20 volts in 10 μs and the current into it is 0.4 ampere, what is the capacity?
10. Refer to Problem 9. If the capacitor is replaced with a 1000 pF capacity, what is the new current?

Reference Section 13.2

11. Assume two parallel plates are spaced 0.15 mm apart with an air dielectric. If each plate is 4 cm by 5 cm, what is the capacity of the device?
12. If the air dielectric in Problem 11 is replaced with porcelain, what is the new capacity?
13. If two parallel plates of 0.3 cm^2 area are to develop a capacity of 100 pF, what spacing is required? Assume an air dielectric.
14. Assume the capacitor shown in Figure 13.16(a) used spacers of mica; what is the total capacity? (The spacers fit exactly within the plates and each spacer is 0.1 mm thick by 4 mm by 2 mm.)
15. The capacitor of Figure 13.16(b) is constructed of two cylinders, each 20 mm high. If the inside diameter of the outer cylinder is 3 mm and the outside diameter of the inner cylinder is 2.5 mm, what is the capacity if the dielectric spacer is paper? What is the maximum voltage rating?
16. Assume two long sheets of foil 2 cm by 0.9 meter are rolled into a capacitor similar to Figure 13.16(c). If the dielectric material is 0.1 mm paper, what is the capacitance and what is the voltage breakdown rating?
17. What thickness of paper would be required for Problem 16 for a capacity of 0.001 μF? What would be its voltage rating?

Reference Section 13.3

18. If a 200 pF and a 300 pF capacitor are connected in parallel, what is the total capacity of the combination?
19. If the two capacitors of Problem 18 were connected in series, what would be the total capacity?
20. If a 0.01 $\mu F(C_1)$ and a 0.05 $\mu F(C_2)$ are connected in series with 20 volts applied to the circuit, how much voltage is developed across each capacitor?

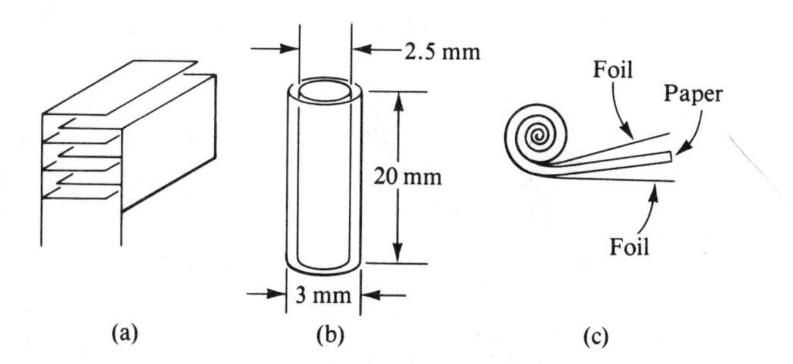

Figure 13.16 Figure for Problems 14, 15, and 16

21. Refer to Problem 20. If the value of C_2 is doubled, what is the change in voltage across it?
22. If a 0.01 μF capacitor at 200 WVDC and a 0.02 μF capacitor at 600 WVDC are connected in series, what maximum dc voltage may be applied to the circuit?
23. Refer to Problem 22. If the voltage rating of the 0.02 μF capacitor is 200 WVDC, what maximum dc may be applied to the circuit?
24. Refer to the circuit of Figure 13.17. (a) What is the total capacity seen by the generator? (b) What is the charge on the 5 μF capacitor? (c) What is the voltage on the 20 μF capacitor?
25. Refer to Problem 24. If a 30 μF capacitor is connected between points A and B, what is the total capacity?

Reference Section 13.4

26. Which will store more energy: a 20 μF capacitor with 30 volts across it, or a 30 μF capacitor with 20 volts across it?
27. If a capacitor has 6×10^{-6} joule of energy stored when 40 volts is applied, what is the capacity?
28. Refer to the circuit of Figure 13.17. How many joules of energy are stored in the 5 μF capacitor?
29. Refer to Figure 13.18. How much applied voltage, V, would be required to store 2×10^{-6} joule of energy in the 0.033 μF capacitor?
30. Refer to Figure 13.18. What is the total charge stored? Use the value of V found in Problem 29.

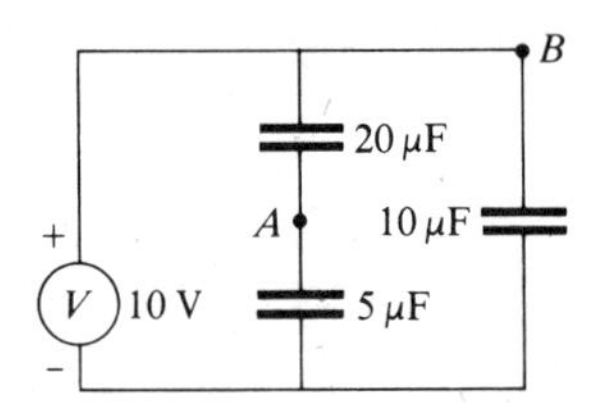

Figure 13.17 Figure for Problems 24 and 25

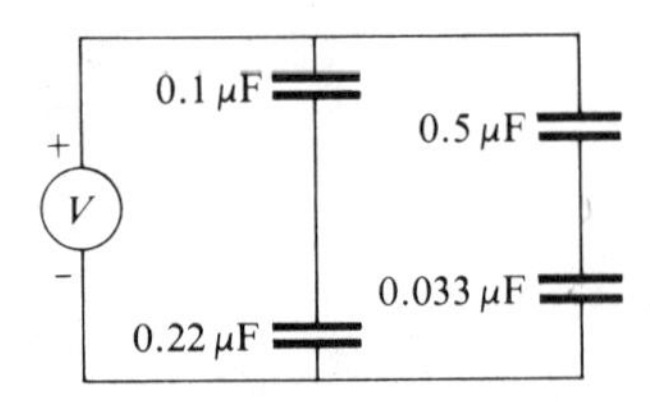

Figure 13.18 Figure for Problems 29 and 30

Reference Section 13.5

31. In the series RC circuit of Figure 13.19(a), what is (a) the maximum current flow? (b) the time constant of the circuit?
32. Refer to Figure 13.19(a). How long will it take for the capacitor to become fully charged?
33. Refer to Figure 13.19(b). If the switch is placed into position 1 for 5 time constants, then switched to position 2, how long will it take for the current to diminish to zero?
34. Assume the capacitor in Problem 33 is paralleled by a 0.1 μF capacitor. How long will it take for the current to diminish to zero?
35. A series circuit consists of two capacitors, 0.1 μF and 0.5 μF, and a 15 kΩ resistor. The applied voltage is 30 volts. What is the current in the circuit 1.2 time constants after the voltage is applied?
36. Refer to the circuit of Figure 13.20. What is the time constant of the circuit? What is the total current 3.1 time constants after the voltage is applied?
37. Refer to Problem 36. How long will it take for the current into the 0.1 μF capacitor to reach zero?

Reference Section 13.6

38. If a 2 MΩ resistor is in series with a 0.1 μF capacitor, what is the current in the circuit 120 ms after 10 volts dc is applied?

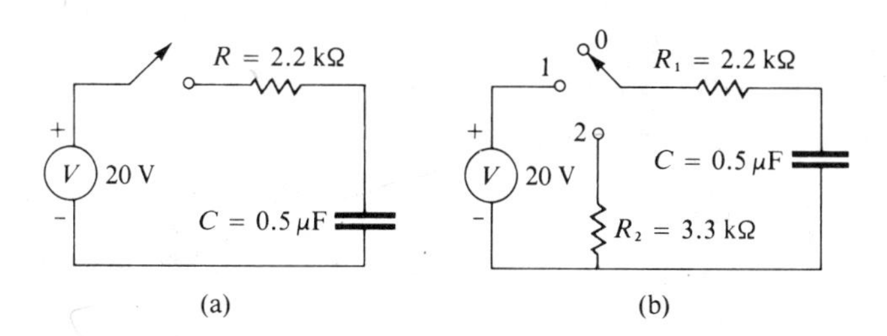

Figure 13.19 Figure for Problems 31, 32, and 33

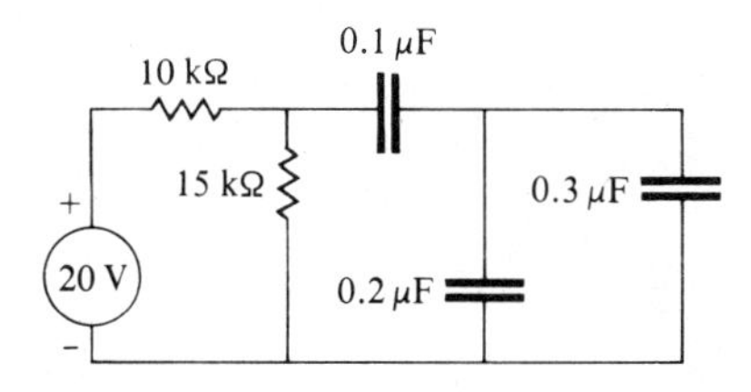

Figure 13.20 Figure for Problems 36 and 37

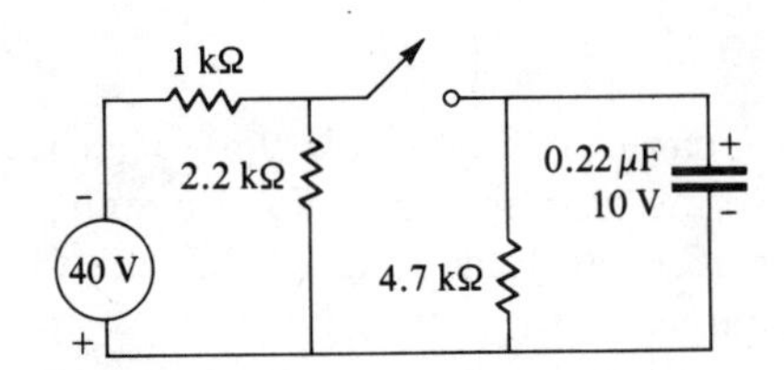

Figure 13.21 Figure for Problem 43, 44, and 45

39. If two capacitors, 0.1 μF and 0.5 μF, are series connected with a 22 kΩ resistor, what is the voltage across the 22 kΩ resistor 1 ms after 30 volts dc is applied? What is it 5 ms after 30 volts is applied?
40. If a series circuit consists of 1 MΩ resistor and a 0.002 μF capacitor, how much voltage will be developed across the resistor 1 ms after 10 volts is applied to the circuit?
41. Refer to the circuit of Figure 13.19(b). What is the voltage across C, 1.5 ms after the switch is placed to position 1 from position 0?
42. Refer to Figure 13.19(b). Assume the capacitor has no charge on it and the switch is in position 0. If the switch is placed at position 1 for 2 ms and then switched to position 2 for 2 ms, what will be the current through R_1?
43. Refer to the circuit of Figure 13.21. Assume the capacitor has 10 volts of charge. After the switch closes, how long will it take for C to become fully charged. What will be the maximum charge in volts?
44. Refer to Figure 13.21. Assume C is charged to 10 volts as shown. If the 2.2 kΩ resistor is removed from the circuit, how long after the switch is closed will the capacitor voltage be zero?
45. Refer to Problem 44. How long will it take the capacitor to charge to 30 volts of the opposite polarity?

CHAPTER 14

ALTERNATING CURRENT

14.1 INDUCED VOLTAGE

Faraday's law states that a conductor moving through a magnetic field will develop potential at the terminal ends of the conductor. The amount of that potential depends upon the flux density, the length of the conductor, and the velocity with which the conductor is moving through the flux. In this chapter we will expand upon these basic concepts and establish some fundamental ideas about alternating current.

A conductor of length ℓ passing through flux density B has an induced voltage equal to:

$$V = B\ell s \tag{14.1}$$

where V is the voltage in volts, B is the flux density in webers per square meter, and s is the velocity of the conductor in meters per second. Equation (14.1) is somewhat limited, however, since it requires that the conductor must cut the magnetic flux perpendicular to the lines of flux. Figure 14.1(a) is a cross-sectional view that illustrates this case. As the conductor moves down through the flux, electrons are moving away from the reader's view of the wire. If the conductor were immersed in the flux field and moved back and forth parallel to the direction of the flux, no movement of electrons would result since no magnetic lines are intersected. Figure 14.1(b) illustrates the movement within the flux with no resultant induced voltage. If the conductor were moved through the flux at some angle other than 90°, a voltage would be induced but not as great as that induced with a 90° cutting. The actual equation defining the voltage is

$$V = B\ell s(\sin\theta) \tag{14.2}$$

Note in Figure 14.1(c) that the angle of cutting of the flux must be with respect to the flux direction. Equation (14.2) shows that if $\theta = 90°$, then sin 90° is 1 and Eq. (14.1) results. If we assume a 100 mm length of wire passed through a flux density of 0.2 wb/m^2 at a velocity of 2.5 m/s and at an angle of 60° with respect to the flux direction, the voltage developed would be 43.3 mV. Increasing the angle of cutting to 90° would result in a voltage of 50 mV.

Assume the length of wire is inside the magnetic flux and moves in a complete circle as shown in Figure 14.1(d). The dotted line shows the path of the length of conductor. As it moves around and passes the south

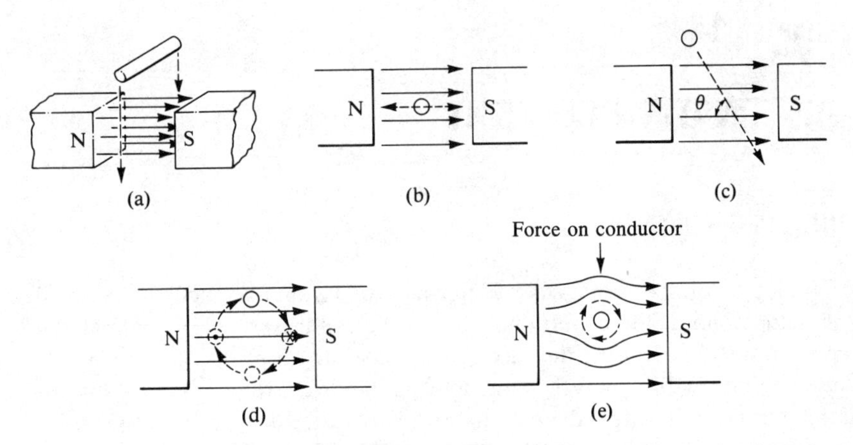

Figure 14.1 Induced Voltage

pole, electron flow moves away from the reader. When it reaches the bottom, the conductor is not cutting flux (moving parallel to), hence no induced voltage. When the length of wire starts up and passes the north pole, electron flow moves toward the reader (creating a negative polarity at the reader's end of the length of wire). Finally it completes 360° and returns to its starting position. It is important to note that through the first 180° the electrons were moving in one direction, then through the next 180° they were moving in the opposite direction.

The development of current and voltage in a length of wire moved through a magnetic flux represents electrical energy ($p = iv$ and $w = pt$). The energy in joules is translated from mechanical energy. This is best seen by analyzing the amount of force required to generate the current in the length of wire. The force in newtons is

$$F = B\ell i \tag{14.3}$$

where B is the flux density in wb/m^2, ℓ is the length of the conductor in meters, and i is the current in amperes. In Figure 14.1(e), assume a length of conductor is not in motion. If for any reason current now moves toward the reader, a magnetic flux surrounds the conductor directly proportional to the current. The flux of the field interacts with the flux of the conductor, resulting in a downward force on the conductor. Since moving a conductor through a field produces a current, a flux is produced that, in turn, interacts with the field to produce a force opposing the original motion of the conductor. Mechanical energy is required to continue the motion of the conductor. This is another example of the consequences of Lenz's law.

Since $w = pt$ and $p = iv$,

$$w = ivt \tag{14.4}$$

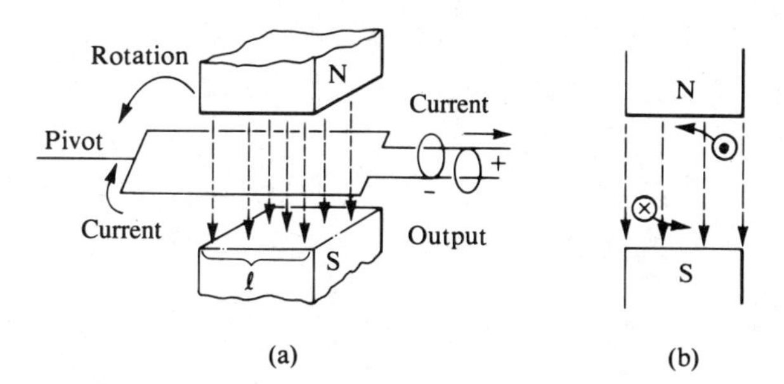

Figure 14.2 Simple AC Generator

Solving Eq. (14.3) for i and substituting Eqs. (14.1) and (14.3) into Eq. (14.4) results in

$$w = Fd \tag{14.5}$$

A force in newtons (F) acting through distance (d) converts mechanical energy (w) to equivalent electrical energy.

Single Loop ac Generator

In Figure 14.1(d), note that the length of wire moving down passes the south pole inducing current in one direction. The length of wire moving upward passes the north pole inducing current in the opposite direction. If a loop of wire is inserted into the magnetic flux and rotated around a central pivot point, the two lengths will simultaneously move past opposite poles.

Figure 14.2 illustrates a single loop rotating in a magnetic flux. As the loop of wire rotates around the pivot point, the current in each leg flows in the direction shown. The polarity at the output terminals reflects the instantaneous charge accumulation. Note that current flows from − to + *inside* the generator. As the two lengths cut the magnetic flux at the 90° point, the induced voltage becomes $V = 2\ B\ell s$. In Figure 14.2(b), a cutaway view shows the direction of the current. An X indicates current moving away from the reader and a dot indicates current moving toward the reader.

Figure 14.3 traces the single loop around 360° of rotation. Assume the maximum voltage produced by this loop is 10 volts. In (a) there is no induced voltage because the cutting angle is zero degrees. The two wires are moving parallel to the magnetic flux. In (b) the loop has moved counterclockwise through 45° of arc. The voltage at the output terminals is 7.07 volts. In (c) the loop has moved through 90° of arc. Here the two wires are perpendicular to the magnetic flux. A maximum of 10 volts is

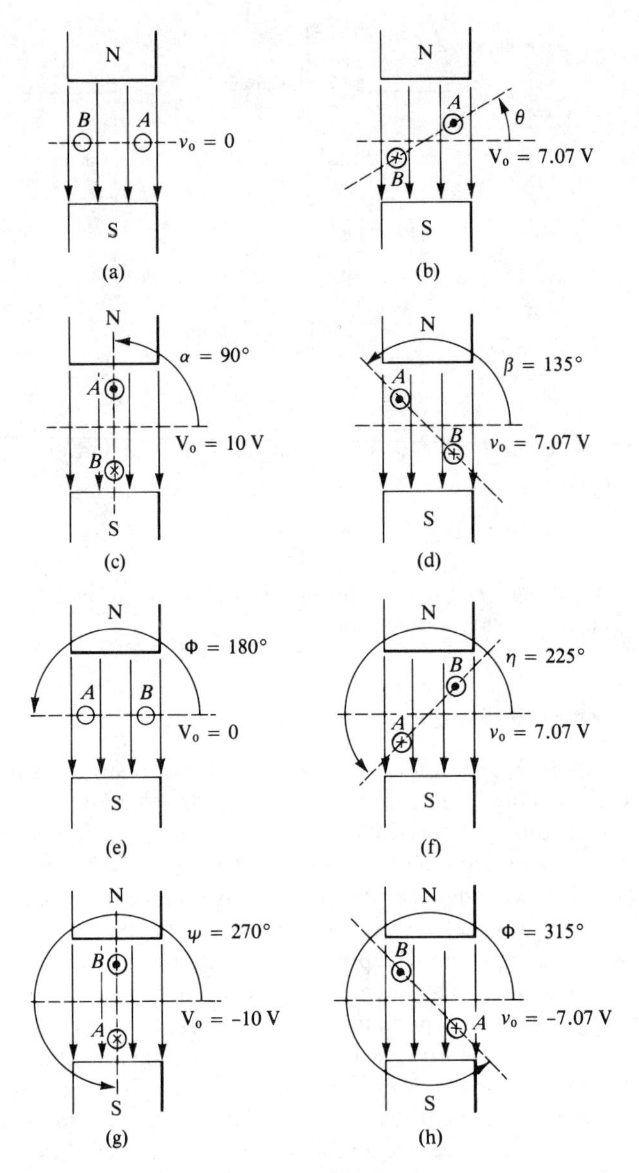

Figure 14.3 Voltages in One Loop

induced. As the loop continues through 135° of arc, note in (d) that the angle is once again 45° with relation to the flux. Once again the output voltage is 7.07 volts. Note in (e) that when the generated angle is 180°, there is no current flowing in either leg. Again the output is zero volts.

In Figure 14.3(f) the angle is 225°. Note that the current in leg *A* has now reversed. It has also reversed in leg *B*. The output voltage is now a

-7.07 volts. In (g) the maximum negative voltage is developed at the output terminals. In (h) the negative voltage is decreasing toward zero as the angle is 315°. When the loop has completed 360°, we return to the starting point of Figure 14.3(a) and no voltage is induced at the output.

Instantaneous Voltage

The characteristics of the rotating loop can be represented by a radius vector shown in Figure 14.4(a). The various reference points around the 360° of rotation are shown. Since the vector rotates, the voltage induced is

$$V = B\ell s\,(\sin\theta)$$

For this reason, the shape of the voltage shown in Figure 14.4(c) is a sine wave. Every generator, single loop or otherwise, has some maximum value of voltage which it can produce. Since that maximum occurs at 90° and 270°, we can solve Eq. (14.2) for V_{max}.

$$V_{max} = \pm B\ell s$$

Ordinarily, B, ℓ, and s are all constants; hence the generator has some predescribed maximum voltage output.

The output voltage may be rewritten as

$$\boxed{V = V_{max}\sin\theta} \qquad (14.6)$$

Note from Figure 14.4(c) that the maximum voltage is also the peak value attained. We often refer to the maximum voltage as the *peak voltage*.

Example 14.1 Assume an alternator produces a peak output of 40 volts. What is the instantaneous output voltage when the loop has rotated from 0° through (a) 30° of arc? (b) 230° of arc? (c) 750° of arc?

a. $V = 40 \sin 30°$ $(\sin 30° = 0.5)$
   $= 40\,(0.5) = 20$ volts

b. 230° occurs in the third quadrant.
   The angle with reference to the real axis occurs in the third quadrant.
   $V = 40\,(-\sin 50°) = -30.64$ volts

c. $V = 40 \sin 750°$
   The rotating vector has completed two full circles (720°) and 30° into the first quadrant of the third cycle. $V = 40 \sin 30° = 20$ volts

Radian Measure

Angle is often expressed in units of *radians* rather than degrees. The relation between radians and degrees is

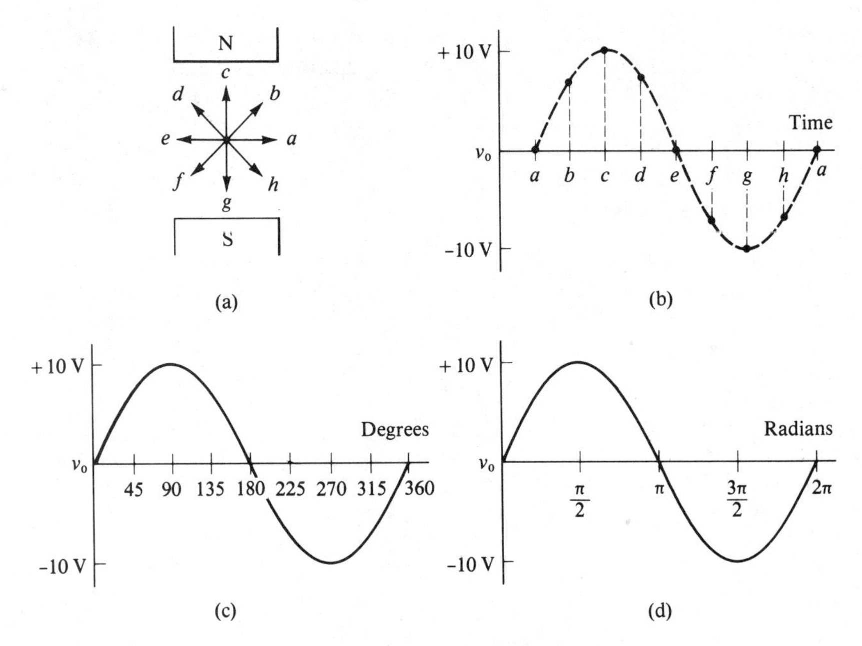

Figure 14.4 Instantaneous Voltages

$$360° = 2\pi \text{ radians} \tag{14.7}$$

Note that the radian is simply a different unit for angle measurement, just as meters and feet are different units for expressing length. Conversions between degrees and radians can be performed in the same way that conversions between other sets of units are performed. For example,

$$(180 \text{ degrees})\left(\frac{2\pi \text{ radians}}{360 \text{ degrees}}\right) = \pi \text{ radians}$$

$$\left(\frac{\pi}{4} \text{ radians}\right)\left(\frac{360 \text{ degrees}}{2\pi \text{ radians}}\right) = 45 \text{ degrees}$$

One radian equals approximately 57.3°. Figure 14.4(d) shows a sine wave plotted against angle expressed in radians. This plot is equivalent to Figure 14.4(c).

14.2 ALTERNATING CURRENT DEFINITIONS

The introduction of sine waves opens a new area of active driving potentials for electric circuits. No longer are we limited to dc energy sources. Several characteristics of sine wave voltages will influence circuit operation.

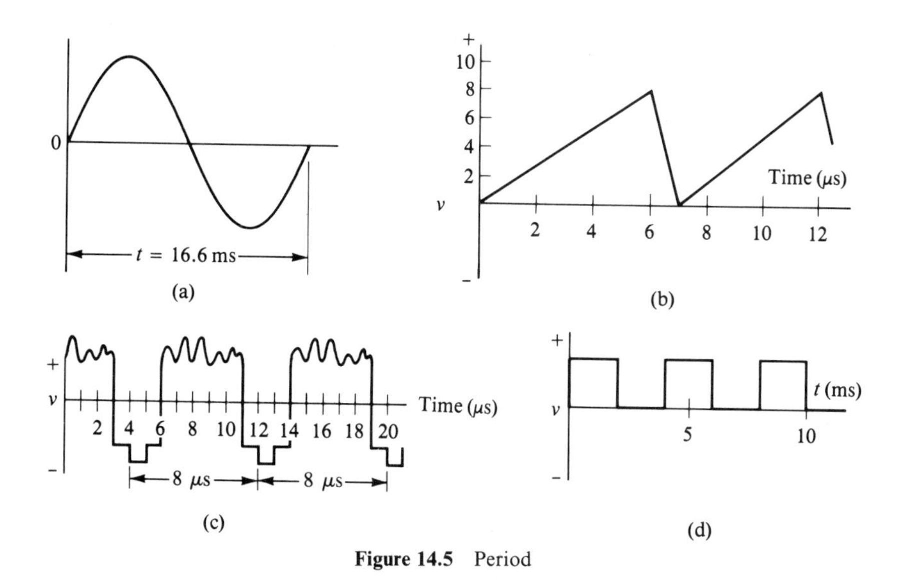

Figure 14.5 Period

Period

In Figure 14.4(b) note that the voltage passed through a series of values within the span of one rotation of the single loop. If the loop continued to rotate within the flux, additional sine waves would be developed. The voltage would oscillate above and below zero in a repeatable pattern. If the loop of the generator is driven with some uniform rotational velocity, the time of one sine wave *cycle* (360° of rotation) will remain fixed. This time (T), the *period*, is measured in seconds.

Period does not necessarily identify a sine wave function. Any sequence of changes recurring at a constant interval can be identified as having a period. Figure 14.5(a) illustrates ordinary household supply voltage. A complete sine wave cycle recurs every 16.6 ms. Note that in Figure 14.5(b) the voltage describes a sawtooth shape. It rises to 8 volts in 6 ms, then decays to zero in 1 ms. The period is 7 ms. The wave shape of Figure 1.45(c) follows a repeatable pattern despite the individual variations. Its period is 8 μs. The square wave of Figure 14.5(d) has a period of 4 ms.

Only the waveforms shown in Figures 14.5(a) and (c) can be classified as *alternating current* (AC) waves, because only these periodically *reverse* direction (go both positive and negative). The periodic reversal of direction is the fundamental characteristic that *defines* the AC wave.

Frequency

Refer again to the rotating loop within the magnetic field of Figure 14.2(a). The rate at which the loop is rotated within the flux will determine the time of the period. The *frequency* is the number of periods (or complete cycles) that occur in 1 second.

$$f = 1/T \qquad (14.8)$$

where f is in hertz and T is in seconds.

Frequency has the units of hertz (Hz), formerly cycles per second (cps) or pulses per second (pps). The sawtooth wave of Figure 14.5(b) has a frequency of $f = 1/(7)(10^{-6}) = 142.8$ kHz. The frequency of the square wave of Figure 14.5(d) is $f = 1/(4)(10^{-3}) = 250$ Hz (pps), while the frequency of the sine wave of Figure 14.5(a) is 60 Hz.

Angular Velocity

Linear velocity has units of length per unit time. An automobile has linear velocity as it completes miles per hour on the highway. In the case of a rotating vector we do not measure length per unit time. A vector rotating about some origin rotates through a certain angle per unit time. The vector, then, has angular velocity. Angular velocity is identified by ω (Greek letter omega). When angle θ is in radians,

$$\omega = \theta/t \text{ rad/sec} \qquad (14.9)$$

Consider the time span to be one period, T. Then $t = T$. The angle θ will be 360° (or 2π radians). The angular velocity then becomes

$$\omega = \frac{2\pi}{T} = 2\pi f \text{ rad/sec} \qquad (14.10)$$

We may now identify the *instantaneous* voltage or current of the sine wave as

$$v = V_p \sin 2\pi ft \qquad (14.11)$$

$$i = I_p \sin 2\pi ft \qquad (14.12)$$

We can find the value of v or i at any instant of time t by substituting t into Eq. (14.11) or (14.12).

Example 14.2 Assume an alternator has a maximum output of 50 volts at a frequency of 400 Hz. (a) What is the period? (b) What is the angular velocity? (c) What is the output voltage when $t = 0.5$ ms after the start of a cycle?

a. $T = 1/f = 1/400 = 2.5$ ms
b. $\omega = 2\pi f = (2)(3.1416)(400) = 2513.28$ rad/s
c. $v = 50 \sin[(2\pi)(400)(0.5)(10^{-3})]$
$= 50 \sin[(360°)(0.2)]$
$= 50 \sin 72° = 50(0.9511) = 47.555$ volts

Sine wave voltages may be applied to resistive circuits with resultant sine wave currents. The voltage and currents in the resistive circuit can be determined by Ohm's law.

Example 14.3 Refer to Figure 14.6(a). What is the peak value of current? At what time t of the cycle will the instantaneous current be 2 mA through R_1?

$$I_p = \frac{V_p}{R_T} = \frac{20}{(1 + 4.7)\ \text{k}\Omega} = 3.51\ \text{mA}$$

$$i = I_p \sin 2\pi f t \text{ (same in } R_1 \text{ and } R_2\text{)}$$

$$(2)(10^{-3}) = (3.51)(10^{-3}) \sin [(360°)(10^3)t]$$

$$\sin^{-1}\left[\frac{(2)(10^{-3})}{(3.51)(10^{-3})}\right] = (360°)(10^3)t$$

$$\sin^{-1} 0.57 = (360°)(10^3)t$$

$$34.8° = (360°)(10^3)t$$

$$96.6\ \mu\text{s} = t$$

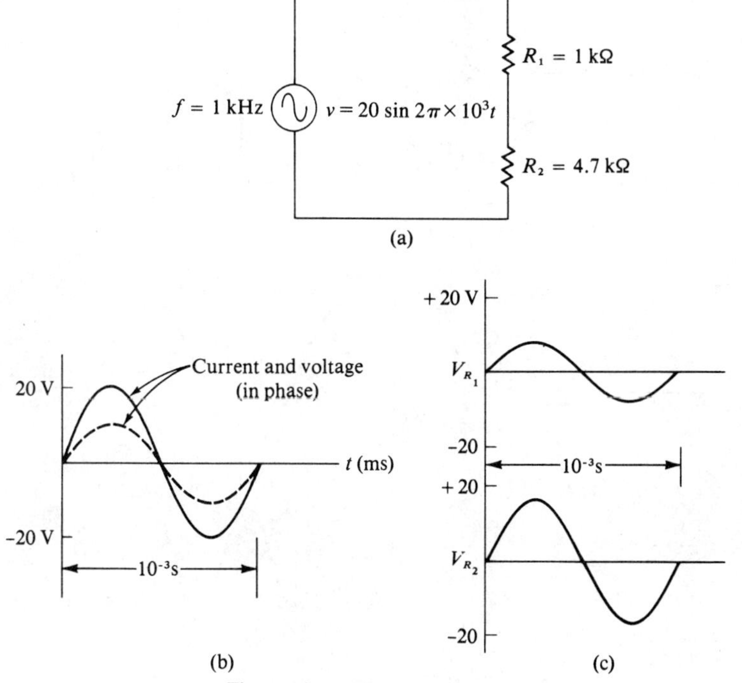

Figure 14.6 AC Series Circuit

Figure 14.6 illustrates how the voltages in the circuit distribute in a sinusoidal relationship. As the applied voltage reaches its peak of 20 volts, the voltage across R_1 reaches its peak of $R_1 20/(R_1 + R_2)$ or 3.51 volts. The voltage across R_2 reaches its peak of 16.49 volts at the same time that the input voltage and circuit current reach their peaks.

14.3 AVERAGE AND EFFECTIVE VOLTAGE

In Section 14.2 it was seen that the circuit conditions involving resistors and ac follow all the rules of Ohm's law. The total ac current in a series circuit is the ac voltage divided by the total resistance. The instantaneous voltages distribute by ratio of the sizes of the resistors.

Return momentarily to dc considerations. In Figure 14.7(a), note that the voltage V causes current to flow through R. The current flowing through R represents electrical energy in joules. This energy is then converted into heat energy (calories). We can assume that all the electrical energy is converted to heat energy. If the switch is moved to position 2 at time t_1, then back to position 1 at time t_2, observe that power rating is a fixed value equal to $P = I^2R$. For a resistance of 1 kΩ and a voltage of 10 volts, the power rating is 0.1 watt. If the interval from t_1 to t_2 is 1

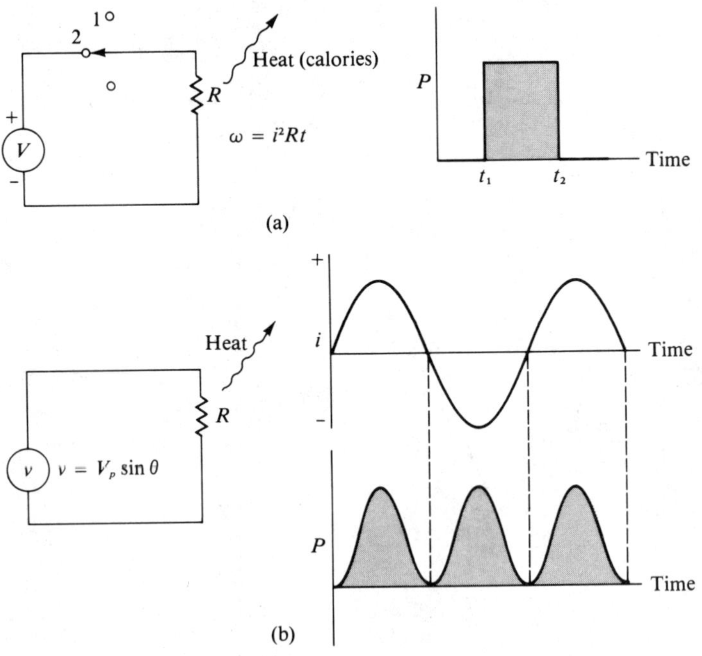

Figure 14.7 Power in a Resistor

second, then 0.1 joule of electrical energy is converted to heat energy. The heat energy is 0.02388 cal. (1 joule equals 0.2388 calories).

In Figure 14.7(b) the voltage source has been replaced by an ac source. Several differences must be considered. The current through R, which represents the quantity charge moving through the resistor, is in one direction for half a period, then drops to zero. The current reverses during the second half of the period. Electrical energy is again being converted to heat energy. The energy conversion is independent of the direction in which the electrons are moving through R. The total energy dissipated in the resistor during the positive and negative halves of the input voltage will be the same. The shaded areas in (a) and (b) of Figure 14.7 indicate the power characteristics of the two circuits. The transfer of energy to heat in the ac case depends upon several factors, such as the peak value of applied voltage, the frequency, and the value of the resistance.

Effective (RMS) Values

The effective value of a sine wave is that value which produces the same heating effect as a direct current. For example, in Figure 14.8(b) note that for a period of 10 ms, the 10 volts dc produces a given heating effect in R. Assume the dc supply in (a) was replaced by a sine wave voltage whose period is 10 ms. It would require a V_p value of 14.14 volts, as shown in Figure 14.8(c), to produce and equivalent heating effect. The effective value of 14.14 volts is 10 volts. If the dc supply is replaced with a sawtooth wave, as in Figure 14.8(d), the voltage would have to build to 17.32 volts in 10 ms, then drop to zero in order to produce the same heating effect as 10 volts dc.

The *effective* value of a voltage or current is the dc value that would cause the same heating in a fixed resistor. Calculation of the effective value requires the use of advanced mathematics. In our study of alternating current theory, we are most interested in the effective value of a sine wave. From the calculus we may compute the effective value of a sine wave. From that derivation,

$$\boxed{V_{eff} = V_p/\sqrt{2}} \qquad (14.13)$$

$$V_{rms} = V_p/\sqrt{2} \cong 0.707V_p \qquad (14.14)$$

where V_{rms} is the effective voltage and V_p is the peak value of the voltage. rms stands for "root mean square" and means the same as "effective."

Generally, the rms or effective value is indicated without any subscript. For example, if an alternator has an output of 50 volts, it is understood that this represents the effective value. If it were the peak value, it would be

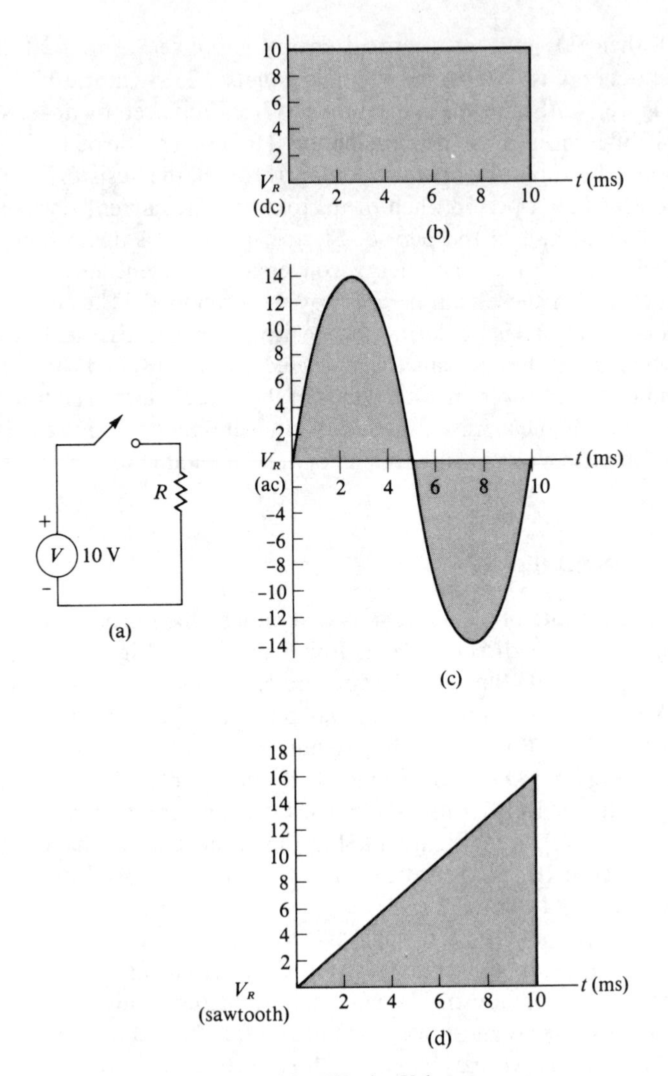

Figure 14.8 Effective Voltage

indicated as such. The unit VRMS is also used to indicate rms values, e.g., 50 VRMS.

Example 14.4 If an alternator has an output of 110 volts at 60 Hz, what are the peak and rms values of the sine wave?

$$V_{\text{rms}} = 110 \text{ volts (the given information)}$$

$$V_p = V_{\text{rms}}/0.707 = 155.58 \text{ volts, pk}$$

The effective value of current can be found using the same general form of Eq. (14.13).

$$I_{eff} = I_p/\sqrt{2} \tag{14.15}$$

Given the peak current I_p, the same multiplying constants can be used.

$$I_p \cong I_{rms}(1.414) \qquad \text{and} \qquad I_{rms} \cong I_p(0.707)$$

Example 14.5 Assume the series circuit of Figure 14.6 has the following circuit conditions. The applied ac is 50 volts, $R_1 = 2.2\ \text{k}\Omega$, and $R_2 = 3.3\ \text{k}\Omega$. What is the peak voltage across R_2? What is the effective current through R_1?

$$I = \frac{V}{R_1 + R_2} = \frac{50}{(5.5)(10^3)} = 9.09\ \text{mA}$$

This is the effective current everywhere in the circuit.

$$V_p = V(1.414) = 50(1.414) = 70.7\ \text{volts, pk}$$

$$V_{p(R_2)} = \frac{R_2 V_p}{R_1 + R_2} = \frac{(3.3)(70.7)}{2.2 + 3.3} = 42.42\ \text{volts, pk}$$

Average Voltage

If a meter which measures direct current is connected across a battery, the mechanical part of the meter converts electrical energy into mechanical energy (except for the case of a digital meter). The meter records the constant current drawn from the source. If the meter were rapidly switched in and out of the circuit, the inertia of the mechanical assembly would read neither the maximum current nor the minimum current. The meter, in fact, would read the *average* value of current. (Refer to Figure 14.9.) If the switch were turned on at t_1, off at t_2, on at t_3, off at t_4, and so on, the dc meter would not read the maximum current of 10 mA but would instead read the average current. If the time spacings are equal, the average current

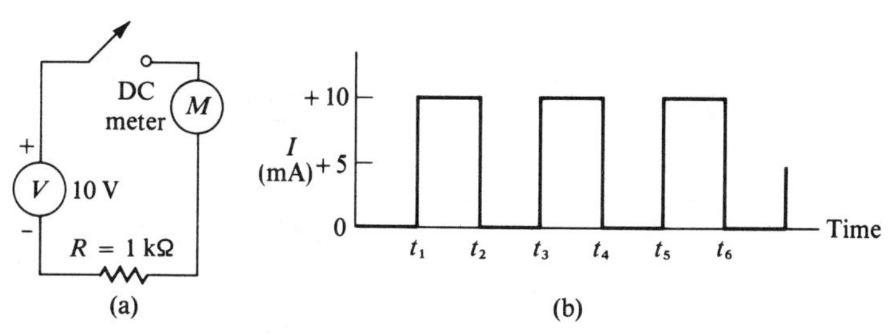

Figure 14.9 Average Current

would be 5 mA. The average value of any waveform is the value that a dc instrument would indicate for it, so the average value is often called the *dc value* or *dc component* of the wave.

One method of determining the average value is through the use of areas. In Figure 14.10(a), it is fairly easy to see that if area (A_1) is spread out over one period, a rectangle results with one side equal to t_2 and the height equal to $I_m/2$ (Figure 14.10(b).

The area method of calculating average current can again be seen in Figure 14.11. The applied voltage is a sawtooth whose period is 0.5 ms. The voltage rises to 10 volts in 0.5 ms, then drops to zero in a negligible time. The current rises to 10 mA in a sawtooth pattern. The average value of any waveform is the total area occupied by the wave in one period, divided by the period. In this case, the area is that of a triangle (A = ½bh) with base equal to the period ($.5 \times 10^{-3}$ sec) and height equal to the peak value, 10×10^{-3} A. Therefore, the average value is

$$\frac{A}{T} = \frac{\frac{1}{2}(.5 \times 10^{-3})(10 \times 10^{-3})}{(.5 \times 10^{-3})} = 5 \text{ mA}$$

We should note that any area falling below the horizontal axis (corresponding to negative voltage or current) is counted as negative area and

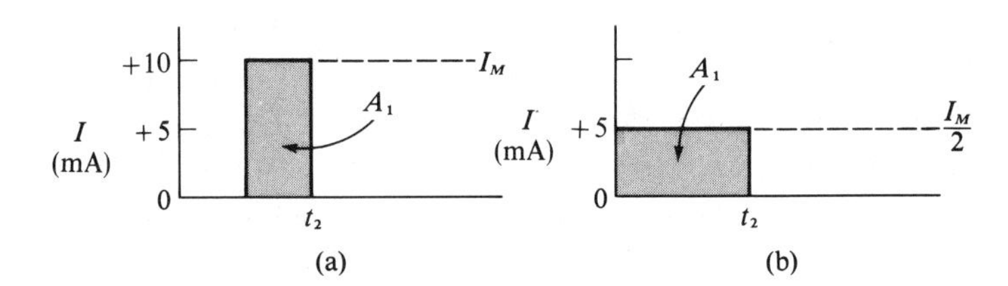

Figure 14.10 Average Current Equivalency

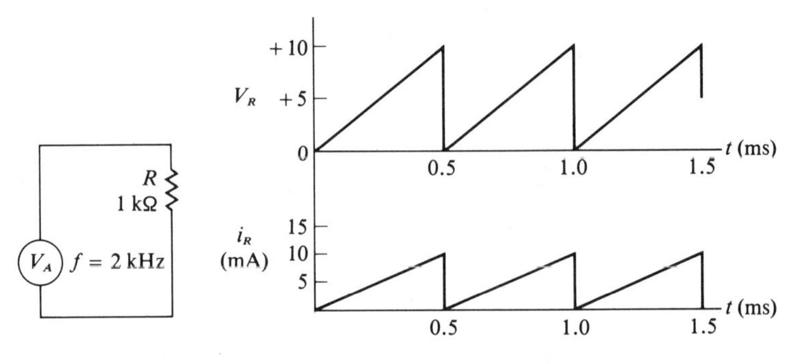

Figure 14.11 Average Current (Sawtooth)

subtracts from the total area. Thus any wave that is symmetrical about the horizontal (zero) axis, such as a sine wave, has *zero* average value.

Figure 14.12 shows a waveform consisting of 10V pulses recurring at 10 ms intervals. The area in one period is the same as the area of one pulse, A = (10)(2 × 10^{-3}) = 20 × 10^{-3} V-sec. Therefore, the average value is

$$\frac{A}{T} = \frac{20 \times 10^{-3}}{10 \times 10^{-3}} = 2V$$

Examine half of a sine wave as shown in Figure 14.13 and note how the average value is determined graphically using areas. Assume the sine wave is "squashed" down into a rectangle with one side being the half period. The other side, or the average value, would be 0.637 of the peak value. This value can be accurately determined using advanced mathematics.

$$V_{\text{ave}} = \frac{2V_p}{\pi} = 0.637V_p \qquad (14.16)$$

The average value of current can be similarly derived. Figure 14.14 shows the relational aspects of the peak, effective, and average voltages. Note that the figure also uses a $V_{\text{p-p}}$ notation. Often when using various test

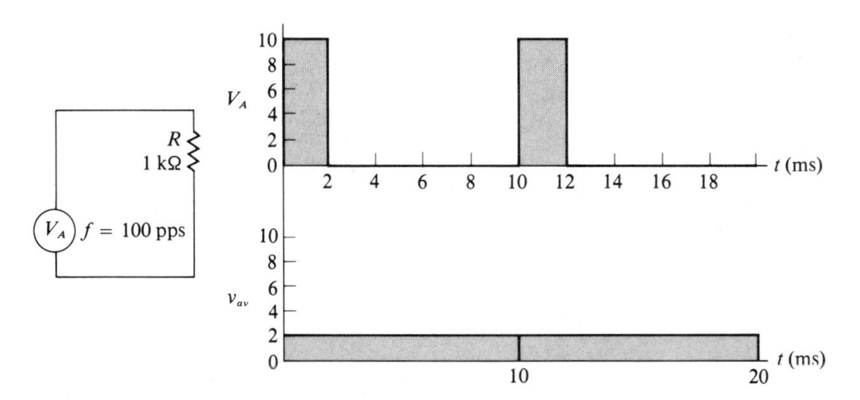

Figure 14.12 Average Voltage (Pulse)

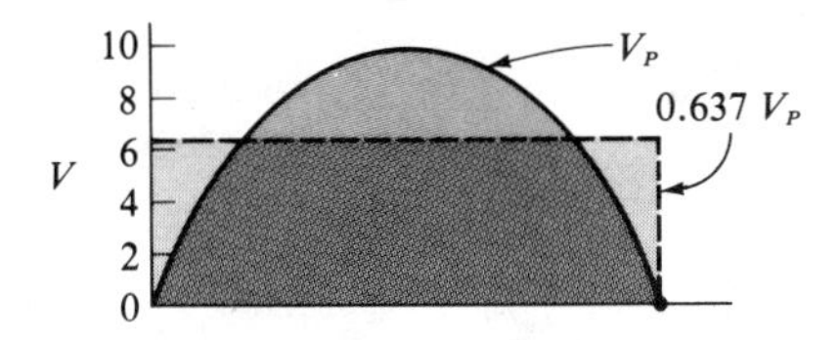

Figure 14.13 Half Sine Wave (Average Value)

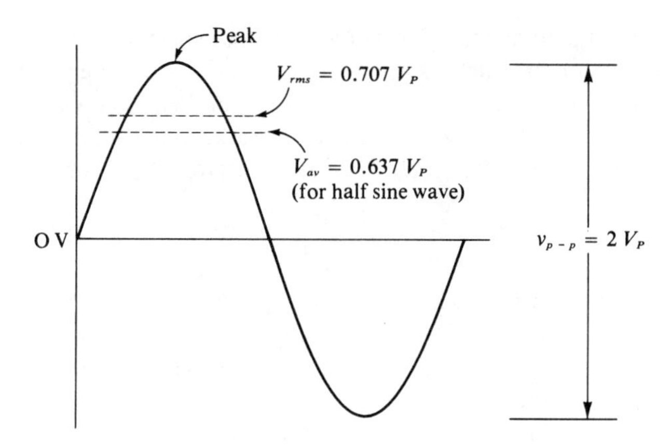

Figure 14.14 Peak, RMS, and Effective Voltages

equipment (such as an oscilloscope), many voltage relationships are required. The peak-to-peak voltage is $2V_p$ for a wave that is symmetrical about the zero axis. *Form factor* (*FF*) is another such relationship that is used in the design of power circuits and ac measurements. It is defined as the ratio of effective value to the half period average value.

$$FF = \frac{V_{\text{rms}}}{V_{\text{ave}}} = 1.11 \text{ (for half sine wave)}$$

Table 14.1 shows the multiplying factor when converting from one form of voltage or current to another.

Example 14.6 Refer to the circuit of Figure 14.15. Assume $V = 100$ volts, $R_1 = R_2 = 20\ \text{k}\Omega$, and $R_3 = 30\ \text{k}\Omega$. What is the peak voltage across R_1? What is the peak-to-peak current through R_2? What is the effective voltage across R_3?

TABLE 14.1 AC Conversions (Sine Wave)

From	To	Multiplier
Peak	Effective	0.707106781
Peak	Average	0.636619772
Peak	Peak-to-peak	2.0
Effective	Average	0.900316316
Effective	Peak	1.414213562
Effective	Peak-to-peak	2.828427124
Average	Effective	1.110720732
Average	Peak	1.570796326
Average	Peak-to-peak	3.141592653
Peak-to-peak	Peak	0.5
Peak-to-peak	Effective	0.353553390
Peak-to-peak	Average	0.318309886

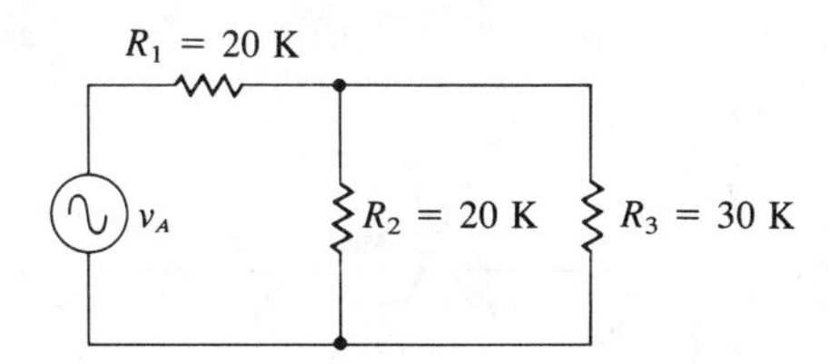

Figure 14.15 Figure for Example 14.6

R_2 in parallel with R_3 is 12 kΩ. The peak voltage across R_1 is

$$V_p = \frac{(20)(100)(1.414)}{20 + 12} = 88.39 \text{ volts}$$

The total current is $V/R_T = 100/32\ \text{k}\Omega = 3.125$ mA. By ratio the peak-to-peak current through R_2 is

$$I_{\text{pp}R_2} = \frac{(R_3)(3.125)(2.828)}{R_3 + R_2}$$

$$I_{\text{p-p}} = 5.3 \text{ mA (in } R_2)$$

Since V_{R_3} is the same as V_{R_2},

$$V_{R_3} = \frac{(100)(12)}{12 + 20} = 37.5 \text{ volts}$$

14.4 AVERAGE POWER

Return momentarily to a direct current passing through a resistor. The power is $p = i^2R$ or $p = iv$. We can extend the idea that the effective values of i and v are constant and, therefore, the average power. The notion of equivalent heat conversion results in a nonmathematical conclusion that the average ac power must be i effective times v effective.

$$P_{\text{ave}} = (I_{\text{rms}})(V_{\text{rms}}) \text{ watts} \qquad (14.17)$$

We may also state the average power as one-half the peak power.

$$P_{\text{ave}} = \frac{I_p}{\sqrt{2}}\frac{V_p}{\sqrt{2}} = \frac{V_pI_p}{2} \text{ watts} \qquad (14.18)$$

$$P_{\text{ave}} = \frac{P_p}{2} \text{ watts} \qquad (14.19)$$

Note in Figure 14.16 that the power is the shaded area. Note also that the frequency is twice the frequency of the applied sine wave voltage. This fact does not influence the average power since we consider the average of one or more complete cycles. The average power is the dc component in the plot, which is $P_p/2$.

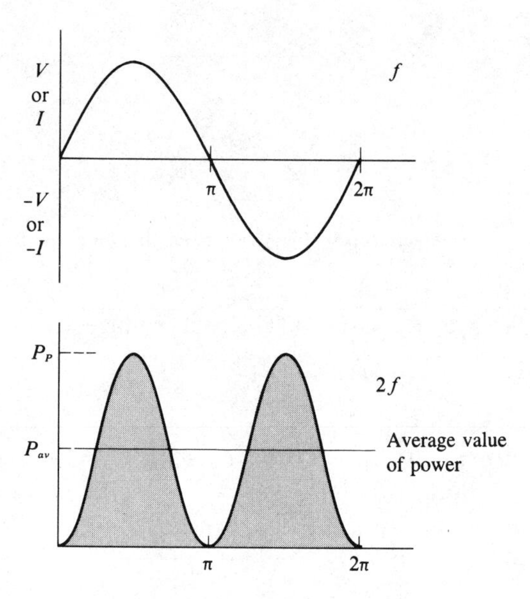

Figure 14.16 Average Power

The equation just discussed deals with the average power developed at a resistive load. We generally refer to this power as the *true power* or the power across a resistance.

$$V_{\text{rms}} = I_{\text{rms}} R$$
$$P = I^2 R \text{ watts} \tag{14.20}$$

where P is the true power, I is the effective current in amperes, and R is the resistance in ohms. Note the similarity of Eq. (14.20) to the dc case.

Example 14.7 Two parallel resistors $R_1 = 10\text{ k}\Omega$ and $R_2 = 40\text{ k}\Omega$ are connected across a 600 Hz source (Figure 14.17). If the average power across R_1 is 10 mW, what is the peak current through R_2? What is the equation for the applied voltage?

$$P = I_1^2 R_1,\ I_1 = \sqrt{P/R} = \sqrt{10^{-2}/10^4} = 10^{-3} \text{ watts}$$

The effective current is 1 mA through R_1. The effective voltage across R_1 is 10 volts. The peak voltage across R_1 and R_2 is 14.14 volts.

$$I_{2p} = 14.14/(40)(10^3) = 0.3535 \text{ mA}$$
$$\omega = 2\pi f = (6.28)(600) = 3770$$
$$v = V_p \sin 2\pi f t = 14.14 \sin 3770t$$

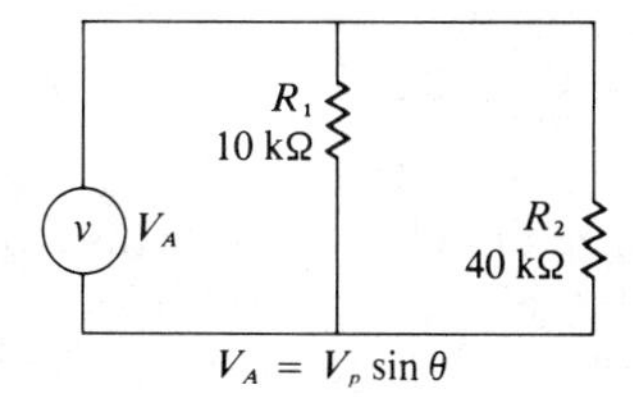

Figure 14.17 Figure for Example 14.7

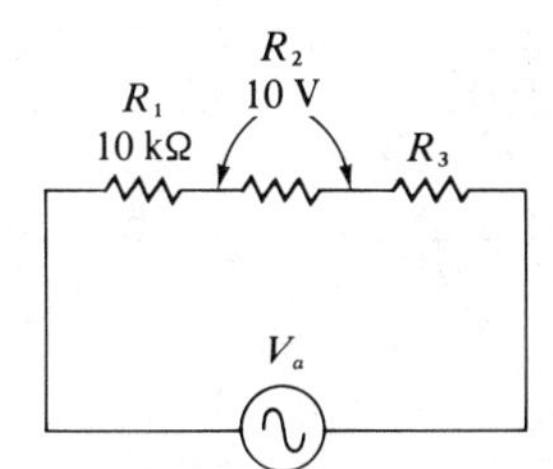

Figure 14.18 Figure for Example 14.8

We may now utilize all of the various known ac circuit characteristics to determine unknown values. From Ohm's law, we know that voltage divided by current equals resistance. In ac, care must be taken to use like units. V_p/I_p will yield resistance—so will V_{rms}/I_{rms}. But V_p/I_{rms} yields nonsense. Perhaps the best illustration is through the following example. The student should try to work the problem without looking ahead to the answers. Then an analysis of the solution will point up areas of misunderstanding.

Example 14.8 Three resistors are connected in series across an ac source as shown in Figure 14.18. The resistance of R_1 is 10 kΩ, the effective voltage across R_2 is 10 volts, and the peak current through R_3 is 10 mA. What is the value of R_2, R_3, and V_a if the total power is 2.47 watts?

The peak current throughout the circuit is the same as the peak current in R_3.

$$
\begin{aligned}
I_{rms} &= (10)(0.707) = 7.07 \text{ mA} \\
R_2 &= V_2/I_2 = 10/7.07 \text{ mA} = 1.414 \text{ k}\Omega \\
V_a &= P/I = 2.47/7.07 \text{ mA} = 349 \text{ volts} \\
V_1 &= IR_1 = (7.07 \text{ mA})(10^4) = 70.7 \text{ volts} \\
V_3 &= 349 - 70.7 - 10 = 268.3 \text{ volts} \\
R_3 &= V_3/I = 268.3/7.07 \text{ mA} = 38 \text{ k}\Omega
\end{aligned}
$$

14.5 AC MEASUREMENTS

Several methods are employed for measuring alternating current. The most common is to utilize only half of the sinusoidal wave. Figure 14.13 illustrated that the average value of half a sine wave is 0.637 V_p. If a series of sine waves similar to Figure 14.19(a) have their negative voltage portions removed (using an electronic device known as a rectifier), the resultant waveshape will be as shown in Figure 14.19(b). Now the half sinusoids have an average value of 0.318 V_p. This positive-only voltage can be measured by a dc meter whose digital readout (or dial face) can be calibrated to read the effective or peak value of the ac voltage. Figure 14.19(c) illustrates the basic schematic.

There are currently several other methods for measuring alternating current.

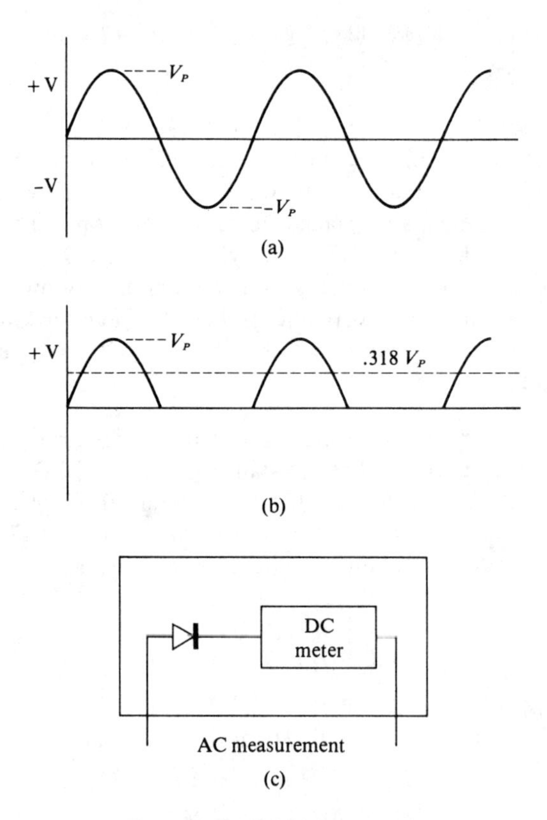

Figure 14.19 AC Measurements

Thermocouple

When two dissimilar metals are heated at their junction, there is a movement of electric charges in the metals. In some metals electrons move toward the source of heat, whereas in others they move away. This principle is illustrated in Figure 14.20(a). The alternating current is passed through a small wire which heats in proportion to the ac magnitude. The heated junction causes a change in Q at the ends of the two metals. The difference in Q results in a dc voltage that is again measured by

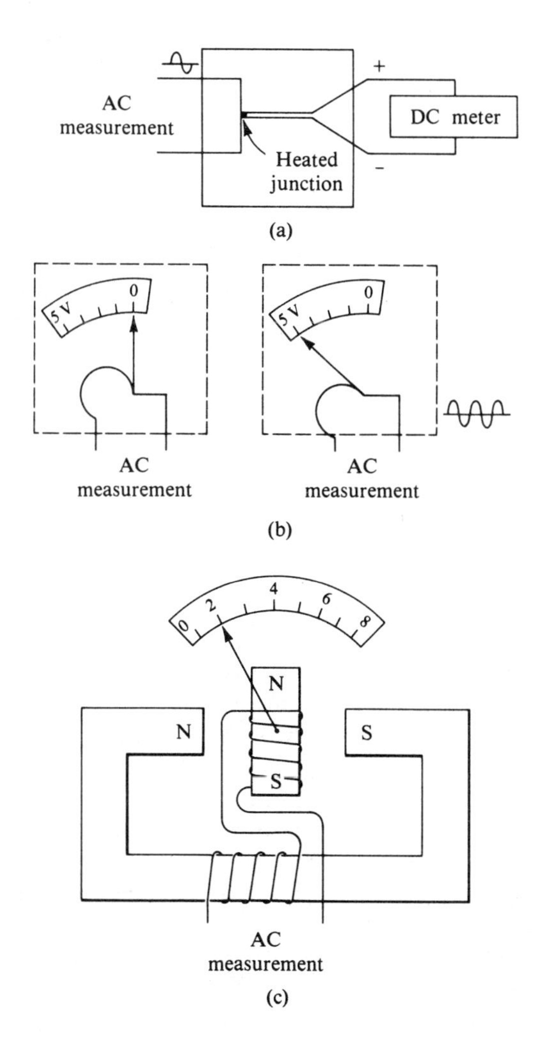

Figure 14.20 Alternating Current Meters

a dc meter. This meter too is calibrated to read ac. Some additional electronic circuitry is required to compensate for the nonlinear characteristics of the thermocouple device.

Hot-Wire Meter

The hot-wire meter does not convert ac to dc for its basic measurement. Its principle of operation is similar to the switching mechanism in an electric toaster. As an element heats, it changes shape. Figure 14.20(b) illustrates the meter. The curved springlike element carries the current through it. The greater the current, the greater the change in the shape of the spring. The rigid needle can be used to point to a scale for calibration of the ac. (Such a meter may be used for dc as well, though scale calibrations must be accounted for.)

Moving Iron Meter

Several types of meters that use magnetic principles are capable of measuring ac. The most common type is the electrodynamometer. Its principle is illustrated in Figure 14.20(c). As current passes through the windings, magnetic poles are set up which cause the center iron to deflect. As the polarity of the ac changes, all the magnetic pole polarities change in unison. The greater the current, the greater the deflection.

SUMMARY

The voltage induced across a conductor moving through a magnetic flux is dependent upon several factors. The density of the flux, the length of the conductor, the angle, and the velocity are all directly proportional. The force acting upon a conductor, in which current is flowing, is proportional to the flux density, the length of the conductor, and the current. For this reason, the continued generation of voltage at the output of a conductor requires an external force in dynes.

When a closed loop of wire is rotating in a magnetic flux, the voltage developed at the loop terminals is a sinusoidal waveshape. The frequency, amplitude, and period of the voltage waveshape depend upon the rotation of the loop.

The effective value of an ac is equivalent to the dc which produces an equivalent heating effect. This value is 0.707 of the peaks of the ac wave. The average value over a half cycle is given as 0.637 of the peak.

In alternating current circuits, the power is determined using the basic equation $p = iv$, where the i and v are the effective values of the sine wave.

The time base calculations for sine wave voltage functions can be determined using time in conjunction with either radians or degrees. In either case, the instantaneous value of either current or voltage may be determined using the relationships $v = V_p \sin 2\pi ft$, $i = I_p \sin 2\pi ft$, where v, i, and V_p, I_p, are instantaneous and peak values, respectively.

PROBLEMS

Reference Section 14.1

1. How much voltage is induced in a 100 mm length of conductor passing through 0.8 wb/m^2 at 20 m/s at an angle of 40° with respect to the magnetic flux?
2. If the angle in Problem 1 is doubled, will the induced voltage double?
3. Assume a 200 mm wire at rest in 0.8 wb/m^2 flux field suddenly conducts 200 mA. What force in newtons is exerted on the wire?
4. Refer to Problem 3. If the current is increased to 300 mA, what new force will be exerted upon the wire?

Reference Section 14.2

5. If a single loop of wire is rotated in a 1.2 wb/m^2 flux, what will be the output frequency in Hz if the loop is rotated 100 rev/min (RPM)? What is the period?
6. If the lengths of each leg of the loop are 200 mm for Problem 5, what is the peak voltage developed? (Loop radius is 50 mm.)
7. Refer to Problem 5. What is the angular velocity in degrees per second? In radians per second?
8. An angle of 104° represents how many radians? How many degrees does 0.87 radians equal?
9. How many degrees are represented by the following radians?
   (a) 1.2, (b) π, (c) $2/3$, (d) 5.63
10. How many radians are equivalent to the following angles?
    (a) 37.5°, (b) 270° (c) 114.6°, (d) 800°
11. An alternator produces a maximum voltage of 100 volts. What is the voltage at 60° of the cycle?
12. Refer to Problem 11. What is the voltage at 160° of the cycle? What is it at 270° of the cycle?
13. If the alternator of Problem 11 has a period of 20 ms, what is the frequency? What is the angular velocity?
14. If a sawtooth wave similar to Figure 14.5(b) has a voltage rise time equal to 50 μs and a decay time of 4 μs, what is its frequency?

15. An alternator has an output frequency of 400 Hz and a maximum output of 20 volts. What is the instantaneous output voltage when $t = 1$ ms? What is it when $t = 2.5$ ms? What is it when $t = 50$ ms?
16. The maximum output of a 60 Hz generator is 150 volts. What instantaneous output voltage will result when $t = 5$ ms of the cycle?
17. Refer to Problem 16. At what time t will the output voltage be half of the maximum input?
18. A series circuit consists of three resistors, 1 kΩ, 2.2 kΩ, and 3.3 kΩ. If the applied voltage is $v = 50 \sin 377t$, what is the peak current in the circuit? What is the voltage across the 2.2 kΩ resistor when $t = 2$ ms?
19. Refer to the circuit of Problem 18. What is the period? What is the voltage across the 1 kΩ resistor when $t = 4$ ms?
20. An ac voltage is applied across a series combination of 2.2 kΩ and R_2. If the frequency is 2 kHz and the voltage across the 2.2 kΩ resistor is 12 volts when $t = 0.1$ ms, what is the peak current in R_2?

Reference Sections 14.3 and 14.4

21. If an oscilloscope shows a sine wave of 56.8 volts peak-to-peak, what is the effective value of the sine wave?
22. What are the average and the peak voltages of an ac wave rated as 20.7 volts rms?
23. If an ac generator produces an rms voltage of 90 volts, what is the instantaneous output voltage when the output cycle is at 50°, 90°, 190°, and 390°?
24. If the instantaneous output voltage of a sine wave is 47 volts at 70°, what is the maximum output voltage of the generator?
25. A sine wave at 160° produces an output voltage of 12 volts, what is the output voltage at 260°?
26. If an ac generator produces an instantaneous output of 80 volts at 30°, what is the effective value of the ac voltage?
27. An oscilloscope shows two complete cycles of the output of an ac source. If the time of the two cycles is 300 μs, what is the frequency of the ac source?
28. A given 1 kHz sine wave produces 0.8 volt when t equals 0.2 ms; what is the average value of the ac wave?
29. The instantaneous value of a sine wave is 15 volts when t equals 10 μs. If the maximum is 45 volts, what is the frequency of the wave?
30. A given circuit consisting of a 5 kΩ resistor and a 15 kΩ resistor in series has a 400 Hz ac wave applied. If the instantaneous current in the circuit is 0.04 ampere when t equals 0.1 ms, what is the effective voltage value of the ac wave applied? What is the peak voltage across the 5 kΩ resistor?

31. What is the peak-to-peak value of 117 volts ac? What is the peak value of this voltage? What is the average value of this voltage over one half cycle?
32. If two resistors, 2.2 kΩ and 3.3 kΩ are connected in series and a generator of 40 volts is applied to the circuit, what voltages will be read across the 2.2 kΩ resistor with an ac voltmeter? If a dc ammeter were connected into the circuit, what current would it read?
33. A heating element is designed to control the temperature in an electric oven. The element is rated at 40 volts dc, but will operate at either ac or dc. What peak-to-peak ac voltage would be required to operate the oven properly?

Optional Computer Problems

34. Write a program in BASIC which will solve for the voltage v_0 for the circuit of Figure 14.21 for values of R_2 from 100 ohms to 1 kΩ. Assume $V_a = 25 \sin \omega t$. Assume 100 ohm increments of R_2.
35. Using BASIC, generate a table for the instantaneous voltages for the function $v = 20 \sin 2\pi f t$, where $f = 1$ kHz and t is in increments of $\frac{1}{40}$th of a period.

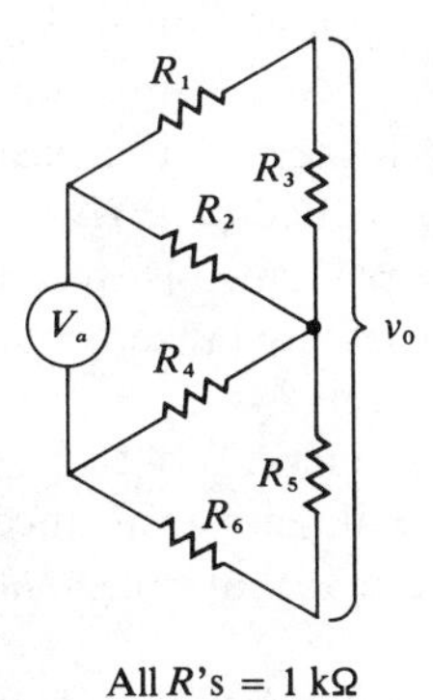

All R's = 1 kΩ

Figure 14.21 Figure for Problem 14.34

CHAPTER 15

INDUCTIVE REACTANCE

15.1 PURE INDUCTANCE

When a dc voltage is applied to a coil of wire, the current builds up to some maximum value. In Figure 15.1(a) note that the current in the circuit will be limited by whatever wire resistance makes up the coil. If the resistance of the wire is 2 ohms, a maximum current of 5 amperes will flow in the circuit. Assume now that the dc supply is replaced by an ac voltage source and the dc meter is replaced by an ac meter. The applied ac voltage is $v = 14.14 \sin \omega t$. Figure 15.1(b) illustrates the circuit. When the switch is closed, 5 amperes of current does not flow. Instead, only 40 mA of current flows in the circuit. We must conclude, therefore, that there is some other opposition to ac that did not exist for the dc case. Through Ohm's law, we can identify the amount of the opposition.

$$\frac{V_{\text{rms}}}{I_{\text{rms}}} = \text{opposition} = X_L \text{ ohms} \tag{15.1}$$

where X_L is the opposition to ac and is measured in ohms. For the example of Figure 15.1(b), X_L (called *inductive reactance*) is $10/(40)(10^{-3}) = 250$ ohms. The inductive reactance is over 100 times greater than the opposition offered by the wire resistance. In this case we can neglect the wire (or dc) resistance of the coil.

The term *pure inductance* refers to a coil of wire whose resistance is negligible. In reality, a coil with zero resistance does not exist. It is often so small that we can make many approximations neglecting the small resistance.

Assume that the applied voltage is a sine wave; therefore, the resultant current is also a sine wave. Faraday identified the voltage developed across an inductance as

$$\boxed{v_L = L\frac{\Delta i}{\Delta t}} \tag{15.2}$$

If we substitute $i = I_p \sin \omega t$, it can be shown (using calculus) that the result is

$$v_L = LI_p(\cos \omega t)\omega$$

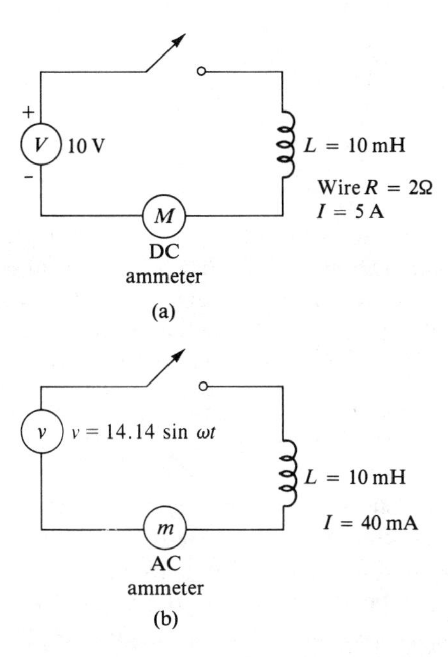

Figure 15.1 Inductive Opposition to Current

The applied voltage is directly across the coil. When $\cos \omega t = 1$, v_L is equal to V_p. Hence,

$$V_p = I_p \omega L$$

$$\frac{V_p}{I_p} = \omega L$$

However, V_p/I_p is volts per ampere which is ohmic opposition. As previously stated, this inductive reactance is X_L,

$$\boxed{X_L = \omega L = 2\pi f L} \qquad (15.3)$$

where f is the frequency in hertz, L is the inductance in henries, and X_L is reactance in ohms.

Note from Eq. (15.3) that the amount of opposition is directly proportional to the frequency of the applied signal and the inductance.

Example 15.1 What current will flow through a 200 mH coil when 10 volts at 600 Hz is applied? What current will flow when the frequency is doubled?

$$X_L = 2\pi f L = (6.28)(600)(0.2) = 754 \text{ ohms}$$

$$I = V/X_L = 10/754 = 13.26 \text{ mA}$$

If the frequency doubles, $X_L = 1.5\ \text{k}\Omega$.

$$I = 10/1.5\ \text{k}\Omega = 6.66\ \text{mA}$$

AC Ohm's Law

We have already seen in the introductory section that a pure inductance creates a reactive opposition in ohms. The principles of dc circuit analysis can be used with inductive-reactive devices when they are connected in various circuit configurations. For example, a series connection of several inductors with negligible resistance has total inductive reactance

$$\boxed{X_{L_T} = X_{L_1} + X_{L_2} + \cdots + X_{L_N}} \qquad (15.4)$$

If the inductors are connected in parallel, their reactances can be treated just as resistors connected in parallel. These conditions, however, require that there be no mutual interactive flux between the coils. We can extend these notions to include all other circuit calculations (except power). For instance, the voltage across an inductor is the product of the current and the inductive reactance. $V_L = I_L X_L$. These principles are illustrated in the following example.

Example 15.2 Two inductors, $L_1 = 100$ mH and $L_2 = 400$ mH, are connected in series across a 20 volt, 1.2 kHz source. What is the current through L_1, and what is the voltage across L_2?

$$X_{L_1} = 2\pi f L_1 = (6.28)(1.2)(10^3)(0.1) = 754\ \text{ohms}$$
$$X_{L_2} = 2\pi f L_2 = (6.28)(1.2)(10^3)(0.4) = 3.014\ \text{k}\Omega$$
$$X_{L_T} = X_{L_1} + X_{L_2} = 0.754\ \text{k}\Omega + 3.014\ \text{k}\Omega = 3.77\ \text{k}\Omega$$
$$I_{L_1} = V/X_{L_T} = 20/3.77\ \text{k}\Omega = 5.3\ \text{mA}$$
$$V_{L_2} = (5.3\ \text{mA})(3.014)\ \text{k}\Omega) = 15.99\ \text{volts}$$

An alternate method for solving for V_{L_2} is through voltage ratios.

$$V_{L_2} = \frac{X_{L_2} V}{X_{L_1} + X_{L_2}}$$

Phase Angle

We have seen in Chapter 12 that when a dc current is applied to an inductive circuit, the current does not rise to its maximum value instantly. In fact, it takes five time constants for the current to reach the maximum

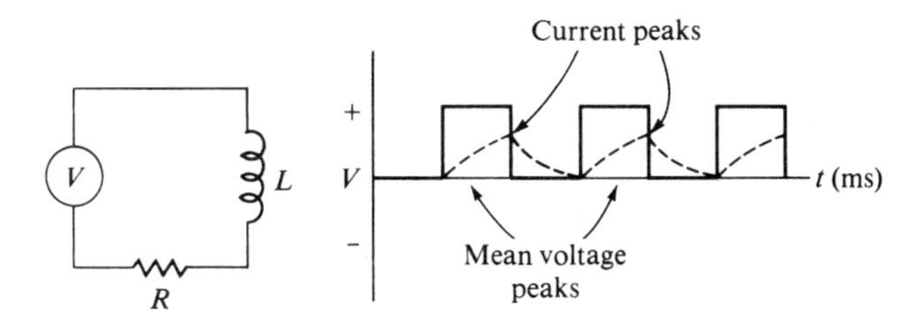

Figure 15.2 Square Wave Applied to L/R Circuit

value. In Figure 15.2, assume that the applied voltage is a square wave. As the voltage holds at its maximum value, the current is building to its maximum value. We can see that the current reaches its peak value at the end of the peak time of the voltage. We say that the current peaks lag behind the voltage peaks.

We have already shown that, when a sinusoidal current $i = I_p \sin \omega t$ flows through a pure inductance L, the voltage across the inductor is $v_L = LI_p\omega \cos \omega t$. Note that $LI_p\omega$ is constant. Since $\cos \theta = \sin(\theta + 90°)$, we have $\cos \omega t = \sin(\omega t + \pi/2)$. (Note that ωt always has the units angle in radians.) Therefore,

$$v_L = LI_p\omega \sin(\omega t + \pi/2) \tag{15.5}$$

The positive angle $\pi/2$ added to ωt means that v_L reaches its peak before $i = I_p \sin \omega t$ reaches its peak. The angle ($\pi/2$ in this case) is called the *phase angle*. We say that v_L *leads* i by $\pi/2$ radians, or 90 degrees. Equivalently, i *lags* v_L by 90 degrees.

The concept of this lagging factor is extremely important. For this reason we will explore what is happening in more detail. Figure 15.3 illustrates the lagging factor in two different ways.

In Figure 15.3(a), note that the applied voltage rises to positive and negative peak values. If this voltage is applied to a resistor, the current through the resistor rises to positive and negative peaks at the same time. The two are said to be *in phase*. Since the sine wave can be illustrated as a rotating vector, the two sine waves in Figure 15.3(b) are shown as two rotating vectors.

In a purely inductive circuit, the current lags the voltage by 90°. Figure 15.3(c) illustrates this relationship using sine waves. Along the time axis, events to the right of a given point occur after events to the left of the point. These characteristics are shown in Figure 15.3(d) using vectors. Note that in the last case the applied voltage has changed from a plus value to a minus value, yet the current, lagging by 90°, is still a positive value.

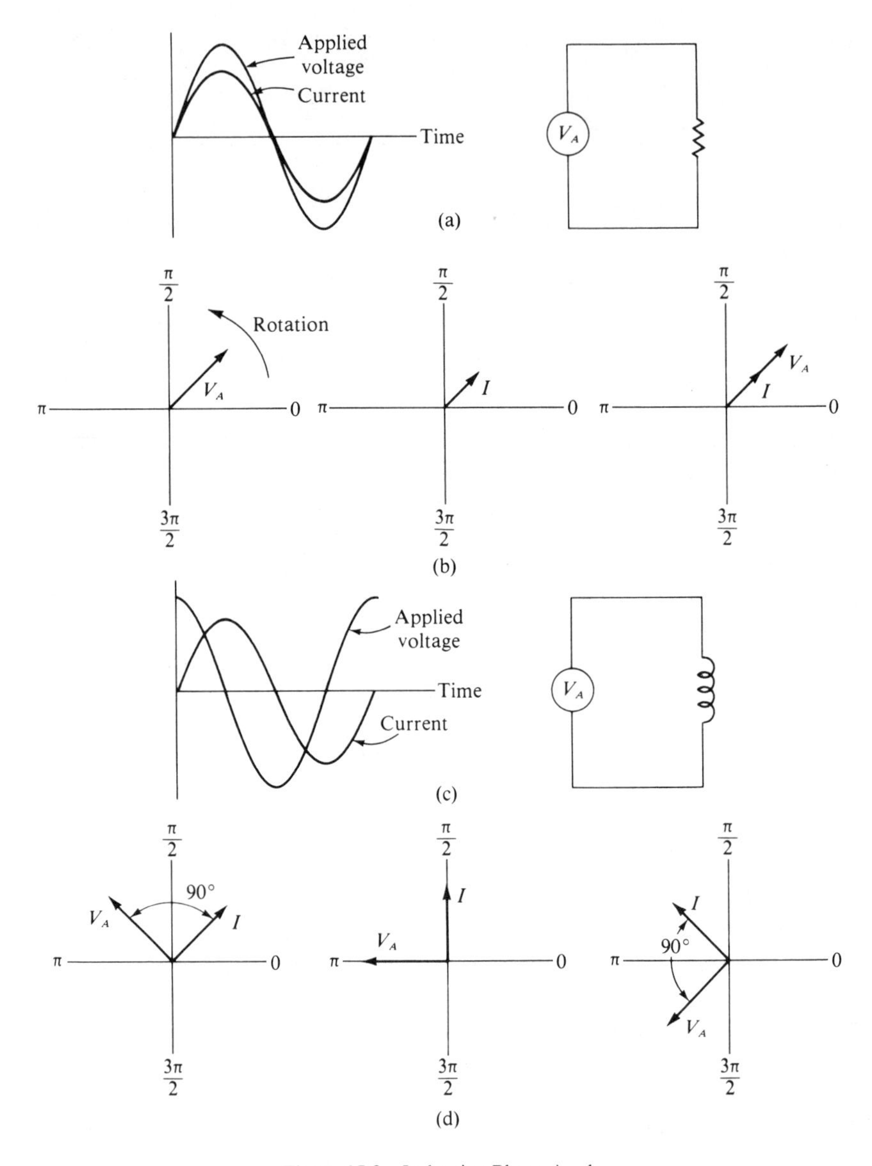

Figure 15.3 Inductive Phase Angle

j Operator

The 90° lagging current shown in Figure 15.3(d) can be identified using complex plane notation. Instead of using the lowercase i, the letter j is used for $\sqrt{-1}$ to avoid confusing this quantity with current. We associate the inductive reactance with $+j$. A 2 henry coil will offer 750 ohms of inductive reactance to 60 Hz. $X_L = 750$ ohms, or $+j750$.

The j means two things: A rotation of 90° and $\sqrt{-1}$ (an imaginary number). Note in Figure 15.4(a) that a vector of 10 units in length is shown extending along the *real* axis. $A = 10$. If the vector is multiplied by j, a rotation of 90° in the positive direction results. In this case (Figure 15.4(b)), $A = j10$. If A is multiplied by $-j$, the rotation is in the negative direction. The $A = -j10$ is shown in Figure 15.4(d). If either the vector shown in (b) or (d) is again multiplied by j or $-j$, respectively, the vector becomes $A = (\sqrt{-1})^2(10) = -10$. This $A = -10$ indicates a rotation of 180° in either the positive or the negative direction.

Before applying these concepts, review the algebra of the complex plane. A, shown in Figure 15.4, is a magnitude. This magnitude may be given direction by the addition of an angle. In part (b), for example: $A\underline{/\theta}$, where A = magnitude and $\underline{/\theta}$ = angle. $j10 = 10\underline{/90^\circ}$. In part (d) the angle is $-90°$. $A\underline{/-\theta} = -j10 = 10\underline{/-90^\circ}$. When these vectors are involved in algebraic manipulations,

$$(A\underline{/\theta})(B\underline{/\phi}) = AB\underline{/\theta + \phi} \tag{15.6}$$

$$(A\underline{/\theta})/B\underline{/\phi} = (A/B)\underline{/\theta - \phi} \tag{15.7}$$

$$(A\underline{/\phi})^N = A^N\underline{/N\phi} \tag{15.8}$$

Note in Eq. (15.6) that when two vectors are multiplied, we multiply their magnitudes and add their angles. In the case of Eq. (15.7), the magnitudes are divided and the angle is the difference of the angles. Since $+j$ is associated with inductive reactance (X_L), we write $X_L\ \underline{/90^\circ}$.

Example 15.3 If a 50 volt 1 kHz signal is applied to a 2 henry inductor, what is the current and its phase angle?

$$X_L = 2\pi fL = (6.28)(10^3)(2) = 12.56\text{ k}\Omega$$

$$X_L = j12.56\text{ k}\Omega = 12.56\text{ k}\Omega\underline{/90^\circ}$$

$$I = V\underline{/\theta}/X_L\underline{/\phi} = 50\underline{/0^\circ}/12.56\text{ k}\Omega\underline{/90^\circ}$$

$$I = 3.98 \times 10^{-3}\underline{/0^\circ - 90^\circ} = 3.98\text{ mA}\underline{/-90^\circ}$$

Example 15.3 illustrates how the current lags the voltage by 90° since the voltage was taken as a reference (0 degrees).

15.2 INDUCTIVE REACTANCE AND IMPEDANCE

In the previous discussion of inductive reactance, the resistive property of the coil could be neglected. At extremely low frequencies, however, that resistive property may contribute significantly to circuit conditions.

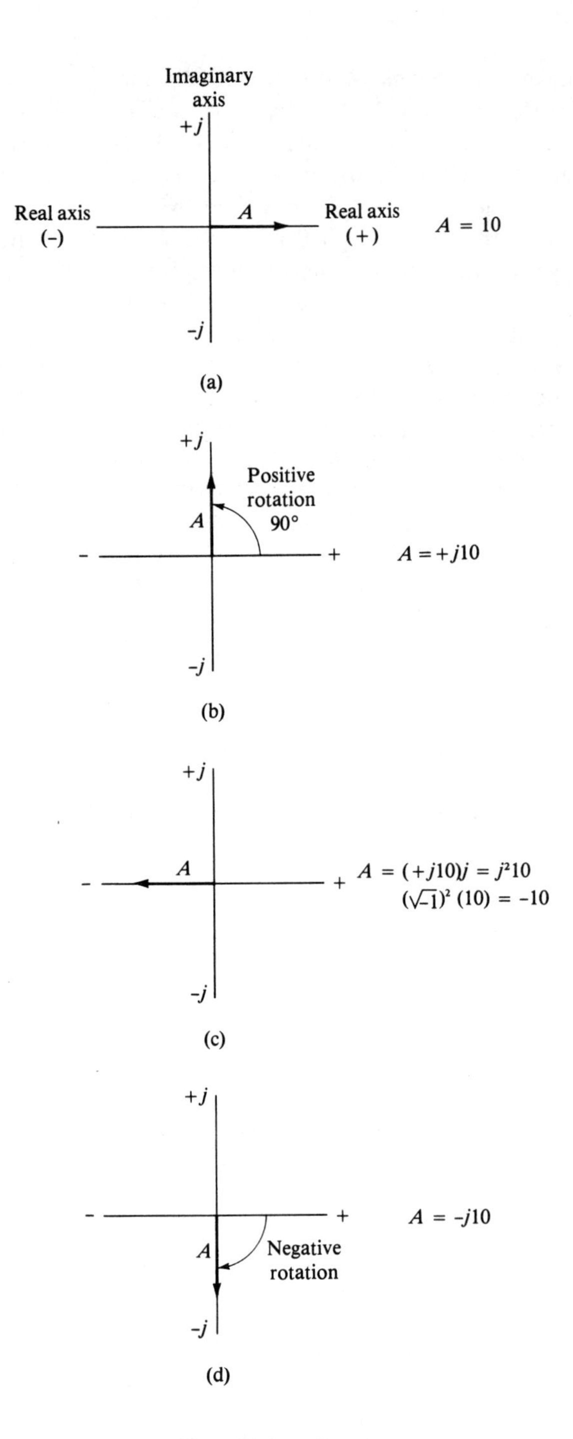

Figure 15.4 *J*-Operator

We must consider, therefore, the influence of X_L combined with a resistance of the device or combined with a resistance in the circuit.

Impedance

The circuit of Figure 15.5 represents the R as a circuit component. It may, in actual practice, represent a combination of circuit resistance and wire resistance of the inductor. An ac relay, for example, could be shown as a series L/R circuit, even though there may not actually be a resistor in the circuit. The opposition consists of the resistance R and the inductive reactance X_L acting at j displacement (90° of positive rotation). Figure 15.5(b) illustrates the two vectors. The equivalent vector, Z, has a length equal to the hypotenuse of the triangle (shown as a dotted line). *The Z represents the total opposition offered to current caused by an ac voltage V_A*. This opposition is identified by Z, the impedance, and is measured in ohms. According to the Pythagorean theorem, the magnitude of the impedance is

$$\boxed{|Z| = \sqrt{R^2 + X_L^2}} \tag{15.9}$$

where Z, R, and X_L are in ohms. The absolute value signs are used to denote magnitude.

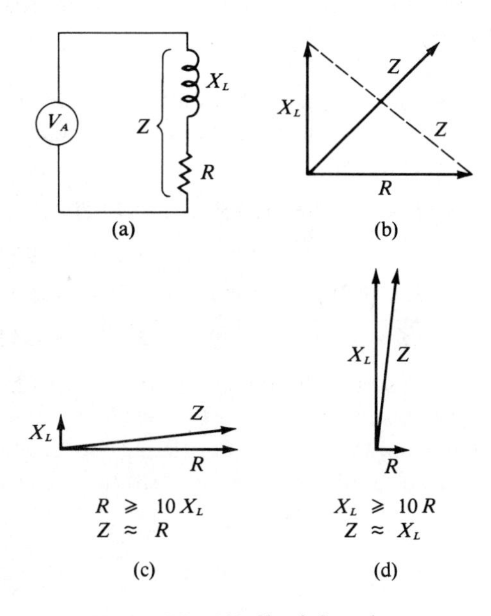

Figure 15.5 L/R Circuit Impedance

Equation (15.9) defines the impedance as an opposition which is the effect of the oppositions of R and X_L. With no inductance in the circuit, the impedance becomes the resistance.

$$|Z| = \sqrt{R^2} = R$$

Similarly, with no resistance, the impedance is the reactance.

$$|Z| = \sqrt{X_L^2} = X_L$$

If a 60 ohm resistor is connected in series with a coil whose reactance is 80 ohms, the impedance will be:

$$|Z| = \sqrt{(60)^2 + (80)^2} = \sqrt{3600 + 6400}$$
$$= \sqrt{10000} = 100 \text{ ohms}$$

The impedance will always be greater than the larger of the two, but less than the sum of the two. In this case the impedance would be greater than 80 ohms but less than 140 ohms.

Equation (15.9) shows that the impedance of the circuit varies with resistance, inductance, and frequency. Since increasing the frequency causes an increase in the inductive reactance (Eq. 15.3), the impedance increases as well.

When the resistance is much larger than the inductive reactance, the impedance value approaches the resistive value. Figure 15.5(c) shows a resistance 10 times greater than the reactance. Similarly, when X_L is greater than 10 times the resistance, the impedance approaches the value of X_L (Figure 15.5(d)).

Example 15.4 Assume a series L/R circuit similar to Figure 15.5(a). If $R = 2.2\text{ k}\Omega$ and $L = 500\text{ mH}$, what is the total impedance if the applied voltage has a frequency of 600 Hz? What is it if the frequency is 1.2 kHz?

$$X_L = 2\pi fL = (6.28)(600)(0.5) = 1.884\text{ k}\Omega$$
$$|Z| = \sqrt{(2.2 \times 10^3)^2 + (1.884 \times 10^3)^2} = 2.896\text{ k}\Omega$$
$$X_L = 2\pi fL = (6.28)(1.2)(10^3)(0.5) = 3.768\text{ k}\Omega$$
$$|Z| = \sqrt{(2.2 \times 10^3)^2 + (3.768 \times 10^3)^2} = 4.363\text{ k}\Omega$$

The series circuit illustrated by the circuit of Figure 15.5(a) adheres to the conditions of Ohm's law similar to that of dc circuits. For example, the total current flowing in the circuit is the applied voltage divided by the circuit's opposition. In this case the opposition is the total impedance of the circuit.

$$I = \frac{V_A}{Z} \tag{15.10}$$

Similarly, the voltage across R and X_L is

$$V_R = IR \qquad \text{and} \qquad V_L = IX_L$$

A study of an example problem illustrates some very important aspects of series L/R circuits.

Assume that a series circuit consists of a 4 kΩ resistor and an inductor whose reactance is 3 kΩ. Let us assume that 50 volts is applied to the circuit. The total impedance, according to Eq. (15.9), is 5 kΩ.

$$|Z| = \sqrt{(3 \times 10^3)^2 + (4 \times 10^3)^2} = 5 \text{ k}\Omega$$

The current in the circuit is 50 V/5 kΩ = 10 mA. The 10 mA is the same everywhere in a series circuit; hence the voltage across R is V_R $(10)(10^{-3})(4)(10^3) = 40$ volts. Similarly, the voltage across L is

$$V_L = (10)(10^{-3})(3)(10^3) = 30 \text{ volts}$$

It would appear that the total voltage should be 70 volts (30 + 40) instead of 50 volts. The two voltages are phase displaced by 90°. Figure 15.6(a) shows the series L/R circuit, and part (b) illustrates the phase displacement. When these two voltages are added as vectors, the magnitude becomes

$$|V_A| = \sqrt{V_R^2 + V_L^2} = \sqrt{40^2 + 30^2} = 50 \text{ volts}$$

Phase Angle

In a pure inductive circuit the current in the circuit lags the applied voltage by 90°. In a pure resistive circuit the phase angle is 0°. When both resistance and inductance are included, the phase angle is between 0° and 90°.

If we assume some arbitrary value of X_L and R as vectors, the triangle of Figure 15.7 results. From right triangle trigonometry the angle θ is the *arc tangent* of the ratio of the opposite to the adjacent side.

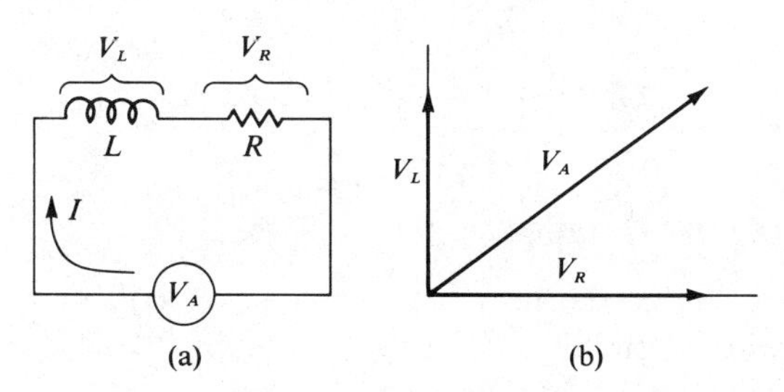

Figure 15.6 L/R Circuit Voltages

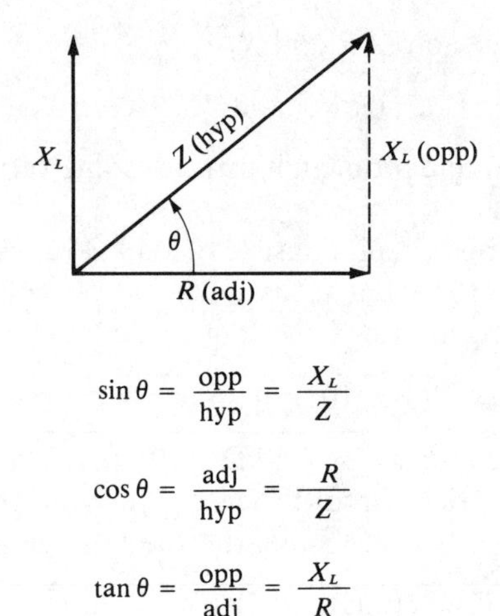

$$\sin\theta = \frac{\text{opp}}{\text{hyp}} = \frac{X_L}{Z}$$

$$\cos\theta = \frac{\text{adj}}{\text{hyp}} = \frac{R}{Z}$$

$$\tan\theta = \frac{\text{opp}}{\text{adj}} = \frac{X_L}{R}$$

Figure 15.7 *L*/*R* Phase Angle

$$\theta = \text{arc tangent } (X_L/R)$$
$$= \tan^{-1}\left(\frac{X_L}{R}\right) \quad (15.11)$$

Example 15.5 Assume a series *L*/*R* circuit consists of a 6.8 kΩ resistor and a 400 mH coil. If the applied voltage is 15 volts at 1.6 kHz, what is the current and its lagging phase angle?

$$X_L = 2\pi f L = (6.28)(1.6)(10^3)(0.4) = 4\text{ k}\Omega$$
$$Z = \sqrt{(4 \times 10^3)^2 + (6.8 \times 10^3)^2} = 7.89\text{ k}\Omega$$
$$I = \frac{15}{(7.89)(10^3)} = 1.9\text{ mA}$$
$$\theta = \tan^{-1}\left[\frac{(4)(10^3)}{(6.8)(10^3)}\right]$$
$$= \tan^{-1}(0.588) = 30.5^\circ \text{ (lagging)}$$

Polar-Rectangular Notation

We may characterize the impedance of a series *L*/*R* circuit using the *j* operator. A series circuit consisting of real (resistance) and imaginary (inductive reactance) components has a total impedance of

$$Z = R + jX_L \quad (15.12)$$

The representation using the j operator expresses the impedance in rectangular notation. (The two vectors are two sides of a triangle.) This same impedance can be expressed in polar notation

$$Z = |Z|\underline{/\theta} = \sqrt{R^2 + X_L^2}\ \underline{/\theta} \tag{15.13}$$

where the angle is Eq. (15.11). Equations (15.12) and (15.13) are two ways of identifying the same circuit impedance.

Example 15.6 Show how the impedance of a series L/R circuit can be represented in both polar and rectangular notations if $R = 1\ \text{k}\Omega$ and $L = 2$ henries. Assume the applied frequency is 60 Hz.

$$X_L = 2\pi fL = (6.28)(2)(60) = 0.754\ \text{k}\Omega$$

$$Z = R + jX_L = (1 + j0.754)\ \text{k}\Omega$$

$$= \sqrt{R^2 + X_L^2}\ \underline{/\tan^{-1}(X_L/R)}$$

$$1.253\ \text{k}\Omega\ \underline{/37^\circ}$$

Polar and rectangular notation may be used even though there may only be one component in the circuit. For example, the impedance of a pure inductance in rectangular notation is

$$Z = 0 + jX_L = jX_L$$

In polar notation it is

$$Z = X_L\underline{/90^\circ}$$

Similarly, a resistance can be represented as

$$Z = R + j0 = R$$

$$= R\underline{/0^\circ}$$

The concepts just studied can be used in actual practice. The following example illustrates the case.

Example 15.7 Assume an inductor has no markings on it. We would like to determine the value of the inductance. Assume the dc resistance of the coil is measured with an ohmmeter and its value is 300 ohms. When 50 volts at 60 Hz is applied to the coil, a current of 0.1 ampere flows. Calculate the inductance.

$$|Z| = \frac{V}{I} = \frac{50}{0.1} = 500\ \text{ohms}$$

From Eq. (15.9) solve for X_L.

$$X_L = \sqrt{Z^2 - R^2} = \sqrt{500^2 - 300^2} = 400\ \text{ohms}$$

From Eq. (15.3) solve for L.

$$L = \frac{X_L}{2\pi f} = \frac{400}{(6.28)(60)} = 1.061 \text{ henries}$$

One additional example will illustrate that other right triangle trig functions may be used to evaluate L/R circuit conditions. Assume the inductor of Example 15.7 is to be used in a circuit where the 0.1 ampere current must lag the applied voltage by 10°. Assume 50 volts at some fixed frequency is applied. Some additional resistance would have to be added in series with the inductor. We must also determine the frequency of the 50 volts source. The impedance must still be 500 ohms since 0.1 ampere must flow. From the trig relationship shown in Figure 15.7,

$$\boxed{R = |Z| \cos \theta} \qquad (15.14)$$

$$R = 500(0.9848) = 492.4 \text{ ohms}$$

Since the wire resistance of the coil is 300 ohms, an additional 192.4 ohms must be connected in series. The frequency of the applied signal can be determined from X_L.

$$\boxed{X_L = |Z| \sin \theta} \qquad (15.15)$$

$$X_L = 500(0.1736) = 86.8 \text{ ohms}$$

Utilizing the L from Example 15.7,

$$f = \frac{X_L}{2\pi L} = \frac{86.8}{(6.28)(1.061)} = 13.02 \text{ Hz}$$

15.3 SERIES L/R CIRCUITS

To understand how we may combine several series resistors and inductors into one equivalent L/R circuit, review some of the concepts of complex plane algebra.* Equations (15.6), (15.7), and (15.8) allow us to manipulate polar representations of impedance values. Certain rules govern the addition and subtraction of rectangular notations of impedances.

$$(A \pm jB) \pm (C \pm jD) = (A \pm C) \pm j(B \pm D) \qquad (15.16)$$

Equation (15.16) states that the sum of two rectangular coordinates is the algebraic sum of the real components and the algebraic sum of the imaginary components. For example, $5 + j3$ added to $6 + j4$ is equal to $11 + j7$. If $2 + j3$ is added to $4 - j3$, the result is $6 + j0$.

*See, for example, "Review of Complex Numbers and Phasor Algebra" in Bogart, *LaPlace Transforms and Control Systems Theory for Technology* (Wiley, 1982).

Though it is easier to multiply and divide polar forms, it is also possible to do so with rectangular forms.

$$(A + jB)(C + jD) = (AC - BD) + j(AD + BC) \tag{15.17}$$

$$(A + jB)/(C + jD) = \frac{(AC + BD) + j(BC - AD)}{C^2 + D^2} \tag{15.18}$$

The circuit of Figure 15.8 illustrates five components connected in series.

The total impedance of the circuit is

$$Z_t = Z_1 + Z_2 + \cdots + Z_5 \tag{15.19}$$

Each impedance can be expressed in its rectangular notation, so

$$\begin{aligned} Z_t &= R_1 + jX_{L_1} + R_2 + jX_{L_2} + R_3 \\ &= (R_1 + R_2 + R_3) + j(X_{L_1} + X_{L_2}) \end{aligned}$$

The equivalent circuit is represented as a series circuit of one resistance $(R_1 + R_2 + R_3)$ and one inductive reactance $(X_{L_1} + X_{L_2})$. From this equivalent circuit we can determine the phase angle and the current.

Example 15.8 Assume a series circuit consists of two resistors and two inductors, $R_1 = 1\ \text{k}\Omega$, $R_2 = 2.2\ \text{k}\Omega$, $L_1 = 1$ mH, $L_2 = 4$ mH. If 20 volts at 50 kHz is applied, what are the phase angle and the voltage across R_1?

$$\begin{aligned} X_{L_1} &= 2\pi f L_1 = (6.28)(50)(10^3)(10^{-3}) = 0.314\ \text{k}\Omega \\ X_{L_2} &= 2\pi f L_2 = (6.28)(50)(10^3)(4)(10^{-3}) = 1.256\ \text{k}\Omega \\ Z &= (1 + 2.2)\ \text{k}\Omega + j(1.256 + 0.314)\ \text{k}\Omega \\ &= (3.2 + j1.57)\ \text{k}\Omega \\ &= \sqrt{3.2^2 + 1.57^2}\ \text{k}\Omega = 3.56\ \text{k}\Omega \\ \theta &= \tan^{-1}(1.57/3.2) = 26.13^\circ \text{(voltage leads current)} \end{aligned}$$

$$I = V_A/Z = \frac{20\ \angle 0^\circ}{3.56 \times 10^3\ \angle 26.13^\circ} = 5.61\ \angle -26.13^\circ\ \text{mA}$$

$$V_1 = IR_1 = (5.61 \times 10^{-3}\ \angle -26.13^\circ)(10^3\ \angle 0^\circ) = 5.61\ \angle -26.13^\circ\ \text{volts}$$

A series L/R circuit may be represented in its polar form. From this polar form we can convert to an equivalent R and X_L.

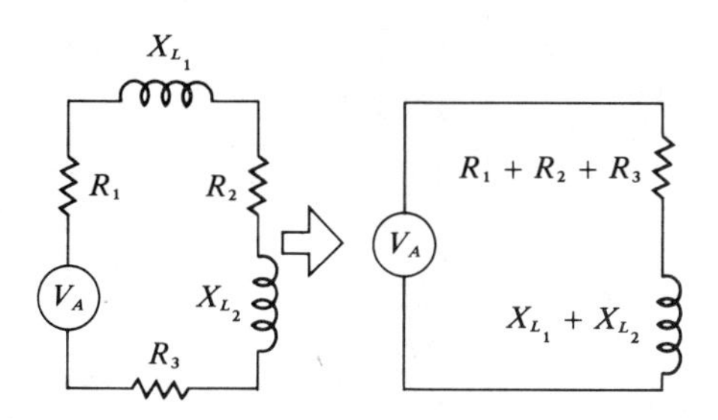

Figure 15.8 Series L/R Circuit

$$Z \angle\theta = R + jX_L$$
$$= Z\cos\theta + jZ\sin\theta \qquad (15.20)$$

Equation (15.20) is essentially a restatement of Eqs. (15.14) and (15.15).

Example 15.9 When an ac voltage of 150 volts is applied to an inductive-resistive load, a current of 2 mA flows with a lagging angle of 40°. What are the resistance and the inductance of the load? Assume $f = 255$ kHz.

$$Z = V \angle 0° / I \angle -40°$$
$$= \frac{150 \angle 0°}{(2)(10^{-3}) \angle -40°} = 75 \text{ k}\Omega \angle 40°$$

According to Eq. (15.20),

$$R = (75)(10^3)\cos 40° = 57.5 \text{ k}\Omega$$
$$X_L = (75)(10^3)\sin 40° = 48.2 \text{ k}\Omega$$
$$L = \frac{X_L}{2\pi f} = \frac{(48.2)(10^3)}{(6.28)(255)(10^3)} = 30 \text{ mH}$$

It is now appropriate to review a methodology of circuit analysis which was first encountered in the latter chapters of dc circuits, namely, computer applications to the solution of complex circuits. As was the situation in dc analysis, a minimal amount of theory must first be presented before problem complexity warrants computer application. The reader has now achieved sufficient background knowledge of ac circuitry that the "tradeoff" between the time it takes to program a problem versus the time it takes for manual calculations renders computer applications more appropriate. As the initial transition into computer solutions of ac circuits, consider Figure 15.9.

Assume the ac voltage source has an internal resistance R_i. This voltage source outputs a sine wave over a wide range of frequencies. The load connected to the source is a simple series L/R circuit. For any frequency, the actual voltage magnitude delivered to the load can be calculated by applying the proportionate voltage ratio law.

$$V_L = \frac{Z_L}{Z_t}(V) = \frac{(R + jX_L)V}{(R + R_i) + jX_L}$$

$$= \frac{V\sqrt{R^2 + X_L^2} \angle \tan^{-1}\left(\frac{X_L}{R}\right)}{\sqrt{(R + R_i)^2 + X_L^2} \angle \tan^{-1}\frac{X_L}{R + R_i}}$$

$$\therefore \quad |V_L| = \frac{V\sqrt{R^2 + X_L^2}}{\sqrt{(R + R_i)^2 + X_L^2}}$$

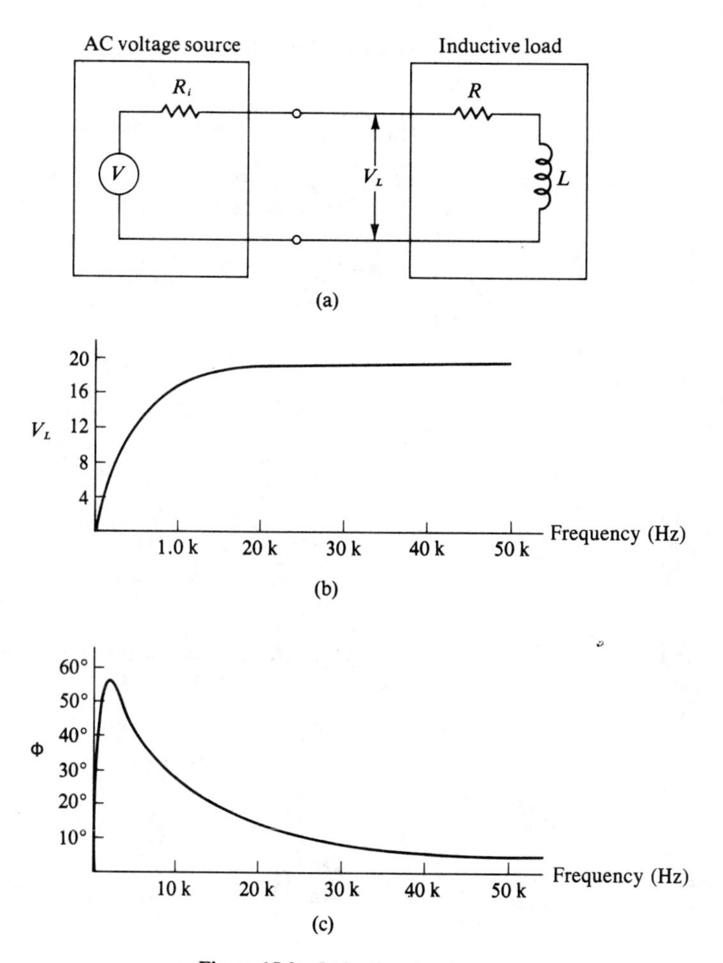

Figure 15.9 Inductive Load

$$\phi = \angle \tan^{-1}\left(\frac{X_L}{R}\right) - \angle \tan^{-1}\left(\frac{X_L}{R + R_i}\right)$$

Example 15.10 With reference to Figure 15.9, V equals 20 volts, R_i equals 1500 ohms, R equals 150 ohms, and L equals 50 mH. Assume it is desired to know the voltage across the load (V_L), and the phase angle (ϕ) between V_L and the input voltage as frequency increases. To facilitate graphic plots, let f increment in 100 Hz steps from 0 Hz to 1400 Hz, and then increment in 2000 Hz steps to 53.5 kHz. Write a computer program in BASIC to (a) provide the values to plot V_L versus f, and (b) provide the

L/R Circuit Printout

```
10   P1=3.14159 : V=20 : R1=1500 : R=150 : L=.05 : X=0
20   PRINT "F IN HZ", "X IN OHMS", "V(LOAD)", "PHASE IN DEG."
25   PRINT
30   FOR F=0 TO 1400 STEP 100 : GOSUB 60 : NEXT F
40   FOR F=1500 TO 53500 STEP 2000 : GOSUB 60 : NEXT F
50   STOP
60   X=(2*P1*F*L)
70   V1=V*(SQR(R↑2+X↑2)/SQR((R1+R)↑2+X↑2))
80   A=(180/P1)*(ATN(X/R)-ATN(X/(R1+R)))
90   PRINT F , X , V1 , A
100  RETURN
110  END
```

F IN HZ	X IN OHMS	V(LOAD)	PHASE IN DEG.
0	0	1.81818	0
100	31.4159	1.85729	10.7382
200	62.8318	1.96982	20.547
300	94.2477	2.1438	28.8727
400	125.664	2.36505	35.5996
500	157.08	2.62083	40.8826
600	188.495	2.90108	44.9709
700	219.911	3.19835	48.1108
800	251.327	3.50726	50.5093
900	282.743	3.82388	52.3296
1000	314.159	4.14531	53.6972
1100	345.575	4.46939	54.7073
1200	376.991	4.79447	55.4331
1300	408.407	5.11923	55.9304
1400	439.823	5.44265	56.2426
1500	471.238	5.76391	56.404
3500	1099.56	11.1936	48.5524
5500	1727.87	14.5187	38.7178
7500	2356.19	16.4156	31.3603
9500	2984.51	17.5253	26.059
11500	3612.83	18.2082	22.169
13500	4241.15	18.6508	19.2327
15500	4869.46	18.9511	16.9544
17500	5497.78	19.163	15.1428
19500	6126.1	19.3176	13.6717
21500	6754.42	19.4335	12.4554
23500	7382.74	19.5225	11.4343
25500	8011.05	19.5922	10.5655
27500	8639.37	19.6479	9.81781
29500	9267.69	19.693	9.16778
31500	9896.01	19.7299	8.59767
33500	10524.3	19.7606	8.09373
35500	11152.6	19.7864	7.64513
37500	11781	19.8083	7.24332
39500	12409.3	19.827	6.88137
41500	13037.6	19.843	6.55368
43500	13665.9	19.857	6.25562
45500	14294.2	19.8692	5.98335
47500	14922.6	19.8799	5.7337
49500	15550.9	19.8893	5.50398
51500	16179.2	19.8976	5.29188
53500	16807.5	19.9051	5.09546

values to plot ϕ versus f. (Note the extensive amount of manual calculations that would be necessary.)

Study the printout on page 440. Note how the voltage increases rapidly to its maximum value of 20 volts. The graph of Figure 15.9(b) illustrates the change. The phase angle starts at zero, rises to a peak value of about 56°, then slowly decays. This function can be seen in Figure 15.9(c).

15.4 INSTANTANEOUS VALUES OF AN *L/R* CIRCUIT

In Figure 15.3 we observed that in a purely inductive circuit, the current lags the voltage by 90°. If the circuit contains resistance as well, the lagging phase angle of the current can vary depending upon how much resistance is in the circuit and the frequency of the applied voltage. It is often necessary to compute the instantaneous current in a given L/R circuit, as well as the instantaneous voltage at any point in the circuit.

If the applied voltage is assumed as a reference

$$v(t) = V_p \sin \theta \tag{15.21}$$

where θ is $2\pi ft$ and $v(t)$ is the voltage at a given instant. Then the instantaneous value of the circuit current is

$$i(t) = I_p \sin (\theta - \phi) \tag{15.22}$$

The instantaneous voltage across the resistor can be found by using Ohm's law.

$$v_R(t) = i(t)R$$

Similarly, the voltage across L can be determined using:

$$v_L(t) = i(t)X_L$$

However, the voltage across a coil always leads the current through it by 90°.

Example 15.11 Assume the series circuit of Figure 15.10 has a resistance of 2.2 kΩ. The inductance L is 200 mH and the applied voltage is 50 volts at 2 kHz. (a) What is the current in the circuit when the input voltage is at 0.1 ms of the period? (b) What is the voltage across R? (c) What is the voltage across L?

a. $X_L = 2\pi fL = (6.28)(2)(10^3)(0.2) = 2.5137 \text{ k}\Omega$

$\theta = 2\pi ft = (360°)(2)(10^3)(0.1)(10^{-3}) = 72°$

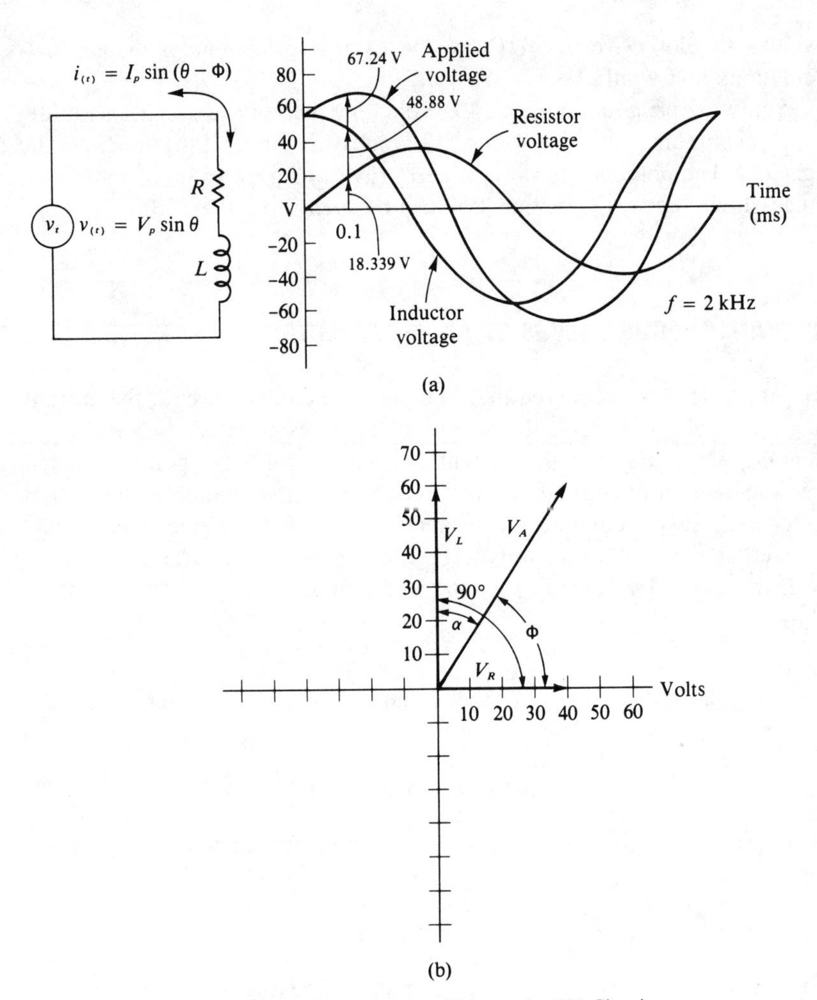

Figure 15.10 Instantaneous Voltages in *R*/*L* Circuit

$$\phi = \tan^{-1}\left(\frac{X_L}{R}\right) = \tan^{-1}\left(\frac{2.5137}{2.2}\right) = 48.8^\circ$$

$$Z = \sqrt{R^2 + X_L^2} = \sqrt{(2.2 \times 10^3)^2 + (2.5137 \times 10^3)^2}$$

$$= 3.34\ \text{k}\Omega \underline{/48.8^\circ}$$

$$I_p = \frac{V_p}{Z} = \frac{(50)(1.414)}{(3.34)(10^3)} = 21.16\ \text{mA}$$

The instantaneous current equation is

$$i(t) = 21.16 \sin(\theta - 48.8^\circ)\ \text{mA}$$

Since $\theta = 72°$ when $t = 0.1$ ms,

$$\begin{aligned}
i &= (21.16)(10^{-3}) \sin 23.2° \\
&= (21.16)(10^{-3})(0.3939) = 8.3358 \text{ mA} \\
V_R &= (8.3358)(10^{-3})(2.2)(10^3) = 18.339 \text{ volts} \\
V_L &= [(21.16)(10^{-3}) \sin (23.2° + 90°)](2.5137)(10^3) \\
&= (21.16)(10^{-3})(2.5137)(10^3) \sin 113.2° \\
&= (21.16)(2.5137) \sin 66.8° \\
&= 48.888 \text{ volts}
\end{aligned}$$

We may verify this voltage by determining the voltage applied when $t = 0.1$ ms.

$$v(t) = 70.7 \sin 72° = 67.239 \text{ volts}$$

The instantaneous sum of 18.3358 and 48.888 is 67.227 volts. This relationship can be seen in Figure 15.10.

In the waveshapes of V_L and V_R in Figure 15.10 the peaks are displaced by 90°. We could solve graphically for the input voltage at a time when the voltage across the resistor is maximum and the voltage across the inductor is minimum. Note that this condition occurs at about 0.125 ms. We can also relate the time base to radians or degrees. There is a direct ratio

$$\frac{t}{\theta} = \frac{T}{360°} \tag{15.23}$$

where T is the period in seconds, t the instant of time in seconds, and θ the lead or lag angle in degrees.

The relationships between the sine waves of Figure 15.10(a) are graphically difficult to visualize. If, however, we presume that each sine wave is a rotating vector, we may represent each voltage in relationship to each other.

Phasor diagrams shown in Figure 15.10(b) illustrate the interrelationship of the voltages. The voltage across the resistor is lagging the applied voltage by ϕ degrees. The voltage across R and L are 90° displaced, while the voltage across L leads the applied voltage by α degrees.

15.5 POWER IN *L*/*R* CIRCUITS

We have already learned that a pure inductor does not convert energy to heat, light, or any other form. It merely stores the energy delivered to it. This energy is returned to the circuit. A resistor, on the other hand, converts all of the electrical energy delivered to it. There is no storage.

For circuits containing both resistance and inductance, we can correctly conclude that not all the energy delivered to such a circuit is converted to another form, nor is all of it stored. Only a percentage is converted to heat. The percentage converted to heat can be determined using the circuit's power relationships.

Substituting $i = I_p \sin(\theta - \phi)$ and $v = V_p \sin\theta$ in $p = iv$, we obtain

$$P = I_p V_p \sin\theta \sin(\theta - \phi) \tag{15.24}$$

where ϕ is the phase angle of the lagging current. Substituting trigonometric identities and solving for the average power:

$$P_{\text{ave}} = \frac{V_p I_p}{2} \cos\phi \tag{15.25}$$

Since the phase angle for a purely inductive circuit is 90°, Eq. (15.25) applied to that special case results in

$$P_{\text{ave}} = \frac{V_p I_p}{2}(\cos 90°) = 0$$

In a purely resistive circuit:

$$P_{\text{ave}} = \frac{V_p I_p}{2}(\cos 0°) = \frac{V_p I_p}{2}$$

Recall that

$$\frac{V_p I_p}{2} = \frac{V_p}{\sqrt{2}} \times \frac{I_p}{\sqrt{2}} = V_{\text{rms}} \times I_{\text{rms}}$$

Since only the resistor converts energy to heat, we call that power the *true power*. In purely resistive circuits it becomes $P_t = I^2R$, where I is the rms value of the current.

The true power in an inductance is zero, yet the inductance draws a current when an ac voltage is connected across it. The quantity

$$P_A = I_{\text{rms}} V_{\text{rms}} \text{ volt-amps} \tag{15.26}$$

is known as the *apparent power*. The *power factor* is the ratio of the true power to the apparent power.

$$\text{PF} = \frac{\text{true power}}{\text{apparent power}} \tag{15.27}$$

In a series L/R circuit the power factor is

$$\text{PF} = \frac{I^2R}{IV}$$

$$= \frac{IR}{V} = \frac{R}{V/I} = \frac{R}{|Z|} = \cos\phi$$

The parts of Eq. (15.25), therefore, are:

$$\text{True power} = (\text{apparent power}) \cos \phi$$

The cos ϕ is the percentage of apparent power that is converted to heat.

Example 15.12 A series L/R circuit consists of 2.2 kΩ and 200 mH. What are the apparent power, the true power, and the power factor if the applied voltage is 50 volts at 2 kHz?

$$X_L = 2\pi fL = (6.28)(2)(10^3)(0.2) = 2.51 \text{ k}\Omega$$
$$|Z| = \sqrt{R^2 + X_L^2} = 3.34 \text{ k}\Omega$$
$$\text{PF} = R/|Z| = 0.658$$
$$I = V/|Z| = 50/\sqrt{2.2^2 + 2.51^2} \text{ k}\Omega = 14.98 \text{ mA}$$
$$P_A = IV = (14.98)(10^{-3})(50) = 0.75 \text{ volt-amps}$$
$$P_t = (P_A)(\text{PF}) = (0.75)(0.658) = 492.5 \text{ mW}$$

SUMMARY

An inductance offers an opposition to an applied alternating current that is directly proportional to the applied frequency and the inductance. In addition, a pure inductance introduces a 90° phase angle which causes the current to lag the applied voltage. The individual reactance can be used in Ohm's law calculations for determining circuit voltages and total reactance.

When a resistance is introduced into the inductive circuit, an L/R results and the total opposition to ac becomes impedance. The phase angle varies with relation to the X_L and the R. When several resistive and reactive components are connected in series, parallel, or series-parallel combinations, polar-rectangular transformations facilitate reducing the circuit to a single L/R series circuit.

Instantaneous voltage and current conditions may be determined using the basic relationship, $v = V_p \sin 2\pi ft$. The current may be determined after the lagging phase angle has been determined, that is, $i = I_p \sin (2\pi ft - \phi)$.

Power in inductive reactive circuits depends upon the rms values of the current and voltage. Power factor, the percentage of power actually generating heat energy, is cos ϕ. The power factor is also the ratio of true power (I^2R) to apparent power (IV).

PROBLEMS

Reference Section 15.1

1. How much opposition (X_L) will a 5 henry coil offer to an ac voltage at 60 Hz and at 400 Hz?

2. If a given coil offered a specific inductive reactance to 1000 Hz, would the inductive reactance increase or decrease if the diameter of the coil were increased?
3. A 12 henry coil is connected in series with a 10 henry coil. If the applied signal is at a frequency of 400 Hz, what is the total X_L offered by the circuit?
4. If the two coils of Problem 3 were connected in parallel and the frequency were increased to 1000 Hz, what would be the total circuit inductive reactance?
5. An ac ammeter is connected in series with a coil of 60 mH. If 10 volts at 10 kHz is applied, what current will flow in the circuit?
6. In Problem 5, if the applied voltage remains the same but the frequency is decreased to 5 kHz, by how much will the current increase or decrease?
7. If a 10 mH coil develops 37.7 ohms of inductive reactance, what is the applied frequency?
8. The inductive reactance (X_L) of a certain unknown coil is 400 ohms. If the applied voltage is 15 volts at 1600 Hz, what is the inductance of the coil?
9. A given coil 3 cm long consists of 600 turns of #12 wire (assume negligible wire resistance) wound on an air core bakelite coil form whose cross-sectional area is 6 cm^2. When 10 volts ac is applied, 6 mA of current flows. When a ferrite core is inserted into the coil, 2 mA of current flows. What is the permeability of the ferrite core?
10. An inductive circuit has 50 volts applied. If the coil is 20 mH and the current is 200 mA, what is the frequency of the applied signal?
11. Refer to the circuit of Problem 10. At what applied frequency will the current be 100 mA?
12. Which coil will draw more current: a 10 mH connected to 5 volts at 100 kHz, or 50 henries connected to 75 volts at 60 Hz?
13. Two coils are connected in series with no interaction between their fields. When 20 volts at 1 kHz is applied to the circuit, 5 mA of current flows. If one-fourth of the total applied voltage appears across L_1 what is the inductance of L_2? What is the inductance of L_1?

Reference Sections 15.2 and 15.3

14. A given series circuit consists of an inductance of 106.1 mH in series with a 3 kΩ resistor. What impedance does the circuit offer to 60 Hz?
15. What is the impedance of the circuit described in Problem 14 to 600 Hz? What is it at 6 kHz?
16. If 14 volts at 1 kHz is applied to a series circuit consisting of a 638 mH coil and a 5 kΩ resistor, what current will flow in the circuit?

17. If the frequency of the voltage generator in Problem 16 doubles but the applied voltage remains the same, what would be the new current flow?
18. Assume a series circuit consists of a 12 kΩ resistor and a 1.28 henry coil. When 40 volts at 2 kHz is applied, what voltage is developed across the resistor? What voltage is developed across the coil?
19. If in a given series circuit (L/R) a 20 volt 12 kHz signal produces 4 mA of current, what is the inductive reactance of the coil if the resistance is 2 kΩ?
20. Refer to Problem 19. What is the inductance of the coil?
21. A coil with an unknown inductance is connected in series with a variable resistor. A 60 Hz voltage is applied and the resistor is adjusted until the voltage across it is 10 volts. The voltage across the coil is measured at 15 volts. The R is measured to be 4.7 kΩ. What is the value of L? What is the applied voltage?
22. When 10 volts dc is applied to a solenoid (coil), it is found that it takes 8 ms for the current to build up to its maximum value of 4 mA. What current would flow into the coil if 10 volts at 400 Hz were applied?
23. If a series circuit consists of a coil of 2 henries and a resistor of 1 kΩ, what is the phase angle of the circuit when 10 volts at 60 Hz is applied? What is the current flow in the circuit?
24. In the circuit of Problem 23, if the inductance is changed to 1 henry, by how many degrees will the current lag the voltage?
25. If a series circuit consists of a coil and a 10 kΩ resistor, what is the inductance of the coil if the phase angle in the circuit is 20° when 4 volts at 400 Hz is applied? What is the current in the circuit?
26. A series L/R circuit consists of a 3.3 kΩ resistor and an 0.8 henry coil. What is the impedance at 700 Hz? What is the phase angle? What are $|Z|$ and θ at 7 kHz? What are $|Z|$ and θ at 70 kHz?
27. A series circuit consists of a 200 ohm resistor and a 0.5 henry coil. If the voltage applied to the circuit is 50 at 60 Hz, answer the following questions:
    a. What is the total impedance of the circuit (polar and rectangular forms)?
    b. What is the total current flow and what is its phase angle?
    c. What are the voltage and phase angle across the resistor?
    d. What are the voltage and phase angle across the coil?

Reference Section 15.4

28. Assume in a given circuit that the current lags the applied voltage by 30°. The period is 475 ms, the maximum voltage is 30 volts and the peak current is 4 mA. What is the instantaneous current when t of

the voltage cycle is 40 ms? What is the current (instantaneous) when t of the voltage cycle is 100 ms?

29. Refer to Problem 28. At what time of the input voltage cycle will the current be 1 mA? What will be the applied voltage (instantaneous) when the current is 1 mA?
30. Refer to Problem 28. When the instantaneous current is 0, what are the values of the instantaneous voltage across R and L?
31. A given ac generator is operating into a load such that the current peaks lag the voltage peaks by 10 μs. If the frequency of the applied voltage is 10 kHz and the maximum voltage is 20 volts, what is the lagging phase angle?
32. Assume the current in a given circuit is lagging the voltage by 70°. The maximum current is 2 mA. The instantaneous voltage of 10 volts occurs when t is 4 ms. If the maximum voltage is 80 volts, what is the instantaneous current when the voltage is 10 volts?
33. Refer to Problem 32. What is the instantaneous voltage when the current is at its zero crossing points?

Reference Section 15.5

34. A series circuit consists of a 2 henry coil and a 1 kΩ resistor. If the applied voltage is 20 volts at 60 Hz, what is the power factor? What is the true power?
35. A given L/R circuit has a power factor of 0.8. The circuit consists of a 1 kΩ resistor and has 20 volts at 400 Hz applied. What is the phase angle?
36. Refer to the circuit of Figure 15.11. What are the power factor and the phase angle?
37. A given circuit consists of a 2.2 kΩ resistor and a 100 mH coil. At what input frequency will the circuit be consuming 50 percent of the apparent power?
38. A given L/R circuit has a phase angle of 60° and a total impedance of 4.8 kΩ. If the dc time constant of the circuit is 100 μs, what is the value of L?

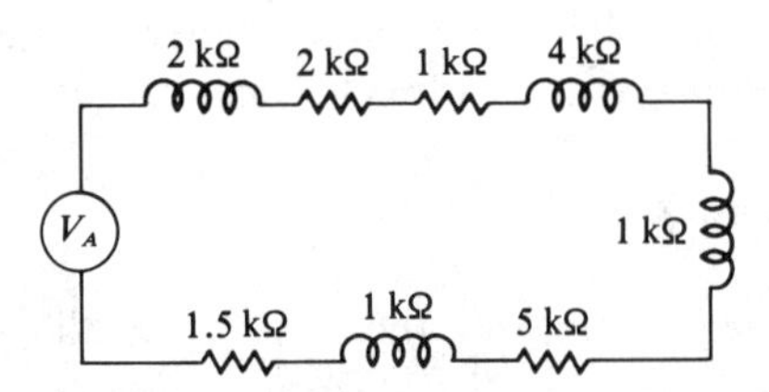

Figure 15.11 Circuit for Problem 15.33

Optional Computer Problems

39. Rewrite the program of Example 15.10. Assume a fixed frequency of 1400 Hz. Assume it is desired to show the voltage across the load and the phase angle as L is varied from 5 mH to 100 mH in 2 mH steps.
40. A series L/R circuit consists of a 4.7 kΩ resistor and a 250 mH coil. The applied voltage is 20 volts over a range of frequencies from 0.5 kHz to 40 kHz. Write a program in BASIC which will provide a printout of how power factor, the voltage across L, and the total impedance vary over the frequency range.

CHAPTER **16**

CAPACITIVE REACTANCE

16.1 PURE CAPACITANCE

The series resistive-capacitive circuit is probably the most universal equivalent circuit used in all areas of electrical science. Almost all devices—resistors, integrated circuits, transistors, relays, coils, and even wires—have, or introduce, some capacity to a given circuit. That capacity will have an effect on the circuit operation. The effect might only be noticed at some high frequency or at some extremely fast switching time. The influence can be significant or trivial. In either case the student should have a solid understanding of capacitance and capacitive reactance.

In this chapter we will explore the behavior of a capacitor in an ac circuit. How the pure capacitor's opposition to alternating current varies with frequency is the prime focus of the chapter. We will also study the characteristics of series and parallel circuits involving capacitors and resistors. Finally, the power factor of an *RC* circuit will be determined.

The ac phase angle will be developed for various applied frequencies, and several computer programs involving varied circuit configurations will be analyzed.

Chapter 15 showed that when an ac current was applied to an inductance, the coil exhibited an opposition to current flow. In a similar manner, when a voltage is impressed across a capacitor, the current is limited by an opposition. In Figure 16.1(a), note that the opposition can be identified using Ohm's law.

$$\frac{V_{\text{rms}}}{I_{\text{rms}}} = \text{opposition} = X_C \text{ ohms} \tag{16.1}$$

where X_C is the capacitive reactance in ohms when V_{rms} is in volts and I_{rms} is in amperes. (The rms selection is arbitrary. Peak units or peak-to-peak units can be used as well.)

In a capacitor the amount of leakage current through the dielectric is generally quite small. The dielectric material along with its physical dimensions determines the amount of leakage. If this leakage is quite small, the amount of leakage resistance that parallels the capacitive plates is very large (refer to Figure 16.1b). The actual value of this leakage can be determined using a simple dc test. If, for example, 50 volts dc is applied to the capacitor, a small current will flow through the dielectric. It is generally measured with a sensitive dc ammeter. If 20 nA flows, the

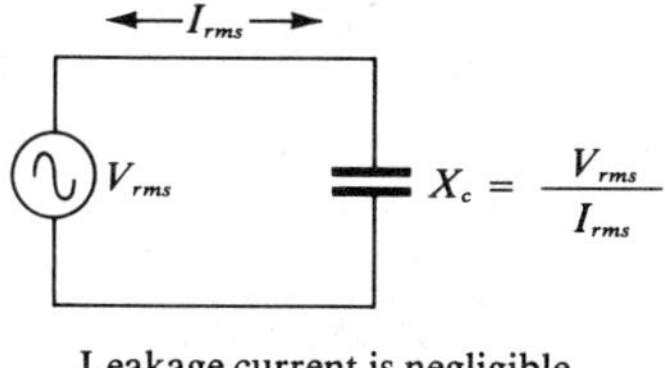

Leakage current is negligible
(a)

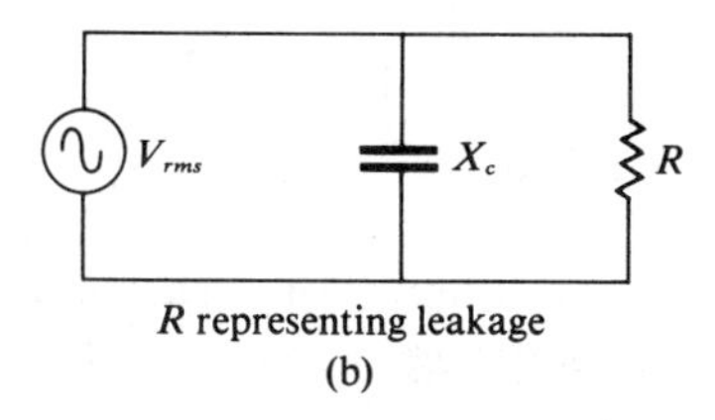

R representing leakage
(b)

Figure 16.1 Capacitive Opposition

leakage resistance is $50/(20)(10^{-9}) = 2500$ MΩ. This extremely large value paralleling X_C is negligible. For practical purposes, we can neglect the effect of R and assume the C is a pure capacitance.

Capacitive reactance can be determined using Coulomb's law. Coulomb stated that the current flowing into the plates of a capacitor is dependent upon the capacity and the rate of change of voltage across the capacitor plates.

$$i_C = C\frac{\Delta v}{\Delta t} \tag{16.2}$$

In Figure 16.1(a), note that the voltage is a sine wave. Substituting $v = V_p$ sin ωt in Eq. (16.2) and using calculus, it can be shown that:

$$i_C = CV_p(\cos \omega t)\omega$$

The applied voltage is directly across the capacitor, hence, when i is equal to I_p (the peak value), cos ωt is equal to 1. Solving for V_p/i_p,

$$\frac{V_p}{I_p} = \frac{1}{\omega C} = X_C \tag{16.3}$$

However, since ω is equal to $2\pi f$, the capacitive reactance becomes:

$$X_C = \frac{1}{2\pi f C} \text{ ohms} \tag{16.4}$$

where f is in hertz, C is in farads, and X_C is in ohms.

Example 16.1 What current will flow through a 0.05 μF capacitor when 20 volts at 2 kHz is applied? What will the current be when the frequency of the applied voltage is doubled to 4 kHz?

$$X_C = \frac{1}{2\pi fC} = \frac{1}{(6.28)(2)(10^3)(0.05)(10^{-6})}$$

$$= 1.59 \text{ k}\Omega \text{ at } 2 \text{ kHz}$$

$$I = V/X_C = 20/1.59 \text{ k}\Omega = 12.566 \text{ mA}$$

$$X_C = 0.796 \text{ k}\Omega \text{ at } 4 \text{ kHz}$$

$$I = 25.13 \text{ mA}$$

If Eq. (16.4) is solved for either C or f, additional circuit information can be determined. The following example illustrates this case.

Example 16.2 When a 10 volt 60 Hz source is placed across an unknown capacitor, a 2 mA current flows. What is the value of C?

$$X_C = \frac{V}{I} = \frac{10}{(2)(10^{-3})} = 5 \text{ k}\Omega$$

$$C = \frac{1}{2\pi f X_C} = \frac{1}{(6.28)(60)(5)(10^3)}$$

$$= 0.53 \ \mu\text{F}$$

AC Ohm's Law

The basic principles of dc current can be applied to circuits containing pure capacitive reactance. The behavior of X_C is the same as that of resistance, so long as the circuit contains only capacitors. For example, if several capacitors are connected in series, the total ac opposition is the sum of the individual capacitive reactances.

$$X_{C_T} = X_{C_1} + X_{C_2} + \cdots + X_{C_N} \tag{16.5}$$

Similarly, when capacitors are connected in parallel configurations, the total capacitive reactance is the reciprocal of the sum of the reciprocals.

$$X_{C_T} = \frac{1}{\dfrac{1}{X_{C1}} + \dfrac{1}{X_{C_2}} + \cdots + \dfrac{1}{X_{C_N}}} \tag{16.6}$$

If only two capacitors are connected in parallel, the total capacitive reactance is the product over the sum of the individual capacitive reactances.

Of course, the total capacitive reactance can be calculated from the total capacity. This total capacity can be determined first, then the reactance calculated.

Example 16.3 What is the total capacitive reactance of a 0.5 μF and a 0.3 μF capacitor connected in series across a 20 volt 2 kHz source? What is the reactance of the 0.3 μF capacitor?

$$C_T = \frac{C_1 C_2}{C_1 + C_2} = \frac{(0.5)(0.3)}{0.5 + 0.3}\ \mu\text{F} = 0.1875\ \mu\text{F}$$

$$X_{C_T} = \frac{1}{2\pi f C_T} = \frac{1}{(6.28)(2)(10^3)(0.1875)(10^{-6})} = 424\ \text{ohms}$$

$$X_C = \frac{1}{2\pi f C} = \frac{1}{(6.28)(2)(10^3)(0.3)(10^{-6})} = 265\ \text{ohms}$$

The principles of dc current ratios and voltage distribution ratios apply to capacitive series and parallel circuits.

The circuit of Figure 16.2 is not an uncommon configuration of capacitors in a practical circuit. Capacitor C_1 might be a coupling capacitor joining the 25 volt source to some load. Capacitor C_3 could represent some input capacity of the load, while C_2 becomes a combination of circuit and distributed capacities. The amount of current flowing out of the generator is not all delivered to C_3 because of the bypassing effect of C_2. The principle is best illustrated through an example.

Example 16.4 Refer to the circuit of Figure 16.2. How much of the current leaving the generator will flow into C_3?

$$X_{C_1} = \frac{1}{2\pi f C_1} = \frac{1}{(6.28)(3)(10^3)(0.04)(10^{-6})} = 1.326\ \text{k}\Omega$$

$$X_{C_2} = \frac{1}{2\pi f C_2} = \frac{1}{(6.28)(3)(10^3)(0.02)(10^{-6})} = 2.652\ \text{k}\Omega$$

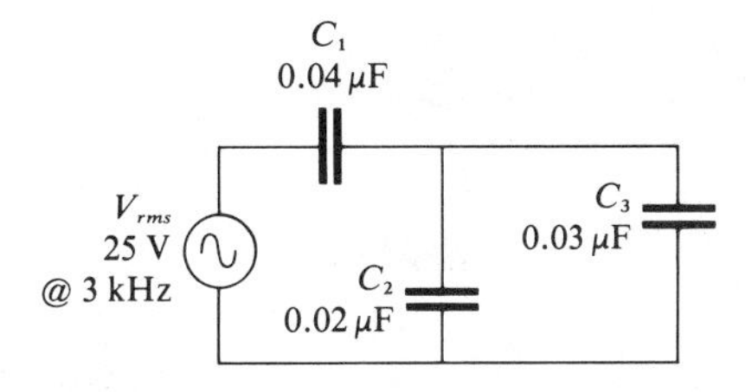

Figure 16.2 Series-Parallel Capacitive Circuit

$$X_{C_3} = \frac{1}{2\pi f C_3} = \frac{1}{(6.28)(3)(10^3)(0.03)(10^{-6})} = 1.768\ \text{k}\Omega$$

$$X_{C_T} = \frac{X_{C_2}X_{C_3}}{X_{C_2} + X_{C_3}} + X_{C_1} = \frac{(2.652)(1.768)}{(2.652) + (1.768)} + 1.326 = 2.387\ \text{k}\Omega$$

$$I_T = \frac{V}{X_{C_T}} = \frac{25}{2.387\ \text{k}\Omega} = 10.474\ \text{mA}$$

The current distributes by ratios:

$$I_3 = \frac{X_{C_2}(I_T)}{X_{C_2} + X_{C_3}} = \frac{(2.652)(10.474)(10^{-3})}{2.652 + 1.768} = 6.28\ \text{mA}$$

(Note that the total current also distributes by the direct ratio of the capacitors as well.)

$$I_3 = \frac{C_3}{C_2 + C_3}(I_T)$$

$$I_3 = \frac{0.03(10.474)(10^{-3})}{0.03 + 0.02} = 6.28\ \text{mA}$$

Phase Angle

The behavior of a capacitor in a series RC circuit was discussed in Chapter 13. We observed that the voltage across a capacitor does not build instantly to its maximum voltage after a dc is impressed across an RC combination. The charge $q = Cv$ cannot be accumulated instantly on the plates of the capacitor. In Chapter 13 we observed that the current began to flow immediately and the buildup of voltage across the capacitor lagged behind.

In the preceding chapter, it was noted that in an inductive circuit the current lagged the voltage. In a capacitive circuit the current leads the voltage. In a purely capacitive circuit the current leads by 90°.

This phase angle relationship can be established using Coulomb's law. From Eq. (16.2),

$$i_c = C\frac{\Delta v}{\Delta t}$$

Assume that current will be used as a reference. If this current is $i = I_p \sin \omega t$, then substituting into Eq. (16.2),

$$I_p \sin \omega t = C\frac{\Delta v}{\Delta t} \tag{16.7}$$

Using advanced mathematics, we can solve this equation for v.

$$v = \frac{I_p}{\omega C} \cos \omega t \tag{16.8}$$

Using the trigonometric identity

$$\sin (\alpha - 90°) = \cos \alpha$$

Equation (16.8) becomes

$$v = V_p \sin (\omega t - 90°) \tag{16.9}$$

Recall that current was used as a reference. Equation (16.9) means that the voltage will lag the current by 90° in a pure capacitive circuit, which is the same as saying that the current leads the voltage by 90°.

This characteristic is illustrated in Figure 16.3(a). Here again events to the right of a given point occur after events to the left. Note that the voltage peak occurs after the current peak. Since the sinusoidal variations are more

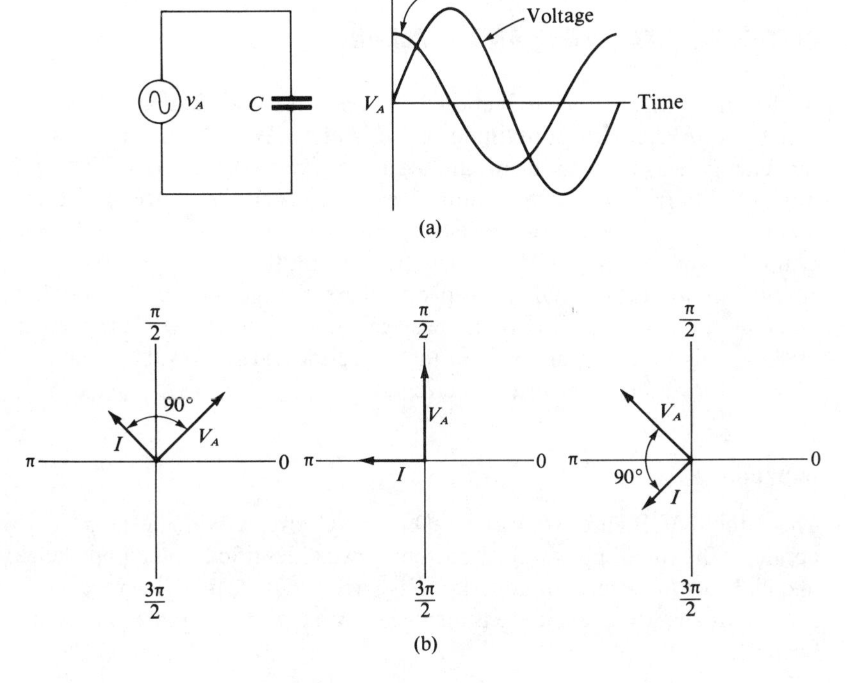

Figure 16.3 Capacitive Phase Angle

difficult to follow, the vector representation of Figure 16.3(b) illustrates the lead-lag relationships at three different times in the cycle.

j Operator

The 90° leading current relationship can be identified using the complex plane notation:

$$-jX_C = X_C \angle -90^\circ$$

Example 16.5 Assume the series circuit of Figure 16.1(a) has 110 volts at 60 Hz applied. If the capacity is 2 μf, what are the current and phase angle?

$$X_C = \frac{1}{2\pi fc} = \frac{1}{(6.28)(60)(2)(10^{-6})} = 1.33 \text{ k}\Omega$$

$$I = \frac{V}{X_C} = \frac{110 \text{ V} \angle 0^\circ}{(1.33)(10^3) \angle -90^\circ}$$

$$= 82.94 \text{ mA} \angle +90^\circ$$

(The plus sign indicates a positive lead angle of the current with respect to the reference voltage.)

16.2 CAPACITIVE REACTANCE AND IMPEDANCE

In the entire discussion of Section 16.1, we presumed that the capacitor acted within a circuit containing no resistance. We also presumed that the leakage was not significant and could be ignored. In actual circuitry, we are rarely involved with circuits where there is no resistance. There are times, however, when the resistive properties can be neglected. For example, when dealing with very low frequencies the reactance approaches nearly infinite values. Often, capacitors are thought of as devices that block dc (infinite opposition) and pass ac (finite opposition). This section explores this blocking property of the capacitor and studies the combined effects of R and C in a series ac circuit.

Impedance

The circuit of Figure 16.4(a) is a basic RC circuit with an ac voltage applied. The total opposition to current flow is identified as the impedance and includes the effects of both the resistance and the capacitive reactance. Since the capacitive reactance is displaced by a factor of $-j$, the impedance becomes

$$|Z| = |R - jX_C| = \sqrt{R^2 + X_C^2} \text{ ohms} \qquad (16.10)$$

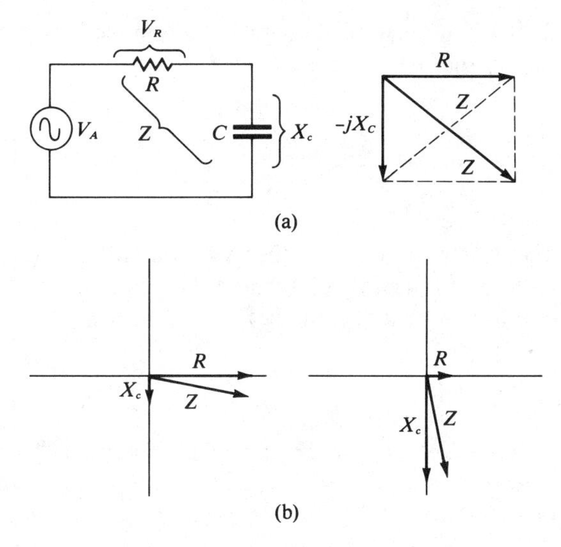

Figure 16.4 *R/C* Circuit Impedance

The impedance is measured in ohms and is influenced by the resistance, the capacitance, and the frequency. Since increasing the frequency causes the capacitive reactance to decrease, the total impedance of an *RC* circuit approaches the value of the resistance at high frequencies.

Figure 16.4(b) illustrates two cases of impedance. The first shows how the impedance approaches the value of R as X_C is equal to or less than 10 percent of R. In the second case the impedance approaches the value of X_C as R approaches 10 percent or less of X_C.

Example 16.6 If a 0.05 μF capacitor is connected in series with a 3 kΩ resistor, what is the total impedance if the applied frequency is 2 kHz and 4 kHz?

$$X_{C_1} = \frac{1}{2\pi f_1 C} = \frac{1}{(6.28)(2)(10^3)(0.05)(10^{-6})} = 1.59 \text{ k}\Omega$$

$$|Z| = \sqrt{R^2 + X_{C_1}^2} = \sqrt{(3^2)(10^3)^2 + (1.59)^2(10^3)^2} = 3.4 \text{ k}\Omega$$

$$X_{C_2} = \frac{1}{2\pi f_2 C} = \frac{1}{(6.28)(4)(10^3)(0.05)(10^{-6})} = 800 \text{ ohms}$$

$$|Z| = \sqrt{R^2 + X_{C_2}^2} = \sqrt{(3^2)(10^3)^2 + 800^2} = 3.1 \text{ k}\Omega$$

In the circuit of Figure 16.4 note that the total impedance is equal to the vector sum of R and X_C. These values determine the voltage across each of these components using traditional Ohm's law calculations.

The current through each device can be determined from the applied voltage divided by the impedance:

$$I = V/|Z| \tag{16.11}$$

Similarly, the voltage across each component is

$$V_R = IR \qquad V_C = IX_C$$

Here, as in the L/R case, observe that the individual voltages do not algebraically add up to the applied voltage. The individual voltages must be added vectorially in order to satisfy Kirchhoff's law.

Phase Angle

Chapter 15 covered the concepts of polar and rectangular notation. The principles are exactly the same with the exception of one sign notation. The impedance of a series RC circuit in rectangular form is given as:

$$\boxed{Z = R - jX_C} \tag{16.12}$$

This same impedance can be represented in polar form.

$$Z = |Z| \angle \theta \tag{16.13}$$

where the $|Z|$ is given as $\sqrt{R^2 + X_C^2}$ and the angle $\theta = \tan^{-1}(-X_C/R)$.

Example 16.7 Assume a 1 kΩ resistance in series with 0.33 μF capacitor is connected across a 28 V, 400 Hz power source. What are the rectangular and polar representations of the impedance? What is the phase angle of the total current?

$$X_C = \frac{1}{2\pi f C} = \frac{1}{(6.28)(400)(0.33)(10^{-6})} = 1.2\text{ k}\Omega$$

$$Z = (1 - j1.2)\text{ k}\Omega \text{ (rectangular coordinates)}$$

$$Z = \sqrt{R^2 + X_C^2} = 1.566\text{ k}\Omega$$

$$\theta = \tan^{-1}\left(\frac{-X_C}{R}\right) = -50.32°$$

$$Z = 1.566\text{ k}\Omega \angle -50.32° \text{ (polar coordinates)}$$

$$I = \frac{V}{Z} = \frac{28 \angle 0°}{(1.566)(10^3) \angle -50.32°} = 17.88\text{ mA} \angle +50.32°$$

The transformation from polar to rectangular form is exactly as described in Chapter 15. Given the polar form of the impedance, the rectangular components of R and X_C are:

$$R = |Z| \cos \theta \tag{16.14}$$

$$X_C = |Z| \sin \theta \tag{16.15}$$

where θ is the known phase angle and $|Z|$ is the absolute value of the impedance.

16.3 SERIES *RC* CIRCUITS

The circuit of Figure 16.5(a) illustrates several resistors and capacitors connected in a series circuit. Any combination of *RC* components can be reduced to an equivalent series circuit which consists of one resistor and one capacitor. The basic equation that describes such a circuit is given as:

$$Z = (R_1 + R_2 + \cdots + R_N) - j(X_{C_1} + X_{C_2} + \cdots + X_{C_N})$$

This equation states that all of the real components can be added together to produce one resistive component, and all of the reactive components can be added to produce one reactive component.

The equivalent circuit of Figure 16.5(a) is shown in Figure 16.5(b).

The voltage across any individual component can be determined using voltage ratios. The voltage across any component N is the resistance or reactance of N, times the applied voltage, divided by the total impedance.

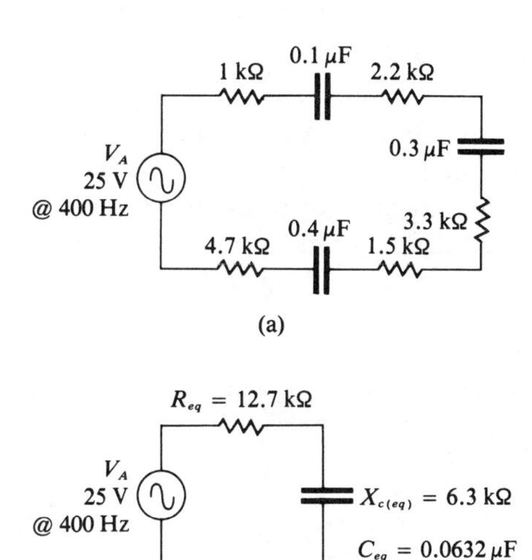

Figure 16.5 Series *RC* Circuit

$$V_N = \frac{R_N V_A}{R_{eq} - jX_{C_{eq}}} \tag{16.16a}$$

$$V_N = \frac{X_{C_N} V_A}{R_{eq} - jX_{C_{eq}}} \tag{16.16b}$$

By changing the denominator of Eq. (16.16a) or (16.16b) to polar form, the phase angle can also be determined. In the case of the reactance, the numerator must also be given in polar form. This principle is best illustrated using a numerical example.

Example 16.8 Refer to Figure 16.5(a). What is the voltage across the 2.2 kΩ resistor? What is the voltage across the 0.3 μF capacitor?

Using Eq. (16.16a),

$$Z_t = (12.7 - j6.3)\ \text{k}\Omega = 14.17\ \text{k}\Omega \angle -26.38^\circ$$

$$\begin{aligned} V_R &= \frac{25 \angle 0^\circ (2.2)(10^3) \angle 0^\circ}{(12.7 - j6.3)(10^3)} \\ &= \frac{55 \angle 0^\circ}{\sqrt{12.7^2 + 6.3^2} \angle \tan^{-1} 6.3/12.7} = \frac{55 \angle 0^\circ}{14.17 \angle -26.38^\circ} \\ &= 3.88\ \text{V} \angle 26.38^\circ \end{aligned}$$

$$X_C = \frac{1}{(6.28)(400)(0.3)(10^{-6})} = 1.326\ \text{k}\Omega \angle -90^\circ$$

$$\begin{aligned} V_C &= \frac{(1.326)(10^3) \angle -90^\circ (25) \angle 0^\circ}{(14.17)(10^3) \angle -26.38^\circ} \\ &= 2.34\ \text{V} \angle -63.62^\circ \end{aligned}$$

When two electronic devices or circuits are coupled together using a capacitor, a series equivalent circuit results. See Figure 16.6(a). Assume the source device is an ac generator whose internal resistance is 800 ohms. This source generates a 20 volt rms signal. To determine how much of the 20 volts is actually transferred to the load resistance, we must evaluate the equivalent series *RC* circuit.

Figure 16.6(b) illustrates the source voltage, the source resistance, the coupling capacitor, and the load resistance. The voltage delivered to the load is

$$V_L = \frac{R_L V_A}{(R_i + R_L) - jX_C} \tag{16.17a}$$

Equation (16.17a) shows that the amount of source voltage transferred to the load depends not only upon the ratio of resistors R_L and R_i, but upon the frequency of the applied signal as well. Note that as the frequency of

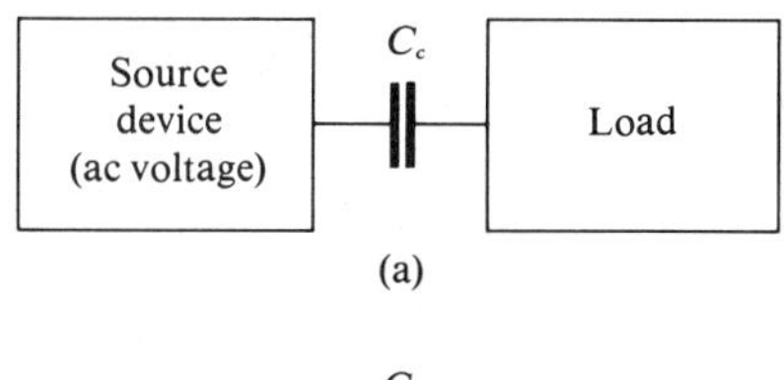

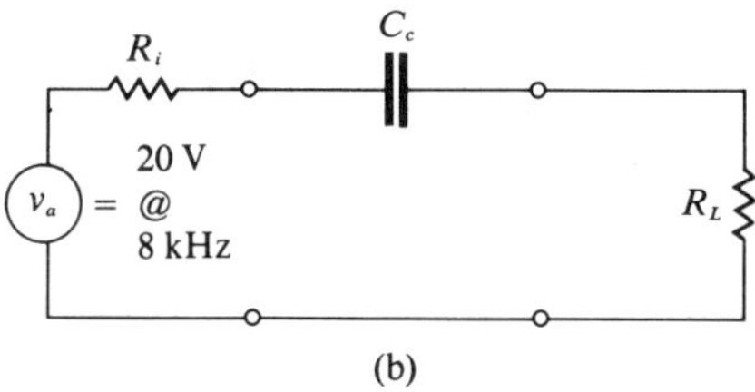

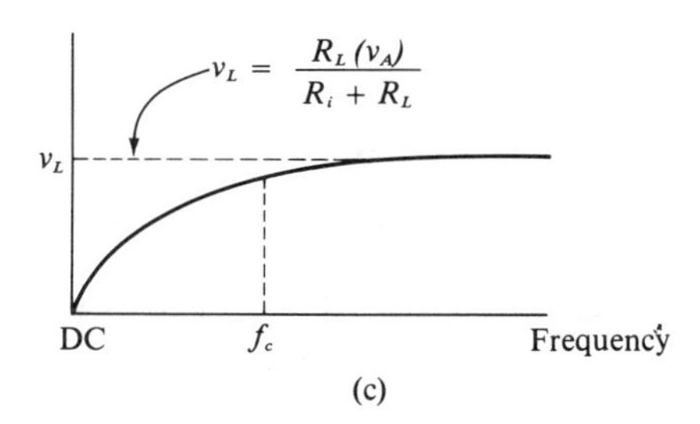

Figure 16.6 Capacitive Coupling

the signal becomes quite large, the value of X_C diminishes to a very small quantity. For large values of f, the X_C decreases to less than $0.1(R_i + R_L)$. The voltage delivered to the load becomes:

$$V_L = \frac{R_L V_A}{R_i + R_L} \tag{16.17b}$$

Conversely, if the frequency drops to a very low value, the capacitive reactance increases to an infinitely large value. The denominator of Eq. (16.17a) approaches an infinitely large value; hence the voltage delivered to the load is zero. Here again we see that a capacitor tends to block dc while passing alternating currents. Figure 16.6(c) illustrates how the voltage delivered to the load varies with frequency.

Above the frequency given as f_C, the voltage at the load depends primarily on the ratio of resistances.

Example 16.9 Refer to Figure 16.6. Assume $R_i = 0.8\ \text{k}\Omega$, $C_C = 0.02\ \mu\text{F}$, and $R_L = 2\ \text{k}\Omega$. What is the voltage delivered to the load when $f_1 = 1$ kHz, when $f_2 = 8$ kHz, and when $f_3 = 100$ kHz? What are the phase displacements?

$$\text{At } f_1, X_C = 7.96 \text{ k}\Omega$$
$$f_2, X_C = 995\ \Omega$$
$$f_3, X_C = 80\ \Omega$$

$$V_{L_1} = \frac{(2)(10^3)(20)}{(2.8)(10^3) - j(7.96)(10^3)} = \frac{40 \angle 0^\circ}{8.43 \angle -70.6^\circ} = 4.74 \angle 70.6^\circ$$

$$V_{L_2} = \frac{(2)(10^3)(20)}{(2.8)(10^3) - j(.995)(10^3)} = \frac{40 \angle 0^\circ}{2.97 \angle -19.56^\circ} = 13.46 \angle 19.56^\circ$$

$$V_{L_3} = \frac{(2)(10^3)(20)}{(2.8)(10^3) - j(0.08)(10^3)} = \frac{40 \angle 0^\circ}{2.8 \angle -1.6^\circ} = 14.28 \angle 1.6^\circ$$

16.4 PARALLEL *RC* CIRCUITS

Many configurations of *RC* circuits involve the parallel combination of two or more *RC* components. We have already seen that the total impedance of a *series* combination of *R* and *C* can be determined using Eq. (16.10). Any parallel *RC* circuit can be reduced to an equivalent series *RC* circuit.

Assume that the parallel circuit of Figure 16.7 has a given applied voltage v at a fixed frequency f. The capacitive reactance of C can be determined. The total impedance, and hence the equivalent R_{eq} and $X_{C_{eq}}$, is

$$Z = \frac{V \angle 0^\circ}{I_t \angle \phi}$$

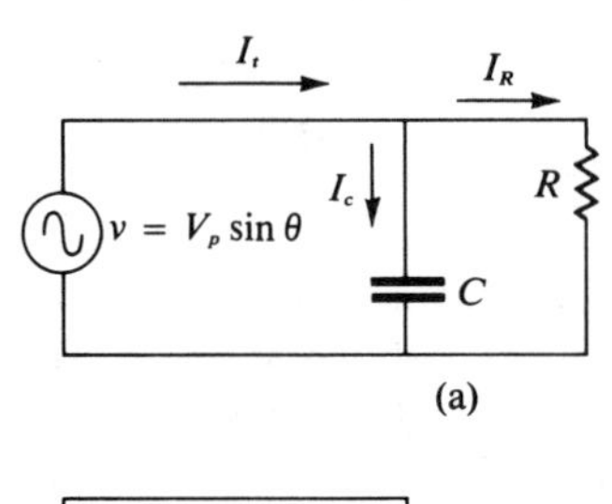

(a)

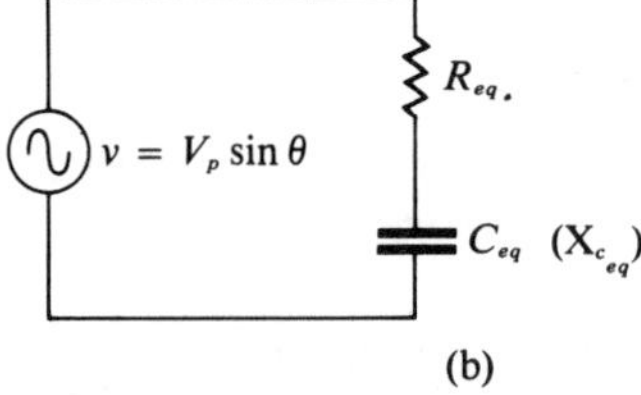

(b)

Figure 16.7 Parallel *RC* Circuit

The total current is the vector sum of the individual currents through each of the branches.

$$I_t = I_C + I_R$$

The voltage across each branch is V; hence,

$$I_C = \frac{V}{-jX_C} = +j\frac{V}{X_C}$$

$$I_R = \frac{V}{R}$$

$$I_t = \frac{V}{R} + j\frac{V}{X_C}$$

If this value of I_t is converted to its polar equivalent, it becomes $I_t \angle\phi$. The equivalent series circuit has

$$R_{eq} = \frac{V}{I_t}\cos\phi \qquad X_{C_{eq}} = \frac{V}{I_t}\sin\phi \tag{16.18}$$

An alternate method for solving for R_{eq} and $X_{C_{eq}}$ directly involves solving the parallel circuit of Figure 16.7(a). Since R and C are in parallel,

$$Z = \frac{R(-jX_C)}{R - jX_C}$$

Multiplying numerator and denominator by the conjugate of the denominator,

$$Z = \frac{R(-jX_C)}{(R - jX_C)}\,\frac{(R + jX_C)}{(R + jX_C)} = \frac{-jRX_C(R + jX_C)}{R^2 + X_C^2}$$

Multiplying out the numerator,

$$Z = \frac{-jR^2X_C + RX_C^2}{R^2 + X_C^2}$$

Separating the real and the j terms,

$$Z = \frac{RX_C^2}{R^2 + X_C^2} - j\frac{R^2X_C}{R^2 + X_C^2}$$

Simplifying,

$$Z = \frac{R}{(R/X_C)^2 + 1} - j\frac{X_C}{(X_C/R)^2 + 1} \tag{16.19}$$

The first term represents the equivalent resistance R_{eq} and the j term represents the equivalent capacitive reactance, $X_{C_{eq}}$. Figure 16.7(b) illustrates the equivalent series circuit.

Example 16.10 What is the equivalent series circuit for a 2 kΩ resistor paralleled by a capacitor whose reactance is 4 kΩ at 8 kHz? What is the equivalent capacity?

$$R_{eq} = \frac{(2)(10^3)}{(2/4)^2 + 1} = 1.6 \text{ k}\Omega$$

$$X_{C_{eq}} = \frac{(4)(10^3)}{(4/2)^2 + 1} = 0.8 \text{ k}\Omega$$

$$C = \frac{1}{2\pi f X_{C_{eq}}} = \frac{1}{(6.28)(8)(10^3)(0.8)(10^3)}$$
$$= 0.025 \ \mu\text{F}$$

Which of the two methods outlined should be used to solve for the equivalent RC circuit depends upon the original circuit. For example, the circuit of Figure 16.8(a) consists of four parallel branches. The easiest method is to sum the individual currents and solve for the total impedance.

The circuit of Figure 16.8(b) consists of a series-parallel combination. Here the voltage across the two parallel components is not known, hence the equivalent circuit approach of Eq. (16.19) best suits the calculation.

Example 16.11 What is the phase angle of the current leaving the generator of Figure 16.8(b). Assume $R_1 = R_2 = 1$ kΩ, $C_1 = C_2 = 0.1$ μF.

$$X_{C_1} = X_{C_2} = \frac{1}{(6.28)(2)(10^3)(0.1)(10^{-6})} = 800 \text{ ohms}$$

$$R_{eq} = \frac{10^3}{(1/0.8)^2 + 1} = 390 \text{ ohms}$$

$$X_{C_{eq}} = \frac{800}{(0.8/1)^2 + 1} = 488 \text{ ohms}$$

$$Z = (1 + 0.39) \text{ k}\Omega - j(0.8 + 0.488) \text{ k}\Omega$$
$$= (1.39 - j1.288) \text{ k}\Omega$$
$$= 1.895 \text{ k}\Omega \underline{/-42.82^\circ}$$

Example 16.12 Refer to the configuration of Figure 16.8(b). Assume $R_1 = R_2 = 1$ kΩ, $C_1 = C_2 = 0.1$ μF, $V = 25$ V. Write a program in BASIC to calculate and print out f, I_{R_1}, V_{C_1}, and Z_t as f varies from 100 Hz to 32.5 kHz. To facilitate graphing, break down the frequency into 100 Hz increments to 1 kHz and then increment in 1500 Hz steps.

Printout for Example 16.12

```
10   P1=3.14159 : R1=1E3 : R2=1E3 : C1=1E-7 : C2=1E-7
12   PRINT "F IN HZ", "IR1 IN MA", "VC1 IN VOLTS", "Z IN OHMS"
15   PRINT
20   FOR F=100 TO 900 STEP 100 : GOSUB 40 : NEXT F
25   FOR F=1000 TO 32500 STEP 1500 : GOSUB 40 : NEXT F
30   STOP
40   X1=1/(2*P1*F*C1) : X2=X1/((X1/R1)↑2+1) : X=X1+X2
45   R2=R1/((R1/X1)↑2+1) : R=R1+R2
50   Z=SQR(R↑2+X↑2) : I=25E3/Z : V=I*X1/1E3
55   PRINT F , I , V , Z
60   RETURN
65   END
```

F IN HZ	IR1 IN MA	VC1 IN VOLTS	Z IN OHMS
100	1.55257	24.71	16102.3
200	3.00425	23.9071	8321.54
300	4.28915	22.7547	5828.66
400	5.38733	21.4355	4640.52
500	6.31285	20.0944	3960.18
600	7.09551	18.8215	3523.35
700	7.76749	17.6605	3218.54
800	8.35687	16.6255	2991.55
900	8.88576	15.7135	2813.49
1000	9.37058	14.9138	2667.92
2500	14.8158	9.43202	1687.39
4000	18.4218	7.32979	1357.09
5500	20.6192	5.96663	1212.46
7000	21.9479	4.99017	1139.06
8500	22.7799	4.26533	1097.46
10000	23.3243	3.71217	1071.85
11500	23.6957	3.27938	1055.05
13000	23.9586	2.93318	1043.47
14500	24.1507	2.65083	1035.17
16000	24.2949	2.41667	1029.02
17500	24.4058	2.2196	1024.35
19000	24.4927	2.05165	1020.71
20500	24.562	1.90691	1017.83
22000	24.6181	1.78095	1015.51
23500	24.6642	1.6704	1013.62
25000	24.7024	1.57261	1012.05
26500	24.7346	1.48552	1010.73
28000	24.7618	1.40749	1009.62
29500	24.785	1.33717	1008.67
31000	24.8051	1.2735	1007.86
32500	24.8224	1.21557	1007.15

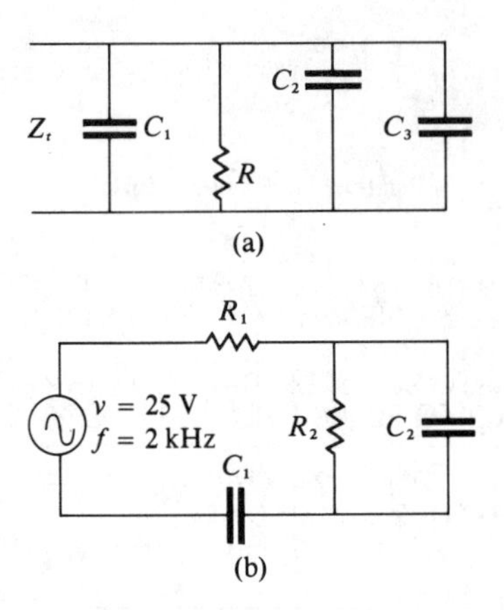

Figure 16.8 Parallel Circuits

16.5 INSTANTANEOUS VALUES OF AN *RC* CIRCUIT

We noted that in a purely capacitive circuit, the current leads the voltage by 90°. This leading current phase angle can vary, depending upon how much resistance is in the circuit and the frequency of the applied voltage. Since the frequency affects the phase angle and the total impedance, the current will be affected as well. It is often necessary to identify the instantaneous current in a given *RC* circuit, as well as the instantaneous voltage at any point in the circuit.

The applied voltage is generally assumed as a reference. Equation (15.21) is repeated here for convenience.

$$v(t) = V_p \sin \theta$$

where θ is $2\pi ft$ and $v(t)$ is the voltage at the given instant. We are able to calculate θ at any instant of time t by substituting t in $2\pi ft$. The frequency also enables us to calculate X_C. Knowing X_C, we can determine the phase angle of the leading current. This phase angle is ϕ. The instantaneous value of the circuit current becomes

$$i(t) = I_p \sin (\theta + \phi) \tag{16.20}$$

Equation (16.20) illustrates that for any instant within the applied voltage cycle the circuit current can be determined. For example, the instantaneous current in an *RC* circuit after the voltage has completed half a period ($t = \pi$) is

$$i(t) = -I_p \sin \phi$$

which is the peak current times the sine of the leading phase angle ϕ.

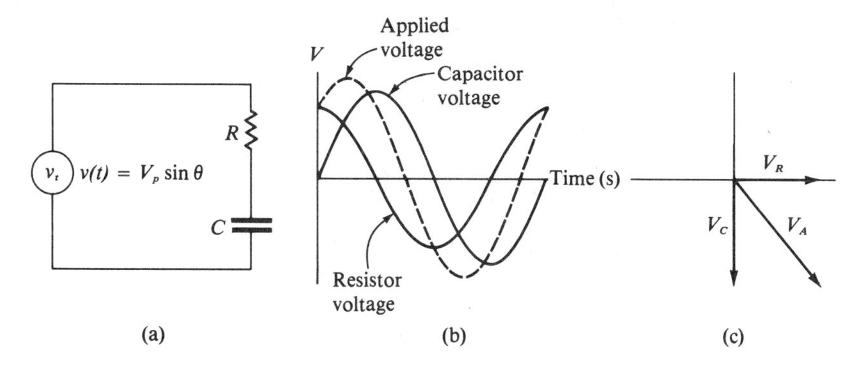

Figure 16.9 Phase Displacement

The instantaneous voltage across the resistor can be found by using Ohm's law:

$$v_R = i(t)R$$

Similarly, the voltage across C can be determined using:

$$v_C = i(t)X_C$$

However, the voltage across a capacitor lags the current through it by 90°. The phase displacement between the capacitor and the resistor voltages will be 90°. Figure 16.9 illustrates this phase displacement.

Example 16.13 Assume the series circuit of Figure 16.9(a) has a resistance of 3.3 kΩ. The capacitance C is 0.02 μF and the applied voltage is 50 volts at 2 kHz. (a) What is the current in the circuit when the input voltage is at 0.1 ms into the period? (b) What is the voltage across R? (c) What is the voltage across C?

a. $$X_C = \frac{1}{2\pi fC} = \frac{1}{(6.28)(2)(10^3)(0.02)(10^{-6})} = 4 \text{ k}\Omega$$

$$\theta = 2\pi ft = (360°)(2)(10^3)(0.1)(10^{-3}) = 72°$$

$$\phi = \tan^{-1}\left(\frac{-X_C}{R}\right) = \tan^{-1}\left(\frac{(4)(10^3)}{(3.3)(10^3)}\right) = -50.48°$$

$$|Z| = \sqrt{R^2 + X_C^2} = \sqrt{(3.3)^2(10^3)^2 + (4)^2(10^3)^2} = 5.186 \text{ k}\Omega$$

$$Z = 5.186 \text{ k}\Omega \angle -50.48°$$

$$I_p = \frac{V_p}{Z} = \frac{(50)(1.414)}{(5.186)(10^3) \angle -50.48°} = 13.63 \text{ mA} \angle 50.48°$$

From Eq. (16.20),

$$i(t) = (13.63)(10^{-3}) \sin (\theta + 50.48°)$$
$$i(t) @ t = 0.1 \text{ ms} \qquad (\theta = 72°)$$
$$i(t) = (13.63)(10^{-3}) \sin (122.48°)$$
$$= 11.5 \text{ mA}$$

b. $v_R = (11.5)(10^{-3})(3.3)(10^3) = 37.94$ V

c. $v_C = [(11.5)(10^{-3}) \sin (122.48° - 90)](4)(10^3)$
$= 24.7$ volts

The waveshapes of v_C and v_R in Figure 16.9(b) illustrate that the peaks are displaced by 90°. We could solve graphically for the input voltage at a time when the voltage across the resistor is maximum and the voltage across the capacitor is minimum. The time can be related to radians or degrees by the same relationship as in L/R circuits. Equation (15.23) shows that the period corresponds to one cycle (360°) by the same ratio as the instantaneous time (t) corresponds to the lead phase angle.

In many electronic circuits it is often necessary to determine a phase angle displacement in advance. For example, in an integrated circuit Wein-bridge oscillator, one leg of the circuit must displace the current by 60°. Assume the circuit of Figure 16.10 is one leg of an RC oscillator circuit. The voltage across R will be phase displaced from the applied voltage. This phase displacement and the current can be tailored to any desired frequency.

Example 16.14 The circuit of Figure 16.10 is one leg of an oscillator which must operate at 8 kHz. If R is 1 kΩ and the applied voltage is 25 volts rms, at what time t of the applied voltage cycle will the instantaneous voltage across R be 16 volts?

$$i_R = \frac{V_R}{R} = \frac{16}{10^3} = 16 \text{ mA}$$

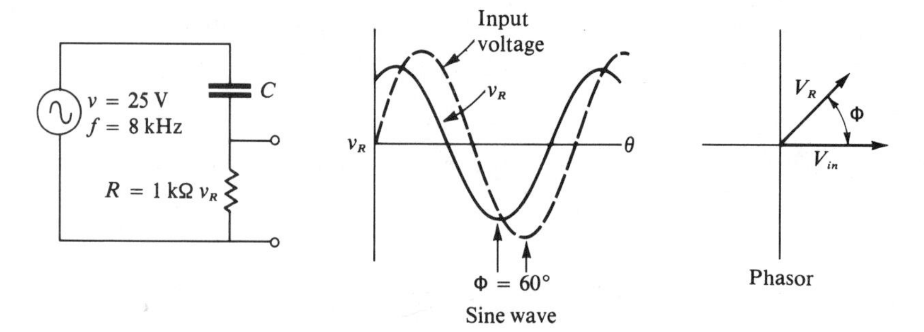

Figure 16.10 Figure for Example 16.14

According to Eq. (16.14)

$$|Z| = \frac{R}{\cos 60^\circ} = \frac{10^3}{0.5} = 2\ \text{k}\Omega$$

$$|I_p| = \frac{|V_p|}{|Z|} = \frac{(25)(1.414)}{(2)(10^3)} = 17.67\ \text{mA}$$

$$i(t) = I_p \sin(2\pi ft + 60^\circ)$$

$$\frac{16}{17.67} = \sin(2\pi ft + 60^\circ)$$

$$\sin^{-1}\left(\frac{16}{17.67}\right) = 2\pi ft + 60^\circ$$

$$64.89^\circ = 2\pi ft + 60^\circ$$

$$64.89^\circ - 60^\circ = 2\pi ft$$

$$4.89^\circ = (360^\circ)(8)(10^3)t$$

$$t = \frac{4.89^\circ}{(360^\circ)(8)(10^3)} = 1.7\ \mu\text{s}$$

16.6 POWER IN *RC* CIRCUITS

A pure capacitor, like a pure inductor, does not convert energy to heat, light, or any other form. It merely stores the energy delivered to it. This energy is returned to the circuit. The energy delivered to the plates of a capacitor is stored in the electric field developed across the dielectric.

Consider now the power relationships of circuits that contain both resistance and capacitance. Not all the energy delivered to such a circuit is converted to another form such as heat, nor is all of it stored. Only a percentage is converted to heat. We can assess what that percentage is using the circuit's power relationships. These relationships are very similar to those already encountered in L/R circuits. Since $p = iv$, these instantaneous values can be substituted: $i(t) = I_p \sin(\theta + \phi)$ and $v(t) = V_p \sin\theta$.

$$p = I_p V_p \sin\theta \sin(\theta + \phi) \tag{16.21}$$

where ϕ is the phase angle of the leading current. By substituting trigonometric identities and solving for the average power:

$$P_{\text{ave}} = \frac{V_p I_p}{2} \cos\phi \tag{16.22}$$

Since the phase angle for a pure capacitive circuit is 90°, Eq. (16.22) is simplified to a product of the apparent power and the power factor (cos ϕ). Cos ϕ, the power factor, is the percentage of apparent power that is converted to heat.

Example 16.15 If a series RC circuit consists of 2.2 kΩ and an 0.08 μF capacitor, what are the apparent power, the true power, and the power factor if the applied voltage is 50 volts at 2 kHz?

$$X_C = \frac{1}{2\pi fC} = \frac{1}{(6.28)(2)(10^3)(0.08)(10^{-6})} = 995 \text{ ohms}$$

$$\theta = \tan^{-1}\left(\frac{-X_C}{R}\right) = -24.34° \qquad |Z| = \sqrt{R^2 + X_C^2} = 2.415 \text{ k}\Omega$$

$$|I| = \frac{|V|}{|Z|} = \frac{50}{(2.415)(10^3)} = 20.7 \text{ mA}$$

$$P_A = IV = (20.7)(10^{-3})(50) = 1.04 \text{ volt-amps}$$

$$P_t = P_A(\text{PF}) = (1.04)(\cos 24.34°) = 0.943 \text{ watt.}$$

16.7 COMPLEX CIRCUIT ANALYSIS

Prior to the advent of the modern digital computer, complex circuit analysis was a time-consuming operation. Often an important concept became clear only after many hours of solving a circuit for one set of conditions and then sequentially solving for other conditions. Today, we need only know the discipline of writing problems in BASIC and the resultant printouts give us immediate insights to important concepts. The following example illustrates how significant circuit concepts can be gleaned from a computer printout.

Example 16.16 Refer to Figure 16.11. Assume R_1 varies from 100 ohms to 4 kΩ in increments of 100 ohms. Write a program in BASIC to calculate the peak-to-peak voltage between points A and B for each value of R_1. Print out V_A (both magnitude and phase angle), and V_{AB} (both magnitude and phase angle).

The reader should immediately recognize that this would be a difficult task (especially considering the complexity of ac circuit analysis) if done manually for each increment of R_1. Looping through a computer program, however, reduces the total time to merely writing the programming steps. There is another attribute to programming that the reader should be developing, namely, an "intuitive feeling" for the approximate range or direction of the answers. This problem is especially appropriate to illustrate this idea.

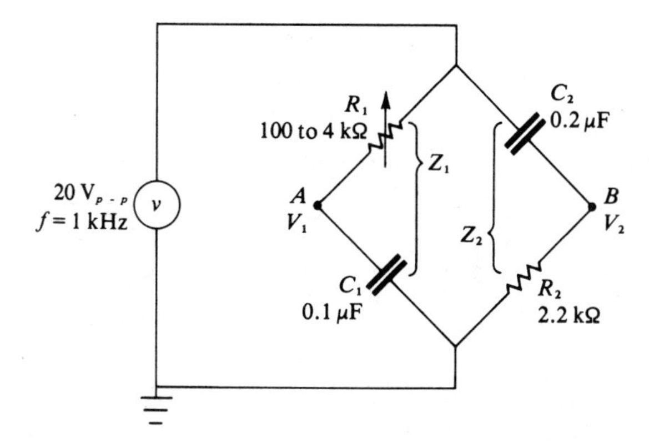

Figure 16.11 Figure for Example 16.16

In studying Figure 16.11, it is seen that V_B (p-p) remains constant, because the voltage input magnitude and input frequency remain constant. Only V_A varies. Since X_{C_1} is in series with R_1, V_A decreases across C_1 as R_1 increases. Since $V_{AB} = V_A - V_B$, V_{AB} should approach and then exceed the constant value of V_B, as R_1 increases. The same is true for the phase angle of V_{AB}. It should also be approaching the constant phase angle of V_B with respect to the reference voltage (V). Since V_{AB} has V_B as one of its vectors, V_{AB} will eventually exceed V_B and its phase angle will approach (but always be greater than) the phase angle of V_B.

A single, manual calculation results in the vector V_B (a constant) being equal to 18.81 $\angle 19.89°$. The following summary may now be made concerning each "educated approximation."

a. V_A starts out reasonably large (approximately equal to V_B) and begins to diminish.
b. Because of the capacitive phase shift, the phase angle of V_B starts out at a small negative value and becomes increasingly negative.
c. V_{AB} starts out larger than V_B, but approaches 18.81 volts (V_B) as a limiting value.
d. The phase angle of V_{AB} starts out larger than the phase angle of V_B, but approaches 19.89° (the phase angle of V_B) as a limiting value.

Note that all the phase angles are with respect to the applied input voltage (20 $\angle 0°$). Observe how the computer printout substantiates these expected results.

Printout for Example 16.16

```
10  P1=3.14159 : C1=1E-7 : C2=2E-7 : R2=2.2E3 : F=1E3 : V=20
15  PRINT "R1 (OHMS)" TAB(11) " V(A)" TAB(24) "ANGLE(A)" ;
20  PRINT TAB(36) " V(AB)" TAB(48) "ANGLE(AB)" : PRINT
25  X1=1/(2*P1*F*C1) : X2=1/(2*P1*F*C2)
30  FOR R1=100 TO 4E3 STEP 100
35  Z1=SQR(R1↑2+X1↑2) : A1=ATN(X1/R1) : V1=V*X1/Z1: A=-P1/2+A1
40  Z2=SQR(R2↑2+X2↑2) : A2=ATN(X2/R2) : V2=V*R2/Z2
45  G1=V1*COS(A) : I1=V1*SIN(A)
50  G2=V2*COS(A2) : I2=V2*SIN(A2)
55  G=G2-G1 : I=I2-I1 : J=SQR(G↑2+I↑2) : K=ATN(I/G)*180/P1
60  IF G2=>G1 THEN 70
65  K=K+180
70  PRINT R1 ; TAB(11) ; V1 ; TAB(24) ; A*(180/P1) ;
75  PRINT TAB(36) ; J ; TAB(48) ; K
80  NEXT R1
85  PRINT : PRINT "V2 IS " V2 " VOLTS"
90  PRINT "AT AN ANGLE OF " A2*(180/P1) " DEGREES"
95  END
```

R1 (OHMS)	V(A)	ANGLE(A)	V(AB)	ANGLE(AB)
100	19.9606	−3.59519	7.96892	106.291
200	19.8439	−7.16238	9.09481	102.723
300	19.6539	−10.6747	10.169	99.211
400	19.3968	−14.1077	11.182	95.7781
500	19.0806	−17.4405	12.1271	92.4452
600	18.7143	−20.6559	13	89.23
700	18.3075	−23.7409	13.7992	86.145
800	17.8695	−26.6865	14.5249	83.1994
900	17.4092	−29.4875	15.1794	80.3984
1000	16.9347	−32.1418	15.7662	77.7441
1100	16.4528	−34.6502	16.2896	75.2357
1200	15.9694	−37.0156	16.7546	72.8704
1300	15.4895	−39.2424	17.1664	70.6435
1400	15.0169	−41.3363	17.5298	68.5496
1500	14.5546	−43.3037	17.8501	66.5822
1600	14.1047	−45.1516	18.1317	64.7343
1700	13.6688	−46.887	18.379	62.9988
1800	13.2479	−48.517	18.5959	61.3689
1900	12.8428	−50.0485	18.786	59.8375
2000	12.4535	−51.4881	18.9525	58.3978
2100	12.0802	−52.8422	19.0981	57.0437
2200	11.7227	−54.1168	19.2255	55.7691
2300	11.3805	−55.3176	19.3368	54.5683
2400	11.0533	−56.4498	19.4339	53.4361
2500	10.7406	−57.5183	19.5187	52.3676
2600	10.4417	−58.5277	19.5925	51.3582
2700	10.1561	−59.4822	19.6567	50.4037
2800	9.8832	−60.3856	19.7124	49.5003
2900	9.62236	−61.2415	19.7607	48.6444
3000	9.37301	−62.0533	19.8024	47.8327
3100	9.13454	−62.8239	19.8383	47.062
3200	8.90643	−63.5561	19.8691	46.3298
3300	8.68811	−64.2525	19.8954	45.6334
3400	8.47907	−64.9155	19.9177	44.9704
3500	8.27882	−65.5473	19.9365	44.3386
3600	8.08691	−66.1499	19.9521	43.736

3700	7.90286	−66.7251	19.965	43.1608
3800	7.72629	−67.2746	19.9754	42.6113
3900	7.55678	−67.8001	19.9837	42.0858
4000	7.39397	−68.303	19.99	41.5829

```
V2 IS  18.8074  VOLTS
AT AN ANGLE OF  19.8859  DEGREES
```

SUMMARY

A capacitor offers an opposition to alternating current identified as capacitive reactance. The reactance is inversely proportional to the frequency of the applied voltage. In ac series circuits this reactance is additive to other capacitive reactances. In parallel these X_C's can be combined similar to resistors in parallel.

A capacitance introduces a phase angle into a circuit. The current leads the voltage by 90° in a purely capacitive circuit. When resistance is introduced in series with the capacitor, the phase angle varies depending upon the ratio of capacitive reactance and resistance.

Parallel *RC* circuits can be reduced to an equivalent series *RC* circuit. If the capacitive reactance is known, the equivalent series X_C is determined by dividing the X_C by one plus the square of the ratio of X_C to R. By a similar method, the equivalent series R can be determined.

The instantaneous value of current and voltage in an *RC* circuit can be determined for an instant within a given cycle. The instantaneous voltage depends upon the frequency and the phase angle to some reference.

Series, parallel, and series-parallel *RC* circuits can be solved using the simplifications resulting from polar-to-rectangular and rectangular-to-polar transformations.

RC circuits may be simplified and analyzed using voltage ratio and current ratio techniques. When the ratio methods are used, care must be taken to effect proper polar and rectangular transformations.

In complex circuits, where several functions may be varied, a computer solution provides the easiest methods. In circuits where the applied frequency, voltage, and resistance can vary, their effects on angle, reactance, and impedance can be solved using BASIC.

PROBLEMS

Reference Section 16.1

1. How much capacitive reactance will a 0.05 μF capacitor offer to 60 Hz? How much to 120 Hz?
2. How much opposition to ac will the capacitor of Problem 1 offer to 1600 Hz? How much to 10 MHz?
3. If a 0.1 μF capacitor offers 1600 ohms of opposition at 1 kHz, how

much change in C would be required to maintain the reactance at 1600 ohms if the frequency were decreased to 60 Hz?

4. A capacitor offers 800 ohms of opposition to 20 volts ac at 600 Hz. What is its capacity? What is the current flow?
5. When 10 volts peak-to-peak at 1000 Hz is applied across a capacitor of 0.05 μF, what rms current will flow?
6. Two capacitors, 0.03 μF and 0.1 μF, are connected in parallel. When 25 volts at 400 Hz is connected across the combination, what current will flow through the 0.1 μF capacitor?
7. Refer to Problem 6. Without working the problem, decide if the current through the 0.03 μF capacitor will be more, less, or the same as the current through the 0.1 μF capacitor.
8. The two capacitors of Problem 6 are connected in series and the same voltage is applied across the circuit. What is the total current in the circuit?
9. Refer to the circuit of Figure 16.12. What will be the voltage across the 0.3 μF capacitor?
10. Refer to the circuit of Figure 16.13. What current will flow through C_3 if $C_1 = 0.1\ \mu$F, $C_2 = C_3 = C_4 = 0.2\ \mu$F, $C_5 = 0.05\ \mu$F, and $V = 10$ volts rms at 16 kHz?
11. Refer to Problem 10. What is the voltage drop across C_1? What is the current through C_2?

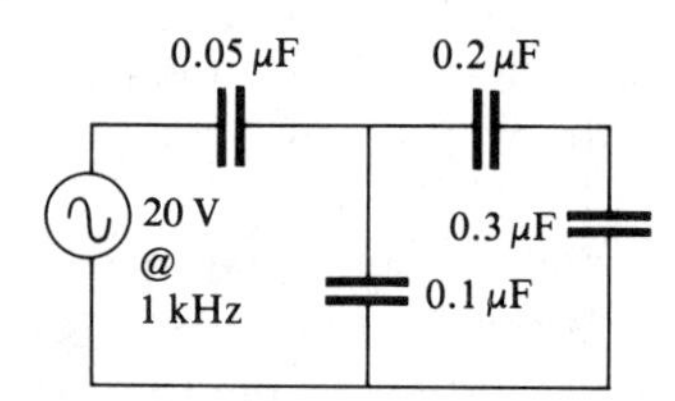

Figure 16.12 Figure for Problem 9

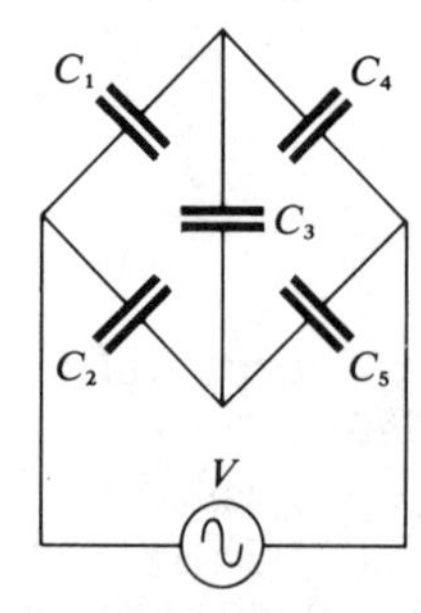

Figure 16.13 Figure for Problems 16.10 and 16.11

Reference Section 16.2

12. If a series RC circuit consists of a 60 kΩ resistor and a capacitive reactance of 80 kΩ, what is the total impedance?
13. An RC circuit consists of $R = 100$ kΩ, $X_C = 155$ kΩ. What is the total impedance? What is the Z if $R = X_C = 100$ kΩ?
14. A series circuit has an impedance of 98.4 kΩ. If R is 12 kΩ, what is X_C? What is the phase angle?
15. By how many degrees does the current lead the applied voltage if a series circuit consists of an R and a C where $R = X_C$?
16. The impedance of a circuit is given as $30.15\ \text{k}\Omega \angle -81.5^\circ$. What are the resistance and reactance?
17. Refer to Problem 16. If the applied frequency is 8 kHz, what is the value of C?
18. An RC circuit is to develop 10 kΩ of impedance to a 2 kHz sine wave. If R is 4.7 kΩ, what is the value of C?
19. If a 35 μF capacitor is connected in series with a 200 ohm resistor and 20 volts at 50 Hz is applied:
    a. What is the total impedance of the circuit?
    b. What is the total current flow and what is its phase angle?
    c. What are the voltage and phase angle across the resistor?
    d. What are the voltage and phase angle across the capacitor?

Reference Section 16.3

20. A given circuit consists of two resistors, 68 kΩ and 30 kΩ in series, with a capacitive reactance of 12 kΩ (20 volts rms is applied).
    a. What is the total impedance?
    b. What is the voltage across the 68 kΩ resistor?
21. A given RC circuit has an impedance of 150 ohms at 60°. What current will flow when a 50 volt 400 Hz voltage is applied to the circuit?
22. Refer to Problem 21. What value of capacity should be added in parallel with the circuit in order to cause the current to lead the applied voltage by 30°?
23. Refer to Problem 21. What type of circuit must be added in series with the circuit to cause the total current leaving the generator to lead the applied voltage by 15° and have a magnitude of 0.1 ampere?
24. A series circuit consists of a 0.05 μF capacitor and a resistance of 3200 ohms. Its current leads the applied voltage by 50°. What is the applied frequency?
25. If, in a series ac circuit, the voltage across the resistor is 60 volts and the voltage across the capacitor is 80 volts, what is the phase angle of the leading current? What is the applied voltage?
26. In a series circuit the resistance is 50 ohms and the capacitive reactance is 40 ohms. With 100 volts applied, find the phase angle change when R varies from 50 to 100 ohms.

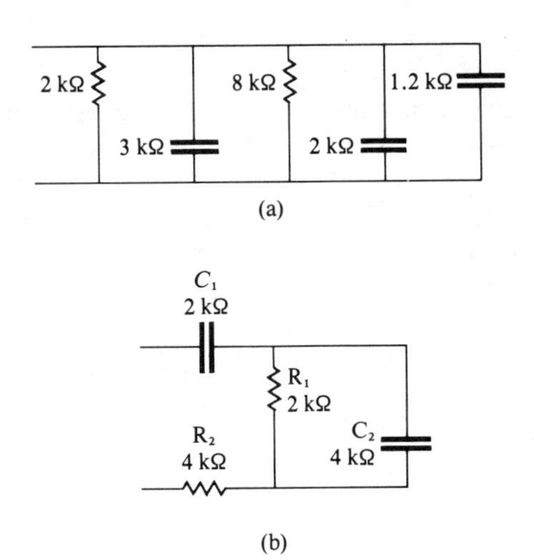

Figure 16.14 Figure for Problems 28 through 30

Reference Section 16.4

27. A resistor and a capacitor are connected in parallel with the applied voltage 20 volts. What would be the total impedance of the circuit and by how many degrees would the current lead the applied voltage if $R = 1$ kΩ and $X_C = 2.4$ kΩ?
28. Refer to the circuit of Figure 16.14(a). What is the total impedance of the circuit?
29. What is the total impedance of the *RC* circuit of Figure 16.14(b)?
30. Refer to Figure 16.14(b). If a capacitor whose reactance is 4 kΩ is connected in parallel with R_2, what is the total impedance of the circuit?
31. A parallel *RC* circuit will have what equivalent series circuits if the applied frequency approaches zero? If the applied frequency approaches an infinitely large value?
32. A 100 ohm resistor and a capacitor, connected in series, have a voltage of 50 volts at 60 Hz applied to the circuit. The voltage across the resistor is 40 volts when the voltage across the capacitor is 25 volts. What is the capacitance and what is the phase angle?

Reference Section 16.5

33. Assume a series *RC* circuit has phase angle of 42° and *R* is 2.2 kΩ. The applied voltage is 30 V(rms) at 4000 Hz. If $C = 0.02\ \mu$F, what is the instantaneous current when the voltage has completed 100° of its cycle?

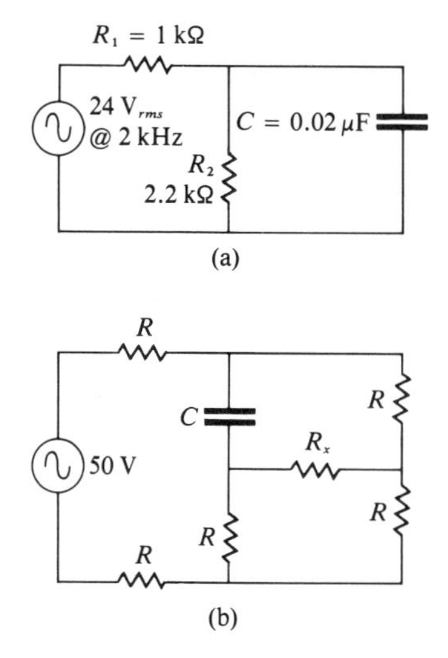

Figure 16.15 Figure for Problems 35 and 36

34. Refer to Problem 33. At what time t of the voltage cycle will the voltage across R be 5 volts? At what time will the voltage be 10 volts?
35. The circuit of Figure 16.15(a) consists of two resistances. What is the current through R_2 when the input voltage has completed 20° of its cycle?
36. Refer to Figure 16.15(a). By how many degrees does the current through C lead or lag the current through R_1?
37. Assume the resistances in the circuit of Figure 16.15(b) are all 1 kΩ. If $C = 0.2\ \mu F$, what value of f is required for the total current of 17.7 mA to lead the applied voltage by 8.5°?
38. Refer to Problem 37. Will the voltage across R_x lead or lag the applied voltage?

Optional Computer Problems

39. Refer to Figure 16.15(b). Assume $C = 0.02\ \mu F$ and all resistors are 1 kΩ. For low frequencies, the voltage phase angle across R_x is large and decreases toward zero degrees as frequency increases. Determine that frequency at which the phase angle of v_x is 25 degrees. Print out a table of F and the phase angle across R_x. Use successively tighter iterative loops, starting upward from 10 kHz in 10 kHz steps.
40. Refer to Figure 16.15(a). With the help of a computer printout, plot the voltage waveshape across R_1 with respect to the applied voltage as C varies from 1000 pF to 0.2 μF.

CHAPTER 17

AC NETWORKS

17.1 SERIES *RCL* CIRCUITS

Complex plane notation with varying frequency conditions leads to fairly cumbersome methods of circuit analysis. To aid in the analysis of various circuits, the concepts of conductance, admittance, and susceptance are introduced. The chapter then goes on to explore how the concepts of network theorems can be applied to ac circuit analysis. Often several orders of simplification can be achieved by applying a basic theorem to the analysis. The chapter concludes with a discussion of series and parallel resonance. Not only are the basic concepts of resonance discussed, but the factors influencing resonance are included.

Chapters 15 and 16 dealt with circuit configurations resulting in either a resistive-inductive or a resistive-capacitive circuit. Most electric circuit configurations include all three circuit components. The analysis of such circuits is not much different from those already outlined. The total impedance of a series *RCL* circuit is the sum of the individual impedances of the circuit.

Figure 17.1(a) illustrates a series circuit with several *RCL* components connected to an ac source. The individual impedances are R_1, $-jX_{C_1}$, R_2, $+jX_{L_1}$, R_3, R_4, $+jX_{L_2}$, and $-jX_{C_2}$. Since the total impedance is

$$Z_t = Z_1 + Z_2 + \cdots + Z_N \tag{17.1}$$

the equation for the circuit of Figure 17.1(a) becomes

$$\begin{aligned} Z_t &= R_1 + R_2 + R_3 + R_4 + jX_{L_1} + jX_{L_2} - jX_{C_1} - jX_{C_2} \\ &= (R_1 + R_2 + R_3 + R_4) + j(X_{L_1} + X_{L_2} - X_{C_1} - X_{C_2}) \\ &= R_{eq} \pm jX_{eq} \end{aligned}$$

Note that the circuit reduces to a series equivalent circuit whose resistance is the sum of the individual resistances and whose reactance is the algebraic sum of the capacitive and inductive reactances. The equivalent circuit is shown in Figure 17.1(b). From this circuit the phase angle of the original circuit can be determined.

Example 17.1 A series *RCL* circuit consists of a 2.2 kΩ resistor, a 0.05 μF capacitor, and a 400 mH coil. When 25 volts at 2 kHz is applied:

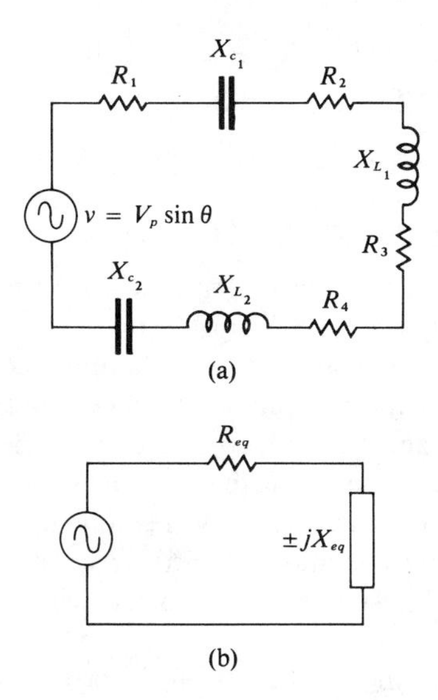

Figure 17.1 Series *RCL* Circuit

(a) What is the equivalent circuit?
(b) What is the phase angle of the current?
(c) What is the voltage across R, L, and C?

a. $X_L = 2\pi fL = (6.28)(2)(10^3)(0.4) = 5.027\ \text{k}\Omega$

$$X_C = \frac{1}{2\pi fC} = \frac{1}{(6.28)(2)(10^3)(0.05)(10^{-6})} = 1.592\ \text{k}\Omega$$

$$X_{\text{eq}} = +j5.027\ \text{k}\Omega - j1.592\ \text{k}\Omega = +j3.435\ \text{k}\Omega$$

$$Z = 2.2\ \text{k}\Omega + j3.435\ \text{k}\Omega$$

$$L = \frac{X_L}{2\pi f} = \frac{(3.435)(10^3)}{2\pi(2)(10^3)} = 273\ \text{mH}$$

The equivalent circuit consists of a 2.2 kΩ resistor and a 273 mH inductor.

b. $\theta = \tan^{-1}\dfrac{X_L}{R} = \tan^{-1}\left(\dfrac{3.435}{2.2}\right) = 57.36°$

c. $Z = 4.079\ \text{k}\Omega\ \underline{/57.36°}$

$$I = \frac{V}{Z} = \frac{25}{(4.079)(10^3)\ \underline{/57.36°}} = 6.129\ \text{mA}\ \underline{/-57.36°}$$

$$V_R = IR = (6.129)(10^{-3})\ \underline{/-57.36°}\ (2.2)(10^3) = 13.48\ \text{V}\ \underline{/-57.36°}$$

$$V_L = IX_L = (6.129)(10^{-3})\ \underline{/-57.36°}\ (5.027)(10^3)\ \underline{/90°}$$
$$= 30.81\text{ V}\ \underline{/32.64°}$$
$$V_C = IX_C = (6.129)(10^{-3})\ \underline{/-57.36°}\ (1.592)(10^3)\ \underline{/-90°}$$
$$= 9.757\text{ V}\ \underline{/-147.36°}$$

Note in Example 17.1 that the voltage across the inductor is 180° out of phase with the voltage across the capacitor. Figure 17.2 shows the vectors that correspond to the voltage across each component.

If in Example 17.1 the frequency is reduced to 236 Hz, the capacitive reactance increases and the inductive reactance decreases. The student should verify that the current in the circuit becomes 10.75 mA $\underline{/+19°}$, which means that the current is now leading the applied voltage. The equivalent circuit now contains capacitive reactance instead of inductive reactance. Figure 17.3 shows the vectors that represent the new circuit voltages.

If the frequency of the voltage applied across a series *RCL* circuit is varied from very low values to extremely high values, a predictable pattern of voltage drops occurs. At very low frequencies the X_C is quite large and X_L small. At very high frequencies the X_L is large and the X_C quite small. Figure 17.4 shows the circle diagram nature of V_R (and hence the circuit current) as the frequency varies over a wide range.

Note the case where $V_R = V_A$. At this frequency the capacitive reactance exactly equals the inductive reactance. This condition is known as *series resonance* and will be explored in greater detail in Section 17.5 of

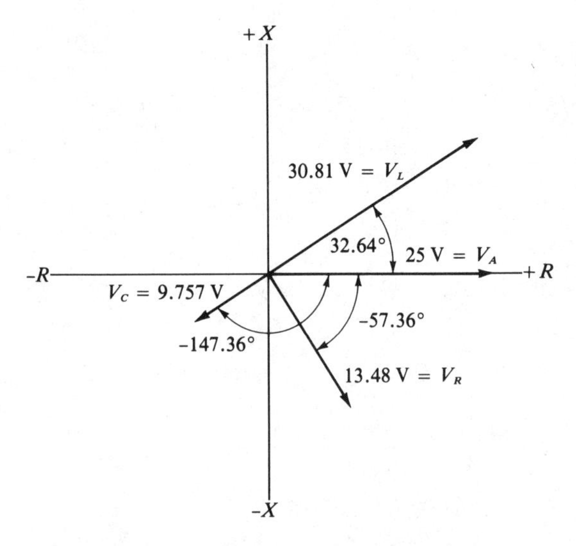

Figure 17.2 Series *RCL* Voltages

this chapter. We should note that at the frequency of resonance (f_0) the inductive and capacitive reactance of the *RCL* circuit cancel each other. The net effect is a purely resistive circuit with the current in phase with the applied voltage. At resonance the impedance of the circuit is equal to the series equivalent resistance.

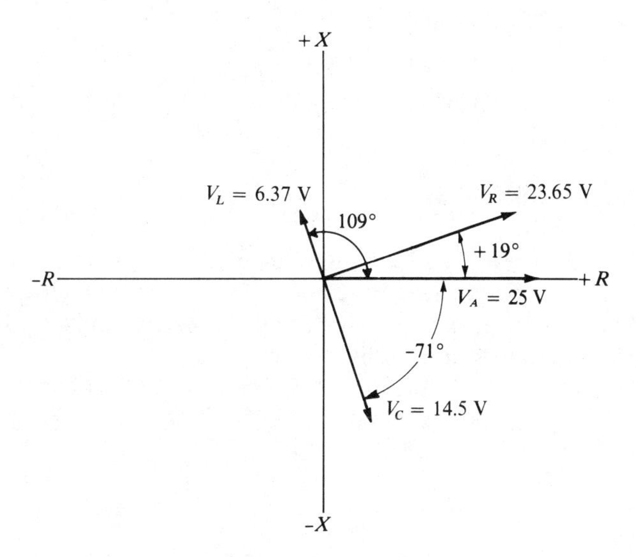

Figures 17.3 Voltage Vectors

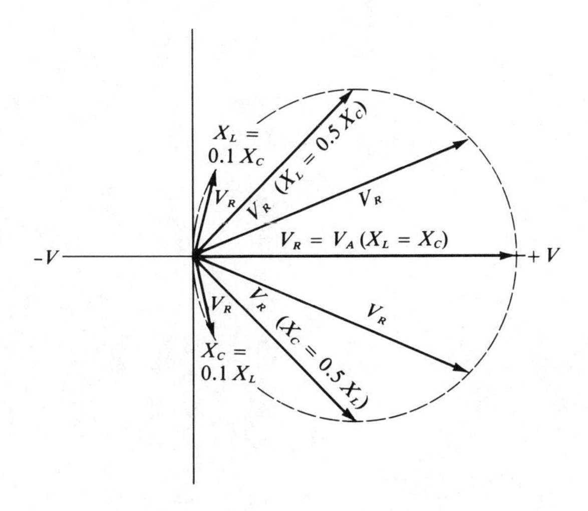

Figure 17.4 Variations of V_R with *f* in *RCL* Circuit

Example 17.2 A series RCL circuit consists of a 3.3 kΩ resistor, an $X_L = 26$ kΩ, and an $X_C = 26$ kΩ. If 30 volts rms is applied to the circuit, what are the voltage drops across R, L, and C?

$$Z = R + jX = 3.3\text{ k}\Omega + j26\text{ k}\Omega - j26\text{ k}\Omega = 3.3\text{ k}\Omega$$

$$I = V/Z = 30/(3.3)(10^3) = 9.09\text{ mA}$$

$$V_R = IR = (9.09)(10^{-3})(3.3)(10^3) = 30\text{ volts}$$

$$V_L = IX_L = (9.09)(10^{-3})(26)(10^3)\,\underline{/+90^\circ} = 236\text{ V}\,\underline{/90^\circ}$$

$$V_C = IX_C = (9.09)(10^{-3})(26)(10^3)\,\underline{/-90^\circ} = 236\text{ V}\,\underline{/-90^\circ}$$

17.2 PARALLEL *RCL* CIRCUITS

We will consider parallel RCL circuits composed of two branch circuits and multiple branch circuits. In the case of two branch circuits, the best approach is to find the equivalent impedance by calculating the product of the branch impedances divided by their sum.

As an example, refer to the circuit of Figure 17.5(a). The two parallel branches Z_A and Z_B are:

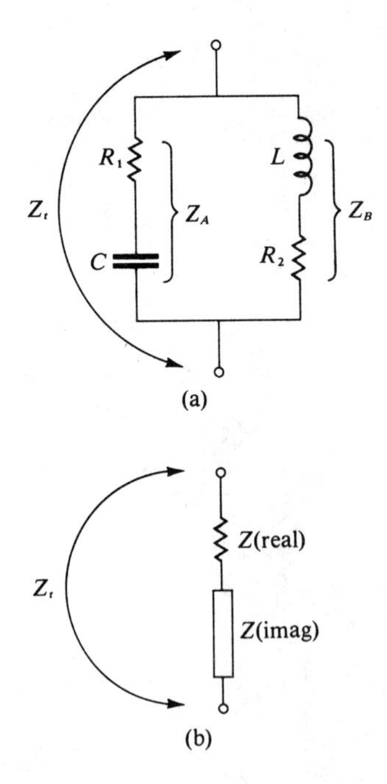

Figure 17.5 Parallel *RCL* Circuits

$$Z_A = Z_A \angle \theta = R_1 - jX_C$$
$$Z_B = Z_B \angle \phi = R_2 + jX_L$$

The total impedance is

$$Z_t = \frac{Z_A Z_B}{Z_A + Z_B}$$

The numerator should be expressed as a product of polar forms, and the denominator as a sum of rectangular forms.

$$Z_t = \frac{Z_A Z_B \angle \theta + \phi}{(R_1 + R_2) + j(X_L - X_C)} \tag{17.2}$$

To obtain an expression for Z_t in polar form, we should now convert the denominator of Eq. (17.2) to polar form and divide the result into the polar form of the numerator. If desired, this polar form can then be converted to its equivalent rectangular form:

$$Z_t = Z_{\text{real}} \pm jZ_{\text{imag}} \tag{17.3}$$

Equation (17.3) represents the equivalent series circuit of the two parallel branches. This equivalent circuit is shown in Figure 17.5(b).

An alternate method for solving this circuit is to solve for the branch currents. These currents are added together to get the total current. Dividing this total current into the applied voltage yields the total equivalent impedance.

Example 17.3 Refer to the circuit of Figure 17.5. Assume $R_1 = R_2 = 3\ \text{k}\Omega$, $X_C = 2\ \text{k}\Omega$, $X_L = 4\ \text{k}\Omega$. What is the total impedance of the circuit? What is the equivalent series R and X?

$$Z_A = (3 - j2)\ \text{k}\Omega = 3.6\ \text{k}\Omega \angle -33.69^\circ$$
$$Z_B = (3 + j4)\ \text{k}\Omega = 5\ \text{k}\Omega \angle 53.13^\circ$$
$$Z_t = \frac{(5)(10^3) \angle 53.13^\circ (3.6)(10^3) \angle -33.69^\circ}{(3 + 3)(10^3) + j(4 - 2)(10^3)}$$
$$= \frac{(18)(10^6) \angle 19.44^\circ}{(6.325)(10^3) \angle 18.43^\circ} = 2.85\ \text{k}\Omega \angle 1.01^\circ$$
$$R = (2.85)(10^3) \cos 1.01^\circ = 2.849\ \text{k}\Omega$$
$$X = (2.85)(10^3) \sin 1.01^\circ = 0.05\ \text{k}\Omega$$
$$Z_{\text{eq}} = (2.849 + j0.05)\ \text{k}\Omega$$

Although the impedance of the circuit of Example 17.3 has a very small angle (1.01°), the circuit is not quite at resonance. Resonance occurs when the j term in the equivalent circuit diminishes to zero. This condition may or may not occur when $X_L = X_C$. The following example problem illustrates this case.

Example 17.4 Assume the circuit of Figure 17.5 has the following values: $R_1 = 10$ kΩ, $R_2 = 1$ kΩ, $X_L = X_C = 4$ kΩ. What is the total impedance across the parallel combination? What is the equivalent series circuit?

$$Z_A = (10 - j4)\ \text{k}\Omega = 10.77\ \text{k}\Omega\ \angle -21.8^\circ$$

$$Z_B = (1 + j4)\ \text{k}\Omega = 4.123\ \text{k}\Omega\ \angle 75.96^\circ$$

$$Z_t = \frac{(10.77)(10^3)\angle -21.8^\circ (4.123)(10^3)\angle 75.96^\circ}{(11 + j0)(10^3)} = 4.036\ \text{k}\Omega\ \angle 54.17^\circ$$

$$Z_{eq} = 2.36\ \text{k}\Omega + j3.27\ \text{k}\Omega$$

Note that Z_{eq} is not purely resistive, even though $X_L = X_C$. If the two resistors in Example 17.4 are reduced by a factor of 10 ($R_1 = 1$ kΩ, $R_2 = 0.1$ kΩ), the parallel equivalent impedance becomes 15 kΩ. Note that the parallel combination produces an impedance that is greater than the impedance of either branch. Reducing the resistance by another factor of 10 causes the parallel combination to have an impedance of 145.5 kΩ!

This increasing parallel impedance can be seen from Eq. (17.2). If R_1 and R_2 equal zero, the equation becomes:

$$Z_t = \frac{(-jX_C)(+jX_C)}{j(X_L - X_C)}$$

Z_t approaches a limit of infinity as R_1 and R_2 approach zero. In reality the resistances can never be zero since the wire that makes up the inductance has resistance.

The following computer program illustrates how a complex parallel circuit can approach infinite impedance even though the resistance in either or both branches approaches zero ohms.

The following program in BASIC calculates Z_t of the circuit of Figure 17.5(a). Assume $X_C = X_L = 4$ kΩ, $R_1 = 0.25$ kΩ. Let R_2 decrease in value from 10 kΩ (2.5 X_L) to 1 kΩ (0.25 X_L) in 1 kΩ increments and then continue to decrease from 900 ohms to 100 ohms ($0.025X_L$) in 100 ohm increments. The printout contains the incremental values of R_2, the corresponding values of Z_t, and the phase angle of Z_t.

RCL Circuit Printout

```
10      P1=3.14159 : X1=4E3 : X2=4E3 : R1=250
15      PRINT "R2(OHMS)" , "Z(T) IN K" , "ANGLE(ZT)"
17      PRINT
20      FOR R2=10E3 TO 1E3 STEP -1E3 : GOSUB 200 : NEXT R2
30      FOR R2=900 TO 100 STEP -100 : GOSUB 200 : NEXT R2
40      STOP
200     Z1=SQR(R1↑2+X1↑2) : A1=ATN(-X1/R1)
210     Z2=SQR(R2↑2+X2↑2) : A2=ATN(X2/R2)
220     R3=R1+R2 : Z=(Z1*Z2)/R3 : A=A1+A2
230     PRINT R2 , Z*1E-3 , A*(180/P1)
240     RETURN
250     END
```

```
R2(OHMS)      Z(T) IN K       ANGLE(ZT)

 10000         4.21126        -64.6223
 9000          4.26728        -62.4612
 8000          4.34508        -59.8587
 7000          4.45682        -56.6788
 6000          4.62411        -52.7337
 5000          4.88809        -47.7639
 4000          5.33449        -41.4237
 3000          6.16586        -33.2936
 2000          7.96598        -22.9887
 1000          13.2197        -10.4599
 900           14.2887        -9.10406
 800           15.5702        -7.73361
 700           17.1314        -6.34992
 600           19.0713        -4.95443
 500           21.5413        -3.54868
 400           24.7865        -2.13426
 300           29.2296        -.71282
 200           35.6695         .713933
 100           45.8179         2.14424
```

17.3 CONDUCTANCE AND SUSCEPTANCE

By calculating the sum of the branch currents of a parallel circuit, we are able to determine the total current. The total impedance can then be found from $Z_t = (V_A/I_t)\angle\theta$. In the case where several branches are connected in parallel, the number of calculations can be reduced by an alternate method. That method requires an understanding of conductance and *susceptance*.

Chapter 3 showed that the reciprocal of resistance is conductance, $G = 1/R$. Resistance is the property that describes an opposition to current flow. Conductance describes the ease with which current passes through the device. Conductance is measured in siemens (S).

The reciprocals of capacitive and inductive reactance can also be calculated. These reciprocal functions are known as susceptance and are also measured in siemens.

Capacitive susceptance is

$$B_C = \frac{1}{-jX_C} \times \frac{+j}{+j} = +j\frac{1}{X_C} \tag{17.4}$$

Similarly, inductive susceptance is

$$B_L = \frac{1}{+jX_L} \times \frac{-j}{-j} = -j\frac{1}{X_L} \tag{17.5}$$

When capacitive, inductive, and resistive devices are connected in parallel, the individual conductances and susceptances can be added. This combination is called *admittance* (Y).

$$Y = (G_1 + G_2 + \cdots + G_N) + j(B_1 + B_2 + \cdots + B_N) \tag{17.6}$$

The phase angle is

$$\theta = \tan^{-1}\left(\frac{B_1 + B_2 + \cdots + B_N}{G_1 + G_2 + \cdots + G_N}\right)$$

Example 17.5 Assume a parallel RCL circuit consists of a 2.2 kΩ resistor, an $X_L = 1$ kΩ, and an X_C of 2.5 kΩ. What is the admittance of the circuit?

$$G = \frac{1}{R} = \frac{1}{(2.2)(10^3)} = (0.4545)(10^{-3}) \text{ siemens}$$

$$B_C = +j\frac{1}{X_C} = +j\frac{1}{(2.5)(10^3)} = +j(0.4)(10^{-3}) \text{ siemens}$$

$$B_L = -j\frac{1}{X_L} = -j\frac{1}{10^3} = -j(10^{-3}) \text{ siemens}$$

$$\begin{aligned} Y &= G + j(B_C - B_L) = [0.4545 + j(0.4 - 1)](10^{-3}) \\ &= (0.4545 - j0.6)(10^{-3}) \\ &= (0.7527)(10^{-3})\angle -52.85^\circ \end{aligned}$$

The reciprocal relationship between admittance and impedance allows us to determine the equivalent circuit for the parallel combination.

$$Y = \frac{1}{Z} \quad \text{and} \quad Z = \frac{1}{Y} \tag{17.7}$$

For example, using the polar form of the answer in Example 17.5, we can determine the impedance and hence the equivalent circuit.

$$Z = \frac{1}{Y} = \frac{1}{(0.7527)(10^{-3})\,\underline{/-52.85^\circ}} = 1.3285 \text{ k}\Omega\,\underline{/52.85^\circ}$$

$$= 802 + j1.058 \text{ k}\Omega$$

$$R_{eq} = 802 \text{ ohms}$$

$$X_{L_{eq}} = 1.058 \text{ k}\Omega$$

The circuit of Figure 17.6(a) is an excellent example of how the use of the reciprocal functions simplifies the total calculation of impedance. The admittance of each individual branch (A through D) can be calculated and then the resultants added to produce a total admittance. This admittance in turn can be converted to an equivalent impedance. To convert branch B (Figure 17.6a) to its equivalent admittance, first determine R and X_L.

$$Z_1 = R_2 + jX_{L1}$$

$$Y_1 = \frac{1}{R_2 + jX_{L1}} \quad \frac{(R_2 - jX_{L1})}{(R_2 - jX_{L1})}$$

$$= \frac{R_2}{R_2^2 + X_{L1}^2} - j\frac{X_{L1}}{R_2^2 + X_{L1}^2}$$

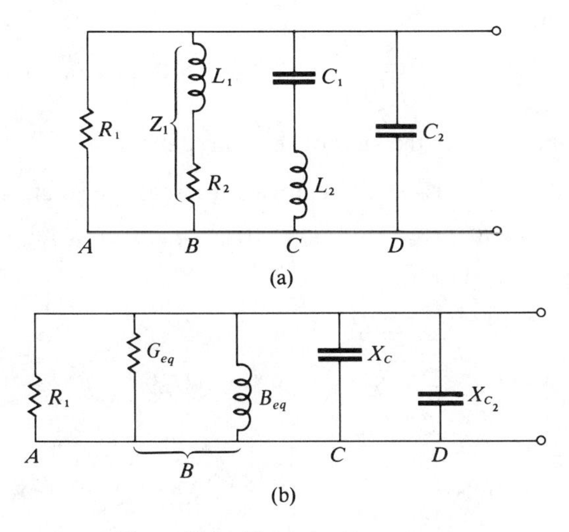

Figure 17.6 Figure for Example 17.6

$$G_{eq} = \frac{R}{R^2 + X_L^2} \tag{17.8}$$

$$B_{eq} = \frac{-jX_L}{R^2 + X_L^2} \tag{17.9}$$

Example 17.6 Refer to the circuit of Figure 17.6(a). If $R_1 = R_2 = 1\ \text{k}\Omega$, $X_{L_1} = X_{L_2} = 2\ \text{k}\Omega$, and $X_{C_1} = X_{C_2} = 3\ \text{k}\Omega$, what is the total circuit impedance? What is the power factor?

For branch A,

$$R = 1\ \text{k}\Omega, \qquad G = \frac{1}{R} = 10^{-3}$$

For branch B, using Eq. (17.8)

$$G_{eq} = \frac{10^3}{(1)^2(10^6) + (2)^2(10^6)}$$
$$= (0.2)(10^{-3})\ \text{siemens}$$

Using Eq. (17.9)

$$B_{eq} = \frac{-j(2)(10^3)}{(1)^2(10^6) + (2)^2(10^6)}$$
$$= -j(0.4)(10^{-3})$$

For branch C,

$$X_{C_1} = (3)(10^3), \qquad X_{L_2} = (2)(10^3)$$

$$B_C = \frac{1}{(-j3 + j2)\ \text{k}\Omega} = +j10^{-3}$$

For branch D,

$$B_C = +j\frac{1}{(3)(10^3)} = j(0.3333)(10^{-3})\ \text{siemens}$$

The total admittance is the sum of the admittances.

$$Y_t = Y_A + Y_B + Y_C + Y_D \text{ (equivalent circuit in Figure 17.6(b))}$$
$$= 10^{-3} + (0.2)(10^{-3}) - j(0.4)(10^{-3}) + j10^{-3} + j(0.333)(10^{-3})$$
$$= (1.2 + j0.933)(10^{-3})$$
$$= (1.52)(10^{-3})\angle 37.87^\circ$$

$$Z_t = \frac{1}{Y_t} = 658\angle -37.87^\circ = 519.4 - j403.9$$

$$R_{eq} = 519.4\ \text{ohms}$$

$$X_{C_{eq}} = 403.9\ \text{ohms}$$

$$\text{PF} = \frac{R_{eq}}{|Z_t|} = 0.789$$

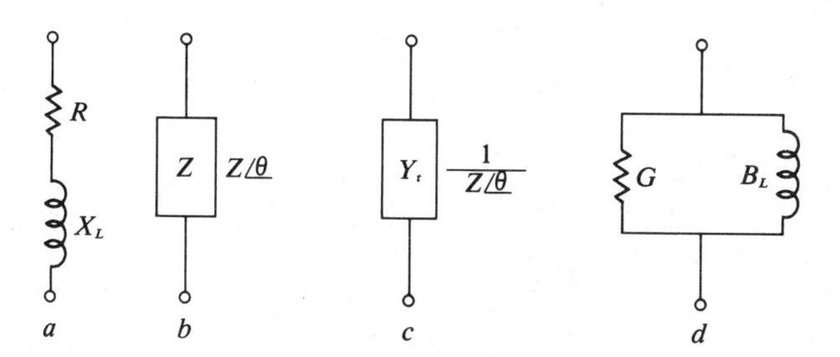

Figure 17.7 *RL* to *GB* Conversion

The student should solve Example 17.6 for its total impedance using the methods outlined in previous sections dealing with parallel impedances. Such a solution will underscore the convenience of the admittance approach. Whenever circuit simplifications require transitions between resistance, impedance, reactance, and susceptance, one major precaution should be noted. This is best illustrated by a detailed example. If a 2 kΩ resistor in series with a 4 kΩ inductive reactance is to be converted to conductance and susceptance, it is not a case of merely taking reciprocals of R and X_L individually. In Figure 17.7(a) note the series RL circuit. If we take reciprocals individually, we get $1/R = 0.5 \times 10^{-3}$ and $1/X_L = 0.25 \times 10^{-3}$. In Figure 17.7(b) we note the polar form of $R + jX_L$. In this case $(2 + j4) = 4.472\ \text{k}\Omega\ \angle 63.43°$. The reciprocal of this polar form impedance is 0.224×10^{-3} siemens. This admittance is shown in Figure 17.7(c). If this value is converted to its equivalent rectangular form, $(0.1 - j0.2) \times 10^{-3}$ results. However, if Figure 17.7(c) is in rectangular form, the equivalent parallel circuit of Figure 17.7(d) results.

17.4 *AC* NETWORKS

In this study of alternating current networks, we demonstrate that the method of analysis of ac circuits is essentially the same as that outlined for resistive circuits. Figure 17.8(a) illustrates a typical resistive series-parallel configuration. Note that R_1 and R_2 are in series and this series combination is in parallel with R_3. The combined parallel resistance is in series with R_4. The total resistance as seen by the generator becomes

$$R_T = \frac{(R_1 + R_2)(R_3)}{R_1 + R_2 + R_3} + R_4$$

We can generalize the circuit of Figure 17.8(a) using the impedance notation instead of the resistive notation. For example, we can replace the

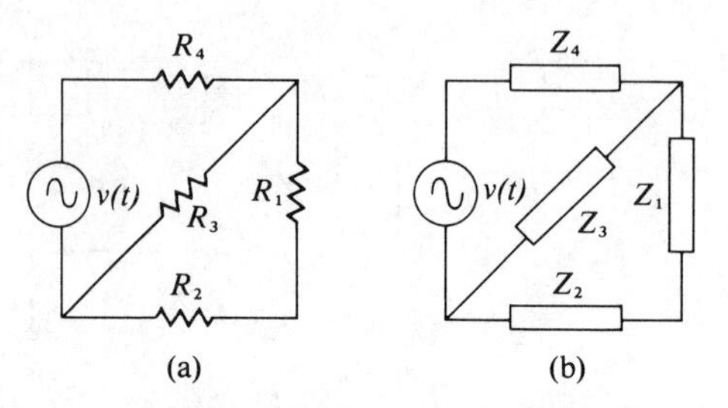

Figure 17.8 Resistive Network

resistances with impedances that represent only real values (resistance). The circuit would appear as in Figure 17.8(b). The equation, therefore, becomes

$$Z_t = \frac{(Z_1 + Z_2)(Z_3)}{Z_1 + Z_2 + Z_3} + Z_4$$

If the branch impedances are nonresistive (that is, the series branch includes inductance and/or capacitance with or without a series resistance), then these respective impedances must be expressed in either polar or rectangular notation depending upon which is most convenient in the arithmetic manipulation. Recall it is easier to add and subtract rectangular forms and to multiply and divide polar forms.

Example 17.7 Assume that an ac source is applied to the circuit of Figure 17.9(a). The frequency is such that $X_L = 4\ \text{k}\Omega$ and $X_C = 3\ \text{k}\Omega$. What is the total impedance of the circuit? What will be the leading or lagging phase angle of the current? What is the equivalent series circuit? What is the power factor?

The circuit of Figure 17.9(a) reduces to three impedance branches: Z_A, Z_B, and Z_C. Their interconnection is shown in Figure 17.9(b). First identify each impedance in both its polar and rectangular form.

$$Z_A = 3\ \text{k}\Omega\ \underline{/0^\circ}, \qquad (3 + j0)\ \text{k}\Omega$$

$$Z_B = 4\ \text{k}\Omega\ \underline{/90^\circ}, \qquad (0 + j4)\ \text{k}\Omega$$

$$Z_C = 7.616\ \text{k}\Omega\ \underline{/-23.2^\circ}, \qquad (7 - j3)\ \text{k}\Omega$$

$$Z_t = \frac{Z_B Z_C}{Z_B + Z_C} + Z_A$$

$$= \frac{(4)(10^3)\ \underline{/90^\circ}\ (7.616)(10^3)\ \underline{/-23.2^\circ}}{(+j4 + 7 - j3)(10^3)} + (3)(10^3) + j0$$

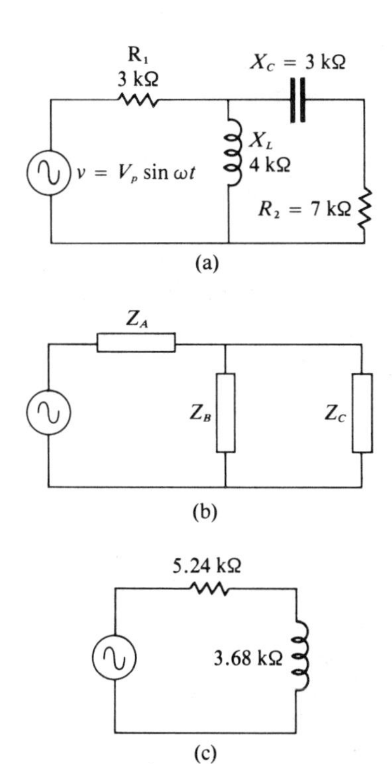

Figure 17.9 Figure for Example 17.7

$$= \frac{(30.46)(10^3)\ \angle 66.8^\circ}{7.071\ \angle 8.13^\circ} + 3\text{ k}\Omega$$

$$= 5.24\text{ k}\Omega + j3.68\text{ k}\Omega$$

Figure 17.9(c) shows the equivalent circuit.

$$Z_t \text{ (in polar form)} = 6.403\text{ k}\Omega\ \angle 35.08^\circ$$

The current will lag the applied voltage by 35.08°.
The power factor, PF, equals the cos θ.

$$\text{PF} = \cos 35.08^\circ = 0.818$$

Delta-Wye Conversions

Delta-wye simplifications are often required to solve large, more complex circuits. In resistive bridge circuits the total resistance could only be found after making a delta-to-wye conversion. The same basic theory may be used for ac conversions as were used for dc.

In Figure 17.10(a) the delta circuit consists of Z_A, Z_B, and Z_C. This three-terminal circuit has terminal points 1, 2, and 3. The wye circuit, which can replace the delta, consists of Z_x, Z_y, and Z_z. The three terminal points must be the same as those of the delta circuit. The equivalent wye is shown in Figure 17.10(b). The equivalency equations are:

$$Z_x = \frac{Z_A Z_B}{Z_A + Z_B + Z_C} \tag{17.10}$$

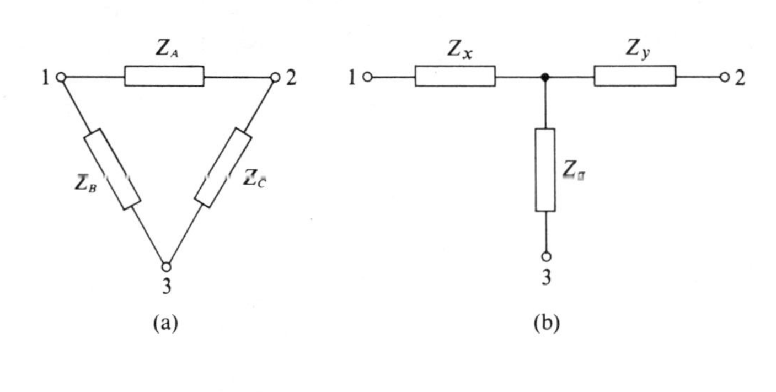

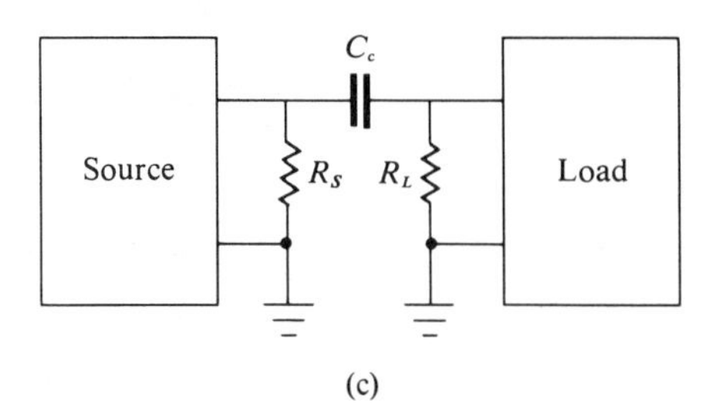

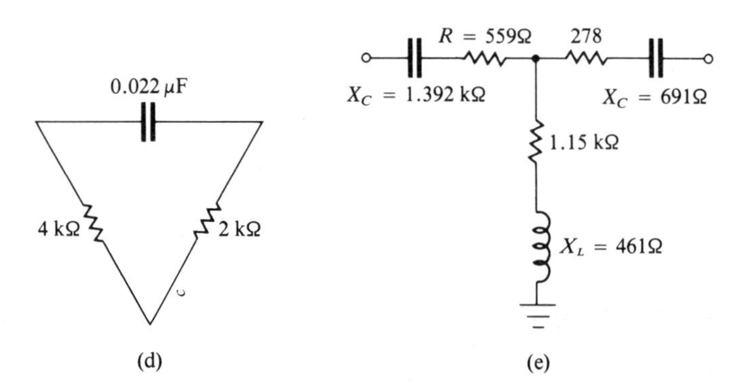

Figure 17.10 Delta-Wye Conversions

$$Z_y = \frac{Z_A Z_C}{Z_A + Z_B + Z_C} \tag{17.11}$$

$$Z_z = \frac{Z_B Z_C}{Z_A + Z_B + Z_C} \tag{17.12}$$

When a wye circuit is to be converted to an equivalent delta, the conversion equations are:

$$Z_A = \frac{Z_x Z_y + Z_x Z_z + Z_y Z_z}{Z_z} \tag{17.13}$$

$$Z_B = \frac{Z_x Z_y + Z_x Z_z + Z_y Z_z}{Z_y} \tag{17.14}$$

$$Z_C = \frac{Z_x Z_y + Z_x Z_z + Z_y Z_z}{Z_x} \tag{17.15}$$

Example 17.8 Assume the circuit of Figure 17.10(c) has capacitive coupling between the two devices. The input resistance of the load is 2 kΩ and the output resistance of the source is 4 kΩ. If the source frequency is to be 3 kHz and C is 0.022 μF, what is the wye equivalent circuit which may be used for coupling?

We must express Z_A, Z_B, and Z_C of Figure 17.10(d) in their polar and rectangular forms.

$$Z_A = -j1/2\pi fC = -j(2.41)10^3 = 2.41 \text{ k}\Omega \underline{/-90^\circ}$$
$$Z_B = 4 \text{ k}\Omega + j0 = 4 \text{ k}\Omega \underline{/0^\circ}$$
$$Z_C = 2 \text{ k}\Omega + j0 = 2 \text{ k}\Omega \underline{/0^\circ}$$

Using Eqs. (17.10), (17.11), and (17.12),

$$Z_x - \frac{(2.41)(10^3) \underline{/-90^\circ} (4)(10^3) \underline{/0^\circ}}{(6 - j2.41)(10^3)} = \frac{(2.41)(4)(10^3) \underline{/-90^\circ}}{6.466 \underline{/-21.88^\circ}}$$
$$= 1.5 \text{ k}\Omega \underline{/-68.12^\circ} = 0.559 \text{ k}\Omega - j1.392 \text{ k}\Omega$$

$$Z_y = \frac{(2.41)(10^3) \underline{/-90^\circ} (2)(10^3)}{(6.466)(10^3) \underline{/-21.88^\circ}}$$
$$= 0.745 \text{ k}\Omega \underline{/-68.12^\circ} = 278 - j691$$

$$Z_z = \frac{(4)(10^3)(2)(10^3)}{(6.466)(10^3) \underline{/-21.88^\circ}}$$
$$= 1.237 \text{ k}\Omega \underline{/21.88^\circ} = 1.15 \text{ k}\Omega + j461$$

The equivalent coupling circuit is shown in Figure 17.10(e). It is important to note that this equivalent coupling circuit is valid at one and only one frequency of 3 kHz. Any variation of the source frequency will result in an altered coupling arrangement.

Example 17.9 The bridge circuit of Figure 17.11(a) is a typical bridge arrangement. The circuit is at resonance. Prove that the phase angle of the total circuit is zero.

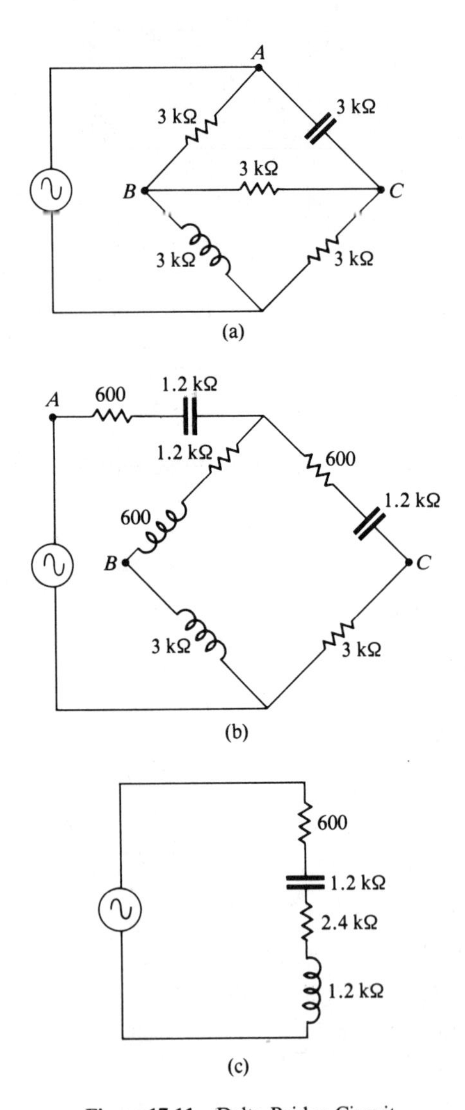

Figure 17.11 Delta Bridge Circuit

The circuit of Figure 17.11(a) has a delta circuit which can be replaced with its wye equivalent. The equivalent is shown in Figure 17.11(b). When the parallel circuit portion is reduced to its equivalent circuit, the series circuit of Figure 17.11(c) results. The two reactive properties cancel each other and the total impedance is resistive. $R_{eq} = 3\ \text{k}\Omega$.

Thevenin's Theorem

In Chapter 9 we learned that the Thevenin equivalent circuit is a simplified equivalent of a complex network. The Thevenin resistance was identified as the resistance looking back into the open circuit terminals with all voltage and current sources replaced by their respective internal resistances. For many ac circuits, this same technique can be used without difficulty. In this, as in all ac network analysis techniques, it must be remembered that changes in source frequency change the impedance values of reactive elements and therefore change the equivalent circuit. An alternate method for determining the Thevenin impedance is to divide the open circuit voltage by the short circuit current.

$$Z_{Th} = \frac{V_{oc}}{I_{sc}} \qquad (17.16)$$

In this manner the Thevenin equivalent impedance of a single or multiple source circuit may be determined without the removal of the active sources. This principle is best illustrated with an example.

Example 17.10 Assume the circuit of Figure 17.12(a) is to be analyzed using Thevenin's equivalent circuit. If the 2 kΩ load is to be removed, solve for the equivalent Thevenin circuit.

If the 2 kΩ load resistor is removed the circuit of Figure 17.12(b) results. V_{oc} can now be determined. According to voltage ratios:

$$V_{oc} = \frac{(3)(10^3)\ \angle 90^\circ\ (25)}{(3 + j3)(10^3)} = 17.68\ \text{V}\ \angle 45^\circ$$

The short circuit current can be found using Ohm's law (Figure 17.12(c)).

$$I_{sc} = \frac{25}{(3)(10^3)} = 8.333\ \text{mA}\ \angle 0^\circ$$

$$Z_{Th} = \frac{V_{oc}}{I_{sc}} = \frac{17.68\ \angle 45^\circ}{(8.333)(10^{-3})} = 2.122\ \text{k}\Omega\ \angle 45^\circ$$

$$= 1.5\ \text{k}\Omega + j1.5\ \text{k}\Omega$$

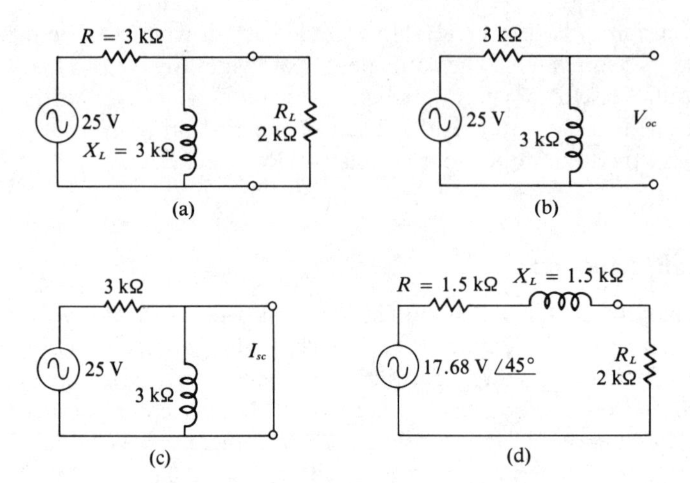

Figure 17.12 Figure for Example 17.10

The Thevenin equivalent circuit with R_L connected is shown in Figure 17.12(d). Note in the preceding example that the total impedance is $(3.5 + j1.5)$ kΩ. The current flow will be the Thevenin voltage divided by this impedance:

$$I = \frac{17.68 \angle 45^\circ}{3.807 \text{ k}\Omega \angle 23.198^\circ} = 4.64 \text{ mA} \angle 21.8^\circ$$

If the circuit is solved using conventional series-parallel methods, the total current out of the 25 volt generator is 5.577 mA $\angle 11.89^\circ$. When this current divides by ratio, the current through the 2 kΩ resistor is 4.643 mA $\angle 21.8^\circ$.

Superposition Theorem

The principle of superposition finds wide application, and not only in the field of electrical science. It is especially useful in networks that contain more than two sources. It is beyond the scope of this text to present the varied number of possibilities of circuit configurations. The example presented here illustrates a general approach. The principles can be applied to most circuit configurations.

Example 17.11 Assume the circuit of Figure 17.13(a) has a 20 volt ac source applied and another 20 volt ac source (displaced by 90°) applied as shown. Both sources operate at the same frequency. What is the current through the 2 kΩ resistor?

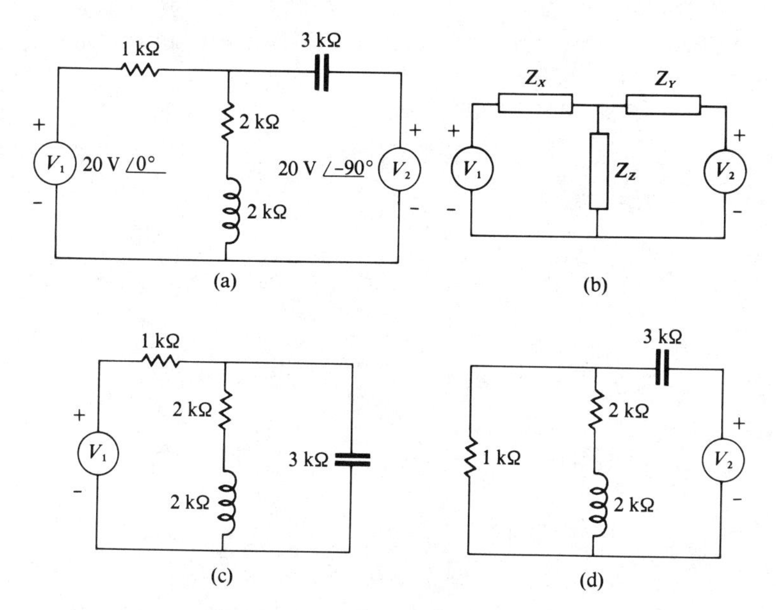

Figure 17.13 Superposition Theorem

The three branch impedances are:

$$Z_x = 1\text{ k}\Omega + j0 = 1\text{ k}\Omega \angle 0^\circ$$
$$Z_z = (2 + j2)\text{ k}\Omega = 2.828\text{ k}\Omega \angle 45^\circ$$
$$Z_y = 0 - j3\text{ k}\Omega = 3\text{ k}\Omega \angle -90^\circ$$

These impedances are shown in Figure 17.13(b).

Replacing V_2 with a short circuit results in the equivalent circuit of Figure 17.13(c).

The total impedance seen by V_1 is

$$Z_{t_1} = Z_x + \frac{Z_y Z_z}{Z_y + Z_z}$$
$$= (1\text{ k}\Omega + j0) + \frac{(2.828\text{ k}\Omega \angle 45^\circ)(3\text{ k}\Omega \angle -90^\circ)}{(2 + j2 - j3)\text{ k}\Omega}$$
$$= 4.754\text{ k}\Omega \angle -14.62^\circ$$
$$I_{t_1} = \frac{20 \angle 0^\circ}{(4.754)(10^3) \angle -14.62^\circ} = 4.207 \angle 14.62^\circ\text{ mA}$$
$$I_1 = \frac{(3)(10^3) \angle -90^\circ (4.207 \angle 14.62^\circ)(10^{-3})}{(2 - j)(10^3)} = 5.644 \angle -48.81^\circ\text{ mA}$$
$$= 3.716 - j4.247\text{ mA}$$

Replacing V_1 with a short circuit results in the equivalent circuit of Figure 17.13(d).

The total impedance seen by V_2 is

$$Z_{t_2} = Z_y + \frac{Z_x Z_z}{Z_x + Z_z}$$

$$= -j(3)(10^3) + \frac{10^3 \angle 0^\circ (2.828)(10^3) \angle 45^\circ}{(3 + j2)(10^3)} = 2.95 \text{ k}\Omega \angle -74.88^\circ$$

$$I_{t_2} = \frac{20 \angle -90^\circ}{(2.95)(10^3) \angle -74.88^\circ} = 6.78 \angle -15.12^\circ \text{ mA}$$

The current through R_2 is the ratio

$$I_2 = \frac{(10^3)(6.78)(10^{-3}) \angle -15.12^\circ}{(3 + j2)(10^3)} = 1.88 \angle -48.81^\circ \text{ mA}$$

$$= 1.238 - j1.415 \text{ mA}$$

The actual current through R is the sum of the two superposition currents.

$$I_R = I_1 + I_2$$
$$= (3.716 - j4.247 + 1.238 - j1.415) \text{ mA}$$
$$= (4.954 - j5.662) \text{ mA}$$
$$= 7.52 \text{ mA} \angle -48.81^\circ$$

The circuit of Example 17.11 can be solved by many methods. For example, the 2 kΩ resistor could be removed and the remaining circuit solved for its Thevenin equivalent. Then the 2 kΩ resistor could be connected across the Thevenin equivalent circuit and the current could be determined. There is no simple rule which can suggest which theorem or which network solution to use for calculating specific circuit conditions.

Perhaps one of the most effective methods of solution involves determinants. In multiple loop networks it is generally convenient to identify each branch impedance in both its polar and rectangular forms. The determinant method then involves several products which can be readily solved in polar form.

Example 17.12 Solve for the current through the 2k resistor in Example 17.11 using simultaneous loop equations.

From Figure 17.13(b) we may write two loop equations (all Z's assumed to be kilohms). A clockwise direction is assumed for the left loop and a counterclockwise direction for the right loop.

$$V_1 = (Z_x + Z_z)I_1 + Z_z I_2$$
$$V_2 = Z_z I_1 + (Z_z + Z_y)I_2$$
$$20\angle 0^\circ = (3 + j2)I_1 + (2 + j2)I_2$$
$$20\angle -90^\circ = (2 + j2)I_1 + (2 - j)I_2$$
$$20\angle 0^\circ = 3.605\angle 33.69^\circ I_1 + 2.828\angle 45^\circ I_2$$
$$20\angle -90^\circ = 2.828\angle 45^\circ I_1 + 2.236\angle -26.56^\circ I_2$$

Solving for I_1 using determinants:

$$I_1 = \frac{\begin{vmatrix} 20\angle 0^\circ & 2.828\angle 45^\circ \\ 20\angle -90^\circ & 2.236\angle -26.56^\circ \end{vmatrix}}{\begin{vmatrix} 3.605\angle 33.69^\circ & 2.828\angle 45^\circ \\ 2.828\angle 45^\circ & 2.236\angle -26.56^\circ \end{vmatrix}}$$

$$= \frac{(20\angle 0^\circ)(2.236\angle -26.56^\circ) - (2.828\angle 45^\circ)(20\angle -90^\circ)}{(3.605\angle 33.69^\circ)(2.236\angle -26.56^\circ) - (2.828\angle 45^\circ)(2.828\angle 45^\circ)}$$

$$= \frac{(40 - j20) - (40 - j40)}{8 + j1 - j8} = \frac{+j20}{8 - j7}$$

$$= \frac{20\angle 90^\circ}{10.63\angle -41.18^\circ} = 1.88 \text{ mA } \angle 131.18^\circ$$

Solving for I_2 using determinants

$$I_2 = \frac{(3.605\angle 33.69^\circ)(20\angle -90^\circ) - (20\angle 0^\circ)(2.828\angle 45^\circ)}{8 - j7}$$

$$= \frac{72.1\angle -56.31^\circ - 56.56\angle 45^\circ}{8 - j7} = \frac{(40 - j60) - (40 + j40)}{8 - j7}$$

$$= \frac{-j100}{8 - j7} = \frac{100\angle -90^\circ}{10.63\angle -41.18^\circ} = 9.407 \text{ mA } \angle -48.82^\circ$$

The current through R is $I_1 + I_2$.

$$I_R = I_1 + I_2 = (-1.237 + j1.415) + (6.19 - j7.08)$$
$$= 4.953 - j5.665$$
$$= 7.52 \text{ mA } \angle -48.83^\circ$$

Note that this answer agrees with the answer using superposition.

Networks Approach

In some circuit problems considerable simplification can be achieved by rearranging circuit components. In the circuit of Figure 17.14(a) the circuit can be made to appear symmetrical, hence Bartlett's theorem may be used. If, for instance, we wish to determine the current flowing out of generator V_1, we can readily rearrange the circuit as shown in Figure 17.14(b). The 6 kΩ inductive reactance is divided into two separate inductive reactances. Similarly, the capacitive reactance is converted into two parallel reactances whose combined reactance is 2 kΩ. An infinite impedance bisector (dashed line) is passed through the line of symmetry. Figure 17.14(c) illustrates the resultant Bartlett circuit. The actual current leaving the generator is 25 V $\angle 20°$ divided by the impedance seen by the generator: $Z = 5\ \text{k}\Omega \angle -53.1°$, $I_1 = 5\ \text{mA} \angle 73.1°$.

17.5 SERIES RESONANCE

We have already encountered series resonance as a condition in a series *RCL* circuit when $X_L = X_C$. Since this condition can occur at only one frequency for a given *RCL* combination, we can determine that frequency algebraically.

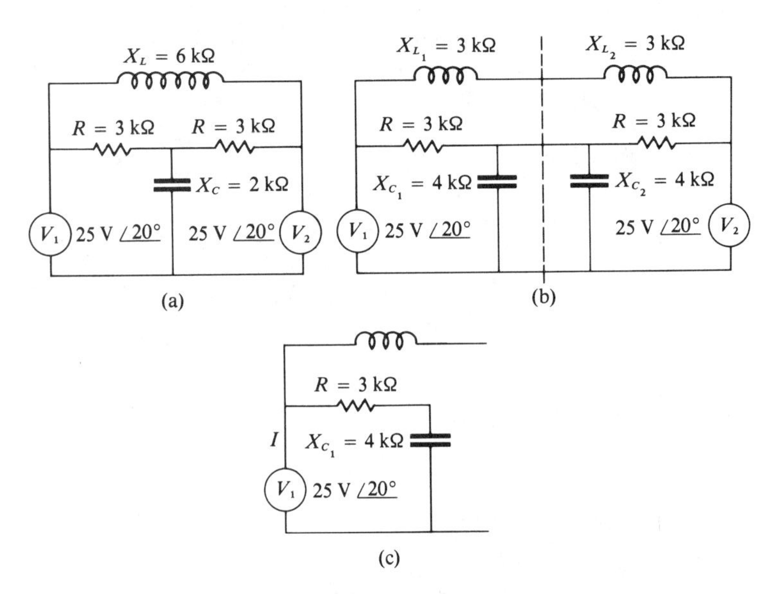

Figure 17.14 AC Network Solutions

$$2\pi fL = \frac{1}{2\pi fC}, \qquad 4\pi^2 f^2 LC = 1$$

$$\boxed{f_0 = \frac{1}{2\pi\sqrt{LC}}} \tag{17.17}$$

where f_0 is the resonant frequency in hertz, L is the inductance in henries, and C is the capacitance in F.

By varying either L or C we can vary the resonant frequency. This can be seen on a graph by observing the capacitive and inductive reactance curves. Refer to Figure 17.15(a). Note that the intersection of X_L and X_C occurs at the resonant frequency point. Increasing L or C will decrease the resonant point. Similarly, decreasing L or C will increase the resonant point. Note in Figure 17.15(a) the basic shapes of the reactance curves. In Figure 17.15(b) the resonant point has decreased because L has been

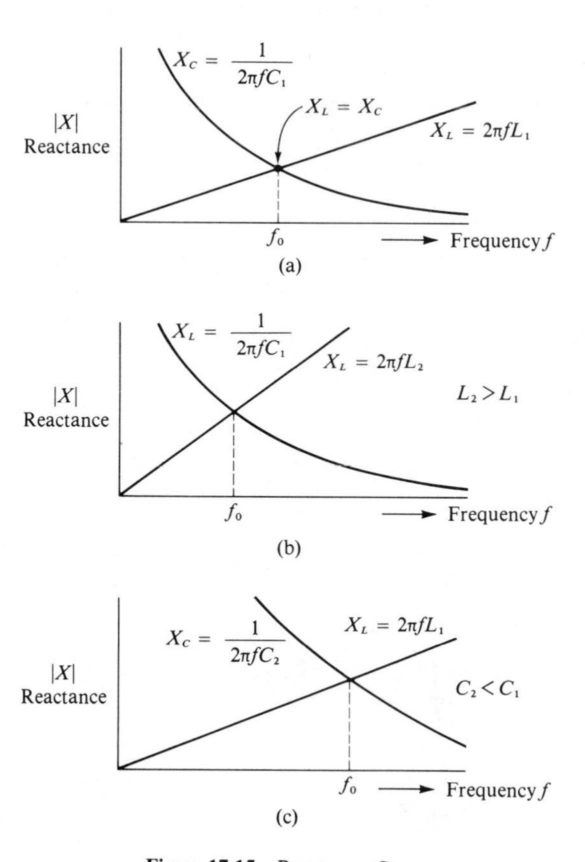

Figure 17.15 Reactance Curves

increased. In Figure 17.15(c) the resonant point has increased because C has been decreased.

The value of the resonant frequency, according to Eq. (17.17), can be changed by changing L or C. Often L and C either individually or collectively are varied in order to produce a different resonant frequency. Sometimes dynamic circuit conditions influence these values and cause an undesirable change in f_o. Usually it is the inductance that is most influenced by dynamic circuit conditions through its magnetic flux. For a given resonant frequency, the choice of L should be as large as practicable since its variations cause the smallest variations of f_o. Note in Figure 17.16 that the change in f_o for small changes in L at large values of L causes a smaller Δf_o than for smaller values of L.

The voltages across the inductor and the capacitor can be determined using Ohm's Law. In a series circuit the current is the same throughout the circuit, hence $V_C = V_L$ at resonance. From this information and from our previous studies, several important conclusions can be drawn concerning series resonant circuits operated at resonance:

1. Since $X_L = X_C$, the reactive properties cancel each other and $Z = R$.
2. Since $Z = R$, the current at resonance is maximum and has a zero phase angle.
3. Since the phase angle is zero, the power factor of the circuit is 1 (cos $0° = 1$).

Refer to the circuit of Figure 17.17(a). Assume the applied voltage V_A is a fixed value whose frequency is varied from zero hertz to a value greater than 10 times the resonant frequency. We can determine the impedance as $|Z| = \sqrt{R^2 + (X_L - X_C)^2}$. At frequencies lower than f_o, the impedance has a capacitive component. At frequencies greater than f_o, the impedance has an inductive component.

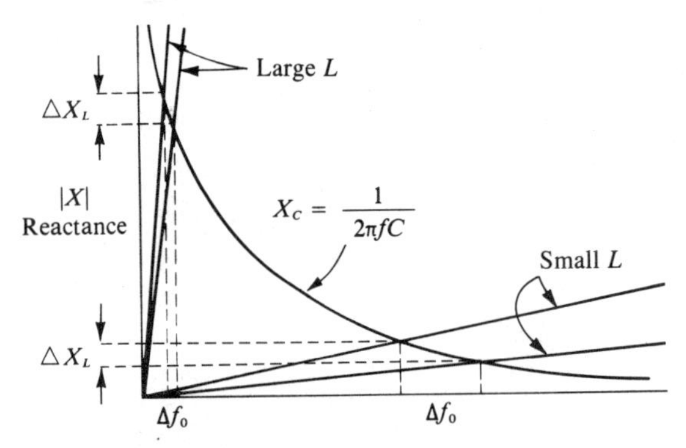

Figure 17.16 Change in f_0 Resulting from Changes in L
(Note that Δf_0 is larger with the small values of L.)

In each case the value of X depends upon the difference between the two reactances. Figure 17.17(b) shows how the impedance varies with frequency. Here we see that at f_o the impedance dips to its lowest value, R.

The phase angle of a series RCL circuit is

$$\theta = \tan^{-1}\left(\frac{X_L - X_C}{R}\right)$$

The greater the net reactance, the greater the phase angle. The further below resonance, the greater the negative phase angle. The maximum phase displacement will be 90° and this condition is approached when $|X| \gg R$. The same conditions prevail above resonance, except that the phase angle is positive since the circuit is inductive and resistive. Figure 17.17(c) illustrates the changing phase angle with variations of frequency.

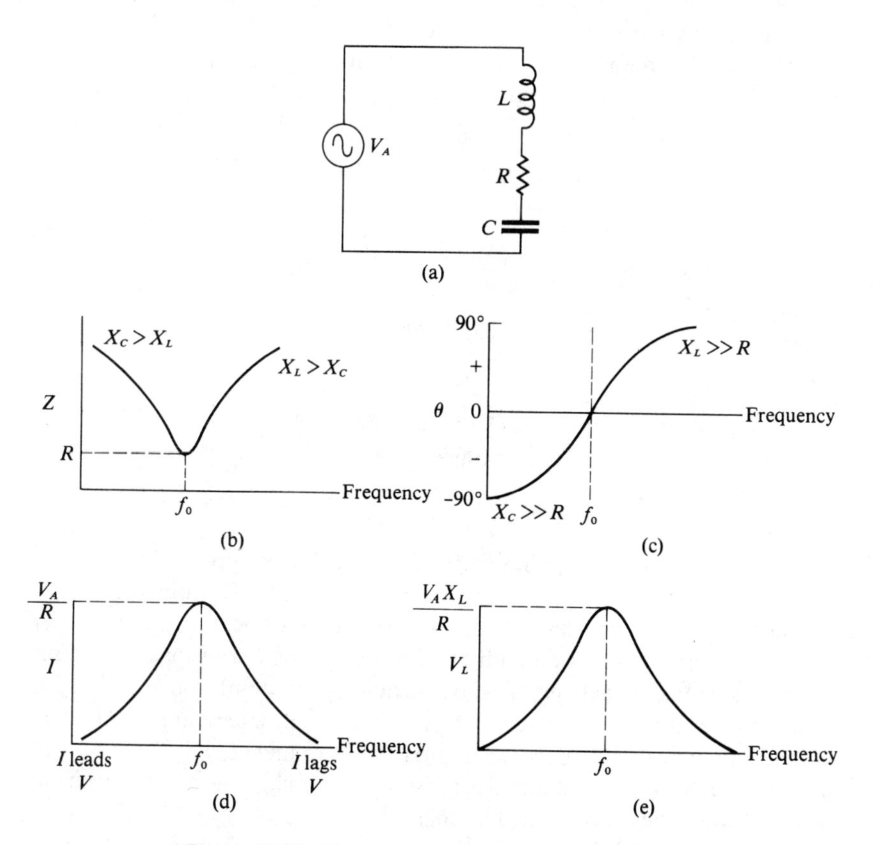

Figure 17.17 Characteristics of Series-Resonant Circuit (Frequency scales are logarithmic.)

The current in a series RCL circuit is $I = V_A/Z$. Note in Figure 17.17(d) that at resonance the current rises to its peak value of V_A/R. This means that at the resonant point the current can be controlled by varying the value of R. Since the voltages across L and C are determined by

$$V_L = IX_L \qquad \text{and} \qquad V_C = IX_C$$

these voltages are dependent on R. Making R small will, at resonance, result in large voltages across L and C. Figure 17.17(e) shows how the voltage across L rises to its peak value IX_L or V_AX_L/R at resonance. The voltage across C is also V_AX_L/R since $X_L = X_C$ at resonance.

The voltage across the resistor is solely dependent upon the current in the circuit (Ohm's law). Its variation, therefore, is the same as that of the current. It rises to a peak value at resonance.

Example 17.13 Assume a series RCL circuit consists of $R = 50$ ohms, $L = 4$ mH, and $C = 0.1$ μF. If 20 volts is applied to the circuit, determine the following values at resonance: f_0, I, V_L, and V_C.

The resonant frequency can be determined using Eq. (17.17).

$$f_0 = \frac{1}{(6.28)\sqrt{(4)(10^{-3})(10^{-7})}} = \frac{0.159}{(2)(10^{-5})}$$

$$= 7.95 \text{ kHz}$$

$$I = \frac{V_A}{Z} = \frac{V_A}{R} = \frac{20}{50} = 0.4 \text{ ampere}$$

$$X_L = 2\pi fL = (6.28)(7.95)(10^3)(4)(10^{-3})$$

$$= 200\ \Omega$$

$$V_L = IX_L = (0.4)(200) = 80 \text{ volts}$$

$$V_C = V_L \text{ at resonance}$$

$$= 80 \text{ volts}$$

The preceding example again illustrates that voltages across L and C can rise to values much greater than the applied voltage. Recall that inductive and capacitive devices do not convert energy to heat. They store the energy. At resonance, the condition known as *oscillation* occurs where the energy transfers between L and C reach a peak condition.

It is useful for us to reexamine the preceding example with one change. If the inductance is changed to 16 mH, what influence will that change have on the current in the circuit at resonance? None, since $I = V/R$! The resonant frequency, however, has changed. The new f_0 is 3.975 kHz. The reactances of X_L and X_C are now changed to 400 ohms. The voltages across L and C are now 160 volts.

Series Resonant *Q* Factor

In the preceding example note that the current supplied to the circuit was 0.4 ampere at resonance. The apparent power in the circuit is $P_a = IV$. For the example studied, this power becomes (0.4)(20) = 8 volt-amps. Similarly, the true power is $P_t = I^2R$. Calculating its value, $P_t = (0.4)^2(50) =$ 8 W. We have seen that the 0.4 ampere flows through the X_L and the X_C, both nonzero quantities. We can identify the reactive power as

$$P_Q = I^2X_L \text{ volt-amps, reactive (vars)} \tag{17.18}$$

For the circuit of Example 17.13, this reactive power becomes $(0.4)^2(200)$ = 32 vars. Similarly, the capacitive reactive power is 32 vars; however, the energy transfers related to these power ratings are mutually exchanging. That is, the capacitive energy is transferring to the inductor and then the inductive energy is transferring to the capacitor at the resonant frequency rate. The resonant frequency equations show that if the inductance is doubled and the capacitance reduced in half, the frequency remains the same. At resonance, the reactances have increased and therefore so has the reactive power.

For a given *RCL* circuit, the possibility exists for higher or lower reactive powers at a given resonant frequency. There are many circuit applications where high reactive powers are desirable. In communications, filters, and tuned circuits, we try to achieve very high reactive powers. For these applications a high *L* to *C* ratio is desirable.

A figure of merit, identified as *Q*, is associated with *RCL* circuits. The *Q* of an *RCL* circuit, also referred to as the *Q* of a coil, is the ratio of reactive power to true power.

$$Q = \frac{P_Q}{P_t} = \frac{I^2X_L}{I^2R} = \frac{X_L}{R} \tag{17.19}$$

Also,

$$Q = \frac{I^2X_C}{I^2R} = \frac{X_C}{R}$$

where X_L and X_C are the values computed at f_o. Since $X_L = 2\pi fL$ and f in this equation is the resonant frequency, the *Q* becomes:

$$Q = \frac{2\pi L}{2\pi\sqrt{LC}\,R} = \frac{1}{R}\sqrt{\frac{L}{C}} \tag{17.20}$$

Example 17.14 Refer to Example 17.13. What is the *Q* of the circuit? What is the resonant frequency *Q* if the inductance is increased to 16 mH?

$$Q = X_L/R = 200/50 = 4$$

$$= (1/50)\sqrt{\frac{(4)(10^{-3})}{10^{-7}}} = 4$$

For $L = 16$ mH,

$$Q = \frac{1}{50}\sqrt{\frac{(16)(10^{-3})}{10^{-7}}} = 8$$

Bandwidth

Perhaps one of our earliest curiosities about electrical science is the simple process of receiving radio signals. At the heart of this process is an *RCL* circuit. The circuit can be symbolized as shown in Figure 17.18(a). We can assume that radio signals of many different frequencies cut across the coil. All of these waves induce currents in the inductor. At one particular frequency (f_0), the current is maximum. That current develops a voltage drop, IX_C, which can be amplified and detected. Hence, if a radio station is broadcasting at 1330 kHz and L and C are properly adjusted, that station's signal will be selected. How selective the tuned circuit is depends upon the circuit components *RCL*.

In order to receive the signal properly, only a narrow band of frequencies must be selected. Bandwidth is defined as

$$\Delta f = f_2 - f_1$$

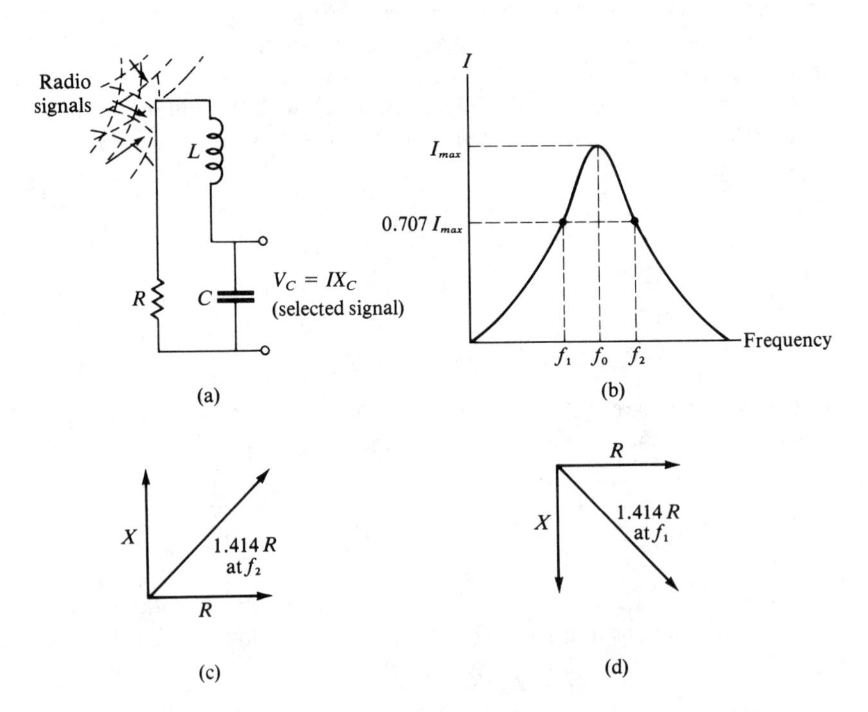

Figure 17.18 Bandwidth Response Curve

where f_1 and f_2 are called the lower and upper *cutoff frequencies*, respectively. From Figure 17.18(b) we can see that the maximum current occurs at f_0 and the current at f_1 and f_2 is $0.707I_{max}$. The bandwidth boundaries are also referred to as the half-power points since at those points $P = I^2R = (0.707I_{max})^2R = 0.5I_{max}R$.

The impedance of the circuit at f_1 and at f_2 is not the same as it is at resonance. Since the current has decreased by a factor of 0.707, the impedance must have increased by a factor of 1.414. The impedance at f_1, therefore, must be $1.414R$. Since Z is the vector sum of two rectangular components, the combined reactances of X_L and X_C at f_1 must be equal to R (Figure 17.18(c) and (d)).

$$X_L - X_C = R$$

In order to state this condition in terms of Q, we may substitute X_L for X_C.

$$X_{L2} - X_{L1} = R$$

$$2\pi f_2L - 2\pi f_1L = R$$

$$f_2 - f_1 = \frac{R}{2\pi L}$$

Dividing by f_0,

$$\frac{\Delta f}{f_0} = \frac{R}{2\pi f_0 L} = \frac{1}{Q}$$

$$\Delta f = \frac{f_0}{Q} \qquad (17.21)$$

Note from Eq. (17.21) that the bandwidth of a series RCL circuit becomes a function of Q. The greater the Q (hence the smaller the resistance), the narrower the bandwidth. A narrow bandwidth results in a highly selective circuit.

Example 17.15 A series RCL circuit is to consist of a 50 mH coil whose wire resistance is 20 ohms, and a 600 pF capacitor. What is the bandwidth? What would be the bandwidth if a 100 mH, 20 ohm coil were used with a 300 pF capacitor?

$$f_0 = \frac{1}{2\pi\sqrt{LC}} = \frac{1}{(6.28)\sqrt{(50)(10^{-3})(0.6)(10^{-9})}} = 29.06 \text{ kHz}$$

$$Q = \frac{2\pi fL}{R} = \frac{(6.28)(29.06)(10^3)(50)(10^{-3})}{20} = 456.47$$

$$\Delta f = \frac{(29.06)(10^3)}{456.47} = 64 \text{ Hz}$$

If $L = 100$ mH, and $C = 300$ pF, $f_0 = 29.06$ kHz

$$Q = \frac{(6.28)(29.06)(10^3)(100)(10^{-3})}{20} = 912.87$$

$$\Delta f = \frac{(29.06)(10^3)}{912.87} = 31.83 \text{ Hz}$$

For high Q circuits ($Q \geq 10$), the cutoff frequencies may be approximated by

$$f_2 = f_0 + \frac{\Delta f}{2}$$

and

$$f_1 = f_0 - \frac{\Delta f}{2}$$

17.6 PARALLEL RESONANCE

Parallel-resonant circuits are used extensively in radio frequency circuits. The basic circuit is shown in Figure 17.19(a). Note that this circuit can be thought of as a series circuit joined into a loop. At resonance the currents in the series loop are high; however, the current in the outside circuit is quite low at resonance. It follows, then, that the impedance between terminals A and B must be high at resonance.

The circuit of Figure 17.19(a) is a representative circuit. In actuality a resistor would not be connected in the loop. The resistance shown represents the wire resistance of the inductor's winding. If the resistance is very low, the resonant frequency is very near the resonant frequency of a series circuit (Eq. 17.17).

A common form of the parallel-resonant circuit is shown in Figure 17.19(b). In a television receiver, for example, it is important to receive a broad bandwidth of frequencies. The parallel R is used to broaden the bandwidth. (The series resistance in the L branch can be neglected).

To evaluate the resonant frequency of the first case, we must first solve the equation for the total impedance. The capacitive branch is in parallel with the L/R branch.

$$Z_t = \frac{(-jX_C)(R + jX_L)}{R + jX_L - jX_C}$$

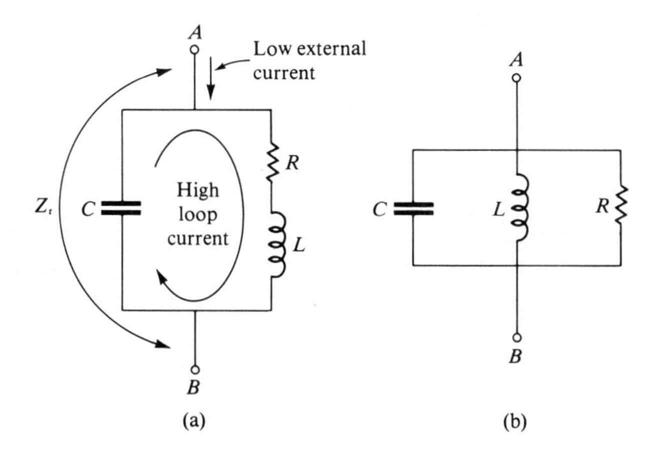

Figure 17.19 Parallel *RCL* Circuit

Multiplying numerator and denominator by the conjugate,

$$Z_t = \frac{-jR^2 X_C - RX_C(X_L - X_C) + RX_L X_C - jX_L X_C(X_L - X_C)}{R^2 + (X_L - X_C)^2}$$

Separating the j and the non-j terms

$$Z_t = \frac{RX_L X_C}{R^2 + (X_L - X_C)^2} - j\frac{R^2 X_C + X_C X_L(X_L - X_C)}{R^2 + (X_L - X_C)^2} \tag{17.22}$$

We may make some fairly reliable approximations from this equation. Since the j term is zero at resonance, the actual impedance is

$$Z_t = \frac{RX_L X_C}{R^2 + (X_L - X_C)^2}$$

However, $(X_L - X_C)$ is near zero, hence

$$Z_t \approx \frac{RX_L X_C}{R^2} = \frac{R2\pi fL}{R^2 2\pi fC} = \frac{L}{RC} \tag{17.23}$$

The resonant frequency can be determined by setting the j portion of Eq. (17.22) to zero and solving for f.

$$0 = \frac{R^2 X_C + X_L X_C(X_L - X_C)}{R^2 + (X_L - X_C)^2}$$

Substituting $X_L = 2\pi fL$ and $X_C = 1/2\pi fC$ results in

$$0 = CR^2 - L + 4\pi^2 f^2 L^2 C$$

Solving f_0,

$$f_0 = \frac{1}{2\pi}\sqrt{\frac{L - CR^2}{L^2C}} \tag{17.24}$$

The characteristics of parallel-resonant circuits are represented in Figure 17.20(a), (b). Note that the impedance reaches its highest value at resonance. Below resonance the parallel "tank" circuit appears as an inductive reactance; while above, it is capacitive. The external current (outside the parallel tank) flowing into the parallel circuit is in phase with the applied voltage at resonance. Below resonance the X_L is less than the X_C, hence the total current is principally inductive. The impedance reflects a positive phase angle. Above resonance the impedance reflects a negative phase angle. Figure 17.20(c) shows the phase angle changes.

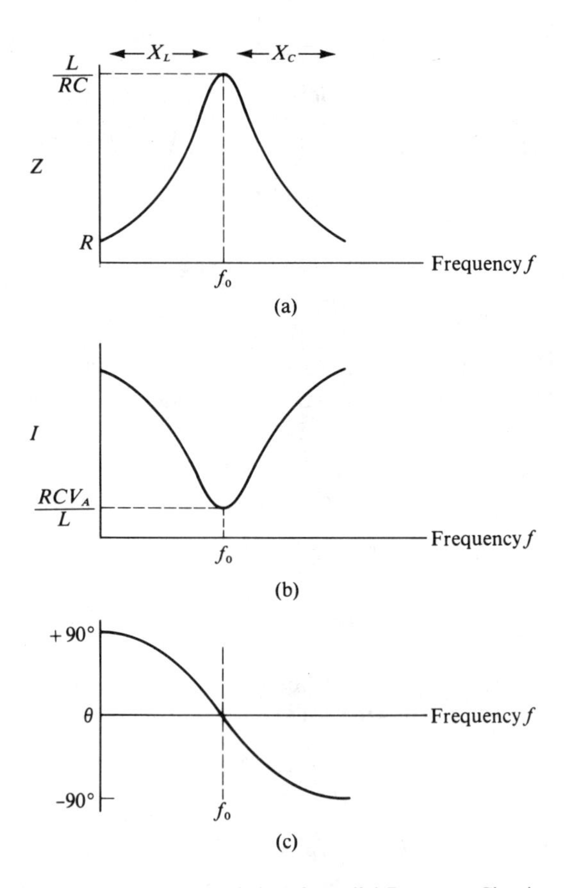

Figure 17.20 Characteristics of Parallel-Resonant Circuits

Example 17.16 Assume the parallel resonant tank circuit of Figure 17.19(a) has the following characteristics: $L = 4$ mH, $C = 0.1\ \mu$F, and the coil's resistance is 100 ohms. What is the impedance between points A and B? What is the resonant frequency?

Z_t, according to Eq. (17.23), is

$$Z_t \approx \frac{L}{RC} = \frac{(4)(10^{-3})}{(100)(10^{-7})} = 400 \text{ ohms}$$

$$f_0 = \frac{1}{(6.28)}\sqrt{\frac{(4)(10^{-3}) - (10^{-7})(100)^2}{(4)^2(10^{-3})^2(10^{-7})}}$$

$$= 6.89 \text{ kHz}$$

(Note that if the equation $f_0 = 1/2\pi\sqrt{LC}$ had been used, the resonant frequency would have calculated to 7.95 kHz—a substantial error!) The student should rework Example 17.16 using the same L and C except assuming the coil's resistance is only 10 ohms. The resonant frequency becomes 7.94 kHz. We may conclude, therefore, that the Q of the coil influences the resonant frequency of a parallel-resonant circuit. Solving Eq. (17.20) for R and substituting into Eq. (17.24) results in

$$f_0 = \frac{1}{2\pi\sqrt{LC}}\sqrt{1 - \frac{1}{Q^2}} \tag{17.25}$$

From Eq. (17.25) we may conclude that for very large values of Q (hence small values of R) the parallel-resonant frequency is calculated using the same equation as that used for the series-resonant frequency condition.

Our final consideration is that of power factor. Since the impedance of a parallel-resonant tank circuit is purely resistive at resonance, the power factor is 1 (same as the series-resonant case). In parallel-resonant circuits the true power is I^2Z, where Z is resistive.

The following example problem illustrates some very important information. The results should be studied carefully and conclusions concerning Q and bandwidth should be observed.

Example 17.17 The digital computer may be used to provide a wide range of information concerning parallel-resonant tanks. A tuned circuit consists of a 20 μH coil and a variable capacitor that ranges from 50 pF to 350 pF. Four 20 μH coils are available, each with resistances of 5, 10, 15, and 20 ohms. For each coil, write a program to compute f_0, Q, Δf, and Z_t as C varies in increments of 50 pF.

Printout for Example 17.17

```
10    P1=3.14159 : L=20E-6 : FOR R=5 TO 20 STEP 5 : PRINT
15    PRINT TAB(15) "FOR A COIL OF " R " OHMS RESISTANCE" :PRINT
20    PRINT "C <PF>" TAB(10) "F(0) <KHZ>" TAB(22) " Q" ;
25    PRINT TAB(34) "BW <HZ>" TAB(46) "Z(T) <KOHMS>" : PRINT
30    FOR C=50E-12 TO 350E-12 STEP 50E-12
35    F=(1/(2*P1))*SQR((L-C*R↑2)/(L↑2*C))
40    Q=(1/R)*SQR(L/C) : F1=F/Q : Z=L/(R*C)
45    PRINT C*1E12  TAB(10)  F*1E-3  TAB(22)  Q ;
50    PRINT TAB(34)  F1  TAB(46)  Z*1E-3
55    NEXT C : NEXT R : END
```

FOR A COIL OF 5 OHMS RESISTANCE

C <PF>	F(0) <KHZ>	Q	BW <HZ>	Z(T) <KOHMS>
50	5032.77	126.491	39787.5	80
100	3558.59	89.4427	39786.3	40
150	2905.49	73.0297	39785.1	26.6667
200	2516.15	63.2456	39783.8	20
250	2250.44	56.5685	39782.6	16
300	2054.3	51.6398	39781.3	13.3333
350	1901.85	47.8091	39780.1	11.4286

FOR A COIL OF 10 OHMS RESISTANCE

C <PF>	F(0) <KHZ>	Q	BW <HZ>	Z(T) <KOHMS>
50	5032.3	63.2456	79567.7	40
100	3557.93	44.7214	79557.7	20
150	2904.67	36.5148	79547.7	13.3333
200	2515.2	31.6228	79537.8	10
250	2249.39	28.2843	79527.9	8
300	2053.14	25.8199	79517.9	6.66667
350	1900.6	23.9046	79507.9	5.71429

FOR A COIL OF 15 OHMS RESISTANCE

C <PF>	F(0) <KHZ>	Q	BW <HZ>	Z(T) <KOHMS>
50	5031.51	42.1637	119333	26.6667
100	3556.81	29.8142	119299	13.3333
150	2903.31	24.3432	119266	8.88889
200	2513.63	21.0819	119232	6.66667
250	2247.62	18.8562	119198	5.33333
300	2051.21	17.2133	119165	4.44444
350	1898.52	15.9364	119131	3.80952

FOR A COIL OF 20 OHMS RESISTANCE

C <PF>	F(0) <KHZ>	Q	BW <HZ>	Z(T) <KOHMS>
50	5030.41	31.6228	159076	20
100	3555.25	22.3607	158996	10
150	2901.4	18.2574	158916	6.66667
200	2511.43	15.8114	158837	5
250	2245.16	14.1421	158757	4
300	2048.51	12.9099	158677	3.33333
350	1895.6	11.9523	158597	2.85714

SUMMARY

Series RCL circuits respond to alternating currents at various frequencies in a unique way. A specific RCL combination exhibits resonance at a point where $X_L = X_C$. At resonance the phase angle is zero and the impedance is minimum. As the frequency drops below the resonant frequency, the circuit becomes capacitive; above, it becomes inductive.

Parallel LC circuits have a specific resonant frequency. At resonance the impedance across the parallel circuit approaches L/RC ohms. The phase angle is zero and below resonance the parallel circuit appears inductive; above, it appears capacitive.

The reciprocal of impedance is admittance. $Y = 1/Z$. The conductance is the reciprocal of resistance, and susceptance is the reciprocal of reactance. A series R and X circuit must be transformed into conductance and susceptance by, first, converting impedance to polar form, second, taking the reciprocal, and then, third, converting the polar admittance to the equivalent rectangular form. The result is a parallel equivalent circuit.

Network theorems may be applied to ac circuits providing care is taken to make all the proper polar and rectangular conversions. When theorems such as Thevenin's is used, the Thevenin impedance should be calculated using V_{oc}/I_{sc}.

All ac circuits involving RCL components reflect a series-resonant Q factor. The Q factor is used to determine the bandwidth of the resonant circuit. $\Delta f = f_0/Q$.

PROBLEMS

Reference Section 17.1

1. If a series circuit consists of a 10 mH coil, a 0.015 μF capacitor, and a 1 kΩ resistor, what is the total current and its phase angle when 10 volts at 6 kHz is applied?
2. A series circuit consists of a resistance of 2 kΩ, an X_L of 3 kΩ, and a capacitive reactance of 4 kΩ. If $f =$ 10 kHz find:
   a. the total impedance
   b. the phase angle
   c. the L and C
3. A series RCL circuit has a total impedance of 3 kΩ $\angle 30°$.
   If the $X_C =$ 1 kΩ, what is:
   a. resistance in the circuit?
   b. the inductive reactance?

4. If a series circuit consists of 1 kΩ resistor and a 0.2 henry coil, how much capacity must be added in series to have a zero phase angle when 10 volts at 5 kHz is applied?
5. If $100 + j200$ is connected in series with $300\underline{/-25^\circ}$, find:
   a. the total impedance
   b. the total phase angle
6. A series circuit consists of a 70 mH coil and a capacitor. What is the capacitance of the circuit if the applied voltage is 20 volts at 5 kHz and the coil's resistance is 600 ohms? The current flowing in the circuit is 33 mA. Assume the *net* reactance of the circuit is inductive at 5 kHz.
7. If the frequency in problem 6 is doubled, what is the new current flow in the circuit?
8. A series circuit consists of a coil and a capacitor. The dc resistance of the coil is 20 ohms. The frequency is 455 kHz. If the inductance of the coil is 0.1 mH, what capacitance is needed in the circuit to cause resonance?
9. What is the impedance of the circuit of problem 8 at resonance?
10. A series circuit consists of a 10 mH coil, a 50 pF capacitor, and a 4 kΩ resistor. If 10 volts 240 kHz is applied across the circuit what is:
    a. the voltage drop across L?
    b. the voltage drop across C?
    c. the voltage drop across R?
11. A coil in a series tuned circuit of a radio receiver has an inductance of 300 μH and a resistance of 15 ohms. What value of capacitance must be connected in series with the coil of the circuit to be series resonant at 840 kHz?
12. If a given *RCL* series circuit has 25 volts rms at the resonant frequency of 2 kHz, a current of 4 mA flows. If the capacitor is 0.5 μF, what is the inductance?

Reference Section 17.2

13. If the three components of problem 1 are connected in parallel, what are the total current flow and phase angle when the same voltage and frequency are applied?
14. If $240\underline{/70^\circ}$ is connected in parallel with $-j150$, what is the series equivalent impedance?
15. If a parallel combination of $(1 - j2)$ kΩ and 3 kΩ has 20 volts applied, what is the total current? What is the current through the 3 kΩ resistor?
16. A parallel *RCL* circuit consists of a 20 mH coil, a 1 kΩ resistor, and an unknown C. What value of C is needed if the current leaving a 20 volt 2 kHz generator is to lead by 20°?

17. The series circuit 2 kΩ − j4 kΩ can be replaced with what parallel circuit?

Reference Section 17.3

18. Refer to Problem 13. What are the conductance and the susceptance of each of the three branches?
19. What are the conductance and the susceptance of $Z = 420\angle -30°$?
20. If a parallel circuit consists of $G = 0.002$, $\mathbf{B} = -j0.015$, what is the equivalent series circuit?
21. Assume a parallel circuit has a total admittance of $0.052 + j0.008$ siemens. What value of reactance must be added in the series equivalent circuit for the circuit to be series-resonant? What is the resonant frequency?
22. A parallel RCL circuit consists of a 1.5 kΩ, an X_L of 2 kΩ, and an X_C of 3.2 kΩ. What is the admittance of the circuit? What is the equivalent series circuit?

Reference Section 17.4

23. Refer to the delta circuit of Figure 17.21. What is the total impedance between points A and B. What is the phase angle?
24. Refer to Figure 17.21. What is the power factor of the circuit?
25. Refer to Figure 17.21. If 20 volts ac is applied across A and B, what is the Thevenin equivalent circuit "seen" by the 5 kΩ resistor?
26. Refer to Figure 17.22. What is the Thevenin voltage for terminals A and B?
27. What is the current flowing through the inductive reactance shown in Figure 17.23?

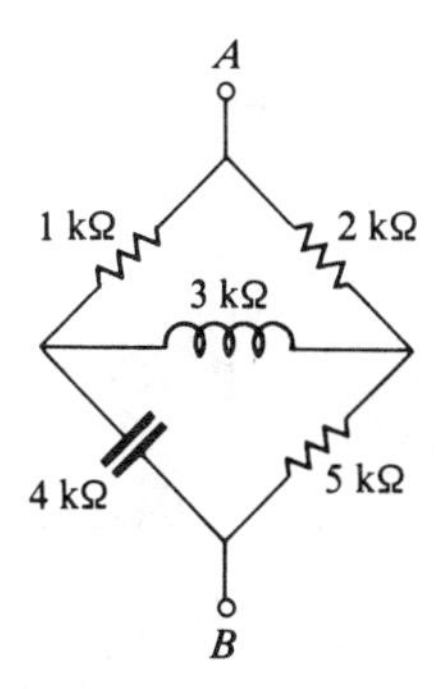

Figure 17.21 Figure for Problems 17.23 through 17.25

28. Refer to Figure 17.23. What are the voltage and phase angle across the inductance?
29. Find Z_T, θ, and i_t when the input frequency to the circuit of Figure 17.24 is 1515 Hz, 1590 Hz, 1675 Hz.
30. If the inductor of Figure 17.24 is changed to 0.2 henry, and the capacitance is changed to 0.05 μF, find Z_T, θ, and i_t when input frequency is 1515 Hz and 1675 Hz.

Reference Sections 17.5 and 17.6

31. A series-resonant circuit has a Q of 120 and the dc resistance of the circuit is 50 ohms. If the resonant frequency is 30 kHz, what is the inductance of the coil? What is the value of capacitance?

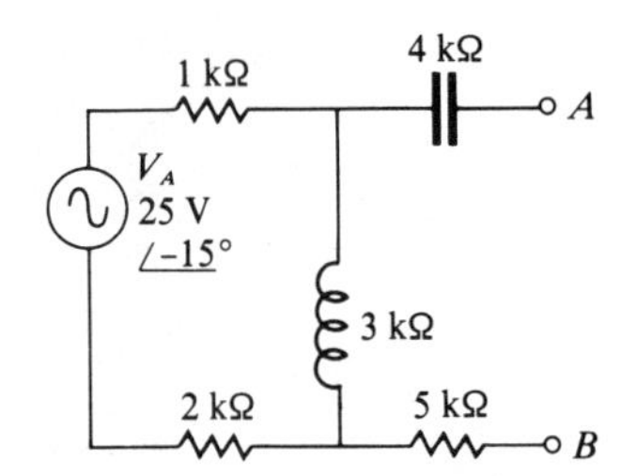

Figure 17.22 Figure for Problem 17.26

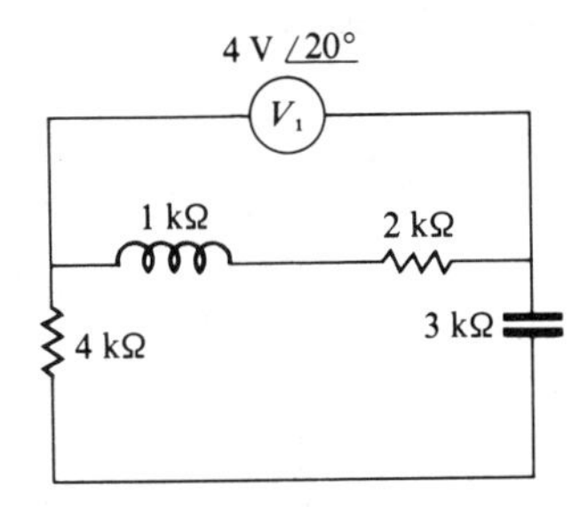

Figure 17.23 Figure for Problems 17.27 and 17.28

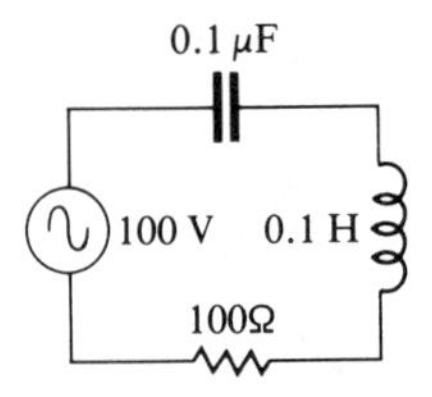

Figure 17.24 Figure for Problems 17.29 and 17.30

32. A given series-resonant circuit has a coil of 1.5 mH and a capacitance of 500 pF. If the impedance of the circuit at resonance is 75 ohms, what is the Q of the circuit? What is the resonant frequency?
33. Consider the following series-resonant circuit: $L = 4$ mH, $C = 0.1$ μF, R = 10 ohms. If 10 volts at resonance is applied, what is:
    a. the current in the circuit?
    b. the Q of the circuit?
    c. the voltage across X_C at resonance?
34. A given series-resonant circuit has a resonant frequency of 1.6 kHz. If $R = 100\Omega$, $L = 400$ mH, and $C = 0.025$ μF, at what frequency is the total impedance $Z_t = 141\angle 45°$? At what frequency is $Z_t = 141\angle -45°$? What is the bandwidth?
35. If 10 volts at 1.6 kHz is applied to the circuit of problem 34, what is the voltage across L? What is it across C?
36. A given series-resonant circuit has a resonant frequency of 250 kHz. If $L = 200$ μH and $R = 40$ ohms, what is the bandwidth?
37. A series circuit is to pass a band of frequencies from 80 kHz to 100 kHz. If a 30 mH coil is selected, what Q should it have? What series capacitance would be required?
38. If the applied voltage in Problem 37 is 120 mV, what would be the voltage across L at 80 kHz?
39. Refer to the circuit of Figure 17.25. Determine the following: f_0, Z_0.
40. A capacitor has a reactance of 10^4 ohms and it is connected in parallel with an inductance whose reactance is also 10^4 ohms. Assume the dc resistance of the coil's windings is 200 ohms. The resonant frequency of the combination is 100 kHz.
    a. What is the Q of the circuit?
    b. What is the bandwidth of the resonant tank?
41. Assume a 4 mH inductance is connected in parallel with a 40 μF capacitance. Assume the 10 ohms of resistance produced by the windings of the coil can be neglected. What is the resonant frequency?

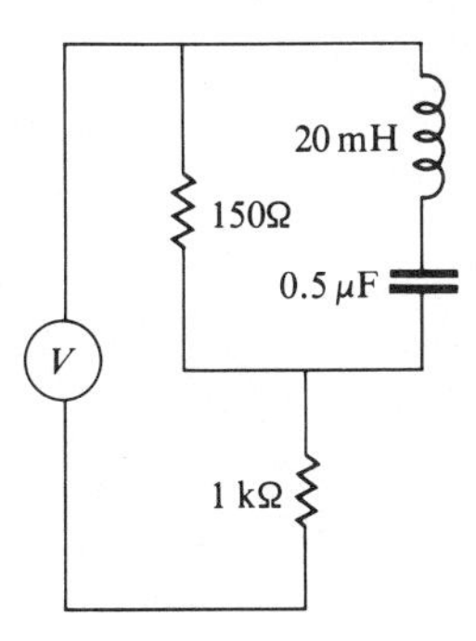

Figure 17.25 Figure for Problem 17.39

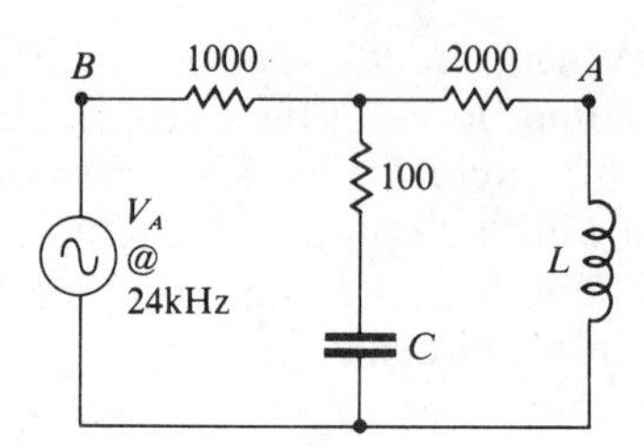

Figure 17.26 Figure for Problem 45

42. Assume that a 300 μH coil is to be used in parallel with a variable capacitor to tune across the AM broadcast band (550 kHz to 1650 kHz). What should be the maximum and minimum values of the variable capacitor?
43. A 200 μH coil with a dc resistance of 500 ohms is to resonate at 30 MHz in a parallel tank configuration. What value of capacity is required?
44. If a 10 kΩ resistor is connected in parallel with the tank circuit of Problem 43, what is the resonant frequency?

Optional Computer Problem

45. Reference Figure 17.26. Let L = 10 mH, C = 0.001 microfarad, V = 20 volts, frequency f is variable. Calculate and print out a table in 2 kHz increments, listing Z_{in} and I_{in} (magnitude and phase angle) for $f = f_0 - 14$ kHz to $f_0 + 26$ kHz. Note: this range will indicate the impedance maximum point, the current minimum point and the phase angle crossover point. Because of the complex nature of the circuit, the three do not occur simultaneously.

PART IV

This concluding segment deals with some applications of the principles just studied. Magnetism and electromagnetism, along with associated magnetic circuits, lead to an understanding of the transformer. Here we see how one electrical circuit may transfer voltages to another circuit without any physical connection between them. This coupling from one circuit to another is further explored in nontransformer circuits also. The losses and tradeoffs of design are analyzed for the general problems of interconnecting one circuit to another. The attenuation, or reduction, of voltages and currents because of the coupling arrangement is introduced.

Often in electrical circuits unwanted voltages and currents introduce disruptive effects for a given operation. Various circuit connections are used to block or eliminate the undesirable signals. Such circuits are identified as *filters*. This text provides only an introduction to filters since the subject could be the material of a complete text.

CHAPTER 18

TRANSFORMERS

18.1 MUTUAL INDUCTANCE

Since transformers are used extensively in power supply circuits, communication systems, and other circuits, the coverage in this chapter will emphasize the basic theory behind these applications. Both power and high frequency transformers are analyzed. The general theory of ideal transformers of both the iron type and the air core type are discussed.

The discussion of mutual inductance in Chapter 11 illustrated that when two inductors share mutual variations in flux, a mutual inductance appears in the circuit of each inductor. This mutual inductance depends upon the percentage of interactive flux called the *coefficient of coupling k*.

In Figure 18.1(a) note that the two windings A and B are wound on an air core cylindrical form. The percentage of interactive flux is quite small. The coefficient of coupling for such a device is generally $0 < k < 0.05$. The symbol for such a transformer is shown in Figure 18.1(b).

The transformer of Figure 18.1(c) shows three windings on a powdered iron toroid. Windings B and C are considered output windings, A an input winding. There is no special convention concerning input and output windings since such a selection is purely arbitrary. The symbol for such a physical arrangement is shown in Figure 18.1(d). The dashed lines

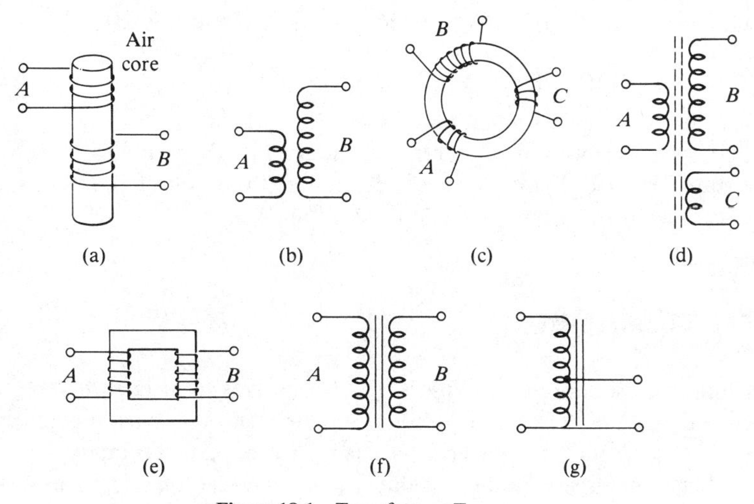

Figure 18.1 Transformer Types

indicate a powdered iron core. Powdered iron core material produces coefficients of coupling $0 < k < 0.2$.

The transformer of Figure 18.1(e) shows an iron configuration which captures almost all of the flux in the core area. The core itself is generally constructed of laminated sheets of iron insulated from each other. The iron core types may embody many windings on the basic core. The schematic symbol for iron core transformers is shown in Figure 18.1(f). The coefficient of coupling range is $0.5 < k < 0.999$. The input of a transformer is called its *primary* winding and the output is called its *secondary* winding.

In all of the transformer types shown in Figure 18.1, the mutual inductance between any two windings is given as

$$\boxed{M = k\sqrt{L_p L_s}} \tag{18.1}$$

where L_p is the inductance of the primary, L_s is the inductance of the secondary, k is the coefficient of coupling, and M is the mutual inductance in henries.

Example 18.1 Assume a transformer has a primary inductance of 2 mH and a secondary of 8 mH. What is the mutual inductance if the coefficient of coupling is 0.85?

$$M = k\sqrt{L_p L_s} = 0.85\sqrt{(2)(10^{-3})(8)(10^{-3})} = (0.85)(4)(10^{-3}) = 3.4 \text{ mH}$$

The mutual inductance can also be considered the link between the two windings of a transformer.

The amount of voltage induced into the secondary winding (V_s) depends upon the rate of change of the primary current ($\Delta i/\Delta t$).

$$V_s = -M\frac{\Delta i}{\Delta t} \tag{18.2}$$

where M is the mutual inductance in henries. If the primary current of the transformer of Example 18.1 is changing at the rate of 50 mA/ms, the voltage across the secondary would be 170 mV.

18.2 IDEAL TRANSFORMER

Assume the transformer of Figure 18.1(e) utilizes an iron core such that the coefficient of coupling is unity (100 percent). With such a condition, all the flux ϕ in the iron core is shared by both the primary and the secondary. Assume winding A is the primary and that it has N_p turns on the iron core. Winding B is the secondary and has N_s windings. The amount of voltage induced across the primary winding is given as:

$$V_p = N_p \frac{\Delta\phi}{\Delta t} \tag{18.3}$$

Similarly, the voltage induced across the secondary is

$$V_s = N_s \frac{\Delta\phi}{\Delta t} \tag{18.4}$$

The flux field is the same for each winding. Solve Eqs. (18.3) and (18.4) for $\Delta\phi/\Delta t$ and set them equal to each other.

$$\frac{V_s}{N_s} = \frac{V_p}{N_p} \quad \text{and} \quad \frac{N_p}{N_s} = \frac{V_p}{V_s}$$

The ratio N_p/N_s is defined as the *turns ratio* of the transformer.

$$\boxed{\alpha = N_p/N_s} \tag{18.5}$$

where α = turns ratio.

Similarly

$$\boxed{\alpha = V_p/V_s} \tag{18.6}$$

Figure 18.2(a) shows a transformer whose primary has 200 turns. The secondary has 1000 turns. When an ac voltage of 25 volts rms is applied at the primary, the voltage at the output or secondary is $V_s = V_p/\alpha$, where the α is 0.2. V_s is 125 volts. If the 25 volt source is connected across the secondary, the voltage across the open primary will be 5 volts. When $\alpha <$ 1, the secondary voltage is greater than the primary voltage, and the transformer is called a *step-up* transformer. When the reverse is true ($\alpha >$ 1), it is called a *step-down* transformer.

Note that the voltage across the output windings occur with no load resistance connected. (The secondary is an open circuit.) When a load resistance is connected across the secondary, current will flow in the secondary. Let us again assume that the transformer is ideal, hence there

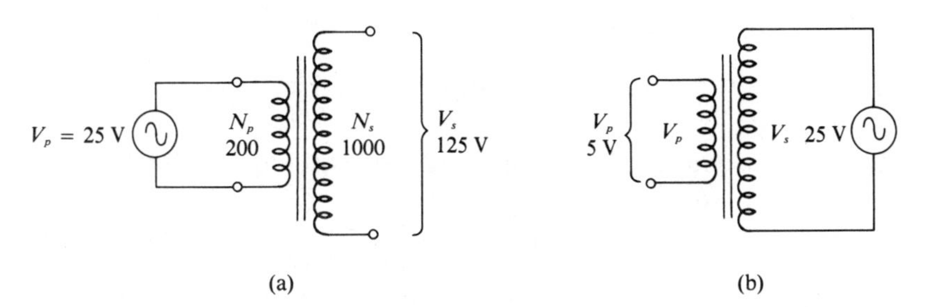

Figure 18.2 Ideal Transformer

are no losses connected with the device and the coefficient of coupling is 1. Since $p = iv$, we may solve for V_p and substitute P/I into Eq. (18.6):

$$\alpha = \frac{P_p/I_p}{P_s/I_s} = \frac{P_pI_s}{P_sI_p}$$

However, since the device is considered ideal, the energy delivered to the primary must equal the energy at the secondary, hence $P_p = P_s$. Therefore,

$$\boxed{\alpha = I_s/I_p} \tag{18.7}$$

The turns ratio can also be identified as the ratio of secondary current to primary current. A transformer can be used to step up the voltage; however, the current is stepped down by the same ratio.

Example 18.2 Assume an ideal transformer has 50 volts applied to a 300 turn primary. The 50 turn secondary is connected to a load resistance of 600 ohms. What is the voltage across the secondary? What is the current in the secondary? What is the current in the primary?

$$\alpha = \frac{N_p}{N_s} = \frac{300}{50} = 6$$

$$V_s = \frac{V_p}{\alpha} = \frac{50}{6} = 8.333 \text{ volts}$$

$$I_s = \frac{V_s}{R} = \frac{8.333}{600} = 13.888 \text{ mA}$$

$$I_p = \frac{I_s}{\alpha} = \frac{13.888(10^{-3})}{6} = 2.315 \text{ mA}$$

We may wish to verify that the power at the primary is the same as the power at the secondary, $I_sV_s = 115.75$ mW and $I_pV_p = 115.75$ mW.

Refer to the transformer of Figure 18.3(a). The impedance of the load Z_L is reflected as an input impedance at the primary terminals. The actual impedance at Z_{in} depends upon the turns ratio and characteristics of the Z_L. The equations for Z_s and Z_p, according to Ohm's law, are:

$$Z_s = \frac{V_s}{I_s} \quad \text{and} \quad Z_p = \frac{V_p}{I_p}$$

Solving these equations for I_s and I_p and substituting into Eq. (18.7),

$$\alpha = \frac{V_s/Z_s}{V_p/Z_p} = \frac{V_sZ_p}{V_pZ_s}$$

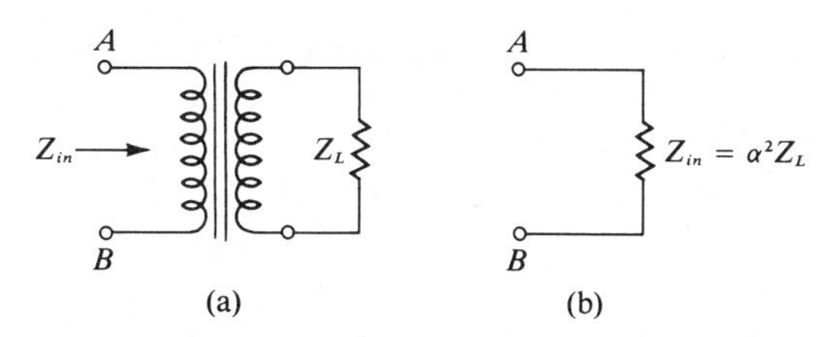

Figure 18.3 Impedance Transformation

Substituting Eq. (18.6) for V_p/V_s,

$$\boxed{\alpha^2 = Z_p/Z_s} \tag{18.8}$$

The impedance reflected back to the primary from the load on the secondary is

$$Z_p = \alpha^2 Z_s \tag{18.9}$$

For a stepup transformer the α is less than 1, hence the load impedance reflected to the primary is less by an amount equal to the square of the turns ratio. If the transformer is a step-down type, that is, fewer turns in the secondary compared to the primary, then the reflected impedance is higher. Figure 18.3(b) shows the equivalent circuit.

Example 18.3 Assume a transformer has 600 turns in the secondary and 3000 turns in the primary. When a 16 ohm load is connected across the secondary, what current will flow in the primary when 24 volts ac is applied across the primary? What is the current in the load?

$$\alpha = \frac{3000}{600} = 5$$

$$Z_{\text{in}} = (5^2)(16) = 400 \text{ ohms}$$

$$I_p = \frac{V_A}{Z_{\text{in}}} = \frac{24}{400} = 60 \text{ mA}$$

$$I_s = \alpha I_p = (5)(60)(10^{-3}) = 300 \text{ mA}$$

Core Losses

The ideal transformer may still be considered arithmetically as an ideal device even though some I^2R (power) loss exists. Since I^2R losses are reflected as energy converted to heat, we can consider the transformer as an energy system. In Figure 18.4(a) the total transformer system has an input energy shown as $I_pV_p = P_p$, the power delivered to the primary. The I^2R losses exist within the core and in the windings. The power delivered to the load is that power which remains. Symbolically this energy relationship is shown in Figure 18.4(b). The energy W_{in} is delivered to the system. W_h represents the lost energy in calories of heat. The energy delivered to the load resistance is $W_o = W_{\text{in}} - W_h$. These energies, expressed as power, can be used to determine the efficiency of the transformer.

If the load resistance of Figure 18.4(a) is removed and a voltage applied at V_p, then the amount of core loss can be determined. For example, with no load connected, assume 25 volts ac is applied at the input. If 2 mA of current flows with a lagging phase angle of 80°, the power factor is cos 80°: PF = 0.174. The apparent power is $IV = (2)(10^{-3})(25) = 50 \times 10^{-3}$ VA. The true power is 50 mW times the power factor, P_t = 8.68 mW, which represents core and winding losses.

When a load resistance is connected across the secondary, the primary and secondary I^2R losses can be determined using power, voltage, and current measuring devices. The primary I^2R, the secondary I^2R, and the core losses all constitute energy transferred to heat. From Figure 18.4(c), note the losses W_C (core), W_p (primary), and W_s (secondary). The total power delivered, therefore, is

$$P = P_c + P_p + P_s + P_L$$

The efficiency, η, is

$$\eta = \frac{P_L}{P_c + P_p + P_s + P_L} \tag{18.10}$$

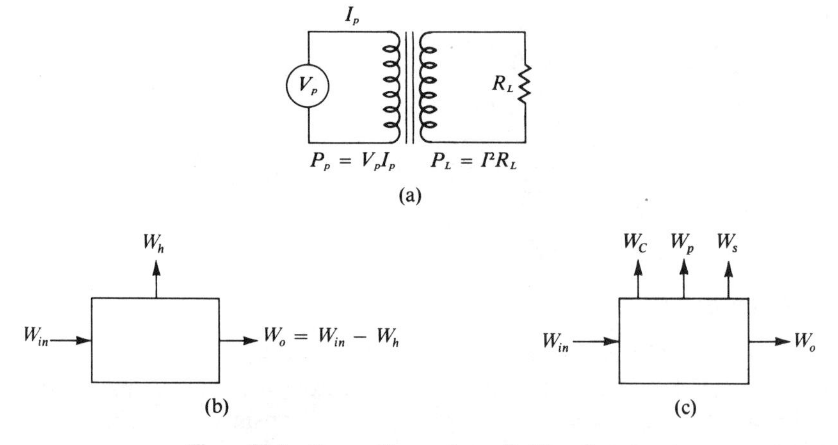

Figure 18.4 Energy Conversions of a Transformer

where P_L is power delivered to the load, P_C is core loss, P_s and P_L are primary and secondary I^2R losses.

Example 18.4 Assume a power transformer is 98 percent efficient. The primary consists of 200 turns and the secondary 50 turns. When a load is connected across the secondary, the current through the load is 1.5 amperes. If the applied voltage source is 110 V, what is the current drawn from the source?

$$\alpha = \frac{200}{50} = 4$$

$$V_s = \frac{V_p}{\alpha} = \frac{110}{4} = 27.5 \text{ volts}$$

$$P_L = IV = (1.5)(27.5) = 41.25 \text{ watts}$$

$$P_A = \frac{P_L}{\eta} = \frac{41.25}{0.98} = 42.092 \text{ volt-amps}$$

$$I_p = \frac{P_A}{V_p} = \frac{42.092}{110} = 382.65 \text{ mA}$$

(Note that if the turns had been used to calculate the current, it would have been 300 mA. The additional 82.65 mA is required for the I^2R losses of the transformer.)

18.3 MUTUAL IMPEDANCE

Equation (18.1) established that there exists between two interactive inductances a mutual inductance. For the transformer case the mutual inductance is seen as an additional inductance in the primary of the transformer. When an ac voltage is applied at the input, the circuit reflects not only any primary impedance but also the impedance due to mutual inductance. This component of impedance is given as:

$$Z_M = 2\pi f M \tag{18.11}$$

Since the mutual inductance M depends upon the coefficient of coupling, the influence of Z_M on the primary and secondary circuits depends upon core material. For example, for an iron core device with $k = 1$ the mutual impedance is

$$Z_M = 2\pi f \sqrt{L_p L_s} \tag{18.12}$$

Example 18.5 Assume a transformer has a primary inductance of 2 henries and a secondary inductance of 7.5 henries. If the coefficient of coupling is 0.95, what is the inductive reactance of the primary and secondary? What is the mutual impedance in the primary and secondary circuits if the applied voltage is 20 volts at 400 Hz?

$$X_{L_p} = 2\pi f L_p = (6.28)(0.4)(10^3)(2) = 5.026 \text{ k}\Omega$$
$$X_{L_s} = 2\pi f L_s = (6.28)(0.4)(10^3)(7.5) = 18.85 \text{ k}\Omega$$
$$M = k\sqrt{L_p L_s} = 0.95\sqrt{(2)(7.5)} = 3.68 \text{ henries}$$
$$Z_M = 2\pi f M = (6.28)(0.4)(10^3)(3.68) = 9.247 \text{ k}\Omega$$

The mutual impedance can be illustrated in an equivalent circuit of the transformer. Assume the transformer of Figure 18.5(a) has a coefficient of coupling near unity. The primary winding not only has inductive reactance, but some resistance as well. The real and imaginary properties of the device can be identified as the primary impedance Z_p. Similarly, the secondary circuit consists of inductive reactance and wire resistance as well as load resistance. The equivalent circuit is illustrated in Figure 18.5(b). Note that the mutual impedance is common to both the primary loop and the secondary loop.

The voltages around the primary loop consist of V_p (the voltage applied to the primary winding). The voltage across Z_p is due to primary current. The voltage across Z_M is the result of the secondary current effect in the primary loop. The voltage in the secondary loop is the result of primary current. The equations for the two loops are

$$V_p = I_p Z_p + I_s Z_M$$
$$0 = I_p Z_M + I_s Z_s$$

Solving the second equation for I_s,

$$I_s = \frac{-I_p Z_M}{Z_s}$$

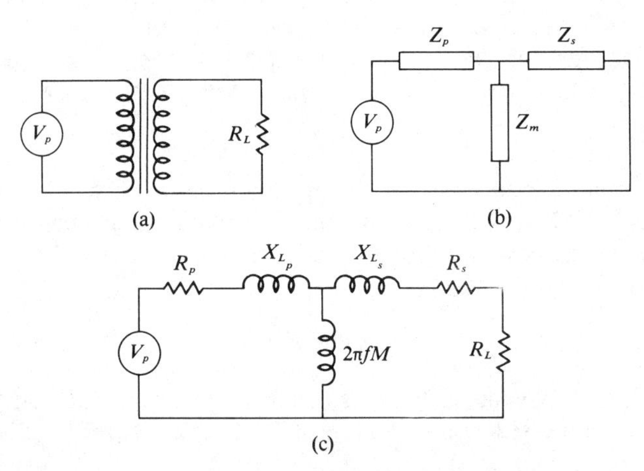

Figure 18.5 Mutual Impedance

Substituting into the first equation,

$$V_p = I_p Z_p - \frac{I_p(Z_M)^2}{Z_s}$$

Dividing by I_p,

$$\frac{V_p}{I_p} = Z_p - \frac{Z_M^2}{Z_s}$$

$$Z_{in} = Z_p - \frac{Z_M^2}{Z_s} \qquad (18.13)$$

The impedances of Eq. (18.13) include the lumped properties of the circuits. Figure 18.5(c) illustrates some of the individual components represented by the impedance.

Example 18.6 Assume an iron core transformer has 25 mV at 10 kHz applied across the primary. The inductance of the primary is 5 mH and the inductance of the secondary is 10 mH. If the R_L across the secondary is 10 kΩ and the coefficient of coupling is 0.9, find (a) the mutual impedance, (b) the current in the primary, and (c) the current in the secondary.

a. $M = k\sqrt{L_p L_s} = 0.9\sqrt{(5)(10^{-3})(10^{-2})} = 6.34 \text{ mH}$

$Z_M = 2\pi f M = (6.28)(10^4)(6.34)(10^{-3}) = +j400 \text{ ohms}$

b. $X_{L_p} = 2\pi f L_p = (6.28)(10^4)(5)(10^{-3}) = 314 \text{ ohms}$

$X_{L_s} = (2\pi)(10^4)(10^{-2}) = 628 \text{ ohms}$

$$Z_{in} = +j314 + \frac{(16)(10^4)}{(10^4 + j628)}$$

since $10^4 \gg 628$

$$Z_{in} \approx +j314 + \frac{(16)(10^4)}{10^4}$$

$$= 16 + j314 \approx 315\angle 90^\circ$$

$$I_p = \frac{V_p}{Z_{in}} = \frac{(25)(10^{-3})}{315\angle 90^\circ} = 80\ \mu\text{A} \angle -90^\circ$$

c. since $I_p X_{L_p} + I_s Z_M = V_p$,

$$(80)(10^{-6})\angle -90^\circ (j314) + I_s(+j400) = 25(10^{-3})$$

$$(25.12)(10^{-3}) + I_s(400\angle 90^\circ) = 25(10^{-3})$$

$$I_s = -(0.3\angle -90^\circ)\ \mu\text{A}$$

(The minus sign for I_s indicates that I_p and I_s are 180° out of phase.) The load resistance in Example 18.6 was given as an extremely large value. Its magnitude was much larger than the inductive reactance of the secondary. This condition causes the denominator of Eq. (18.13) to be a real component, and this large value of load resistance reflects back as a small impedance into the primary circuit. In a similar manner, if the load placed on the transformer secondary is a capacitance, then, depending upon the frequency, that capacitance causes the denominator of Eq. (18.13) to contain a $-j$ component. The effect, then, is to reflect an increased inductive reactance in the primary circuit. The phase angle of the current in the primary is influenced by the resistive and reactive characteristics of the load.

Example 18.7 Assume an isolation transformer has a primary inductance of 10 mH, a secondary inductance of 10 mH, and a coefficient of coupling of 1. If a 20 volt 1 kHz signal is applied at the primary, what is the equivalent circuit seen by the signal source when the secondary sees a 100 ohm resistive load, a 100 ohm capacitive load, and a 100 ohm inductive load. (Assume the primary has a resistance of 20 ohms.)

$$M = k\sqrt{L_p L_s} = \sqrt{(10^{-2})(10^{-2})} = 10 \text{ mH}$$

$$Z_M = X_{L_s} = X_{L_p} = 2\pi f M = (6.28)(10^3)(10^{-2}) = +j62.8$$

With a 100 ohm resistive load:

$$\begin{aligned} Z_{\text{in}} &= 20 + j62.8 + \frac{(62.8)^2}{(100 + j62.8)} \\ &= 20 + j62.8 + 28.28 - j17.76 \\ &= 66 \angle 43^\circ \end{aligned}$$

With a 100 ohm capacitive load:

$$\begin{aligned} Z_{\text{in}} &= 20 + j62.8 + \frac{(62.8)^2}{+j62.8 - j100} \\ &= 20 + j62.8 + j106 \\ &= 170.98 \angle 83.24^\circ \end{aligned}$$

With a 100 ohm inductive load:

$$\begin{aligned} Z_{\text{in}} &= 20 + j62.8 + \frac{(62.8)^2}{+j62.8 + j100} \\ &= 20 + j62.8 - j24.22 \\ &= 20 + j38.58 \\ &= 43.45 \angle 62.6^\circ \end{aligned}$$

Note in the preceding example how the phase angle of the input impedance was increased by the capacitive load. The reflection of the inductive load reduced the phase angle and, for the capacitive load, it increased the phase angle.

In an actual transformer circuit, the resistive properties of the primary and the secondary cannot be ignored at certain operating frequencies. At low frequencies, for example, the reactances of the primary, secondary, and mutual inductances decrease; therefore, the resistive values become more significant.

Example 18.8 Refer to Figure 18.5(a). Assume the transformer has a coefficient of coupling $k = 0.9$. The primary winding is 15 ohms and 10 henries. The secondary is 45 ohms and 40 henries. If 25 volts at 30 Hz is applied, determine the primary current and the phase angle when 7 kΩ is connected across the secondary.

$$M = k\sqrt{L_p L_s} = (0.9)\sqrt{400} = 18 \text{ henries}$$

$$X_{L_p} = 2\pi f L_p = (6.28)(30)(10) = 1.88 \text{ k}\Omega$$

$$X_{L_s} = 2\pi f L_s = (6.28)(30)(40) = 7.54 \text{ k}\Omega$$

$$Z_{\text{in}} = R_p + jX_{L_p} - \frac{(Z_M)^2}{R_L + R_s + jX_{L_s}}$$

$$Z_M = 2\pi f M = (6.28)(30)(18) = j3.38 \text{ k}\Omega$$

$$Z_{\text{in}} = 15 + j(1.88)(10^3) - \frac{[(+j3.38)(10^3)]^2}{(7)(10^3) + j(7.54)(10^3)}$$

$$= +j(1.88)(10^3) + \frac{(11.42)(10^6)}{(10.3)(10^3)\angle 47^\circ}$$

$$= +j(1.88)(10^3) + (1.11)(10^3)\angle -47^\circ$$

$$= +j(1.88)(10^3) + (0.75)(10^3) - j0.812(10^3)$$

$$= 0.75 \text{ k}\Omega + j1.07 \text{ k}\Omega = 1.3 \text{ k}\Omega\angle 55^\circ$$

$$I_p = \frac{V}{Z_{\text{in}}} = \frac{25\angle 0^\circ}{1.3 \text{ k}\Omega\angle 55^\circ}$$

$$= 19.23 \text{ mA}\angle -55^\circ$$

18.4 MULTIPLE WINDING TRANSFORMER

Figure 18.1(d) illustrated a toroid core type transformer with three windings on one core. Many practical transformers utilize more than one winding. In power transformers several windings are used to provide various levels of ac voltage. A television receiver may require one voltage level for its integrated circuits, a different voltage for the high voltage power supply, and still a third voltage for the filament of the picture tube.

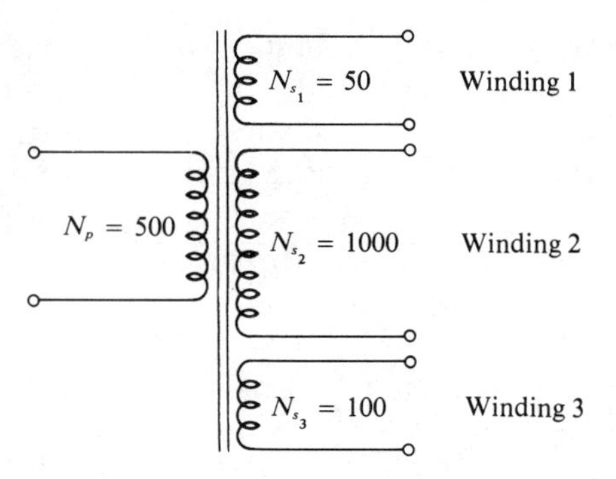

Figure 18.6 Multiple Winding Transformer

In such a transformer there would be one primary and three secondary windings.

Regardless of how many windings are utilized, the system adheres to all the basic theories studied in Sections 18.1 to 18.3. Each secondary winding interacts with the primary winding on a turns ratio basis. Refer to Figure 18.6. This transformer will have three distinct turns ratios:

$$\alpha_1 = \frac{N_p}{N_{s_1}} = 10 \qquad \alpha_2 = \frac{N_p}{N_{s_2}} = 0.5 \qquad \alpha_3 = \frac{N_p}{N_{s_3}} = 5$$

Assume 100 volts ac is applied across the primary. The voltage developed across winding number 1 will be 10 volts; across winding number 2, 200 volts, and across winding number 3, 20 volts. The currents in each of the secondaries will determine how much current must be drawn from the source.

Example 18.9 Refer to Figure 18.7(a). Given the information shown, determine the turns ratio of each secondary.

$$\alpha_1 = \frac{V_p}{V_s} = \frac{120}{12} = 10$$

$$\alpha_2 = \frac{V_p}{V_s} = \frac{120}{600} = 0.2$$

Although the turns ratio relationships may be used to determine the currents and voltages, care must be taken for multiple winding trans-

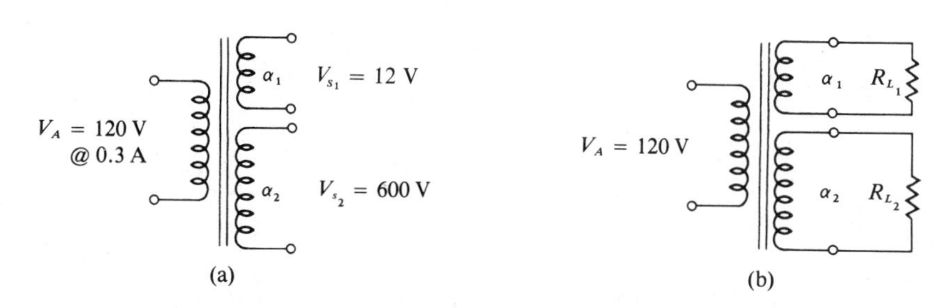

Figure 18.7 Figure for Examples 18.9 and 18.10

formers. Refer to the circuit of Figure 18.7(b). The primary shows a voltage rating of 120 volts. The manufacturer generally rates some maximum current which can flow in the primary circuit. The values of R_{L_1} and R_{L_2} will determine how much current will flow in the primary. Assume α_1 is 2 and α_2 is 0.5. If a maximum of 1.5 amperes may flow in the primary and R_{L_2} is 480 ohms, then the current through R_{L_2} is 0.5 ampere. The power is i^2R, hence 120 watts. The power at the primary is 180 watts. This leaves 60 watts for the other winding. Since 60 volts appears across R_{L_1}, the current may not exceed $i = p/V = 60/60 = 1.0$ amperes. The resistance R_{L_1} may be no less than $60/1.0 = 60$ ohms. Note that the sum of the two currents in the secondaries exceeds the current in the primary.

Example 18.10 Reference Figure 18.7(b). Assuming ideal conditions, let R_{L_2} vary from 1 kΩ to 10 kΩ in 500 ohm increments. Write a program in BASIC to print out the power expended in each secondary winding and the minimum value of R_{L_1} for each incremental change in R_2. Assume V (primary) = 120 volts, I (primary) = 2.5 amperes, $\alpha_1 = 1.75$, $\alpha_2 = 0.25$.

Printout for Example 18.10

```
10    V=120: I=2.5: A1=1.75: A2=0.25: V1=V/A1: V2=V/A2: P=V*I
20    PRINT"R2 IN OHMS   P1(WATTS)  P2(WATTS)    R1(MIN)"
25    PRINT
30    FOR R2= 1E3 TO 10E3 STEP 500
40    P2=(V2**2)/R2: R1=(V1**2)/(P-P2): P1=P-P2
50    PRINT R2, P1, P2, R1: NEXT R2
60    END

R2 IN OHMS      P1(WATTS)        P2(WATTS)        R1(MIN)

 1000            69.6             230.4            67.558
 1500            146.4            153.6            32.1178
 2000            184.8            115.2            25.4439
```

2500	207.84	92.16	22.6234
3000	223.2	76.8	21.0665
3500	234.171	65.8286	20.0795
4000	242.4	57.6	19.3979
4500	248.8	51.2	18.8989
5000	253.92	46.08	18.5178
5500	258.109	41.8909	18.2173
6000	261.6	38.4	17.9742
6500	264.554	35.4462	17.7735
7000	267.086	32.9143	17.605
7500	269.28	30.72	17.4615
8000	271.2	28.8	17.3379
8500	272.894	27.1059	17.2303
9000	274.4	25.6	17.1357
9500	275.747	24.2526	17.052
10000	276.96	23.04	16.9773

The phase relationships of transformer voltages are generally identified by dots. Refer to Figure 18.8(a). The dot indicates the direction of conventional current. The input voltage is not only altered in amplitude by the α of the transformer, but is also changed 180° in phase with respect to the input. In Figure 18.8(b) the dot indicates no phase inversion; however,

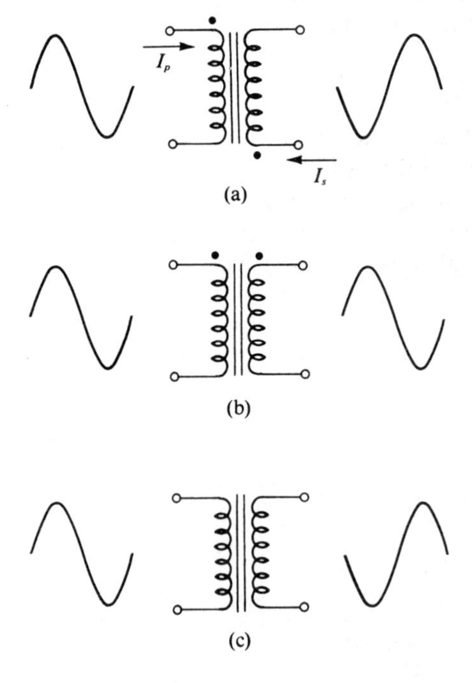

Figure 18.8 Transformer Phase Relationships

the amplitude is altered by the α of the transformer. When there is no dot convention (when no dots are present on the schematic), it is assumed to be a 180° phase reversal. Figure 18.8(c) illustrates the case.

18.5 TRANSFORMER RESONANT CIRCUITS

Chapter 17 discussed the series and parallel resonant circuit. Recall that Figure 17.18(a) illustrated a radio signal "cutting across" the turns of the inductor in the tuned *RCL* circuit. Figure 18.9(a) shows an alternate method of tuning a radio signal. The primary is untuned and the secondary is tuned to resonance with *C*. Since the coefficient of coupling is small (less than 0.1), the secondary inductance and *C* make up the principal factors of resonance. Figure 18.9(b) demonstrates an alternative method in which both primary and secondary are tuned.

It is important to note that the coefficient of coupling not only depends upon the core material, but also on the proximity of the windings. If the primary and the secondary are wound on top of each other, tight coupling results. Conversely, winding the primary and secondary at opposite ends of a cylindrical hollow tube results in loose coupling.

In tuned transformer circuits there exists an optimum (or critical) coupling. Recall that a series-resonant circuit results in a peak current. Figure 18.10 illustrates a functional schematic that includes all of the

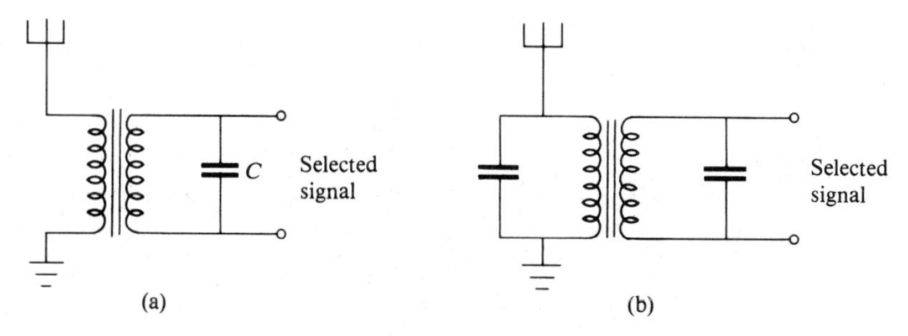

Figure 18.9 Tuned Transformers

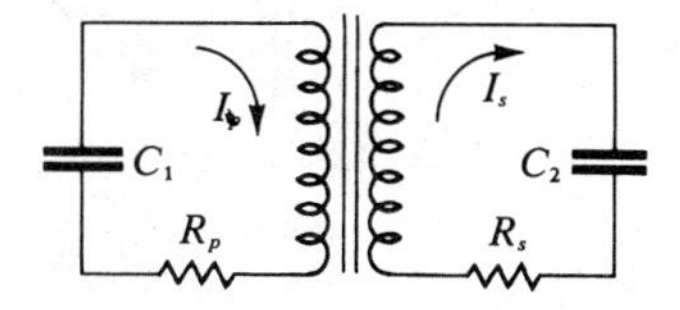

Figure 18.10 Coupling

circuit components. At resonance, I_p seeks its peak value. However, the secondary is also at resonance. Consequently, the circuit impedance is low. This low impedance reflects back to the primary as a high impedance. We can see the conflict. The primary tries to reach a peak current at resonance; the secondary reflects back a high impedance attempting to decrease the primary current. Which of these effects will finally prevail depends upon the degree of coupling between primary and secondary. If loose coupling exists, the reflected impedance effect is minimal.

The primary and secondary currents are shown in Figure 18.11(a). If tight coupling exists, the primary and secondary currents are as shown in Figure 18.11(b). The double hump effect results from the fact that the resonant circuit reflects different reactive properties on each side of resonance.

The waveshape of Figure 18.11(c) shows critical coupling. A given physical placement of primary and secondary results in the optimum conditions.

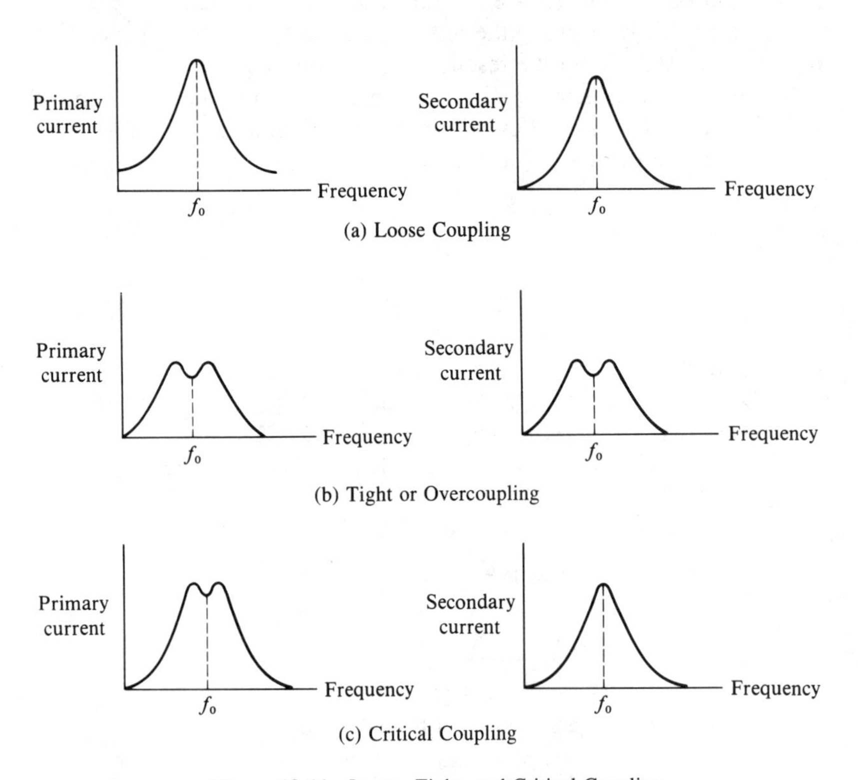

Figure 18.11 Loose, Tight, and Critical Coupling

With ideal coupling conditions, the impedance of the secondary reflected to the primary is equal to the primary impedance. Then, a maximum transfer of energy occurs.

$$R_p = \text{reflected impedance } Z_R$$

However, $Z_R = (2\pi fM)^2/R_s$, therefore

$$R_pR_s = (2\pi fM)^2$$

Substituting for $M = k\sqrt{L_pL_s}$

$$k^2 = \frac{R_pR_s}{(2\pi fL_p)(2\pi fL_s)} = \frac{1}{Q_p} \times \frac{1}{Q_s}$$

$$k_c = \sqrt{1/(Q_pQ_s)} \tag{18.14}$$

where k_c is the critical coupling.

18.6 TYPES OF TRANSFORMERS

Most commercially available transformers fall into three basic groups: power transformers, audio transformers, and radio frequency transformers. Each group has specific applications in the field of electronics. For example, power transformers are generally used for low frequency, high current applications. Figure 18.12 illustrates one type of power transformer. These transformers have heavy laminated cores designed to dissipate a maximum amount of heat and generally have several windings wound on a basic structure. Each winding on a power transformer is isolated from any other winding on the core. One specific case of a power transformer that involves a single winding with a specifically placed tap (often movable) is the autotransformer. See Figure 18.1(g).

Audio transformers are generally similar in construction to the power transformer. They are designed to operate over a wider range of frequency up to 20 kHz. Transformers of these types are used to drive speaker systems, provide crossover networks, and link together radio lines.

The final group of transformers covers a range of frequencies from about 300 kHz up to 1 GHz. These radio frequency transformers (rf) are used in communications equipment. Television receivers employ several rf transformers as intermediate frequency amplifier coupling devices. An rf transformer consists of a core slug that can be moved in and out of the winding. This movable core tunes the transformer by changing the effective permeability of the core.

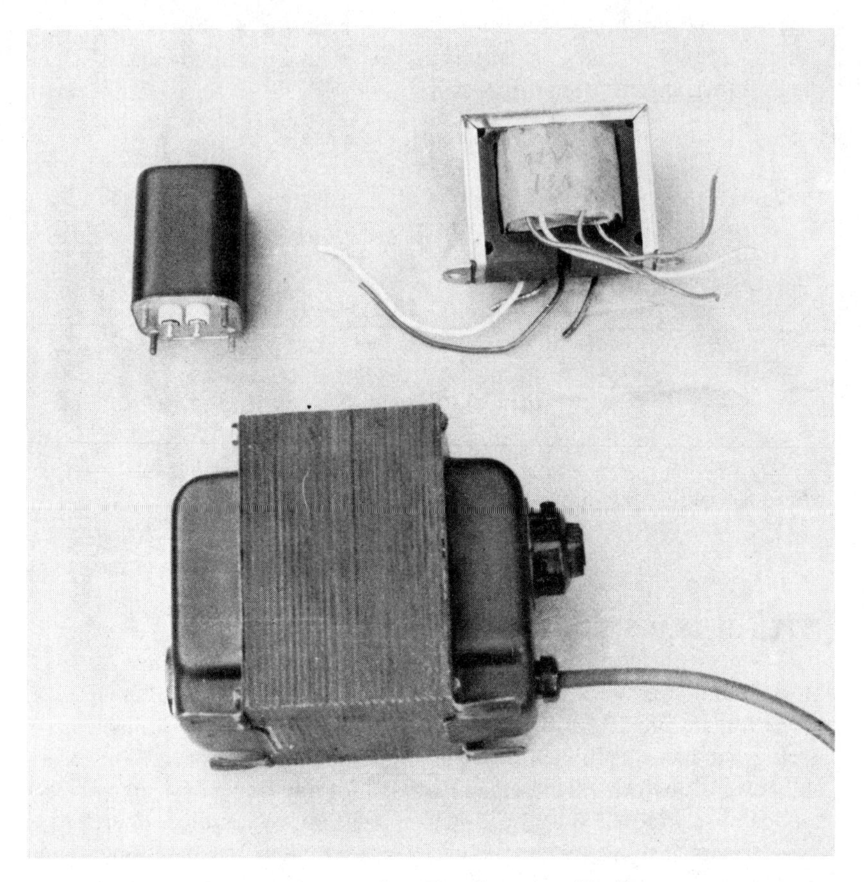

Figure 18.12 Iron Core Transformers and Inductor

SUMMARY

The electrical energy of one circuit may be transferred to another circuit through a transformer. In ideal transformers, the amount of voltage and current transfer depends upon the turns ratio, $N_p/N_s = V_p/V_s = I_s/I_p$. An ideal transformer provides a reflection from its secondary load impedance back to the primary. This reflection is the square of the turns ratio.

Most of the energy losses in the form of heat occur at the iron core of the transformer. The greater the amount of core loss, the lower the efficiency of the transformer. A transformer with an efficiency lower than 1 is considered less than ideal. For these conditions the calculations of primary and secondary currents and voltages must progress through power calculations.

The amount of mutual impedance present between the primary and secondary of a transformer depends upon the coefficient of coupling and the inductances of the windings. The effect of the mutual impedance must be considered part of the primary and secondary circuits. The voltage generated in the primary via the mutual impedance is influenced by the current in the secondary. The voltage generated in the secondary via the mutual impedance is influenced by the current in the primary.

Transformers are classified by the frequency of operation. Power transformers operate below 1 kHz. Audio transformers operate below 50 kHz. Radio frequency transformers operate above 200 kHz.

PROBLEMS

Reference Section 18.1

1. If a transformer has a primary inductance of 10 henries and a secondary of 40 henries, what is the mutual inductance if the coefficient of coupling is 0.6?
2. The coefficient of coupling between the primary and the secondary of a transformer is 0.85. If the mutual inductance is 12 mH and the inductance of the primary is 15 mH, what is the inductance of the secondary?
3. A transformer has a primary inductance of 12 henries and a secondary inductance of 8 henries. If the mutual inductance is 6 henries, what is the coefficient of coupling?
4. If the mutual inductance of a transformer is 6 henries, what voltage will be developed across the secondary if the current in the primary is changing at a rate of 8 mA per second?
5. A transformer consists of a primary of 10 mH and a secondary of 15 mH. If the voltage across the secondary is 10 mV when the current through the primary is changing at the rate of 2 amperes per second, what is the coefficient of coupling between the windings of the transformers?

Reference Section 18.2

6. If the primary of a transformer consists of 150 turns and the secondary of 500 turns, what induced voltage will appear across the secondary when 3 volts is applied to the primary?
7. A given transformer has 120 volts applied at the primary. If 30 volts is measured across the secondary, what is the turns ratio of the transformer?
8. Refer to the transformer of Problem 6. If 2 mA of current is measured in the secondary, what is the current flowing in the primary?

9. If a given audio transformer has 650 turns in its primary and 20 volts is measured across the secondary when 2.5 volts is applied at the primary, how many turns are in the secondary?
10. A given transformer is assumed to be 100 percent efficient. In normal circuit operation 110 volts is applied at the primary. There are two secondary windings: 12 volts (which draws 360 mA) and 30 volts (which draws 0.8 ampere). How much power is being delivered to the primary?
11. A given transformer has a primary voltage of 25 volts and a secondary voltage of 1.5 volts. If the secondary draws 850 mA, how much current is flowing in the primary?
12. A given transformer is 95 percent efficient. The secondary has 15 volts at 250 mA when 10 volts is applied at the primary. What is the current in the primary?
13. Refer to Problem 12. What would the current have been if the transformer were 100 percent efficient?
14. Assume a transformer has a power factor of 0.085. If 20 volts is at the secondary when 30 volts is applied at the primary, and the current in the primary is 100 mA, what are the core losses?
15. Assume a step-down transformer has a turns ratio of 15:1 and the secondary is connected to a load resistance of 4 ohms. What is the impedance at the primary?
16. The secondary impedance of a given transformer is 200 ohms. If the turns ratio of the transformer is $N_p = 600$, $N_s = 40$, what is the impedance of the primary?
17. A given transformer has a matched load to its secondary. When 10 volts is applied to the primary the voltage across the secondary is 60 volts and the current in the secondary is 2 mA. What is the impedance of the primary?
18. A power transformer has a turns ratio (step-up) of 1:12. When 1 kΩ is connected to the secondary, what is the impedance seen at the primary?
19. A 10 volt ac generator is connected to a transformer which has $N_p = 1200$ and $N_s = 400$. When a 500 ohm resistor is connected across the secondary, what current flows in the primary?
20. Refer to Problem 19. If the resistance is increased to 1.5 kΩ (three times greater), what is the new current in the primary?

Reference Section 18.3

21. A given transformer has a primary inductance of 4 henries and a secondary inductance of 12 henries. If the coefficient of coupling is 0.88, what is the mutual impedance seen in the primary and in the secondary when 50 volts at 1 kHz is applied?

22. Assume the transformer of Problem 21 has 1 kΩ connected across the secondary. What current will flow in the primary? What is the power developed at R_L?
23. Assume a transformer has a primary and a secondary each equal to 0.4 henry. The coefficient of coupling is 0.9. Assume 5 volts rms at 8 kHz is applied. If the load resistance on the secondary is 10 kilohms, what is the current and phase angle of the primary current?

Reference Section 18.4

24. A given power transformer has two secondary windings. When 120 volts rms is applied at the primary, the voltage at one secondary is 24 volts, and the voltage across the other secondary is 600 volts. Find the turns ratio for each secondary ($k = 1$).
25. An interstage transformer with a coefficient of coupling $k = 0.95$ is driven by a 50 mV 1 kHz signal. The inductance of the primary is 4 mH and the inductance of the secondary is 8 mH. If the load on the transformer is 10 kΩ, what is:
    a. the mutual inductance?
    b. the mutual impedance?
    c. the primary phase angle?
    d. the primary current?
    e. the current in the load?
26. A transformer has a coefficient of coupling of 0.9. The inductance of the primary is 2.5 henries. The inductance of the secondary is 1.2 henries. If the applied voltage at the primary is 20 volts at 60 Hz and the load on the secondary is 500 ohms, what is the reflected impedance? What is the phase angle of the current in the primary?
27. If the transformer of Problem 26 is connected as shown in Figure 18.13, what is the impedance seen by the 60 Hz generator?

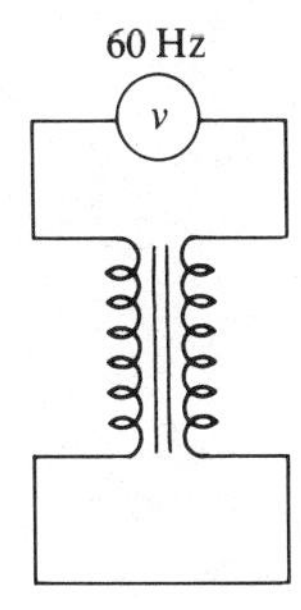

Figure 18.13 Figure for Problem 27

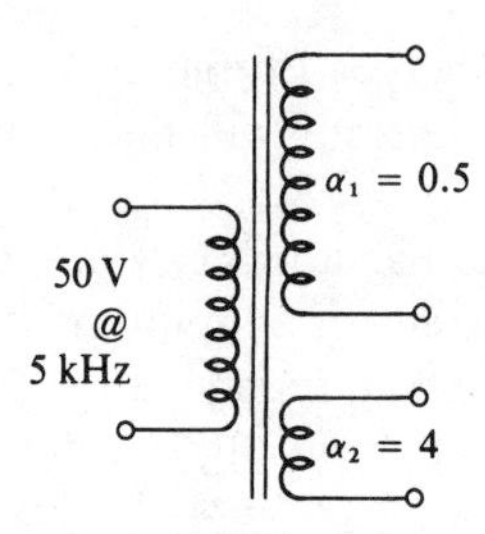

Figure 18.14 Figure for Problem 28

28. Refer to the circuit of Figure 18.14. If a 1 kΩ load is connected to both sets of output terminals, what current will be drawn in the primary circuit?

Reference Section 18.5

29. If a tuned primary and a tuned secondary are to resonate at 450 kHz and the Q of the primary is 150 while the Q of the secondary is 200, what is the value of the critical coupling?
30. A tuned (resonant) primary and a tuned (resonant) secondary have a critical coupling of 0.01. If both coils are identical at 100 μH each, with a dc resistance of 10 ohms each, what capacity is required for resonance?

Optional Computer Problems

31. Write a program in BASIC which will generate a table of values of Z_{in} of the primary for a secondary load resistance ranging from 100 ohms to 10 kΩ in 250Ω steps. Assume $L_p = L_s = 10$ mH, $k = 0.95$, and the applied voltage is 10 volts at 10 kHz.
32. Using the same transformer of Problem 31, generate a table of values for Z_{in} for a fixed value of $R_L = 1$ kΩ and assume the input frequency varies from dc to 10 kHz in 250 Hz steps.

CHAPTER 19

FILTER CIRCUITS

19.1 *RC* FILTER CIRCUITS

Filter circuits in electrical science occur both intentionally and consequentially. A given electric circuit may behave like a filter circuit even though it was designed for another purpose.

The definition of a filter circuit actually embodies several concepts. When an electrical circuit passes or blocks a given band of frequencies, it is considered a filter. If a network passes high frequencies and rejects low (or vice versa), the circuit is a filter. If a network selects a group of frequencies while rejecting all others, it is a filter. These types of filters are frequency selective. There exists another group of filters which are not frequency selective. These filters virtually smooth out variations in voltage waveshapes.

In this chapter we will discuss only frequency selective filters. The output voltage versus the frequency of the input voltage will be analyzed.

In modern high fidelity amplifying systems often a sound system will employ a large speaker (*woofer*) to reproduce the low frequency sounds, and a small speaker (*tweeter*) to reproduce the high frequency sounds. Figure 19.1(a) illustrates the basic system. Note that the two speakers are merely connected in parallel. This means that all frequencies are applied to both speakers. The quality of reproduction can be improved if the frequencies below a certain point are blocked from the high frequency speaker and vice versa. Figure 19.1(b) illustrates the desirable frequency response pattern. Two filters would be required: a highpass filter and a lowpass filter.

Highpass Filter

The basic *RC* highpass filter is shown in Figure 19.2(a). When V_{in} is an ac signal voltage whose frequency varies over a wide range, the output voltage will depend on the frequency of the input. The output voltage V_o is given as the ratio:

$$V_o = \frac{R(V_{in})}{R - jX_c} = \frac{R(V_{in})}{R - \dfrac{j}{2\pi fC}} \qquad (19.1)$$

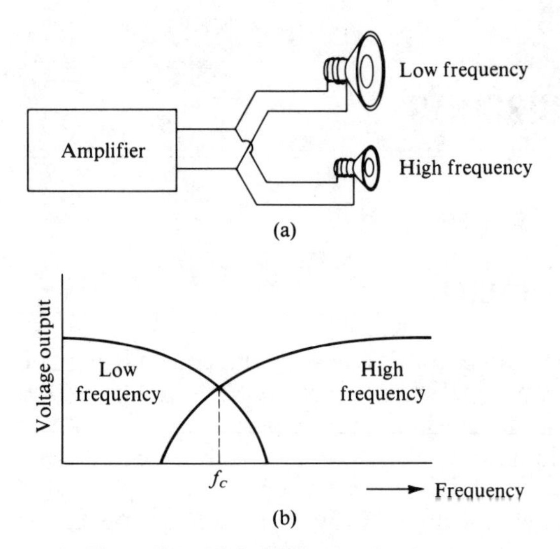

Figure 19.1 High Fidelity Speaker System

Note from Eq. (19.1) that at very low frequencies the denominator is quite large, hence the output quite small. Conversely, as the frequency increases, the output voltage approaches the input voltage.

Figure 19.2(b) illustrates a frequency response characteristic. The point indicated as f_c is the cutoff frequency. By definition the filter is said to pass all frequencies above f_c and attenuate or reduce the amplitude of all frequencies below f_c. The cutoff frequency is identified as that frequency where the voltage across C is equal to the voltage across R (the output voltage). Since R and C are in series with each other and since the voltages are equal, the ohmic oppositions must be equal.

$$R = X_c = 1/2\pi f_c C$$

$$\boxed{f_c = 1/2\pi RC} \tag{19.2}$$

where R is in ohms, C is in farads, and f_c is in hertz. Highpass filters can be constructed using L/R circuits. The circuit of Figure 19.2(c) shows inductive coupling from the generating source. As the frequency decreases, the voltage across the inductor (which shunts the load) decreases since X_L gets smaller than R. The cutoff frequency f_c occurs at a frequency when $V_R = V_L$. For this to occur, X_L must equal R.

$$R = X_L = 2\pi f_c L$$

$$f_c = R/2\pi L \tag{19.3}$$

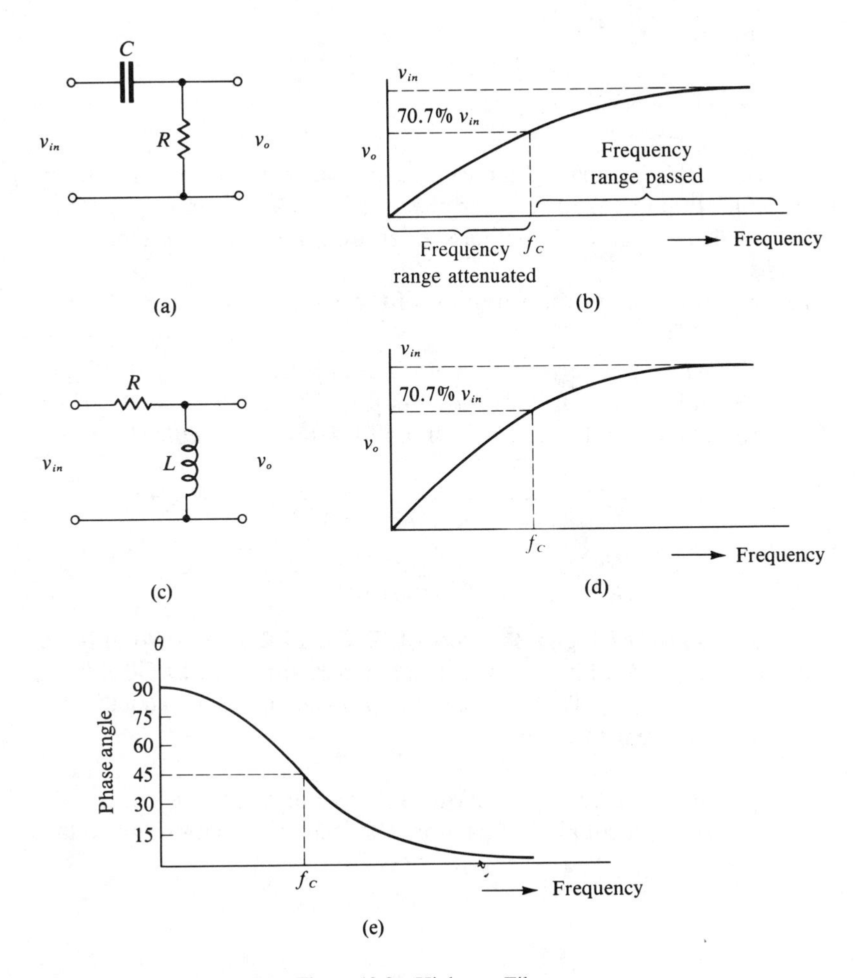

Figure 19.2 Highpass Filters

The cutoff frequency and the response characteristics are shown in Figure 19.2(d). Note that the characteristics are the same as those of the *RC* highpass filter.

Highpass filters introduce phase displacements between the input and the output voltages. The amount of phase displacement depends upon the frequency. For example, at f_c the phase displacement is 45° with the output leading the input. Equation (19.1) can be rewritten to include the phase angle. Solving Eq. (19.2) for *R* and substituting into Eq. (19.1),

$$V_o = \frac{f\,V_{\text{in}}}{f - jf_c}$$

Converting to polar form,

$$V_o = \frac{fV_{\text{in}}}{\sqrt{f^2 + f_c^2}} \angle \tan^{-1} f_c/f = \frac{V_{\text{in}} \angle \tan^{-1}(f_c/f)}{\sqrt{1 + (f_c/f)^2}} \tag{19.4}$$

Note that as the frequency f increases to a value much larger than f_c, the phase angle diminishes to zero and the output equals the input. Equation (19.4) results for an L/R highpass filter through a similar method of analysis.

Figure 19.2(e) shows the variation of the phase angle of an RC filter.

Example 19.1 Assume an RC filter is connected as shown in Figure 19.2(a). What is the cutoff frequency? Assume $R = 2$ kΩ and $C = 0.05$ μF. What capacity would be required for a 5 kHz cutoff frequency?

$$f_c = \frac{1}{2\pi RC} = \frac{1}{(6.28)(2)(10^3)(0.05)(10^{-6})} = 1.59 \text{ kHz}$$

$$C = \frac{1}{2\pi f_c R} = \frac{1}{(6.28)(10^3)(5)(2)(10^3)} = 0.0159\ \mu\text{F}$$

A comparison of the highpass filters of Figure 19.2 shows two different applications. The RC filter will act as a highpass filter and will also block the flow of dc current. The L/R filter will provide highpass characteristics and short dc current to ground.

The circuit of Figure 19.3(a) shows a transistor circuit driving a resistive load through a coupling capacitor. The transistor requires dc voltage

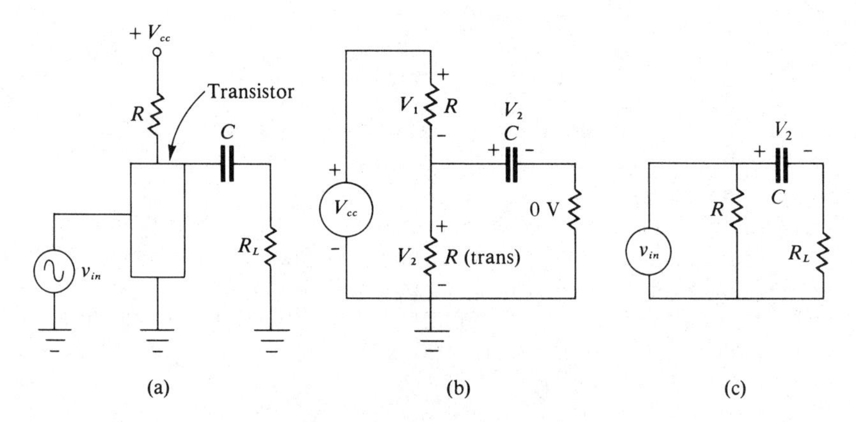

Figure 19.3 Highpass Coupling

for proper operation. The load resistance R_L will not have dc flowing through it because of the coupling capacitor between R_L and the supply voltage V_{cc}. The equivalent circuit of Figure 19.3(b) illustrates the case. The V_{cc} supply voltage distributes across the R and the resistance of the transistor. The capacitor quickly charges to V_2 leaving no dc across the load. The input voltage V_{in} of Figure 19.3(a) sees an equivalent circuit consisting of R, C, and R_L (Figure 19.3c). The ac equivalent circuit is actually a highpass filter circuit.

This ac/dc relationship is best seen through an example.

Example 19.2 The circuit of Figure 19.4 represents a coupling network that also acts as a highpass filter. Assume V_{in} is 5 volts over a wide range of frequencies (5 volts p-p, which varies from +2.5 volts to −2.5 volts). If $R_1 = 5\ k\Omega$, $R_2 = 1\ k\Omega$, $R_3 = 10\ k\Omega$, and $C = 0.022\ \mu F$, what is the output voltage at the cutoff frequency? What is the constant dc voltage across C? What is the dc voltage across R_3?

$$f_c = \frac{1}{2\pi R_3 C} = \frac{0.159}{(10^4)(0.022)(10^{-6})} = 722.7 \text{ Hz}$$

The voltage across R_3 is 70.7 percent of the applied voltage (ac across R_2).

When $R_2 \ll R_3$, this approximation is acceptable.

$$V_{R_2} = \frac{R_2 V_{in}}{R_1 + R_2} = \frac{1(5)}{1 + 5} = 0.833 \text{ volt (p-p)}$$

$$V_o = (0.707)(0.833) = 0.589 \text{ volt (p-p)}$$

The dc voltage across C is the dc voltage across R_2, which is zero since, in this case, there is no dc component in V_{in}. The dc voltage across R_3 is also zero. When R_2 is greater than 10 percent of R_3, the approximation may not be used. A complex ac circuit results which can be simplified using Thevenin's theorem.

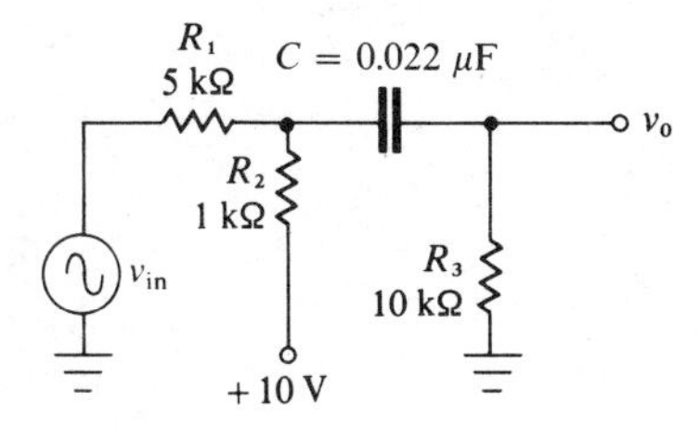

Figure 19.4 Figure for Example 19.2

19.2 LOWPASS FILTER

The *RC* and *RL* circuits of Section 19.1 may be rearranged to produce filter circuits which will pass all frequencies below a given cutoff frequency and block all frequencies above that frequency. The circuit of Figure 19.5(a) illustrates the *RC* lowpass filter. When V_{in} is a constant ac voltage whose frequency varies from zero hertz to an extremely high frequency, the output voltage will be the voltage across the capacitor.

$$V_o = V_c = \frac{-j(1/\omega C)(V_{\text{in}})}{R - j(1/\omega C)} \qquad (19.5)$$

Equation (19.5) simplifies to

$$V_o = \frac{V_{\text{in}}}{1 + jR\omega C} \qquad (19.6)$$

where $\omega = 2\pi f$.

Equation (19.6) shows that, as f approaches zero, ω approaches zero, and the output equals the input. Similarly, as the frequency increases, the denominator increases and the output diminishes toward zero.

Figure 19.5(b) is a graphic representation of the magnitude of V_o in Eq. (19.6). The upper cutoff frequency, f_c, is defined as that frequency where the output voltage has decreased to 70.7 percent of the input voltage. (This is the half-power point.) At this frequency the voltage across R equals the voltage across C. The value of f_c is determined in exactly the same way as it was calculated in the highpass *RC* circuit. Equation (19.2) defines the upper cutoff frequency.

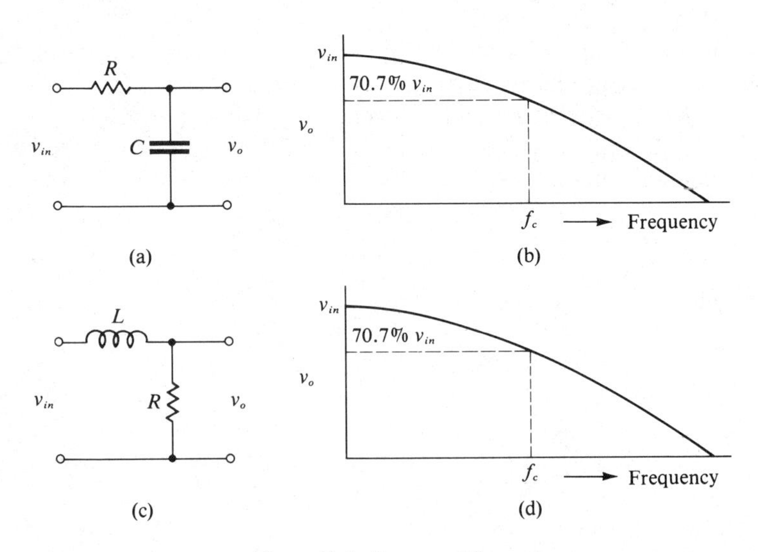

Figure 19.5 Lowpass Filters

V_o lags V_{in} by 45° at the cutoff frequency. The output voltage and phase at any frequency can be determined. Solve Eq. (19.2) for R and substitute into Eq. (19.6). V_o becomes

$$V_o = \frac{V_{in}}{1 + j(f/f_c)} = \frac{V_{in}}{\sqrt{1 + (f/f_c)^2}} \angle -\tan^{-1}(f/f_c) \qquad (19.7)$$

The L/R circuit combination of Figure 19.5(c) is a lowpass filter whose characteristics are the same as those of its counterpart in Figure 19.5(a). As the frequency decreases, the inductive reactance of L decreases, so a larger portion of V_{in} is developed across R. Therefore,

$$V_o = \frac{R(V_{in})}{R + j\omega L}$$

Dividing numerator and denominator by R,

$$V_o = \frac{V_{in}}{1 + j\omega L/R} \qquad (19.8)$$

Figure 19.5(d) illustrates the frequency response characteristics. The upper cutoff frequency is defined by Eq. (19.3). If this equation is solved for R and substituted into Eq. (19.8), the output voltage becomes the same as in Eq. (19.7).

Example 19.3 Using BASIC, write a program which will print out the output voltage and its associated phase displacement of a lowpass RC filter. Assume $R = 4.7$ kΩ, $C = 0.022$ μF. Input voltage equals 1000 mV over a wide range of frequency. Increment the frequency in 10 equal steps between 0 hertz and f_c. Then increment frequency in 9 equal doublings of f_c. (This program is found on page 554.)

Example 19.3 demonstrated how the output voltage diminishes gradually for a single section of an RC filter. If several lowpass filters are cascaded as shown in Figure 19.6(a), the "rolloff" of the voltage can be made to diminish at a faster rate. In this circuit each section has its own cutoff frequency f_c. Linearization has been adapted by the industry as a method of facilitating the analysis of filter graphs. Figure 19.6(b) shows a linearized response of each section of the filter. The plot is along a logarithmic graph. The corner of each graph of a linearized response represents the cutoff frequency. Note that the rolloff has the same slope in each case; only the corner (or cutoff) frequencies are different. Figure 19.6(c) shows the three response curves superimposed on one graph. The actual response characteristic is shown in Figure 19.6(d). The lowpass filter of Figure 19.6(a) provides sharp cutoff characteristics, R_2C_2 continues to diminish the already decreased voltage. R_3C_3 further diminishes the output of the preceding two.

Printout for Example 19.3

```
10   P1=3.14159 : V=1E3 : R=4.7E3 : C=.022E-6
15   F1=1/(2*P1*R*C)
20   PRINT "CUT-OFF FREQUENCY IS " F1 " HZ"  : PRINT
25   PRINT "F IN HZ" , "V(O) IN MV" , "ANGLE IN DEG." : PRINT
30   FOR F=0 TO F1+1 STEP F1/10 : GOSUB 50  : NEXT F : N=1
35   F=(2↑N)*F1 : GOSUB 50   : IF N>8 THEN 45
40   N=N+1 : GOTO 35
45   STOP
50   V1=V/SQR(1+(F/F1)↑2) : A=-ATN(F/F1)
55   PRINT F , V1 , A*(180/P1) : RETURN
60   END
```

```
CUT-OFF FREQUENCY IS  1539.22  HZ

F IN HZ          V(O) IN MV       ANGLE IN DEG.

 0                1000             0
 153.922          995.037         -5.7106
 307.844          980.581         -11.3099
 461.765          957.826         -16.6993
 615.687          928.477         -21.8014
 769.609          894.427         -26.5651
 923.531          857.493         -30.9638
 1077.45          819.232         -34.9921
 1231.37          780.869         -38.6598
 1385.3           743.294         -41.9872
 1539.22          707.107         -45
 3078.44          447.214         -63.435
 6156.87          242.536         -75.9638
 12313.7          124.035         -82.8751
 24627.5          62.3783         -86.4237
 49255            31.2347         -88.2102
 98509.9          15.6231         -89.1049
 197020           7.81227         -89.5525
 394040           3.90622         -89.7763
 788079           1.95312         -89.8882
```

19.3 BANDPASS AND BAND REJECTION FILTERS

By combining the *RC* lowpass and *RC* highpass filter circuits, it is possible to produce a filter which will pass a band of frequencies. Figure 19.7(a) illustrates the basic circuit. R_2 and C_2 make up a highpass filter. R_1 and C_1 make up the lowpass filter. The characteristics of each filter are seen in Figure 19.7(b) and (c). Note that the corner frequency of the lowpass filter is higher than the corner of the highpass filter. The bandwidth of the passed frequencies becomes the difference between the two corner frequencies.

$$\mathrm{BW} = f_{C_1} - f_{C_2} \tag{19.9}$$

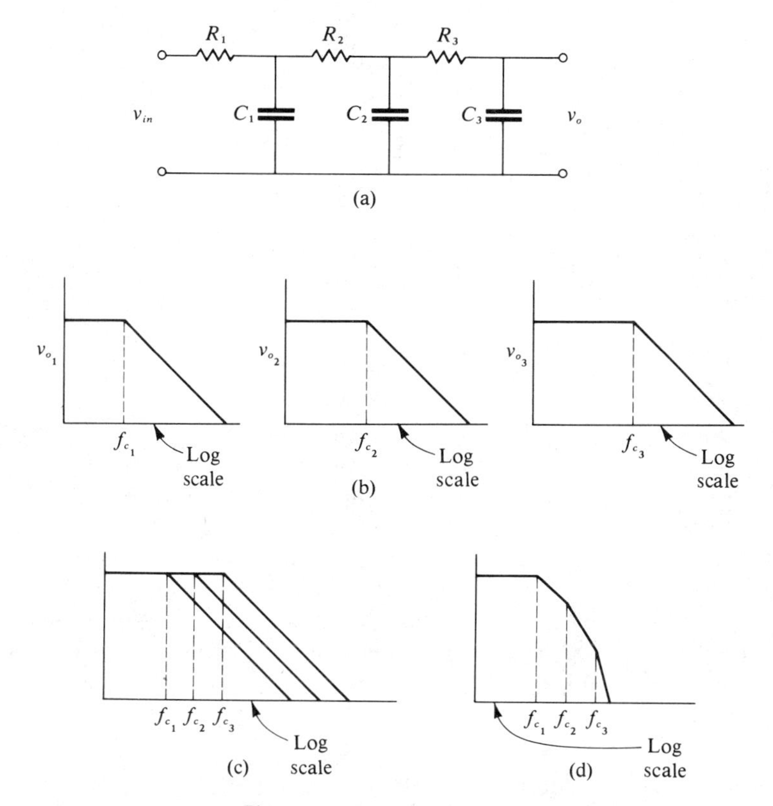

Figure 19.6 Frequency Rolloff

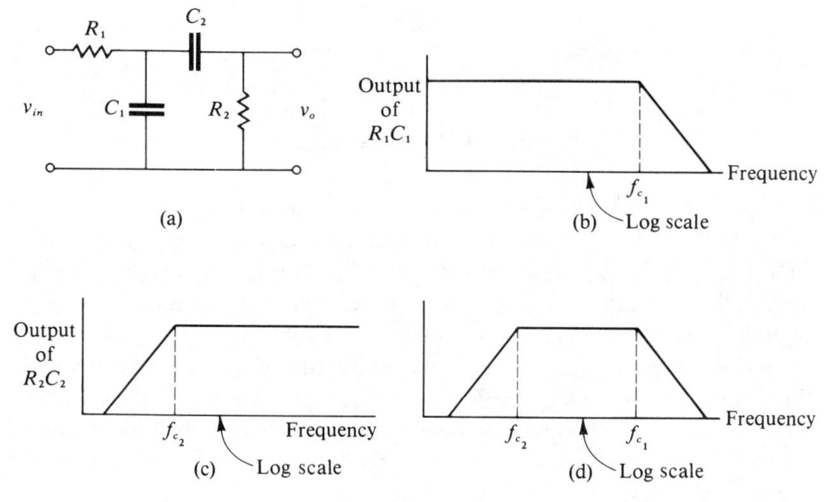

Figure 19.7 Bandpass Filter

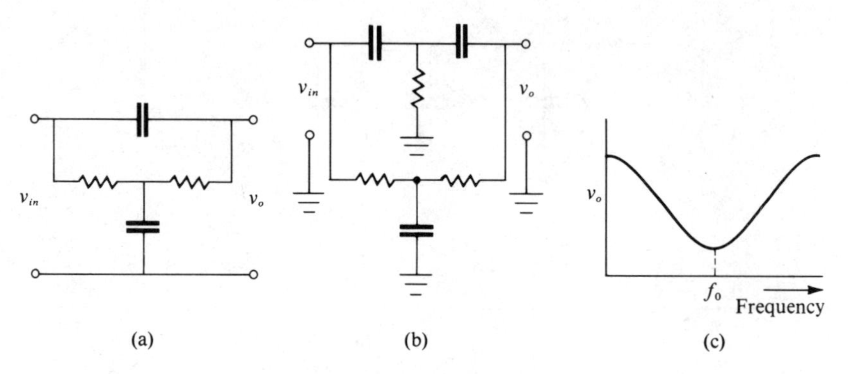

Figure 19.8 Notch Filters

It is important to note, however, that we cannot design the lowpass section and the highpass section independently and then join them together, expecting the cutoff frequencies to be the same. The presence of each section in the overall circuit affects the performance of the other section. Only by *buffering* one section from the other (separating them by an amplifier with a high-input resistance and a low-output resistance) can we expect the two filter sections to perform as they would if isolated from each other.

Example 19.4 Assume the bandpass filter of Figure 19.7(a) is to consist of $R_1 = R_2 = 1\ \text{k}\Omega$, $C_1 = 2000$ pF, and $C_2 = 0.05\ \mu$F. What bandwidth would be (incorrectly) calculated if we analyzed the sections independently?

$$f_{C_1} = \frac{1}{2\pi R_1 C_1} = \frac{0.159}{(10^3)(2000)(10^{-12})} = 79.5\ \text{kHz}$$

$$f_{C_2} = \frac{1}{2\pi R_2 C_2} = \frac{0.159}{(10^3)(0.05)(10^{-6})} = 3.18\ \text{kHz}$$

$$\text{BW} = (79.5 - 3.18)\ \text{kHz} = 76.32\ \text{kHz}$$

This result would be true only if the two sections were separated by a buffer amplifier. Using conventional series/parallel ac circuit analysis, we would find the true value of f_{C_2} in the circuit of Figure 19.7 much closer to 15 kHz. This results in a much smaller bandwidth than that previously calculated. There are many combinations of *RC* band rejection filters. Figures 19.8(a) and (b) illustrate two rejection filters. These are sometimes called *notch* filters. A typical band rejection characteristic is shown in Figure 19.8(c). The theory of their operation is very complex and beyond the scope of this book. Their study is generally included in books on integrated circuits.

Series *RCL* Filters

The *RC* bandpass filters are generally used to pass a wide band of frequencies within the audio range. These filters become impractical at radio frequencies. When a narrow band in the radio frequency range is to be passed from source to load, several *RCL* combinations are possible. The series *RCL* circuit of Figure 19.9(a) may be used. Generally, *R* is quite small and can be neglected, hence the coupling circuit becomes *L* and *C* in series. On either side of resonance the series coupling circuit exhibits a reactance. At resonance, $X_L = X_C$, hence the current is maximum, providing a maximum voltage across R_L.

The transfer characteristic for the circuit is shown in Figure 19.9(b). The band of frequencies is determined by the load being driven. The center frequency is given as:

$$f_0 = 1/2\pi\sqrt{LC} \tag{19.10}$$

The frequencies f_1 and f_2 are the half-power frequencies, where the output is 70.7 percent of the input. Since this condition exists when the impedance is $\sqrt{2}R_L$, we may solve for $f_2 - f_1$. Chapter 17 derives the bandwidth equations. From Eqs. (17.19) and (17.21),

$$\text{Bandwidth} = R/2\pi L \tag{19.11}$$

Note that as *R* gets larger, the bandwidth gets broader. Similarly, decreasing *L* can increase the bandwidth. However, when *L* is varied, the center frequency also varies. If *L* is changed, the *C* must be changed by an appropriate amount in order to maintain f_0.

Example 19.5 A series *RCL* bandpass filter has $L = 10$ mH and $C = .01\ \mu F$. The source resistance is 20Ω and the load resistance is 30Ω. What is the center frequency of the filter? What is its *Q*? What are its lower and upper cutoff frequencies?

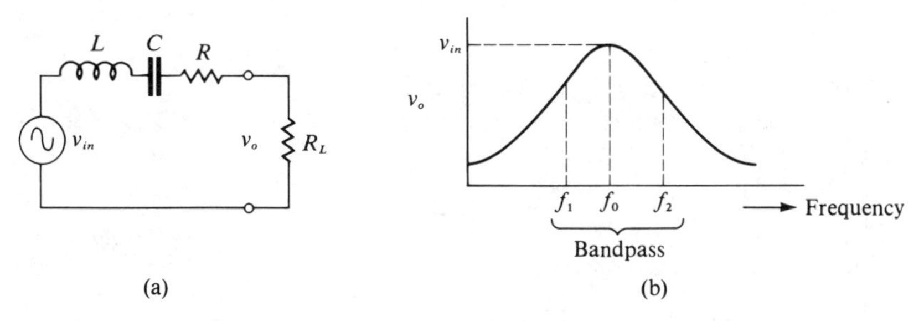

Figure 19.9 Series *RCL* Filter (Series Connection)

$$f_0 = \frac{1}{2\pi\sqrt{LC}} = \frac{1}{2\pi\sqrt{10^{-2} \times 10^{-8}}} = 15.92 \text{ kHz}$$

$$Q = \frac{2\pi f_0 L}{R} = \frac{10^5}{20 + 30} = 20$$

Since $Q > 10$, the center frequency is approximately midway between the lower and upper cutoff frequencies.

$$\text{BW} = f_0/Q = (15.92 \times 10^3)/20 = 776 \text{ Hz}$$
$$f_1 = f_0 - \text{BW} = (15.92 - .776) \times 10^3 = 15.14 \text{ kHz}$$
$$f_2 = f_0 + \text{BW} = (15.92 + .776) \times 10^3 = 16.7 \text{ kHz}$$

The series-resonant circuit may be used to reject a given band of rf when it is connected across the load resistance as shown in Figure 19.10(a). In this configuration the LC combination acts as a large shunt resistance across R_L over a wide range of frequencies. At the resonant frequency the LC combination parallels R_L with a low value of impedance. Above and below resonance the output voltage is approximately:

$$V_o = \frac{R_L V_{in}}{R_L + R_i}$$

The value of R_i may be small compared to R_L, hence the output voltage is equal to the input. At resonance the output voltage drops to

$$V_o = \frac{R V_{in}}{R + R_i}$$

Generally R is small compared to R_i, hence the output is at a minimum.

The transfer characteristic of Figure 19.10(b) illustrates the variation of output over a range of input frequency.

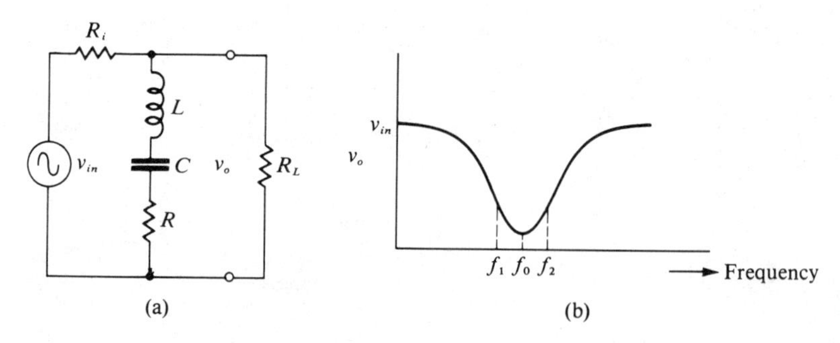

Figure 19.10 Series *RCL* Filter (Shunt Connection)

Parallel *RCL* Filter

Bandpass and band rejection at radio frequencies can be accomplished using parallel resonant tanks as well. Refer to the circuit of Figure 19.11(a). In this circuit the impedance of the parallel-resonant tank is given as

$$Z = L/RC$$

Therefore the output voltage across R_L will be small since L/RC is an extremely large value. On either side of resonance the parallel tank circuit exhibits a reactive property which is ideally much less than R_L.

The transfer characteristics are shown in Figure 19.11(b). The filter will pass all the frequencies below f_1 and above f_2. The band rejection is $f_2 - f_1$. If R is small, the center frequency of the bandpass characteristic is

$$f_0 = 1/2\pi\sqrt{LC}$$

Chapter 17 develops the influence of R on the frequency f_0. The chapter also develops the bandwidth of the circuit.

$$\text{Bandwidth rejected} = f_0/Q \tag{19.12}$$

The value of the load resistance R_L will also influence the band rejection.

Example 19.6 In the filter circuit of Figure 19.11(a), the f_0 is to be 10 kHz. If bandwidth of the rejected band is to be 500 Hz, what must the Q of the inductor be?

$$Q = \frac{f_0}{\text{BW}} = \frac{(10)(10^3)}{(0.5)(10^3)} = 20$$

The parallel-resonant tank can be used in shunt with the load similar to the series-resonant circuit. Figure 19.12(a) illustrates the circuit configuration and the transfer function. Often combinations of parallel- and series-resonant circuits are used to provide different transfer functions. Figure

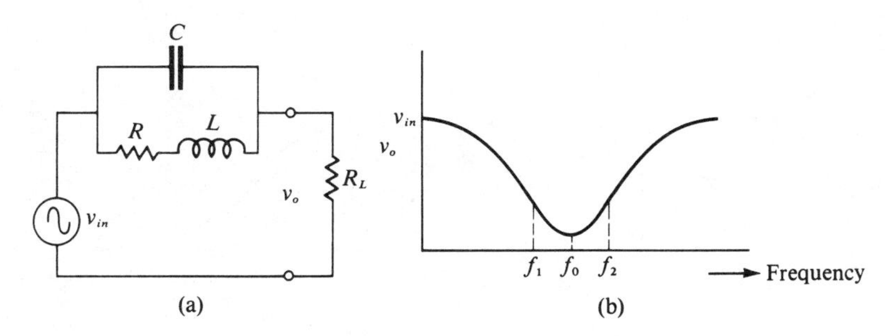

Figure 19.11 Parallel *RCL* Filter (Series Connection)

19.12(b) shows how a narrow band of frequencies can be rejected. All three tank circuits tune to near $f_0 = 1/2\pi\sqrt{LC}$.

Figure 19.12(c) illustrates how a series *LC* circuit can be used as a lowpass filter. Interchanging the two components creates a highpass filter.

19.4 COMBINATION *RCL* FILTERS

Inductors, capacitors, and resistors in various combinations can be used to produce bandpass, lowpass, highpass, and band rejections. Filters are classified in many categories. The two most common are constant-*K* and *m*-derived filters. Quantitative analysis of these filters is so complex that only an introduction is presented.

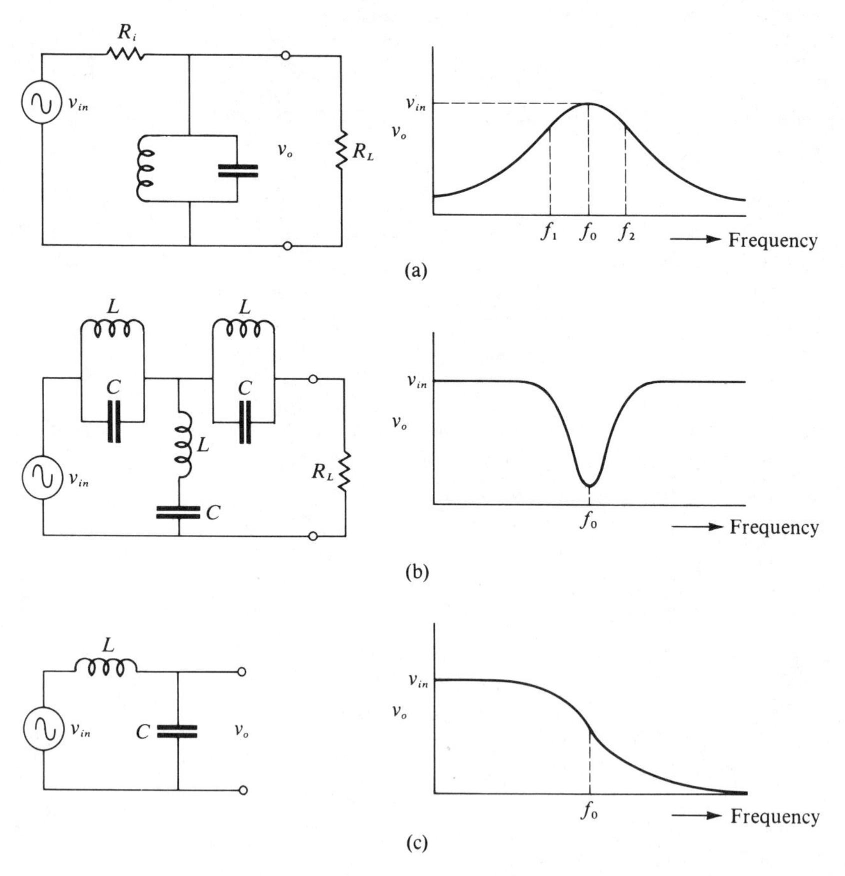

Figure 19.12 Series and Parallel Resonant Filters

Constant-K Filter

The various filters illustrated in Figures 19.9 through 19.12 provide many frequency pass, stop, and bandpass characteristics. The characteristics of these filters are such that the input and output impedances may or may not vary with changes in load resistance or input frequency. Figure 19.13(a) shows a highpass filter consisting of two branches, Z_A and Z_B. Filter sections where the product of Z_A and Z_B is K^2 are called constant-K filters.

$$Z_A Z_B = K^2$$

Solving for K,

$$K = \sqrt{Z_A Z_B}$$

$$= \sqrt{\frac{2\pi fL}{2\pi fC}} = \sqrt{\frac{L}{C}}$$

The cutoff frequency, f_c, is given as the half-power frequency (R_i is assumed negligible). The frequency characteristics are shown in Figure 19.13(b).

$$f_c = \frac{1}{2\pi CR_L} \quad \text{and} \quad f_c = \frac{R_L}{2\pi L}$$

Setting the two f_c equations equal to each other and solving for R_L,

$$R_L = \sqrt{L/C} = K \tag{19.13}$$

Equation (19.13) demonstrates the relationship that must exist for a given cutoff frequency. Various combinations of the basic constant-K filter can be developed.

Example 19.7 Assume an LC circuit is to be used as a highpass constant-K filter similar to the circuit of Figure 19.13(a). R_i can be neglected since its value is low. The load resistance is 2.5 kΩ. (a) What value of L is required for a lower cutoff frequency of 4 kHz? (b) What value of C is required? (c) What is the cutoff frequency?

a. $L = \dfrac{R_L}{2\pi f} = \dfrac{(2.5)(10^3)}{(6.28)(4)(10^3)} = 0.1$ henry

b. From Eq. (19.13), $C = \dfrac{L}{R_L^2} = \dfrac{(0.1)}{(2.5)^2(10^3)^2} = 0.016\ \mu\text{F}$

c. $f_o = \dfrac{1}{2\pi\sqrt{LC}} = 4$ kHz (the cutoff frequency)

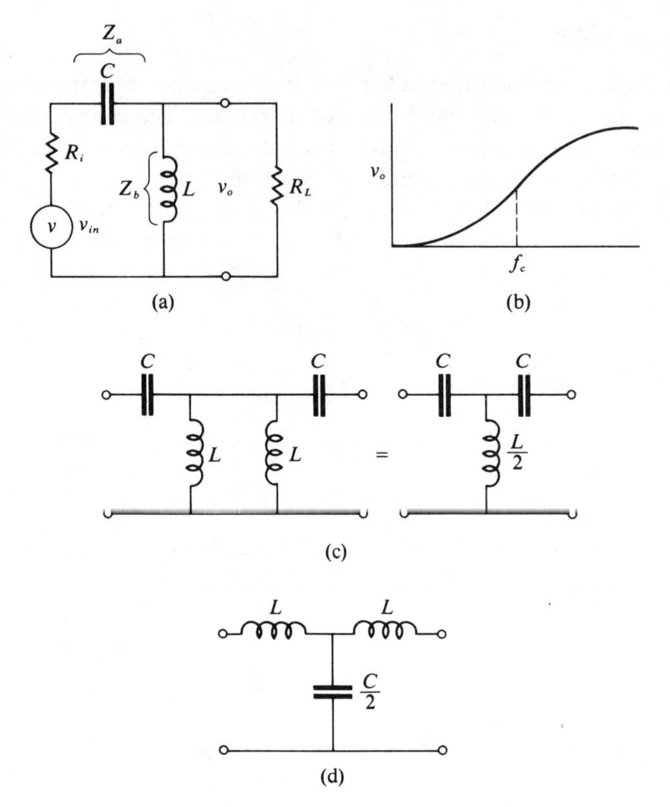

Figure 19.13 Constant-*K* Filter

Figure 19.13(c) illustrates how two sections can be joined to produce the basic T, constant-*K* highpass filter. Interchanging the components produces a *T*, constant-*K* lowpass filter as illustrated in Figure 19.13(d).

The response characteristics of the constant-*K* filter are such that the output beyond the cutoff frequency trails off gradually; that is, there is a slow diminishing of the signal level. When such characteristics are undesirable, a different configuration of filter is required.

m-Derived Filter

The *m*-derived filter is characterized by a capacitor in the series leg and by a shunt arm that can contain a capacitance and an inductance. The opposite configuration is also possible where the series arm contains a parallel *L*/*C* circuit and the shunt arm contains an inductance or capacitance. Figure 19.14(a) illustrates the basic section of an *m*-derived highpass filter. Figure

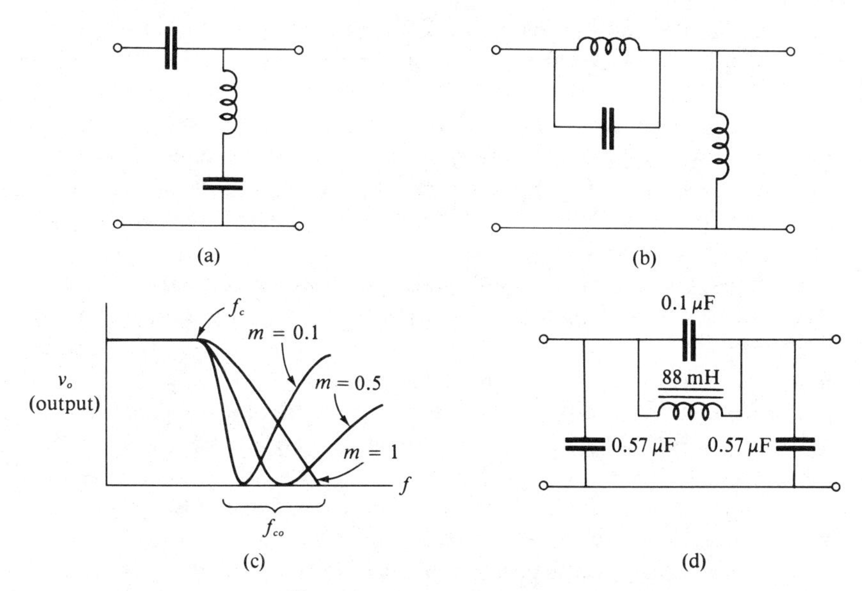

Figure 19.14 *M*-derived Filters

19.14(b) shows an *m*-derived shunt-type highpass filter. The distinguishing feature in both circuits is a resonant circuit, either series or parallel. The resonant frequency and its L/C ratio determine the bandwidth characteristic and its approach to maximum attenuation.

The factor *m* is essentially a ratio which defines cutoff frequency and maximum attenuation frequency, f_{co}. Figure 19.14(c) illustrates the possible characteristics of various lowpass *m*-derived filters. These *m*-derived configurations are used in both high frequency and low frequency applications to improve impedance and phase characteristics. Figure 19.14(d) illustrates an *m*-derived speech filter used in telephone equipment. Its upper cutoff frequency is about 2 kHz.

The series and shunt-type configurations of the constant-*k* and the *m*-derived filters may be used to produce highpass, lowpass, and band attenuation and bandpass characteristics. The actual design of such filters makes use of separate tables and references. The reader should refer to a radio or engineering handbook for filter design.

SUMMARY

Filter circuits are classified by their function. Lowpass filters generally transfer from their input to their output all frequencies below a cutoff frequency. The cutoff frequency is identified as the half-power point which occurs when the output voltage is 70.7 percent of the input. Lowpass filters can be developed from one or more sections of series *RC* or *L*/*R* circuits.

Highpass filters generally transfer from their input to their output all frequencies above a given cutoff frequency. The cutoff frequency is determined in the same way as in the lowpass filter case. In both cases the filter introduces phase angle change of the transferred signal. This angle is 45° at the cutoff frequency.

Bandpass and band rejection filters select a specific range of frequencies. They transfer from their input to the output either all the frequencies of the band or all the frequencies outside of the band. These filters can be constructed from *RC*, *L*/*R*, or *RCL* components.

Resonant filters can pass or reject a narrow band of frequencies. The selectivity of the circuit is dependent upon the *Q* of the circuit.

Constant-*k* and *m*-derived filters include combinations of components to produce almost any transfer characteristic desired. Design of these filters generally requires the calculation of either the *m* or the *k* constant from the frequency characteristics intended. Then the filter's component values are calculated.

PROBLEMS

Reference Sections 19.1 and 19.2

1. A highpass filter consists of a capacitor and a resistor, $C = 0.01\ \mu F$ and $R = 10\ k\Omega$. (a) Draw the schematic showing input and output. (b) What is the lower cutoff frequency?
2. A highpass filter consists of a series capacitor and resistor. If the output is taken across the resistor and its value is 5000 ohms, what is the lowest frequency passed if the value of *C* is 0.2 μF?
3. A series resistor and a capacitor are to be used as a highpass filter. The cutoff frequency is to be 2.4 kHz and the load resistance (*R*) is to be 3.3 kΩ. What value of *C* is required?
4. A "woofer" type speaker is to cover all frequencies below 800 Hz. A series inductor is to act as a lowpass filter. If the impedance of the speaker is approximately 500 ohms (resistive), what value of inductance is required?
5. A series circuit consists of a resistor and an inductor. The value of *R* is 500 ohms and the inductor is 0.6 henry. If the applied voltage is 10 volts at 400 Hz, what voltage will be developed across the resistor? What is the voltage output at twice f_C?
6. Refer to the lowpass filter of Figure 19.15. Assume $R_1 = 22\ k\Omega$, $R_2 = 4.7\ k\Omega$, $C = 0.47\ \mu F$, and the applied voltage is 50 volts at 2 kHz. What is the voltage across *C*? What is the half-power frequency?

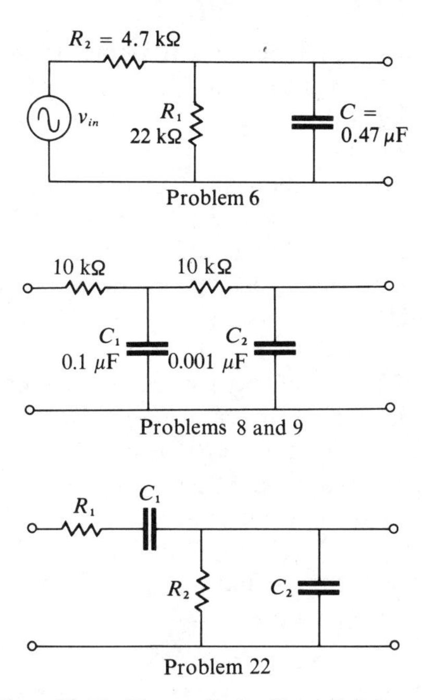

Figure 19.15 Figures for Problems 6, 8, 9, and 22

7. Assume a series *RC* filter (highpass) consists of $R = 20$ kΩ and $C = 1600$ pF. If 15 volts at 10 kHz is applied, what are the output voltage and phase? What is the output voltage at 40 kHz?
8. Refer to the circuit of Figure 19.15. What is the bandwidth of the filter? Assume $R_1 = 10$ kΩ, $R_2 = 10$ kΩ, $C_1 = 0.1$ μF, and $C_2 = 0.001$ μF.
9. Refer to Problem 8. If R_2 is changed to 20 kΩ, what is the new bandwidth?

Reference Section 19.3

10. Refer to Problem 9. If C_1 is removed from the circuit, what is the cutoff frequency of the circuit?
11. Assume that a filter (series resonant) is to be made up from a 2 mH coil whose resistance is 1.2 ohms. What value of *C* is required to filter a center frequency of 3.5 kHz? (R_L is 20 ohms.)
12. What is the bandwidth for the filter of Problem 11? What are the upper and lower cutoff frequencies?
13. What value of resistance would be required to add in series with the circuit of Problem 12 to double the bandwidth?
14. Assume a parallel tank series filter is to couple a source to a load. How much impedance will the parallel-resonant circuit offer if

$L = 25$ mH, $C = 50$ pF, and $R_s = 10$ ohms? If the load is 5 kΩ, what is the output if 10 volts is applied at resonance? What is it at twice the resonant frequency? What is it at half the resonant frequency?

15. A series-resonant filter is connected in a circuit in such a way as to reject a given band of frequencies. If the value of C is 0.01 μF, the L is 200 mH, and the resistance of the load is 500 ohms, what is the band of frequencies rejected by the filter? (R of coil equals 10 ohms)
16. A parallel-resonant filter consists of a capacitor of 0.003 μF, an inductor of 20 mH whose resistance is 40 ohms. What is the center frequency filtered by the circuit? (The tank circuit is in series with the load.) What is the band of frequencies rejected (not passed)?
17. Refer to the band rejection filter of Figure 19.16. Assume a wide range of frequencies is applied at AB, however, all frequencies between 1 kHz and 4 kHz are to be rejected (not appear at CD). If $L = 1$ mH, what values of R and C would be required?
18. The circuit of Figure 19.16 represents an alternate form of the high-pass filter. If $C = 0.005$ μF and $L = 40$ mH, what frequencies will be passed by the filter? Assume $R_L = 1.8$ kΩ.

Reference Section 19.4

19. Assume a constant-K filter similar to Figure 19.13(a) is coupled to a 1 kΩ load. If C is to be 0.022 μF, what is the cutoff frequency?

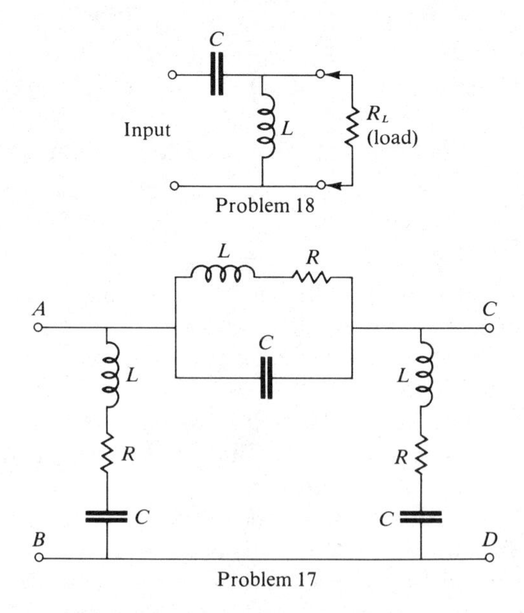

Figure 19.16 Figures for Problems 17 and 18

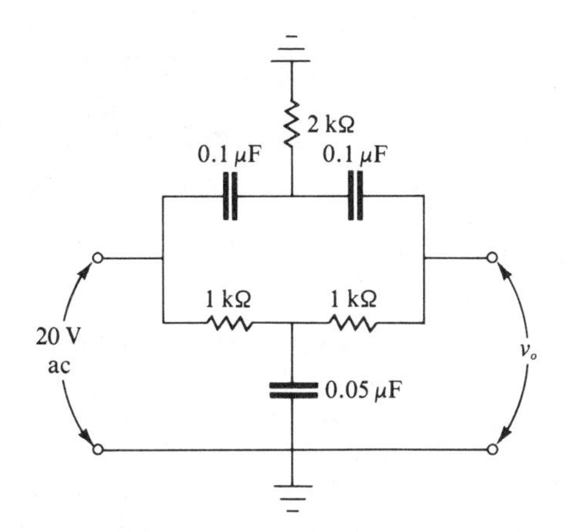

Figure 19.17 Figure for Problem 21

20. Refer to Problem 19. Above the cutoff frequency does the phase angle of the output lead or lag the input wave?

Optional Computer Problems

21. Refer to the notch filter of Figure 19.17. Write a program in BASIC which will print a table of values of v_0 as a function of frequency. From the table what is the center frequency? (*Hint:* Consider the circuit as two *T*-circuits in parallel with each other.)
22. Reference the two-stage *RC* low-pass filter of Figure 19.15. Print out a table of variables relating frequency input, voltage output, and the phase angle between V_{out} and V_{in}. To observe the variables on a logarithmic scale, increment the input frequency in twenty steps between f_c low and f_c high, and then double f_c high 5 times. Assume V_{in} is 10 volts at zero degree phase angle.

CHAPTER 20

ATTENUATORS AND COUPLING NETWORKS

20.1 THREE TERMINAL NETWORKS

Previous chapters have explored various circuit configurations and their application to networks. It is apparent by now that the number of circuit combinations is limitless. For this reason it is not practical to study a long list of circuit configurations. Fortunately there is no need to. We have explored the concepts of dc and ac. We have also studied some basic network theorems and some algebraic manipulations involving the complex plane (j operator). All these concepts can be brought together in the analysis of almost any electric circuit configuration. This chapter will show how the concepts studied may be used in the analysis of coupled networks.

Coupled networks have been chosen for one very important reason. As the field of electronics moves more and more toward miniaturization, the challenges of design and analysis focus on the interconnecting of modular elements. The microintegrated circuit, using large scale integration (LSI) and medium scale integration (MSI) along with the host of logic families, requires a solid understanding of how devices and circuits couple and interconnect (electrically) with each other.

Figure 20.1(a) illustrates a coupling network which interconnects circuit A with circuit B. Note that circuits A and B share the same reference point, ground. In Figure 20.1(b) the coupling network interconnects A and B; however, A and B may or may not have a common reference

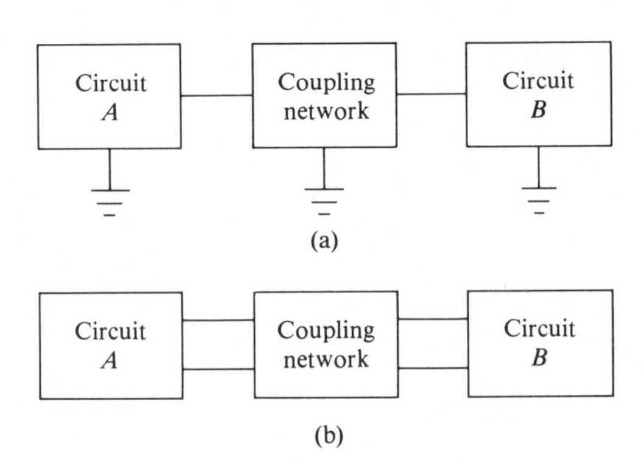

Figure 20.1 Coupling Networks

point. The coupling network of Figure 20.1(a) is said to have two ports with three terminals, while that of part (b) has two ports with four terminals.

A practical example of these two types of networks stems from communication lines. Assume a modem (a modulator/demodulator device which translates computer information into a string of electrical pulses and places them sequentially on telephone lines) is connected to a pair of lines whose impedance is standardized at 600 ohms. One of these lines is ground. The three-terminal arrangement of Figure 20.2(a) may be used. The circuit is known as a *T-pad attenuator*. Although resistors are used here, the T-pad may consist of any combination of LCR components. Note that the sending modem "sees" a 600 ohm impedance "looking" into the T-pad. The receiving modem also sees 600 ohms at the output of the T-pad.

Figure 20.2(b) illustrates an alternate method of coupling the two modems. In this case a four-terminal coupling network is used. The 600 ohm input and output impedances of the modems are "matched" by the H-pad attenuator or balanced-to-ground T-pad.

The three-terminal network of Figure 20.3 is a symbolic representation. It could represent the resistive T-pad of Figure 20.2(a). It could also

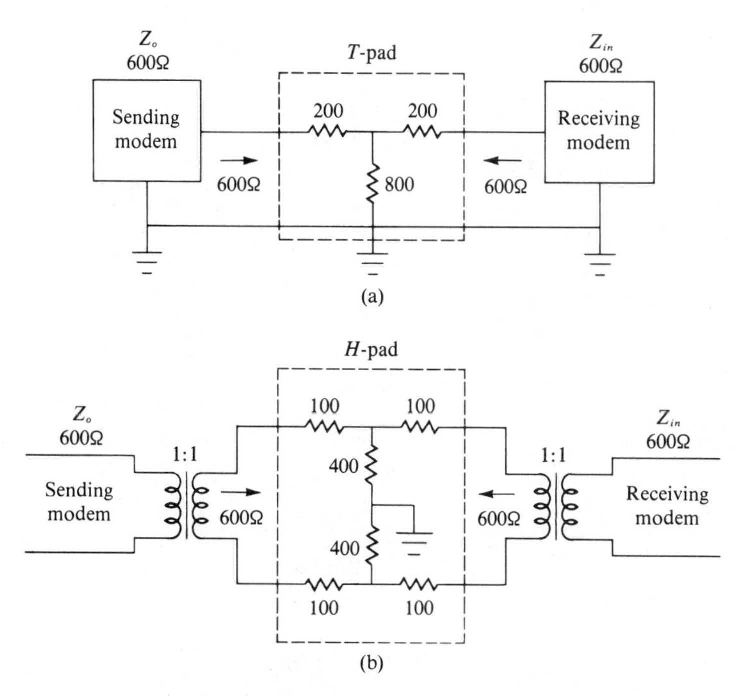

Figure 20.2 Three- and Four-Terminal Pads

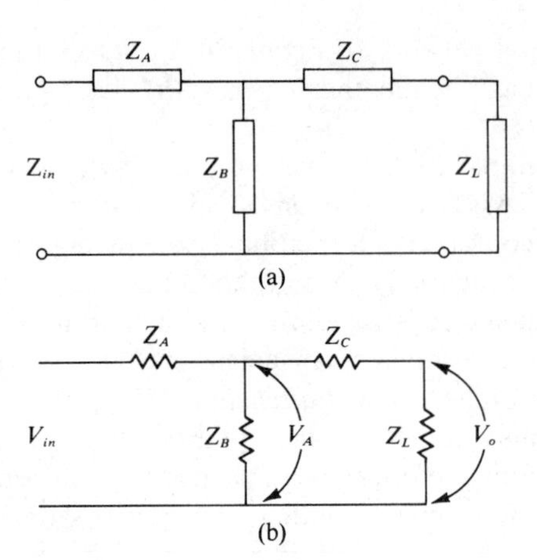

Figure 20.3 Three-Terminal Network

represent the impedance of a section of two wires (similar to the antenna wire used in TV receivers). For the TV line, Z_A and Z_C represent the inductance of the wire, and Z_B represents the capacity between the wires. The network could also represent the impedance equivalencies of a transistor. We will consider the network as a coupling device between a source and a load impedance Z_L. Two parameters of this T network are of interest. The input impedance, Z_{in}, and the voltage attenuation constant, α. The α defines how much the voltage across the load has decreased as a result of the coupling. For example, if 10 volts is applied at Z_{in} and only 3 volts appears across Z_L, the attenuation constant is 0.3. Seventy percent of the applied voltage has been developed across the coupling pad.

The input impedance consists of Z_A in series with the parallel combination of Z_B and $(Z_C + Z_L)$.

$$Z_{\text{in}} = Z_A + \frac{Z_B(Z_C + Z_L)}{Z_B + Z_C + Z_L} \tag{20.1}$$

The attenuation constant is the ratio of V_o to V_{in}. Figure 20.3(b) is an equivalent circuit with T-pad and load. The output voltage is

$$V_o = \frac{Z_L V_A}{Z_C + Z_L} \tag{20.2}$$

The voltage V_A is

$$V_A = \frac{\dfrac{(Z_L + Z_C)Z_B V_{\text{in}}}{Z_B + Z_C + Z_L}}{\dfrac{(Z_L + Z_C)Z_B}{Z_B + Z_C + Z_L} + Z_A}$$

This equation simplifies to

$$V_A = \frac{(Z_L + Z_C)Z_B V_{in}}{(Z_L + Z_C)Z_B + Z_A(Z_B + Z_C + Z_L)} \tag{20.3}$$

Substituting Eq. (20.3) into (20.2) and solving for V_o/V_{in},

$$\alpha = \frac{V_o}{V_{in}} = \frac{Z_L Z_B}{Z_L Z_B + Z_C Z_B + Z_A Z_B + Z_A Z_C + Z_A Z_L} \tag{20.4}$$

In Eq. (20.4) the impedances may be any complex form.

Refer now to the T-pad network of Figure 20.2(a). The attenuation constant of this circuit is 0.5. The process of coupling causes a 50 percent loss of voltage signal.

Example 20.1 Assume a resistive T network is to connect to a source whose output resistance is 1.8 kΩ. If the T network consists of two series leg resistors of 1 kΩ each and a shunt resistor of 2 kΩ, what value of load resistance will match the input resistance of the T-pad to the output resistance of the source? What is the attenuation constant for that load?

Solve Eq. (20.1) for Z_L.

$$Z_L = \frac{Z_{in}(Z_B + Z_C) - Z_A Z_B - Z_C(Z_A + Z_B)}{Z_A + Z_B - Z_{in}}$$

$$= \frac{(1.8)(1 + 2) - (1)(2) - (1)(1 + 2)}{1 + 2 - 1.8} \text{ k}\Omega = 333 \text{ ohms}$$

$$\alpha = \frac{(0.333)(2)}{(0.333)(2) + (2)(1) + (1)(2) + (1)(1) + (1)(0.333)} = 0.11$$

Example 20.1 illustrates that careful selection of component values is required. It is true that this T-pad has matched the output impedance of the source. The maximum-power transfer theorem requires such matching. But the matching process has attenuated the signal to 11.1 percent of the input. Generally, a compromise of design is required. For example, in examining Eq. (20.4), if Z_B is made large and Z_A and Z_C small, the attenuation constant increases. However, these changes influence Z_{in} and the impedance matching range.

In the case of reactive elements in the T-pad, the two equations (20.1 and 20.4) still apply. The added dimension of frequency introduces new considerations. How is the attenuation constant influenced? Does the circuit become resonant within the range of frequencies considered?

Let us consider the case of a source whose impedance is $A \pm jB$. That is, the source driving the T-pad consists of an impedance which is resistive and reactive. For maximum transfer of energy it is important that the source drive an impedance that is the conjugate of the source impedance:

Z_{in} should be $A - jB$ if the source impedance is $A + jB$. If we assume that the load to be driven is $C \pm jD$, then the coupling network must translate the impedance of the load to the impedance of the source.

The modern digital computer becomes a valuable tool in the solution of such a problem. We will explore a fairly simple problem before going on to the more complex. In Figure 20.4(a), assume the value of the load resistance is to vary from zero to infinity (an open circuit). The input impedance Z_{in} will vary between two boundaries. The circuit of Figure 20.4(b) is an equivalent circuit when R_L is zero. When R_L is infinity, the equivalent circuit is shown in Figure 20.4(c).

The general equation for Z_{in} is

$$Z_{in} = \frac{R + R_L}{\omega^2 C^2 (R + R_L)^2 + 1} + R - j\frac{\omega C (R + R_L)^2}{\omega^2 C^2 (R + R_L)^2 + 1} \tag{20.5}$$

If R_L is considered a short circuit, the equation becomes

$$Z_{in} = \frac{R}{\omega^2 R^2 C^2 + 1} + R - j\frac{\omega C R^2}{\omega^2 C^2 R^2 + 1}$$

If R_L is considered an open circuit, the input impedance becomes

$$Z_{in} = R - j\frac{1}{\omega C}$$

These extreme values do not really present much of a problem in terms of calculating the input impedance. If we let R_L *and* f become variables, the calculations become prohibitive. We may instead choose to allow our computer to solve for a series of variables providing us with a printout of tables of solutions. The following program illustrates how Eq. (20.5) may be processed to provide incremental values of Z_{in} for a range of values of R_L.

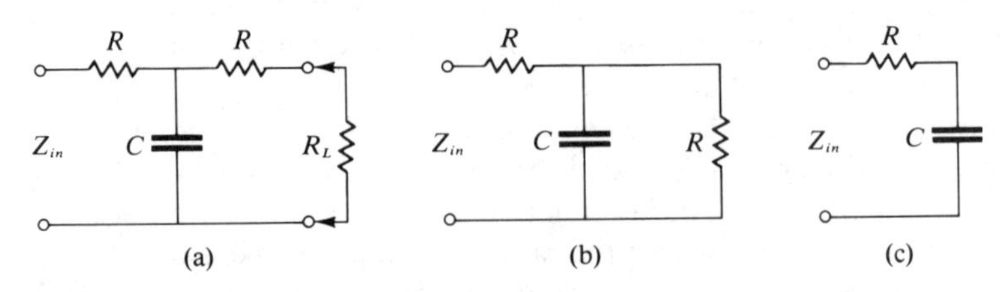

Figure 20.4 *RC* Coupler Pad

Example 20.2 Reference Figure 20.4(a). Let $R = 2.2\ \text{k}\Omega$, $C = 0.015\ \mu\text{F}$, $f = 8.2$ kHz. Write a program in BASIC to calculate Z_{in} as R_L varies in 50 ohm increments from zero ohms to 900 ohms. Print out the values of R_L, Z_{in}, and θ.

The second form of the three-terminal network is the π configuration. Figure 20.5(a) illustrates the basic π network. Here again we are interested in the input impedance and the attenuation constant. The input impedance becomes Z_A in parallel with the combination Z_B in series with the parallel combination of Z_C and Z_L.

$$Z_{\text{in}} = \frac{Z_A Z_C Z_L + Z_A Z_B (Z_L + Z_C)}{(Z_L + Z_C)(Z_A + Z_B) + Z_L Z_C} \tag{20.6}$$

Printout for Example 20.2

```
10   P1=3.14159 : R=2200 : C=0.015E-6 : F=8200 : W=2*P1*F
15   PRINT " R(L)" , "  Z(IN)" , "ANGLE (DEG.)" : PRINT
20   FOR R1=0 TO 900 STEP 50
25   D=(W*C*(R+R1))↑2+1
30   R2=(R+R1)/D+R : X2=(W*C*(R+R1)↑2)/D
35   Z=SQR(R2↑2+X2↑2) : A=ATN(X2/R2)
40   PRINT R1 , Z , -A*(180/P1)
45   NEXT R1
50   END
```

R(L)	Z(IN)	ANGLE (DEG.)
0	2927.78	−19.1695
50	2925.51	−19.4129
100	2923.14	−19.6478
150	2920.67	−19.8745
200	2918.13	−20.0935
250	2915.51	−20.3051
300	2912.85	−20.5095
350	2910.13	−20.7071
400	2907.39	−20.8981
450	2904.6	−21.0828
500	2901.8	−21.2616
550	2898.98	−21.4346
600	2896.15	−21.6021
650	2893.31	−21.7642
700	2890.48	−21.9214
750	2887.64	−22.0736
800	2884.81	−22.2212
850	2881.99	−22.3643
900	2879.18	−22.5031

The attenuation constant becomes:

$$\alpha = \frac{Z_L Z_C}{Z_L Z_C + Z_B(Z_L + Z_C)} \tag{20.7}$$

Example 20.3 Assume that a resistive network similar to Figure 20.5(a) will couple a source to a resistive load which will vary from zero to infinity. (a) What change in resistance is reflected to the source if $R_A = 200$, $R_B = 400$, and $R_C = 200$? (b) What is the range of attenuation constant?

a. With R_L equal to zero, the equivalent circuit is shown in Figure 20.5(b).

$$Z_{\text{in}_1} = R_A \| R_B = \frac{(200)(400)}{200 + 400} = 133 \text{ ohms}$$

With R_L equal to infinity, the equivalent circuit is shown in Figure 20.5(c).

$$Z_{\text{in}_2} = R_A \| (R_B + R_C) = \frac{(200)(600)}{200 + 600} = 150 \text{ ohms}$$

The range of change of Z_{in} is 133 to 150 ohms, or $\Delta Z_{\text{in}} = 17$ ohms.

b. With a short circuit, $\alpha_1 = 0$. With an open circuit,

$$\alpha_2 = \frac{200}{200 + 400} = 0.333$$

The range of α is zero to 0.333 or a $\Delta\alpha$ of 0.333.

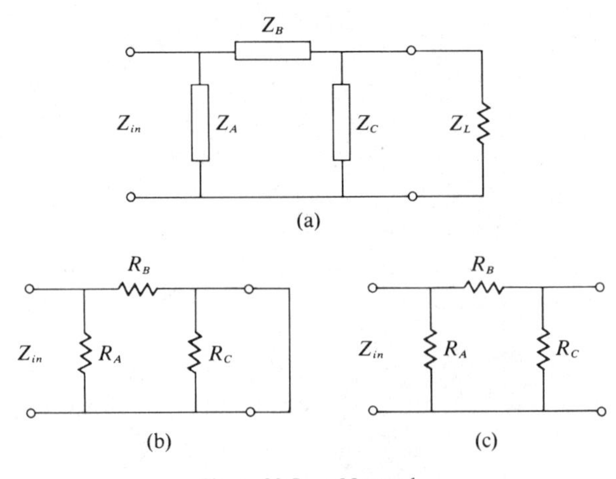

Figure 20.5 π Network

Example 20.4 Repeat Example 20.3 with $R_A = R_C = 400$ and $R_B = 200$.

a. $$Z_{\text{in}_1} = \frac{(200)(400)}{200 + 400} = 133 \text{ ohms}$$

$$Z_{\text{in}_2} = \frac{(400)(600)}{400 + 600} = 240 \text{ ohms}$$

The range of change of Z_{in} is $\Delta Z_{\text{in}} = 107$ ohms.

b. With a short circuit, $\alpha_1 = 0$. With an open circuit,

$$\alpha_2 = \frac{400}{400 + 200} = 0.666$$

The range of α is zero to 0.666 or a $\Delta\alpha$ of 0.666.

Comparison of the two examples illustrates the design tradeoff between isolation of the load and attenuation. The increased isolation results in a sacrifice of signal voltage level.

20.2 FOUR TERMINAL NETWORKS

The four-terminal two-port network of Figure 20.1(b) may consist of any number of resistive, inductive, and/or capacitive elements. We know from network theorems that this circuit can be simplified. Often, it is not known exactly what the elements of the coupling network are, or how they are connected. In such a case we can only determine the nature of the network by how it responds to externally applied voltage and currents.

Consider the circuit of Figure 20.6(a). Assume the circuit within the four terminals is unknown. If the unknown network consists of passive elements, the network can be represented by two groups of parameters.

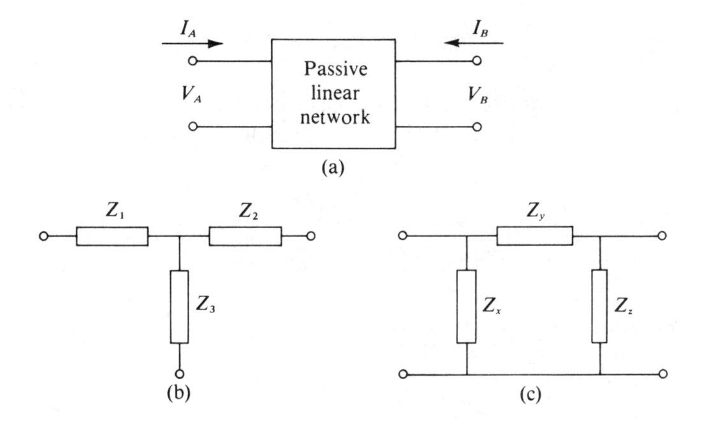

Figure 20.6 Four-Terminal Network

These are identified as the Z and Y parameters. The parameters can be determined by making external measurements such as V_A, V_B, I_A, and I_B.

Once the parameters of an unknown four-terminal network are known, the action of the network can be predicted for any other set of external conditions. The circuit can also be replaced by a known circuit, such as a T or π circuit, with predetermined values of components. The basic equations which define the four-terminal network Z parameters are

$$V_A = Z_{11}I_A - Z_{12}I_B \tag{20.8}$$

$$V_B = Z_{21}I_A - Z_{22}I_B \tag{20.9}$$

The basic equations that define the Y parameters are

$$I_A = Y_{11}V_A - Y_{12}V_B \tag{20.10}$$

$$I_B = Y_{21}V_A - Y_{22}V_B \tag{20.11}$$

Refer to Figure 20.6(b). The T network shown can replace the four-terminal network of Figure 20.6(a). The impedances of the T network are given as:

$$Z_1 = Z_{11} - Z_{12} \tag{20.12}$$

$$Z_2 = Z_{22} - Z_{12} \tag{20.13}$$

$$Z_3 = Z_{12} \tag{20.14}$$

Refer to Figure 20.6(c). The π network shown can replace the four-terminal network of Figure 20.6(a). The impedances of the π are given as:

$$Z_x = 1/(Y_{11} - Y_{12}) \tag{20.15}$$

$$Z_y = 1/Y_{12} \tag{20.16}$$

$$Z_z = 1/(Y_{22} - Y_{12}) \tag{20.17}$$

The functional use of four-terminal network parameters will be examined in more detail.

Z_{11} can be determined by setting I_B in Figure 20.6(a) to zero. This is, in effect, an open circuit at the output. Equation (20.8) reduces to

$$V_A = Z_{11}I_A$$

Hence

$$Z_{11} = V_A/I_A,\ I_B = 0 \tag{20.18}$$

Z_{11} is identified as the open circuit input impedance parameter. Similarly, setting I_B to zero for Eq. (20.9) results in

$$Z_{21} = V_B/I_A,\ I_B = 0 \tag{20.19}$$

Z_{21} is identified as the open circuit forward transfer impedance. It is identified as forward transfer because it is the ratio of output voltage to the input current.

If a voltage is applied at the output terminals and the input terminals are left open circuited the remaining two parameters can be determined. Equation (20.8) reduces to

$$Z_{12} = V_A/I_B, I_A = 0 \tag{20.20}$$

Z_{12} is identified as the open circuit reverse transfer impedance. Similarly, the open circuit output impedance becomes

$$Z_{22} = V_B/I_B, I_A = 0 \tag{20.21}$$

Example 20.5 Assume that the four Z parameters of the network of Figure 20.6(a) are $Z_{11} = 2.1$, $Z_{12} = 0.6$, $Z_{21} = 0.6$, and $Z_{22} = 1.6$. What voltage V_A must be applied in order to produce a current of $\frac{1}{2}$ ampere through a 2 ohm load?

Substituting the known values into Eq. (20.9) and solving for I_A yields (V_B will be IR or $(\frac{1}{2})(2) = 1$ volt):

$$1 = (0.6)I_A - (1.6)(\tfrac{1}{2})$$
$$I_A = 3 \text{ amperes}$$

Substituting 3 amperes of I_A and the known parameters into Eq. (20.8),

$$V_A = (2.1)(3) - (0.6)(\tfrac{1}{2}) = 6 \text{ V}$$

Assume that the four-terminal network of this example is to be replaced with an equivalent T network. Will the 6 volts applied to the equivalent T yield the same $\frac{1}{2}$ ampere through a 2 ohm load? The equivalent T circuit is found by substituting the known parameters in Eqs. (20.12), (20.13), and (20.14).

$$R_1 = (2.1) - (0.6) = 1.5 \text{ ohms}$$
$$R_2 = (1.6) - (0.6) = 1 \text{ ohm}$$
$$R_3 = (0.6) = 0.6 \text{ ohm}$$

The equivalent T circuit is shown in Figure 20.7. Using conventional Ohm's law techniques, the student should solve for the current through the 2 ohm load resistance.

Example 20.6 Refer to the network of Figure 20.8(a). With the values given, determine the Z parameters. What is the equivalent T circuit? As previously stated, any voltages may be assumed to evaluate the Z parameter of the network.

Assume $20\angle 0°$ volts is applied between points A and B. Points C and D are open-circuited. From Eq. (20.18), the current I_A is $V_A/(500 - j300)$.

$$Z_{11} = 500 - j300$$

The voltage V_B in Figure 20.6(a) is the voltage across CD in Figure 20.8(a). By voltage ratio,

$$V_B = \frac{-j300\,V_A}{500 - j300}$$

According to Eq. (20.19),

$$Z_{21} = \frac{-j300\,V_A}{500 - j300} \bigg/ \frac{V_A}{500 - j300} = -j300$$

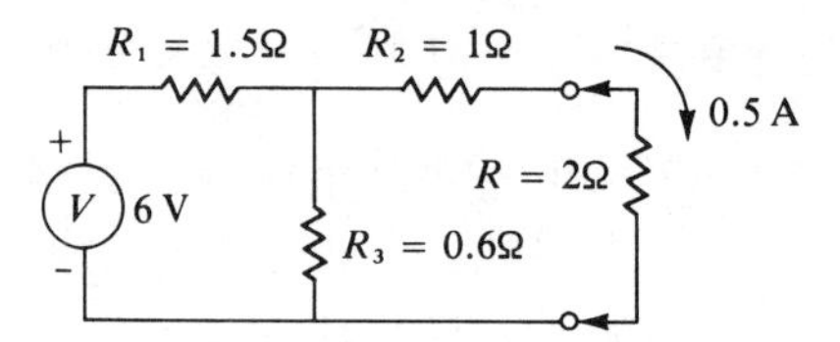

Figure 20.7 Equivalent T Circuit

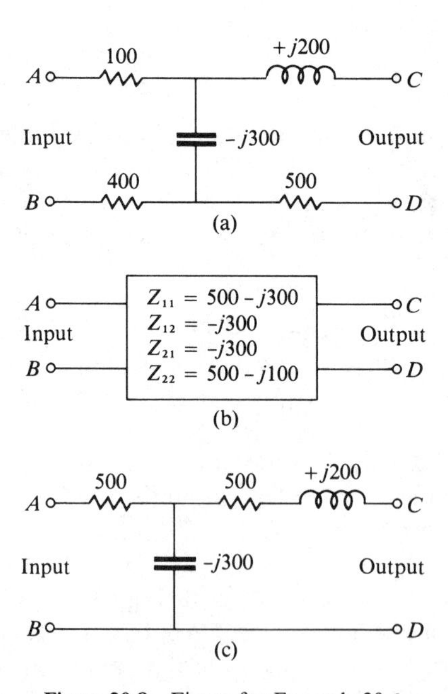

Figure 20.8 Figure for Example 20.6

To find the reverse parameters, again assume $20\,\angle 0^\circ$ volts is applied between points C and D. Similarly, the input terminals are open circuited. According to Eq. (20.21),

$$Z_{22} = \frac{20\,\angle 0^\circ}{20\,\angle 0^\circ/(500 - j100)} = 500 - j100$$

Solving for Z_{12} according to Eq. (20.20),

$$Z_{12} = -j300$$

(Note that $Z_{21} = Z_{12}$. This is true for any network that contains only L, R, or C components.) The equivalent network is shown in Figure 20.8(b).
The equivalent T network results from Eqs. (20.12), (20.13), and (20.14).

$$Z_1 = 500 - j300 - (-j300) = 500$$
$$Z_2 = 500 - j100 - (-j300) = 500 + j200$$
$$Z_3 = -j300$$

The circuit is shown in Figure 20.8(c).

20.3 IMPEDANCE COUPLING

Return now to the circuit of Figure 20.1. We have explored the characteristics of coupler networks and their attenuation characteristics. Often the coupler circuit must "match" two known impedances. In order to effect a maximum transfer of energy from the source to the load, the input impedance of the coupling device must equal the conjugate of the output impedance of the source. In addition, the output impedance of the coupler must equal the conjugate of the input impedance of the load. The resistive networks of Figure 20.2 illustrate this case. However, the design of such circuits often becomes difficult when reactive properties are involved. Two kinds of circuits can be employed: the L-attenuator pad or the H-attenuator pad.

L-type Pad

Figure 20.9 illustrates the L-type pad. The Z_A and Z_B of Figure 20.9(a) form an L configuration, hence the name. Similarly, Figure 20.9(b) shows a variation of the L pad. Z_X is the output impedance of the source. This is generally the known value to be matched. The value of Z_L is the impedance of the load. This too is generally known. The following deriva-

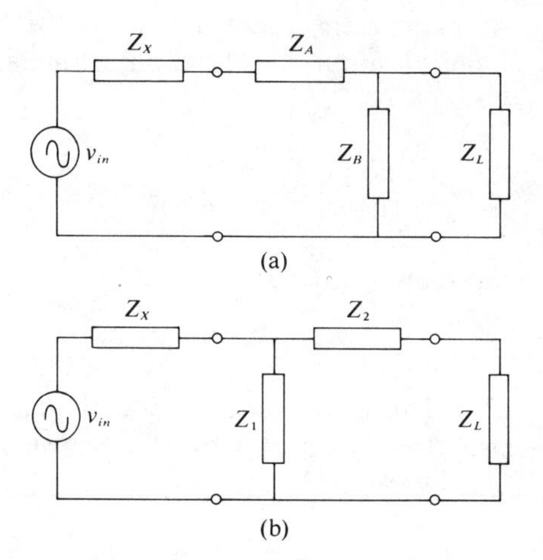

Figure 20.9 *L*-Type Pad

tion presumes that conjugates are used. If the source has an impedance of $200 + j150$, then the Z_X used in the equation should be $200 - j150$. Return now to Figure 20.9(a). By inspection, since Z_L is in parallel with Z_B, we require that

$$Z_X = Z_A + \frac{Z_B Z_L}{Z_B + Z_L} \tag{20.22}$$

The value of Z_X is in series with Z_A. When we view the coupler from its output terminals at Z_L, we require that

$$Z_L = \frac{(Z_A + Z_X)Z_B}{Z_A + Z_B + Z_X} \tag{20.23}$$

Solving Eq. (20.23) for Z_A and substituting into Eq. (20.22), the resultant equation is

$$Z_X(Z_B^2 - Z_L^2) = Z_B Z_L(Z_B + Z_L) + (Z_B + Z_L)(Z_L Z_X + Z_L Z_B - Z_X Z_B)$$

Solving this equation for Z_B,

$$Z_B = \sqrt{\frac{Z_L^2 Z_X}{Z_X - Z_L}} = Z_L \sqrt{\frac{Z_X}{Z_X - Z_L}} \tag{20.24}$$

We note from Eq. (20.24) that an *L*-type coupler cannot be used to match equal impedances. If $Z_X = Z_L$, the denominator approaches zero, and the value of Z_B then becomes undefined. Once determined, we can substitute the value for Z_B into Eq. (20.22) and solve for Z_A. The following example illustrates the case for resistive matching.

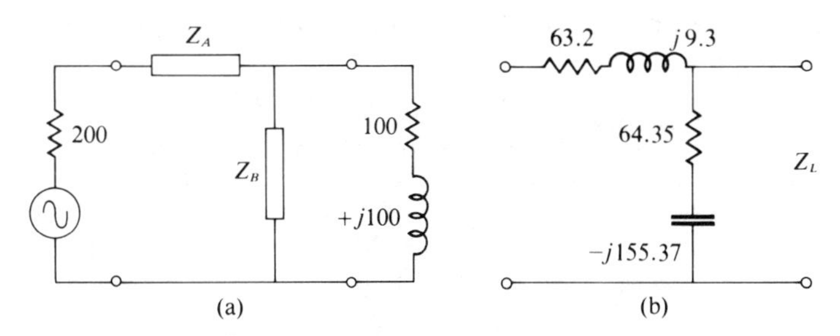

Figure 20.10 Figure for Example 20.8

Example 20.7 Assume a 600 ohm source resistance is to be matched to a 200 ohm load resistance. For maximum power transfer, what values of resistance should be used in an *L*-type attenuator? Assume the circuit of Figure 20.9(a) is to be used.

$$Z_B = \sqrt{\frac{(200)^2(600)}{600 - 200}} = 245 \text{ ohms}$$

$$Z_A = Z_X - \frac{Z_B Z_L}{Z_B + Z_L} = 600 - \frac{(245)(200)}{245 + 200} = 490 \text{ ohms}$$

The attenuation constant of an *L*-type coupler is the ratio of the voltage across Z_L to the input voltage. This factor can be readily developed from the basic circuit.

Example 20.8 Refer to the circuit of Figure 20.10(a). Assume that an *L*-type attenuator is to match the 200 ohm source impedance to a load impedance of $100 + j100$. What values of Z_A and Z_B should be connected into the circuit? Identify the real and the reactive values for each of the branches of the *L*-pad.

The output impedance of the coupler must be the conjugate of the load.

$$Z_o \text{ must be } 100 - j100$$

According to Eq. (20.24),

$$Z_B = \sqrt{\frac{(141.41\,\angle{-45^\circ})^2\,200\,\angle{0^\circ}}{100 + j100}}$$

$$= \sqrt{\frac{141.41^2\,\angle{-90^\circ}(200)}{141.41\,\angle{45^\circ}}}$$

$$= 168.17\,\angle{-67.5^\circ}$$

$$= 64.35 - j155.37$$

Solving for Z_A from Eq. (20.22) (here we do not use the conjugate of Z_L):

$$Z_A = 200 - \frac{(168.17\,\underline{/-67.5^\circ})(141.41\,\underline{/45^\circ})}{64.35 - j155.37 + 100 + j100}$$

$$= 200 - \frac{23780\,\underline{/-22.5^\circ}}{173.43\,\underline{/-18.61^\circ}}$$

$$= 200 - 137.12\,\underline{/-3.89^\circ}$$

$$= 200 - (136.8 - j9.3)$$

$$= 63.2 + j9.3$$

The resultant coupler is illustrated in Figure 20.10(b). This example was chosen to illustrate the design of an *L*-pad. In practice, matching could be achieved more simply by connecting $-j200$ in parallel with the load. The student should verify this.

H-type Attenuator

The *H*-type attenuator pad is shown in Figure 20.11(a). The fundamentals of four-terminal network theory can be applied to this circuit. If we assume that the pad must match an output impedance Z_{in}, then the parameter Z_{11} must be the conjugate of Z_{in}. Similarly, the output parameter Z_{22} must be the conjugate of Z_0. Note in Figure 20.11(b) that the transfer impedance in the forward direction equals the transfer impedance in the reverse direction. Z_2 will equal Z_{21} if Z_1 is the sum of Z_A and Z_B, and Z_3 is the sum of $Z_D + Z_E$. Appropriate substitutions into equations for the four-terminal networks will yield values for the coupler circuit.

SUMMARY

Attenuator and coupling networks are designed to provide coupling from a source to a load. Either a maximum transfer of energy is desired or a controlled amount of attenuation is required.

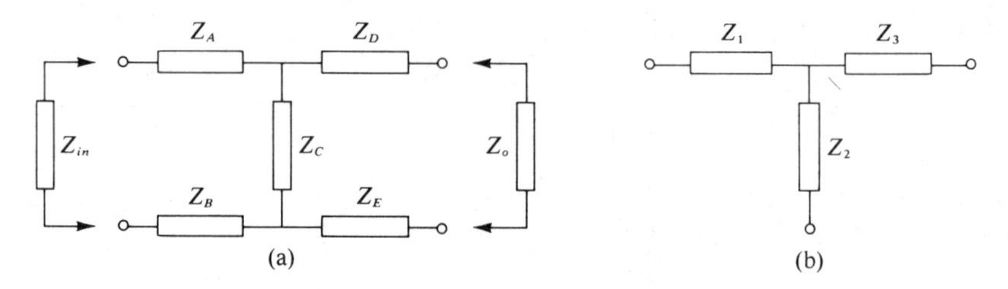

Figure 20.11 *H*-Type Attenuator

Three-terminal and four-terminal networks can be used for coupling. For any given combination of components a fixed attenuation constant, α, results.

Four-terminal network theorems may be used for the analysis of coupler and attenuator circuits in complex solutions. The *L*-pad coupler is used to match impedance for unbalanced lines, while the *H*-pad coupler is used for balanced lines.

PROBLEMS

Reference Section 20.1

1. Assume a *T*-pad coupler consists of three 2.2 kΩ resistors. If the load resistance is 1 kΩ, what is the input resistance? What is the attenuation constant?
2. Refer to Problem 1. What is the range of change of the input resistance if the load resistance is varied from zero to infinity?
3. Assume the *T*-pad of Figure 20.2(a) has two series resistors of 200 ohms each and the shunt resistor is 100 ohms. What is the range of input resistance as the impedance of the load varies from 10 ohms to 1 kΩ?
4. Refer to Problem 3. If the two series resistors are 100 ohms and the shunt resistor is 200 ohms, what is the range of change of input resistance?
5. Assume the *T*-pad of Figure 20.12 has two capacitors of equal value. The capacitive reactance of each is 2 kΩ and the resistance is 2 kΩ. What is the input impedance when the load resistance is 1 kΩ? 2 kΩ? 10 kΩ? What is the attenuation constant for each value of load resistance?
6. Refer to Problem 5. If the 1 kΩ load resistance is replaced with an inductive reactive load of $j4$ kΩ, what is the input impedance? What are the attenuation constant and the phase displacement of the output?

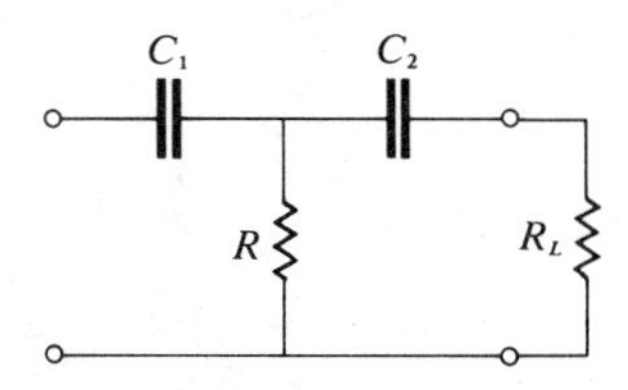

Figure 20.12 Figure for Problem 20.5 and 20.6

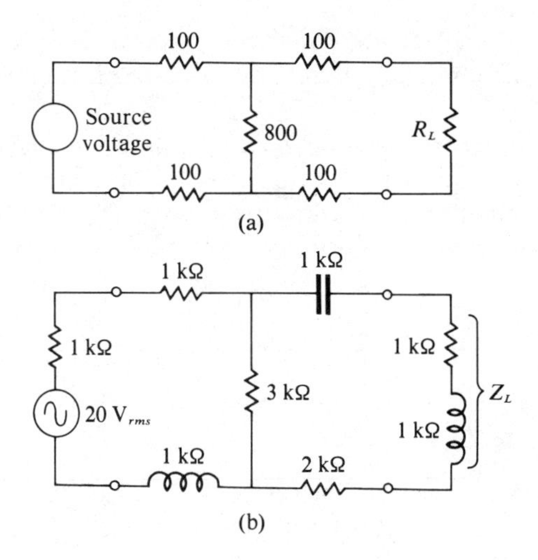

Figure 20.13 Figure for Problems 13 and 18

7. Assume the pi network of Figure 20.5(a) has the following values: $Z_A = 200$, $Z_B = +j200$, $Z_C = +j100$, and $Z_L = 300$. What is the input impedance? What is the attenuation constant?
8. Refer to Problem 7. If the frequency is reduced in half, what is the new attenuation constant?

Reference Section 20.2

9. What are the Z parameters for the pi circuit (without load resistance) of Problem 7?
10. Refer to Problem 9. What is the equivalent T-pad for the circuit?
11. A given network has the following parameters: $Z_{11} = 200 - j300$, $Z_{12} = Z_{21} = -j300$, and $Z_{22} = 100 + j100$. If 20 volts at 0° is applied at the input and 30 volts at +20° is applied at the output terminals, what is the current flowing out of the input generator?
12. Refer to Problem 11. What is the current phase angle flowing out of the 30 volt source?
13. Refer to the H-pad attenuator of Figure 20.13(a). What are the Z parameters of the pad?

Reference Section 20.3

14. Design an L-pad coupler which will match a 300 ohm source resistance to a 72 ohm load resistance. What is the attenuation constant of the coupler?

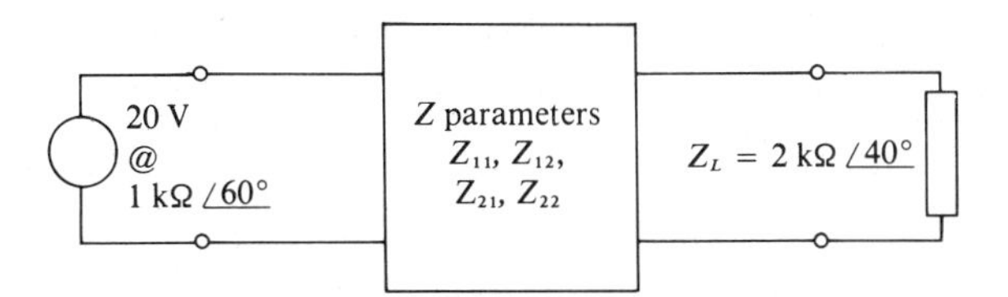

Figure 20.14 Figure for Problem 19

15. If a generator whose internal resistance is 1 kΩ is to drive a load whose impedance is $200\angle 20^\circ$ at a frequency of 4 kHz, what components would be required for an *L*-type coupler circuit?
16. Refer to Problem 14. What is the resonant frequency of the coupler pad?

APPENDICES

APPENDIX A Mathematical and Computer Symbols and Constants

Mathematical Symbol	Computer Symbol	Meaning
$=$	=	*A* equals *B*
$\neq$	< >	*A* does not equal *B*
$\approx$	= =	*A* is approximately equal to *B*
$>$	>	*A* is greater than *B*
$<$	<	*A* is less than *B*
$\geq$	> =	*A* is equal to or greater than *B*
$\leq$	< =	*A* is less than or equal to *B*
$\lvert\ \rvert$	ABS ( )	Absolute positive value of
$\equiv$		By definition, *A* equals *B*
$\sum$		The sum of terms
$\therefore$		Therefore
K		9×10^9 ($\frac{1}{4}\pi\epsilon_0$)
π		3.1415926
e		2.71828
ϵ_0		8.854×10^{-12}

APPENDIX B Electrical Symbols and Constants

Name	Capital	Lower Case	Electrical Meaning
Epsilon	$\mathscr{E}$		Electric field strength
Rho		ρ	Resistivity
Omega	Ω		Ohms
Sigma		σ	Conductivity
Alpha		α	Temperature coefficient
Eta		η	Efficiency
Phi		ϕ	Maxwell (one line of force)
Mu		μ	Permeability
Weber/m^2	B		Flux density
Permeability		μ_0	Permeability of free space
Reluctance	$\mathscr{R}$		Magnetic opposition
Flux intensity	H		Strength of flux
Magnetomotive force	$\mathscr{F}$		Gilbert
Linkages		λ	Flux linkages
Coefficient of coupling		k	Mutual interactive flux
Tau		τ	Time constant
Psi	Ψ		Electric flux
Epsilon		ϵ	Electric force or flux intensity

APPENDIX C System Commands for Microcomputers

This appendix contains the most important system commands for a number of currently popular computers. This is intended to be a convenient reference but does not contain every command, nor every option and variation of every command, available for the computers shown. Consult the programming manual for the individual computer for additional details. Optional entries appear inside brackets { }.

APPLE (Applesoft II)

Command	Description
CLEAR	Sets all numeric variables to zero and all string variables to null
CONT	Causes execution of a program to resume after being halted by a STOP, END, or Control C statement
Control C	Interrupts program execution
DEL n_1, n_2	Deletes lines n_1 through n_2 from a program
FRE(0)	Displays the number of bytes of memory still available to the user
LIST	Causes the entire program to be displayed
LIST n_1, n_2	Causes line numbers n_1 through n_2 to be displayed
LOAD	Loads a program from cassette tape
NEW	Deletes current program and all variables
RUN {n_1}	Begins execution of a program at line n_1—If no n_1 is specified, execution begins at the lowest line number
SAVE	Stores a program on cassette tape
SPEED=n	Sets the rate at which characters are sent to the video display or other I/O device—n is a number between 0 and 255
TRACE	Causes the line number of each statement to be displayed as it is executed

ATARI 400/800

Command	Description
BYE	Causes an exit from BASIC and allows use of keyboard without disturbing any programs in memory
CLOAD	Loads a program from cassette tape
CONT	Causes program execution to resume after a STOP, END, or break
CSAVE	Saves a program on cassette tape

ATARI 400/800—*(contd.)*

Command	Description
ENTER	Causes a cassette tape to play back a program originally recorded using LIST
LIST	Displays entire program
LIST n_1, n_2	Displays line numbers n_1 through n_2
LIST "P"	Prints program on a line printer
LIST "P", n_1, n_2	Prints line numbers n_1 through n_2 on a line printer
LIST "*filename*"	Stores a program on cassette tape under the designation "*filename*"—quotation marks must be used
LOAD "*filename*"	Loads a program designated "*filename*"—quotation marks must be used
RUN {"*filename*"}	Causes current program to be executed if no filename is specified; otherwise, retrieves and runs specified file—quotation marks must be used
SAVE "*filename*"	Saves a program designated "*filename*," which must be enclosed within quotation marks

COMMODORE (CBM and PET)

Command	Description
CLR	Sets all numeric variables to zero and all string variables to null
CONT	Resumes program execution after a STOP or END statement, or after the stop key has been depressed
DLOAD "*filename*"	Loads a file named "*filename*" from disk
DSAVE "*filename*"	Saves a file named "*filename*" on disk
LIST	Displays entire program
LIST n_1-n_2	Displays line numbers n_1 through n_2
NEW	Deletes current program and clears all variables
RENAME "*oldname*" TO "*newname*"	Changes the name of a disk file
Run {n}	Runs a program beginning at line n or from the lowest line number, if n is not specified
SAVE {"*filename*"}	Saves a program on cassette tape or disk—if "*filename*" is used, it must be enclosed by quotation marks
VERIFY {"*filename*" {,device}}	Compares contents of program in memory to file on disk or tape and reports differences—device defaults to cassette number 1; filename defaults to null

HEATH H–8

Command	Description
Auto {*initial sn* , *incr*}	Causes automatic numbering of statements, beginning with initial *sn* and using *incr* as increment; if neither of the latter are specified, then it is assumed that both equal 10
CLEAR	Sets all program variables to zero
CONT	Causes program execution to resume after having ceased due to the execution of a STOP or END statement or a CONTROL/C command.
CONTROL/*ch*	Simultaneous depression of the control key and a *ch* character key, which results in one of the actions listed below
CONTROL/A	Allows use of the EDIT commands on the line currently being typed
CONTROL/C	Interrupts execution of a program
CONTROL/O	Suppresses all output until an INPUT statement is executed
CONTROL/Q	Causes program execution to resume after a CONTROL/S
CONTROL/S	Causes program execution to pause until such time as a CONTROL/Q is entered
CONTROL/U	Erases current line and executes a carriage return
FRE(O)	Causes display of the total number of bytes that are still available in memory
KILL "*filename*"	Deletes file named "*filename*" from disk—quotation marks must be used
LIST	Displays the program currently in memory, statement by statement, beginning with the first statement number
LIST *sn*1 - *sn*2	Displays all statements from statement number *sn*1 to statement number *sn*2, inclusive
LIST *sn*	Lists statement with statement number *sn*
LIST *sn*-	Lists all statements from statement number *sn* through end of program
LIST - *sn*	Lists all statements from the first through statement number *sn*
LOAD "*filename*"	Loads a file named "*filename*" from disk—quotation marks must be used
LOAD "*filename*", R	Same as LOAD "*filename*", except a program loaded from disk is also run
NAME "*oldname*" AS "*newname*"	Assigns the name "*newname*" to a file on disk already having the name "*oldname*"; the "*oldname*" file must exist, and there cannot already be a "*newname*" file
NEW	Deletes program currently in memory and sets all variables to zero
RENUM	Renumbers all the statement numbers in a program; first new statement number is 10, and there is an increment of 10 between each new statement number

HEATH H-8—*(contd.)*

Command	Description
RENUM *sn* , , *incr*	Renumbers all statement numbers in a program using *sn* as the first new statement number and using increment *incr* between the new statement numbers
RENUM *sn*1, *sn*2, *incr*	Same as RENUM *sn* , , *incr*, except renumbering begins with old statement number *sn*2
RESUME	Causes program execution to resume after an error-recovery procedure has been performed; execution resumes at the statement causing the error
RESUME NEXT	Same as RESUME except execution resumes at statement immediately following the error-causing statement
RESUME *sn*	Same as RESUME except program execution resumes at statement number *sn*
RUN {*sn*}	Causes a program to be executed, beginning with statement number *sn*; if *sn* is omitted, then execution begins at lowest numbered statement number
SAVE "*filename*"	Saves a file on disk under the name "*filename*"—quotes must be used
TRON	Causes statement numbers of a program to be displayed as the corresponding statements are executed
TROFF	Disables TRON—i.e., eliminates display of line numbers caused by TRON
WIDTH *int*	Sets the width of the display on the screen of a terminal; *int* is the (integer) number of characters per line that are displayed (*int* must be less than 255)

IBM PERSONAL COMPUTER

Command	Description
AUTO	Generates line numbers automatically
CLEAR	Clears program variables
CONT	Resumes program execution after an interrupt
DELETE	Deletes a range of program lines
LIST	Displays program or portions thereof
LLIST	Prints program on a line printer
LOAD	Loads a program file
NEW	Deletes current program
RUN	Loads and runs a program
SAVE	Saves current program
SYSTEM	Exits BASIC and returns control to operating system

TRS-80

Command	Description
AUTO	Generates line numbers automatically, beginning with 10 and having increment 10
AUTO n_1, n_2	Generates line numbers automatically, beginning with n_1 and having increment n_2
CLEAR	Sets all numeric variables to zero and all string variables to null
CLEAR *n*	Makes *n* bytes available for string storage
CLOAD "*filename*"	Loads a program from cassette tape—quotation marks must be used
CLOAD? "*filename*"	Compares a program stored on cassette with a program in the computer and displays BAD if there are any discrepancies
CONT	Resumes program execution after a STOP instruction or after depression of the break key
CSAVE "*filename*"	Stores current program on cassette tape under "*filename*," which must be enclosed by quotation marks
DELETE n_1-n_2	Deletes line numbers n_1 through n_2
EDIT *n*	Puts computer in edit mode for editing line *n*
LIST n_1-n_2	Displays line numbers n_1 through n_2
LIST *n*-	Displays line number *n* and all higher line numbers
RUN {*n*}	Runs current program beginning at line *n*—if no *n* is specified, run begins at lowest line number
SYSTEM	Causes exit from BASIC and returns control to operating system for loading programs

XEROX SIGMA IX

Command	Description
Break key	Stops current operation and generates > prompt symbol; a second depression of break returns control to system with ! prompt
CATALOG	Lists names of all program files in user's account
CLEAR	Sets numeric variables to zero and string variables to null
CLEAR ARRAYS	Sets only array variables to zero
CLEAR STRINGS	Sets string variables to null
DELETE *n*	Deletes line *n*
DELETE n_1-n_2	Deletes lines n_1 through n_2
DELETE *filename*	Deletes program named *filename*
EXECUTE n_1 {-n_2}	Begins program execution at line n_1—if n_2 is specified, program execution halts at line number preceding n_2

XEROX SIGMA IX—*(contd.)*

Command	Description
EXTRACT *n*	Deletes entire program except line *n*
EXTRACT n_1-n_2	Deletes entire program except line numbers n_1 through n_2
LIST *n*	Displays line *n*
LIST n_1-n_2	Displays line numbers n_1 through n_2
LOAD *filename*	Loads program called *filename*
RENUMBER $\{n_1\{,n_2\{,n_3\}\}\}$	Renumbers lines—n_1 is lowest new line number; n_2 is lowest old line number; n_3 is increment (n_1, n_2, and n_3 are 100, 1, and 10, respectively, by default)
RUN	Begins execution of a program at lowest line number
STATUS	Returns EDITING, COMPILING, or RUNNING
WIDTH *n*	Changes print width from 72 characters to *n* characters, where *n* is a number from 0 to 255

APPENDIX D Solving Simultaneous Equations Using Microcomputer BASIC

Most versions of BASIC used with microcomputers do not have the capability of performing mathematical operations on matrices; we therefore demonstrate in this appendix that two or three equations can be solved simultaneously using only the conventional arithmetic operators. First suppose that we have two equations in the two unknowns x and y and that we have arranged them in the format

$$ax + by = c \tag{D.1}$$

$$dx + ey = f \tag{D.2}$$

where a, b, d, and e are coefficients and c and f are constants. For example, the equations

$$2x - 3y = 5$$
$$x + 4y = -2$$

would give us the following:

$$a = 2 \qquad b = -3 \qquad c = 5$$
$$d = 1 \qquad e = 4 \qquad f = -2$$

Solving Eqs. (D.1) and (D.2) using determinants (see section 8.1) leads to

$$x = \frac{ce - bf}{ae - bd} \tag{D.3}$$

$$y = \frac{af - cd}{ae - bd} \tag{D.4}$$

Note that the denominator is the same in both Eqs. (D.3) and (D.4). The following three BASIC statements could then be used to find x and y.

```
LET D1 = A*E - B*D
LET X = (C*E - B*F)/D1
LET Y = (A*F - C*D)/D1
```

Example D.1 Write a BASIC program to find the currents I_1 and I_2 in the network shown in Figure D.1.

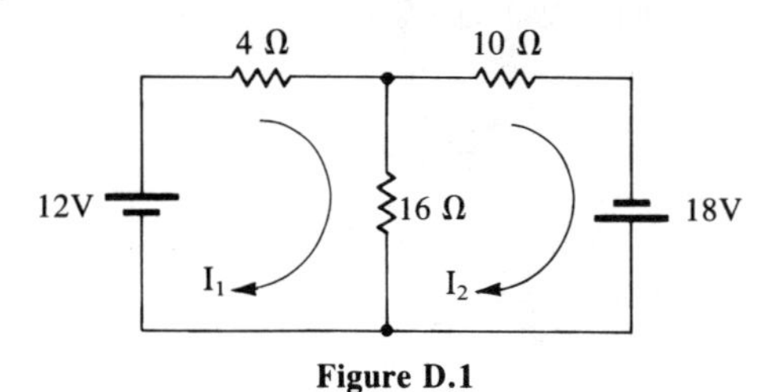

Figure D.1

Writing Kirchhoff's voltage law around loop 1, we obtain

$$4I_1 + 16(I_1 - I_2) = 12$$

or

$$20I_1 - 16I_2 = 12 \tag{D.5}$$

Similarly, for loop 2 we find

$$10I_2 + 16(I_2 - I_1) = -18$$

or

$$-16I_1 + 26I_2 = -18 \tag{D.6}$$

From Eqs. (D.5) and (D.6), we have

$$a = 20 \qquad b = -16 \qquad c = 12$$
$$d = -16 \qquad e = 26 \qquad f = -18$$

The following program finds and prints the values of I_1 and I_2:

```
10      PRINT  "ENTER A,B,C,D,E,F"
20      INPUT  A,B,C,D,E,F
30      LET D1 = A*E - B*D
```

```
40        PRINT  "I1 = ";  (C*E - B*F)/D1;   "AMPS"
50        PRINT  "I2 = ";  (A*F - C*D)/D1;   "AMPS"
60        END
RUN
? 20, -16, 12, -16, 26, -18
I1 = 9.09091E-02 AMPS
I2 = -.636364 AMPS
```

The minus sign associated with the result for I_2 means that the loop current I_2 is actually flowing in the opposite direction from that assumed in Figure D.1.

Now suppose that we have three equations in three unknowns (x, y, and z), which we arrange in the following format:

$$ax + by + cz = d$$
$$ex + fy + gz = h$$
$$ix + jy + kz = l$$

Using determinants, the solutions for these equations can be shown to be

$$x = \frac{(dfk + bgl + chj) - (cfl + dgj + bhk)}{D}$$

$$y = \frac{(ahk + dgi + cel) - (chi + agl + dek)}{D}$$

$$z = \frac{(afl + bhi + dej) - (dfi + ahj + bel)}{D}$$

where

$$D = (afk + bgi + cej) - (cfi + agj + bek)$$

Example D.2 Write a BASIC program that can be used to solve and print the solutions to the following equations:

$$x + 2y - 3z = 4$$
$$2x \quad + z = 1$$
$$-x + 7y + 4z = -12$$

```
10        READ A,B,C,D,E,F,G,H,I,J,K,L
20        LET D1=(A*F*K+B*G*I+C*E*J)-(C*F*I+A*G*J+B*E*K)
30        LET X=((D*F*K+B*G*L+C*H*J)-(C*F*L+D*G*J+B*H*K))/D1
40        LET Y=((A*H*K+D*G*I+C*E*L)-(C*H*I+A*G*L+D*E*K))/D1
50        LET Z=((A*F*L+B*H*I+D*E*J)-(D*F*I+A*H*J+B*E*L))/D1
60        PRINT "X", "Y", "Z"
70        PRINT X,Y,Z
80        DATA 1,2,-3,4,2,0,1,1,-1,7,4,-12
90        END
RUN

X              Y              Z
1.20896        -.731343       -1.41791
```

In circuit analysis involving capacitors and inductors, a transitional state exists between the initial and the final values of the circuit parameters. This in-between state is referred to as the *transient state.* It is the purpose of this section of the appendix to analyze these conditions.

Although complex mathematics could be employed, analysis ultimately reduces to one simple, algebraic approach. Regardless of whether the circuit analysis requires capacitive, inductive, or resistive voltages, or whether one is seeking the current value through the circuit, all these parameters assume this basic form:

$$e_R,\ e_C,\ e_L,\ i_t = A + Be^{-t/\tau} \qquad \text{(E.1)}$$

It should be noted that there are two unknowns in Eq. (E.1). These unknowns are symbolized as A and B. The evaluation of these unknowns requires that two arbitrary choices of time be selected, and that the circuit conditions prevalent at these times be substituted into Eq. (E.1). Although any two times can be chosen, logic dictates that t_0 (conditions at time zero) and t_∞ (conditions after the circuit has reached its steady state) are most easily analyzed. Examples will illustrate this simple methodology. Before beginning, however, it is appropriate to recall the following:

$$e^0 = 1 \qquad \text{(E.2)}$$

$$e^{-\infty} = 0 \qquad \text{(E.3)}$$

Example E.1 Reference Figure E.1. Calculate the voltage across C 1.2 ms after switch 1 is closed. Assume C is initially uncharged.

$$\tau = RC = 10^3 \times 10^{-6} = 10^{-3} = 1 \text{ ms}$$

The voltage across the capacitor the instant switch 1 is closed is zero volts, therefore, from Eq. (E.1),

$$e_C(0) = A + Be^{-0/1\text{ ms}} = 0 \text{ volts}$$

$$A + B(1) = 0$$

$$A + B \quad = 0$$

$$e_C(\infty) = A + Be^{-\infty/1\text{ ms}} = 10 \text{ volts}$$

$$A + B(0) = 10 \text{ volts}$$

$$A = 10$$

If $A + B = 0$ and $A = 10$, then $B = -10$.

Substituting these values into Eq. (E.1) results in

$$e_C = 10 - 10e^{-t/\tau} \qquad \text{(E.4)}$$

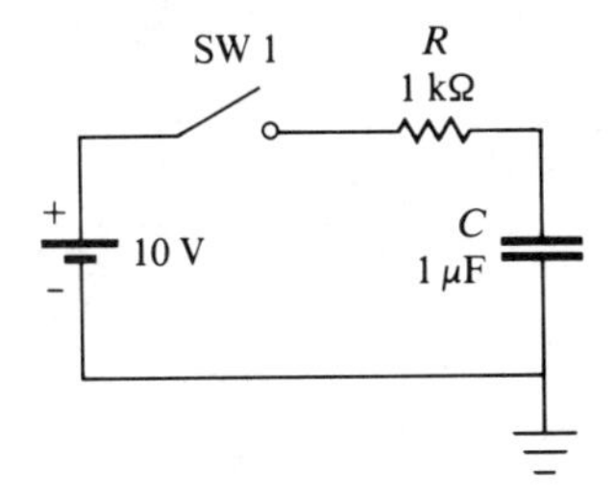

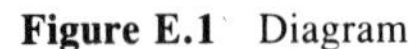

Figure E.1 Diagram

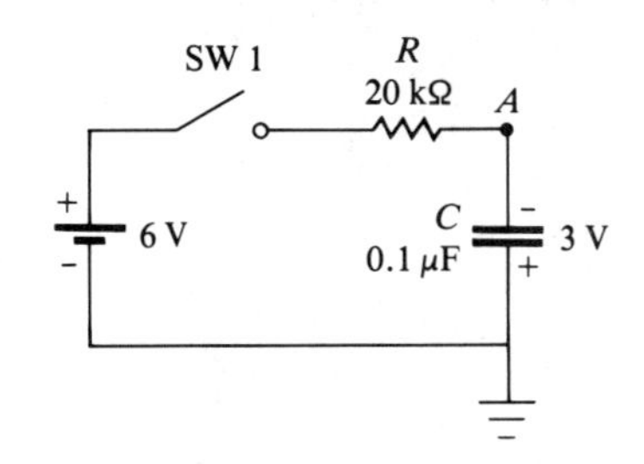

Figure E.2 Diagram

Equation (E.4) states the circuit conditions for any time t after switch 1 is closed. In particular, when $t = 1.2$ ms,

$$e_C = 10 - 10e^{-1.2/1} = 10 - 10e^{-1.2}$$

$$e^{-1.2} = 0.301$$

$$\therefore \quad e_C = 10 - 10(0.301) = 10 - 3.01 = 6.99 \text{ volts}$$

Example E.2 Assume the circuit of Figure E.2. Capacitor C is initially charged to 3 volts with the polarity shown. After switch 1 is closed, point A will eventually charge to 6 volts in a fashion similar to Figure E.3. How long will it take for e_C to equal zero volts?

$$\tau = RC = (20 \times 10^3) \times (0.1 \times 10^{-6}) = 2 \times 10^{-3} = 2 \text{ ms}$$

$$e_C(0) = A + Be^{-0/2 \text{ ms}} = -3 \text{ volts}$$

$$A + B(1) = -3$$

$$A + B \quad = -3$$

$$e_C(\infty) = A + Be^{-\infty/2 \text{ ms}} = 6$$

$$A + B(0) = 6$$

$$A = 6$$

$$\therefore \quad B = -3 - 6 = -9$$

From Eq. (E.1),

$$e_C = 6 - 9e^{-t/2 \text{ ms}}$$

Specifically, if e_C is to equal zero volts, then the following results:

$$0 = 6 - 9e^{-t/2 \text{ ms}}$$

$$e^{-t/2 \text{ ms}} = \frac{6}{9}$$

$$e^{t/2 \text{ ms}} = \frac{9}{6} = 1.5$$

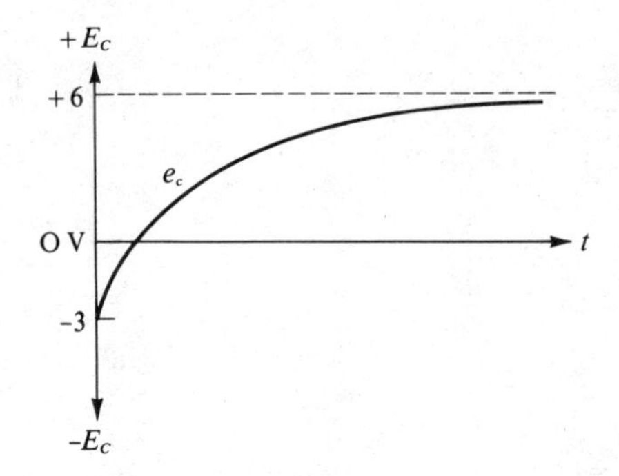

Figure E.3 Diagram

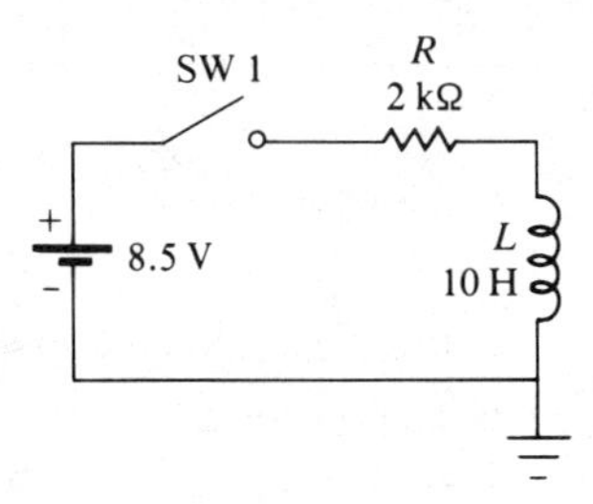

Figure E.4 Diagram

$$t/2 \text{ ms} = \ln 1.5$$
$$t/2 \text{ ms} = 0.405$$
$$t = 0.81 \text{ ms}$$

Example E.3 How long will it take for the current in Figure E.4 to equal 1 mA after the switch is closed?

Initially, at time zero $i_t = 0\,\text{mA}$. After approximately five time constants, $i_t = 8.5 \text{ V}/2 \text{ k}\Omega = 4.25 \text{ mA}$.

The curve i_t is represented in Figure E.5.

$$\tau = \frac{L}{R} = \frac{10}{2 \times 10^3} = 5 \text{ ms}$$

$$i_t(0) = A + Be^{-0/5\text{ ms}} = A + B = 0$$

$$i_t(\infty) = A + Be^{-\infty/5\text{ ms}} = A = 4.25 \text{ mA}$$

$$\therefore \quad B = -4.25 \text{ mA}$$

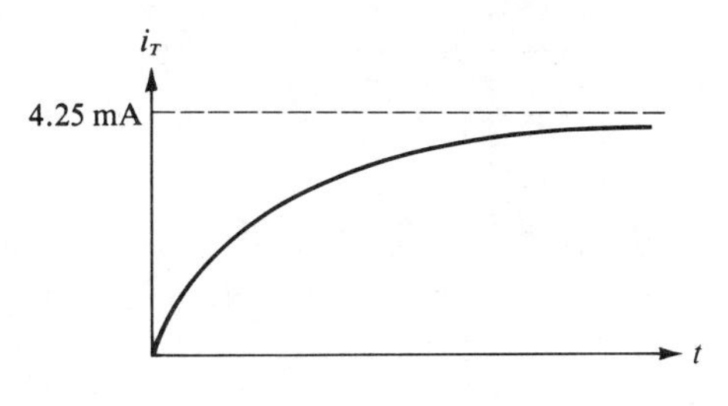

Figure E.5 Diagram

From Eq. (E.1)

$$i_t = 4.25 - 4.25e^{-t/5\text{ ms}}$$

Specifically, when $i_t = 1$ mA,

$$1 = 4.25 - 4.25e^{-t/5\text{ ms}}$$
$$4.25e^{-t/5\text{ ms}} = 3.25$$
$$e^{t/5\text{ ms}} = \frac{4.25}{3.25} = 1.31$$

$$t/5\text{ ms} = \ln 1.31$$
$$t/5\text{ ms} = 0.268$$
$$t = 1.34\text{ ms}$$

APPENDIX F Periodic Chart of the Elements

Shells		IA	IIA	IIIB	IVB	VB	VIB	VIIB	VIII	VIII	VIII	IB	IIB	IIIA	IVA	VA	VIA	VIIA	0
1	K	1 **H** 1.0079																1 **H** 1.0079	2 **He** 4.00260
2	L	3 **Li** 6.941	4 **Be** 9.01218											5 **B** 10.81	6 **C** 12.011	7 **N** 14.0067	8 **O** 15.9994	9 **F** 18.99840	10 **Ne** 20.179
3	M	11 **Na** 22.98977	12 **Mg** 24.305											13 **Al** 26.98154	14 **Si** 28.086	15 **P** 30.97376	16 **S** 32.06	17 **Cl** 35.453	18 **Ar** 39.948
4	N	19 **K** 39.098	20 **Ca** 40.08	21 **Sc** 44.9559	22 **Ti** 47.90	23 **V** 50.9414	24 **Cr** 51.996	25 **Mn** 54.9380	26 **Fe** 55.847	27 **Co** 58.9332	28 **Ni** 58.70	29 **Cu** 63.546	30 **Zn** 65.38	31 **Ga** 69.72	32 **Ge** 72.59	33 **As** 74.9216	34 **Se** 78.96	35 **Br** 79.904	36 **Kr** 83.80
5	O	37 **Rb** 85.4678	38 **Sr** 87.62	39 **Y** 88.9059	40 **Zr** 91.22	41 **Nb** 92.9064	42 **Mo** 95.94	43 **Tc** 98.9062	44 **Ru** 101.07	45 **Rh** 102.9055	46 **Pd** 106.4	47 **Ag** 107.868	48 **Cd** 112.40	49 **In** 114.82	50 **Sn** 118.69	51 **Sb** 121.75	52 **Te** 127.60	53 **I** 126.9045	54 **Xe** 131.30
6	P	55 **Cs** 132.9054	56 **Ba** 137.34	57 **La*** 138.9055	72 **Hf** 178.49	73 **Ta** 180.9479	74 **W** 183.85	75 **Re** 186.207	76 **Os** 190.2	77 **Ir** 192.22	78 **Pt** 195.09	79 **Au** 196.9665	80 **Hg** 200.59	81 **Tl** 204.37	82 **Pb** 207.2	83 **Bi** 208.9804	84 **Po** (210)	85 **At** (210)	86 **Rn** (222)
7	Q	87 **Fr** (223)	88 **Ra** 226.0254	89 **Ac**** (227)	104 (260)	105 (260)													

*Lanthanum Series

6	P	58 **Ce** 140.12	59 **Pr** 140.9077	60 **Nd** 144.24	61 **Pm** (147)	62 **Sm** 150.4	63 **Eu** 151.96	64 **Gd** 157.25	65 **Tb** 158.9254	66 **Dy** 162.50	67 **Ho** 164.9304	68 **Er** 167.26	69 **Tm** 168.9342	70 **Yb** 173.04	71 **Lu** 174.97

**Actinium Series

7	Q	90 **Th** 232.0381	91 **Pa** 231.0359	92 **U** 238.029	93 **Np** 237.0482	94 **Pu** (244)	95 **Am** (243)	96 **Cm** (247)	97 **Bk** (247)	98 **Cf** (251)	99 **Es** (254)	100 **Fm** (257)	101 **Md** (258)	102 **No** (255)	103 **Lr** (256)

APPENDIX G Element Abbreviations and Electron Shell Distributions

	Symbol	Atomic Number	Atomic Weight	Shells K	L	M	N	O	P	Q
Actinium	Ac	89	(227)	2	8	18	32	18	9	2
Aluminum	Al	13	26.9815	2	8	3				
Americium	Am	95	(243)	2	8	18	32	25	8	2
Antimony	Sb	51	121.75	2	8	18	18	5		
Argon	Ar	18	39.948	2	8	8				
Arsenic	As	33	74.9216	2	8	18	5			
Astatine	At	85	(210)	2	8	18	32	18	7	
Barium	Ba	56	137.34	2	8	18	18	8	2	
Berkelium	Bk	97	(247)	2	8	18	32	26	9	2
Beryllium	Be	4	9.01218	2	2					
Bismuth	Bi	83	208.9804	2	8	18	32	18	5	
Boron	B	5	10.81	2	3					
Bromine	Br	35	79.904	2	8	18	7			
Cadmium	Cd	48	112.40	2	8	18	18	2		
Calcium	Ca	20	40.08	2	8	8	2			
Californium	Cf	98	(251)	2	8	18	32	28	8	2
Carbon	C	6	12.011	2	4					
Cerium	Ce	58	140.12	2	8	18	19	9	2	
Cesium	Cs	55	132.9054	2	8	18	18	8	1	
Chlorine	Cl	17	35.453	2	8	7				
Chromium	Cr	24	51.996	2	8	13	1			
Cobalt	Co	27	58.9332	2	8	15	2			
Copper	Cu	29	63.546	2	8	18	1			
Curium	Cm	96	(247)	2	8	18	32	25	9	2
Dysprosium	Dy	66	162.50	2	8	18	28	8	2	
Einsteinium	Es	99	(254)	2	8	18	32	29	8	2
Erbium	Er	68	167.26	2	8	18	30	8	2	
Europium	Eu	63	151.96	2	8	18	25	8	2	
Fermium	Fm	100	(257)	2	8	18	32	30	8	2
Fluorine	F	9	18.99840	2	7					
Francium	Fr	87	(223)	2	8	18	32	18	8	1
Gadolinium	Gd	64	157.25	2	8	18	18	25	9	2
Gallium	Ga	31	69.72	2	8	18	3			
Germanium	Ge	32	72.59	2	8	18	4			
Gold	Au	79	196.9665	2	8	18	32	18	1	
Hafnium	Hf	72	178.49	2	8	18	32	10	2	
Helium	He	2	4.00260	2						
Holmium	Ho	67	164.9304	2	8	18	29	8	2	
Hydrogen	H	1	1.0079	1						
Indium	In	49	114.82	2	8	18	18	3		
Iodine	I	53	126.9045	2	8	18	18	7		
Iridium	Ir	77	192.22	2	8	18	32	15	2	
Iron	Fe	26	55.847	2	8	14	2			
Krypton	Kr	36	83.80	2	8	18	8			
Lanthanum	La	57	138.9055	2	8	18	18	9	2	
Lawrencium	Lw	103	(256)	2	8	18	32	32	9	2
Lead	Pb	82	207.2	2	8	18	32	18	4	
Lithium	Li	3	6.941	2	1					
Lutetium	Lu	71	174.97	2	8	18	32	9	2	
Magnesium	Mg	12	24.305	2	8	2				
Manganese	Mn	25	54.9380	2	8	13	2			
Mendelevium	Md	101	(258)	2	8	18	32	31	8	2
Mercury	Hg	80	200.59	2	8	18	32	18	2	

(continued)

Element Abbreviations and Electron Shell Distributions—(*contd.*)

	Symbol	Atomic Number	Atomic Weight	Shells K	L	M	N	O	P	Q
Molybdenum	Mo	42	95.94	2	8	18	13	1		
Neodymium	Nd	60	144.24	2	8	18	22	8	2	
Neon	Ne	10	20.179	2	8					
Neptunium	Np	93	237.0482	2	8	18	32	22	9	2
Nickel	Ni	28	58.70	2	8	16	2			
Niobium	Nb	41	92.9064	2	8	18	12	1		
Nitrogen	N	7	14.0067	2	5					
Nobelium	No	102	(255)	2	8	18	32	32	8	2
Osmium	Os	76	190.2	2	8	18	32	14	2	
Oxygen	O	8	15.9994	2	6					
Palladium	Pd	46	106.4	2	8	18	18			
Phosphorus	P	15	30.9738	2	8	5				
Platinum	Pt	78	195.09	2	8	18	32	17	1	
Plutonium	Pu	94	(244)	2	8	18	32	24	8	2
Polonium	Po	84	(210)	2	8	18	32	18	6	
Potassium	K	19	39.098	2	8	8	1			
Praseodymium	Pr	59	140.9077	2	8	18	21	8	2	
Promethium	Pm	61	(147)	2	8	18	23	8	2	
Protactinium	Pa	91	231.0359	2	8	18	32	20	9	2
Radium	Ra	88	226.0254	2	8	18	32	18	8	2
Radon	Rn	86	(222)	2	8	18	32	18	8	
Rhenium	Re	75	186.207	2	8	18	32	13	2	
Rhodium	Rh	45	102.9055	2	8	18	16	1		
Rubidium	Rb	37	85.4678	2	8	18	8	1		
Ruthenium	Ru	44	101.07	2	8	18	15	1		
Samarium	Sm	62	150.4	2	8	18	24	8	2	
Scandium	Sc	21	44.9559	2	8	9	2			
Selenium	Se	34	78.96	2	8	18	6			
Silicon	Si	14	28.086	2	8	4				
Silver	Ag	47	107.868	2	8	18	18	1		
Sodium	Na	11	22.9898	2	8	1				
Strontium	Sr	38	87.62	2	8	18	8	2		
Sulfur	S	16	32.06	2	8	6				
Tantalum	Ta	73	180.9479	2	8	18	32	11	2	
Technetium	Tc	43	98.9062	2	8	18	13	2		
Tellurium	Te	52	127.60	2	8	18	18	6		
Terbium	Tb	65	158.9254	2	8	18	26	9	2	
Thallium	Tl	81	204.37	2	8	18	32	18	3	
Thorium	Th	90	232.0381	2	8	18	32	18	10	2
Thulium	Tm	69	168.9342	2	8	18	31	8	2	
Tin	Sn	50	118.69	2	8	18	18	4		
Titanium	Ti	22	47.90	2	8	10	2			
Tungsten	W	74	183.85	2	8	18	32	12	2	
Uranium	U	92	238.029	2	8	18	32	21	9	2
Vanadium	V	23	50.9414	2	8	11	2			
Xenon	Xe	54	131.30	2	8	18	18	8		
Ytterbium	Yb	70	173.04	2	8	18	32	8	2	
Yttrium	Y	39	88.9059	2	8	18	9	2		
Zinc	Zn	30	65.38	2	8	18	2			
Zirconium	Zr	40	91.22	2	8	18	10	2		

ANSWERS TO SELECTED PROBLEMS

Chapter 1

1. a. 1.56×10^{-4}
   b. 2.35×10^{-7}
   c. 1.685×10^{3}
   d. 1.52×10^{-1}
   e. 3.65×10^{6}
   f. 1.23×10^{-2}
   g. 1.2×10^{1}
   h. 1.24×10^{1}
   i. 1×10^{-1}
   j. 1×10^{1}
   k. 1×10^{2}
   l. 1.01×10^{1}
3. a. 12.2×10^{6}
   b. 25.024×10^{-7}
   c. 2.239×10^{6}
   d. 13.29×10^{7}
   e. 2×10^{-4}
   f. 2.41×10^{4}
   g. 9.47×10^{3}
   h. 2.26×10^{-1}
   i. 512×10^{-18}
   j. 2×10^{-1}
   k. 1
   l. 1
   m. $a^{5/6}$
   n. 64×10^{-24}
5. a. 96.56 km/h
   b. 64 cm/s
   c. 2.79×10^{3} cm^2
   d. 0.07 m^3
   e. 9.46 liters
   f. 76.81 decimeters
   g. 10.74×10^{3} grams
   h. 1.344×10^{6} cm/h
7. 113°F
9. −52.44°C
11. 40.23 meters

Chapter 2

1. a. 33 electrons, 33 protons, 42 neutrons
   b. K(2 electrons), L(8 electrons), M(18 electrons), N(5 electrons)
3. 3350.01×10^{-30} kg, 1 proton, 1 electron, 1 neutron
5. 46892×10^{-30} kg
7. 1.24×10^{-10} meter
9. 8.08×10^{22} atoms
11. 5.86×10^{22} atoms/cm^3
13. 13.1×10^{18} electrons
15. 0.293×10^{18} electrons
17. 78×10^{15} electrons
19. 3.75 V

Chapter 3

1. 7.02 ohms
3. 0.024 inch
5. 110.87 meters
7. 0.129 siemens
9. 0.046 mm/edge
11. 0.28 in/edge
13. 0.089 ohm
15. 1.66×10^{-3}
17. 265.1×10^{3} ohm · cm/ft
19. −364°C
21. 0.06 percent
23. 2.36×10^{-3}
25. 4.72 ohms
27. 0.08 mm^2
29. Gray, black, black
31. 635.5 kΩ – 703.5 kΩ

Chapter 4

1.

```
10        PRINT "BASIC"
15        PRINT
20        PRINT "B": PRINT "A": PRINT "S": PRINT "I": PRINT "C"
25        PRINT
30        PRINT "B": PRINT, "A": PRINT, , "S": PRINT, , , "I"
35        PRINT, , , , "C"
40        END
```

```
BASIC

B
A
S
I
C

B
              A
                            S
                                          I
                                                        C
```

3.

```
10      READ  A , B , C
20      INPUT "ENTER VALUE OF X " ; X
30      D=A*X↑2+B*X+C
40      PRINT "FOR THIS EQUATION, WHEN X=  " X " THEN D=  " D
50      DATA 17.15 , -3.61 , 2.8
60      END
```

```
ENTER VALUE OF X ? 7.569
FOR THIS EQUATION, WHEN X=  7.569   THEN D=   957.996
```

5.

```
10      P1=3.14159 : D=.0175 : L=1E4 : A=(P1*D↑2)/4
20      V=A*L : N=8.4E22*V
30      PRINT "THE NUMBER OF FREE ELECTRONS IS " N
40      END
```

```
THE NUMBER OF FREE ELECTRONS IS  2.02043E+23
```

7.

```
10      P1=3.14159
20      A=10.35*3/1.35E-6 : A1=(P1/4)*A : L=SQR(A1)*1E-3
30      PRINT "LENGTH OF EACH SIDE OF THE SQUARE IS " L " INCHES"
40      END
```

```
LENGTH OF EACH SIDE OF THE SQUARE IS  4.25019  INCHES
```

9.

```
10        REM ASSUME AN INITIAL RESISTANCE OF 100 OHMS
20        R1=100: A=.006E-3: T2=-65: T1=20
25        R2=R1*(1+A*(T2-T1))
30        PRINT"PERCENTAGE OF CHANGE IS"R1-R2"%"
40        END

PERCENTAGE OF CHANGE IS .510025E-1 %
```

Chapter 5

1. 39.6 volts
3. 866 ohms
5. 1.36 kΩ
7. 1.86 mA, 2.81 mA
9. 0.41 watt
11. 306.6 volts
13. 40 watts
15. 31.8¢
17. 70.6 watts
19. 21.76 amperes
21. 98,472 N·m/min
23. $32.76
25. $1.14
27. 0.254 hp
29. 3.43 amperes

31.

```
10        INPUT "WHAT IS INPUT PWR IN K-WATTS" P
20        P1= P/.746: D1= P1*.875: D2= D1*.9125
30        PRINT "PWR OUT OF #1 IN HP IS" D1
35        PRINT "AND PWR OUT OF #2 IN HP IS" D2
40        END

WHAT IS INPUT PWR IN K-WATTS
? 16.345
PWR OUT OF #1 IN HP IS 19.1714
AND PWR OUT OF #2 IN HP IS 17.4939
```

Chapter 6

1. 12.3 kΩ
3. R_1 = 4.7 kΩ
   R_2 = 9.4 kΩ
   R_3 = 18.8 kΩ
5. $5.55/R_1$
7. 14.85 volts
9. 24.44 volts
11. 6.92 volts
13. −7.75 volts
15. 8.9 kΩ
17. 1.11 kΩ
19. 4 kΩ
21. 34 mA, 68 mA
23. R_1 = 250 Ω, R_2 = 238 Ω,
    R_3 = 176 Ω, R_4 = 800 Ω
25. I (10 kΩ) = 142.24 mA
    I (22 kΩ) = 64.66 mA
    I (33 kΩ) = 43.10 mA

27. $G_1 = 2.128$ mS
    $G_2 = 1.471$ mS
    $G_3 = 0.667$ mS
    $G_T = 4266\ \mu$S
29. 1.51 mA
31. 4.35 V
33. 90.9 μA
35. 19.7 mA
37. 10.05 mA
39. −3.98 V
41. 1 mA
43. −9 V
45. Yes
47. $V_A = 20$ V, $V_B = 13.75$ V,
    $V_C = 6.75$ V, $V_D = 2.28$ V,
    $V_E = 6.08$ V, $V_F = 5.51$ V,
    $V_G = 1.6$ V, $V_H = \phi$ V

49.

```
1Ø      INPUT"WHAT IS THE VALUE OF THE SOURCE VOLTAGE"V
2Ø      INPUT"WHAT ARE THE VALUES OF THE THREE SERIES
        RESISTORS"R1,R2,R3
3Ø      D=R1+R2+R3: V1=R1*V/D: V2=R2*V/D: V3=R3*V/D
4Ø      PRINT"VOLTAGE ACROSS R1 IS" V1
45      PRINT"VOLTAGE ACROSS R2 IS" V2
5Ø      PRINT"VOLTAGE ACROSS R3 IS" V3
7Ø      END
```

```
WHAT IS THE VALUE OF THE SOURCE VOLTAGE
? 56.985
WHAT ARE THE VALUES OF THE THREE SERIES RESISTORS
? 15Ø,47ØØ,67Ø
VOLTAGE ACROSS R1 IS 1.54851
VOLTAGE ACROSS R2 IS 48.5198
VOLTAGE ACROSS R3 IS 6.91666
```

Chapter 7

1.

```
Ø5      PRINT"      R                   I(MA)
1Ø      FOR R= 1E3 TO 1ØE3 STEP 1E3
2Ø      R1= (3.3E3*R)/(3.3E3+R): I=1Ø/R1
3Ø      PRINT R,I*1E3: NEXT R
4Ø      END
```

```
   R                   I(MA)

 1ØØØ                 13.Ø3Ø3
 2ØØØ                 8.Ø3Ø3
 3ØØØ                 6.36364
 4ØØØ                 5.53Ø3
 5ØØØ                 5.Ø3Ø3
 6ØØØ                 4.69697
 7ØØØ                 4.45887
 8ØØØ                 4.28Ø3
 9ØØØ                 4.14141
 1ØØØØ                4.Ø3Ø3
```

3.

```
1Ø    PRINT" R3(OHMS)    V3(VOLTS)    I3(MA)    P3(MW)"
12    PRINT
15    R1= 2.2E3: R2=4.7E3: R4=1.5E3
2Ø    FOR R3 = IE3 TO 1ØE3 STEP IE3
3Ø    R5= R2+R3: R6= (R4*R5)/(R4+R5): R7= R1+R6: I=25/R7
35    I3=(R4*I)/(R4+R5)
4Ø    V3= I3*R3:   P3= I3*V3
5Ø    PRINT R3, V3, I3*1E3, P3*1E3: NEXT R3
6Ø    END
```

R3(OHMS)	V3(VOLTS)	I3(MA)	P3(MW)
1ØØØ	1.53752	1.53752	2.36395
2ØØØ	2.66999	1.33499	3.56442
3ØØØ	3.53885	1.17962	4.17448
4ØØØ	4.22654	1.Ø5664	4.46592
5ØØØ	4.78438	.956877	4.578Ø7
6ØØØ	5.24598	.87433	4.58671
7ØØØ	5.63246	.8Ø4894	4.53498
8ØØØ	5.9654	.745675	4.44825
9ØØØ	6.25116	.694573	4.34189
1ØØØØ	6.5ØØ26	.65ØØ26	4.22534

5.

```
1Ø    PRINT" V(VOLTS)    R2(OHMS)    P3(MW)"
12    PRINT
15    R3=4.7E3: R1=1.5E3
2Ø    FOR V= 5 TO 25 STEP 5: FOR R2= 1E3 TO 1ØE3 STEP 1E3
3Ø    R4= (R2*R3)/(R2+R3): R5= R4+R1
35    I=V/R5: I3= (R2*I)/(R2+R3)
4Ø    P3= I3**2*R3*1E3
5Ø    IF P3> 32.75 THEN PRINT V, R2, P3
6Ø    NEXT R2: NEXT V
7Ø    END
```

V(VOLTS)	R2(OHMS)	P3(MW)
2Ø	6ØØØ	34.5648
2Ø	7ØØØ	36.1936
2Ø	8ØØØ	37.4919
2Ø	9ØØØ	38.55Ø7
2Ø	1ØØØØ	39.43Ø3
25	3ØØØ	4Ø.1833
25	4ØØØ	46.3318
25	5ØØØ	5Ø.7234
25	6ØØØ	54.ØØ75
25	7ØØØ	56.5525
25	8ØØØ	58.5812
25	9ØØØ	6Ø.2355
25	1ØØØØ	61.6Ø99

7.

```
5       PRINT"R3(OHMS)        V3(VOLTS)          I3(MA)"
10      PRINT: V=50: R1= 2200: R2= 1500: R4= 4700: R5= 6800
15      FOR R3= 200 TO 2200 STEP 200
20      R6= R3+R4: R7= (R6*R2)/(R6+R2): R8= R7+R1+R5
30      I=V/R8: I3= (R2*I)/(R6+R2): V3= I3*R3
40      PRINT R3, V3, I3*1E3
45      NEXT R3
50      END
```

```
R3(OHMS)          V3(VOLTS)          I3(MA)

 200               .230947           1.15473
 400               .447427           1.11857
 600               .650759           1.0846
 800               .842105           1.05263
 1000              1.02249           1.02249
 1200              1.19284           .994036
 1400              1.35397           .967118
 1600              1.50659           .94162
 1800              1.65138           .917431
 2000              1.78891           .894454
 2200              1.91972           .8726
```

Chapter 8

1. a. 24 b. −4 c. 55
   d. −747 e. 164 f. −624
3. $x = bfg + aei + cdh$
   $- bdi - ceg - afh$
5. $I_{R_1} = I_{R_4} = 2.59$ mA,
   $I_{R_2} = I_{R_5} = 1.45$ mA,
   $I_{R_3} = 1.14$ mA
7. −25.01 V
9. For loops *ABCEA*, *CBDFC*, *FDECF*
   $I_1 = 2.49$ mA,
   $I_2 = 2.48$ mA,
   $I_3 = 1.15$ mA
   $\therefore I_{R_1} = 0.01$ mA,
   $I_{R_2} = 2.48$ mA,
   $I_{R_3} = 1.33$ mA
   $I_{R_4} = 1.34$ mA,
   $I_{R_5} = 1.15$ mA,
   $I_{R_6} = 2.49$ mA
11. −10 V
13. $a = -6$, $b = 3$, $c = -2$
15. $I_1 = 2.49$ mA, $I_2 = 2.48$ mA, $I_3 = 1.15$ mA

17.

```
10      REM LOOPS ABDA, ACDA, AND CDBC WERE CHOSEN
20      DIM A(3,3), B(3), C(3), D(5)
30      PRINT "COEFFICIENTS OF MATRIX 'A' ARE"; :MAT INPUT A
37      MAT A= INV(A)
42      FOR V=10 TO 17.5 STEP .5: B(1), B(2)=V: MAT C=A*B
43      IF V= 10 AND C(3)<0 THEN PRINT"REVERSE CHOICE OF I3
        DIRECTION"
44      IF V=10 THEN PRINT"ALL RESISTOR CURRENTS IN MICROAMPS"
45      D(1)=C(1): D(2)=C(2): D(3)=C(3)
50      IF C(3)>=0 THEN 65 ELSE 60
60      D(3)=ABS(C(3)): D(4)=C(1)+D(3): D(5)=C(2)-D(3): GOTO 70
65      D(4)=C(1)-C(3): D(5)=C(2)+C(3)
70      PRINT"FOR V ="V
72      PRINT"I1="D(1)*1E3, "I2="D(2)*1E3, "I3="D(3)*1E3,
73      PRINT"I4="D(4)*1E3, "I5="D(5)*1E3:NEXT V:END

COEFFICIENTS OF MATRIX 'A' ARE? 37,0,-22,0,43,33,-22,33,61.8
REVERSE CHOICE OF I3 DIRECTION
ALL RESISTOR CURRENTS IN MICROAMPS
FOR V = 10
I1= 226.337  I2= 289.262  I3= 73.8874  I4= 300.225  I5= 215.375
```

Note: Above answer is edited for $V = 10$ only.

19.

```
10      REM LOOPS ARE ABDA, ACDA, AND CBDC
        PROGRAM ALLOWS ARBITRARY CHOICE OF I3 DIRECTION
        STATEMENTS 160 AND 165 USED TO ALIGN COLUMNS
30      DIM A(3, 3), B(3), C(3), D(3, 3), E(5)
40      PRINT"COEFFICIENTS OF MATRIX 'A' ARE";:MAT INPUT A
45      PRINT"ALL CURRENTS LISTED IN MICROAMPS": PRINT
50      PRINT"  V";" R2(KOHMS)";"   I1   ";"   I2   ";"   I3   ";
60      PRINT"   I4   ";"   I5   ":PRINT
70      FOR V=1 TO 10
80      FOR R2=1 TO 10
90      A(2, 2)=33+R2: B(1), B(2)=V: MAT D=INV(A): MAT C=D*B
100     IF V=1 AND C(3)<0 THEN PRINT"REVERSE THE CURRENT
        DIRECTION OF I3"
110     E(1)=C(1): E(2)=C(2): E(3)=C(3)
120     IF C(3)>=0 THEN 130 ELSE 140
130     E(4)=C(1)+C(3): E(5)=C(2)-C(3): GOTO 160
140     E(3)=ABS(C(3)): E(4)=C(1)-E(3): E(5)=C(2)+E(3)
160     A$="  ##  ##  ###.###  ###.###  ###.###  ###.###  ###.###"
165     PRINT USING A$, V, R2, E(1)*1E3, E(2)*1E3, E(3)*1E3,
        E(4)*1E3, E(5)*1E3
180     NEXT R2: NEXT V: END
```

Note: A typical sample answer is:
$V = 3$, $R_2 = 5$ (kΩ), $I_1 = 56.733$, $I_2 = 114.508$,
$I_3 = 40.949$, $I_4 = 97.682$, $I_5 = 73.559$

21.

```
1Ø      DIM A(4,4),B(4),C(4)
2Ø      MAT READ A: MAT READ B
3Ø      MAT A= INV(A): MAT C=A*B
4Ø      DATA 37.7,4.7,-1Ø,Ø,4.7,14.8,Ø,-6.8,-1Ø,Ø,14.8,Ø,Ø,-6.8,Ø,38.8
45      DATA 1Ø,22.5,-2.5,12.5
5Ø      FOR K=1 TO 4
55      PRINT"I"K"="C(K)*1E3"MICROAMPS": NEXT K: END

I 1 =-7.39568 MICROAMPS
I 2 =  1816.95 MICROAMPS
I 3 =-173.916 MICROAMPS
I 4 =  64Ø.599 MICROAMPS
```

Chapter 9

1. -8.68 V
3. -0.516 V
5. -20.64 V
7. 2.98 mW
9. 3.84 kΩ
11. $I_N = 1.59$ mA, $R_N = 4.89$ kΩ
13. $I_N = 2.62$ mA, $R_N = 2.98$ kΩ, $I_{R_1} = 1.96$ mA
15. $R = 2.95$ kΩ, $P_{max} = 10.74$ mW
17. 1.48 mA
19. ($R_1 = 1.01$ kΩ, $R_2 = 0.33$ kΩ, $R_3 = 2.23$ kΩ) for Δ *ABC*; ($R_1 = 1.01$ kΩ, $R_2 = 0.71$ kΩ, $R_3 = 1.52$ kΩ) for Δ *BCD*
21. $R_T = 1.67$ ohms
23. $R_{in(1)} = 3R$, $R_{in(2)} = 5.5R$, $R_{in(3)} = 8.619R$, $R_{in(4)} = 11.863R$
25. 0.779 mA
27. $I_{(10\ k\Omega)} = 147.058\ \mu A$

29.

```
Ø5  REM THE METHOD OF SUCCESSIVE TIGHTER LOOPS IS USED
    ASSUME V IS INITIALLY 1Ø VOLTS.  USE LOOPS ABDA, ACDA
    AND ACBA
Ø8  PRINT"  V","  P(MW)": PRINT
1Ø  DIM A(3,3),B(3),C(3): MAT READ A: MAT A= INV(A)
12  R=4.7: S=1Ø: T=1ØØ
15  FOR V= S STEP T/1Ø UNTIL P>35 OR ABS(35-P)<=.ØØØ1
17  B(1),B(2)=V: MAT C= A*B: P=(C(2)**2)*R: PRINT V,P: NEXT V
22  IF ABS(35-P)<=.ØØØ1 THEN 9Ø ELSE PRINT"CROSSOVER POINT"
24  S= V-(T/1Ø)-(T/1ØØ): T=T/1Ø
35  FOR V =S STEP -T/1Ø UNTIL P<35 OR ABS(35-P)<=.ØØØ1
4Ø  B(1),B(2)=V: MAT C= A*B: P=(C(2)**2)*R: PRINT V,P: NEXT V
42  IF ABS(35-P)<=.ØØØ1 THEN 91 ELSE PRINT"CROSSOVER POINT"
44  S= V+(T/1Ø)+(T/1ØØ): T=T/1Ø: GOTO 15
5Ø  DATA 3.7,Ø,-1.5,Ø,14.7,1Ø,-1.5,1Ø,14.8
9Ø  PRINT"V(SOURCE) FOR 35MW ACROSS R IS"V-T/1Ø: GOTO 95
91  PRINT"V(SOURCE) FOR 35MW ACROSS R IS"V+T/1Ø
95  END
```

Note: Solution of program is $V = 29.2383$.

Chapter 10

1. 25×10^6 maxwells
3. 0.25 weber
5. 0.15 weber
7. $\pi \times 10^{-4}$ wb/Amp-m
9. 0.064×10^{-1} At/m
11. 10 cm^2
13. 6250 turns
15. 5.2 At
17. 4 At/m
19. 5.0 A/m
21. 0.6×10^{-3}
23. 2.3×10^{-6} j/cycle
25. 0.8 wb/m^2
27. 8×10^{-6} wb

Chapter 11

1. 0.06 V
3. 0.48 V
5. 0.15×10^2 m/s
7. 4 V
9. 5 H
11. 0.313 μH
13. 0.179 μH
15. (b)
17. 9 mH
19. (a) 6 H; (b) 62 H
21. 20 H, 25 H
23. 20 mH
25. 16 mH
27. 258 mH
29. 10 mH
31. 312.5×10^{-8} joule
33. 40.8 H
35. 4 times greater
37. 33.35 μs
39. 393 mA
41. 20 V

Chapter 12

1. 0.1 newton attracting
3. 0.015×10^{-3} newton attracting Q_1 toward Q_2
5. 1.1×10^{-4} coulomb
7. Not changed
9. 500 V/m;
   17.7×10^{-12} q;
   88.54×10^{-12} joule
11. 2877.55×10^{-10} coulomb/m^2;
    16 kV
13. 5.755×10^{-10} joule
15. 1.4 MV
17. 8.854×10^{-4} coulomb/m^2
19. 56 kV

Chapter 13

1. 0.5 μF
3. 274.56×10^{12} electrons
5. 1 MV/s
7. 4.4 amperes
9. 0.2 μF
11. 118 pF
13. 2.66×10^{-6} meter
15. $C = 45$ pF;
    $V = 2675$ volts
17. 3.46 kV

19. 120 pF
21. 1.51 volt decrease
23. 300 volts maximum
25. 14.545 μF
27. 7500 pF
29. 11.74 volts
31. (a) 9.09 mA; (b) 1.1 ms
33. 13.75 ms
35. 0.6 mA
37. 2.5 ms
39. $V_N = 17.388$ volts; 1.962 volts
41. 14.89 volts
43. $5\tau = 0.66$ ms; C will charge to 24 V of the opposite polarity
45. $t = 483$ μs

Chapter 14

1. 1.03 volts
3. 0.032 newton
5. $f = 1.667$ Hz; $T = 0.6$ second
7. 600°/s; 10.47 rad/s
9. (a) 68.75°, (b) 180°, (c) 38.2°, (d) 322.58°
11. 86.6 volts
13. 50 Hz; 314.16 rad/s
15. 11.76 volts; 0 volts; 0 volts
17. 1.389 ms
19. 7.68 volts
21. 20.08 volts
23. At 50°, $V = 97.5$ volts
    At 90°, $V = 127.28$ volts
    At 190°, $V = -22.1$ volts
    At 390°, $V = 63.64$ volts
25. −34.55 volts
27. 6.67 kHz
29. 5.4 kHz
31. $V_{p\text{-}p} = 330.92$ V,
    $V_p = 165.46$ V,
    $V_{ave} = 105.33$ V
33. 113.14 $V_{p\text{-}p}$

Chapter 15

1. 1885 Ω at 60 Hz; 12.57 kΩ at 400 Hz
3. 55.3 kΩ
5. 2.653 mA
7. 600 Hz
9. 3
11. 3.98 kHz
13. $L_1 = 0.159$ H; $L_2 = 0.477$ H
15. 3027 Ω at 600 Hz
    5 kΩ at 6 kHz
17. 1.484 mA
19. 4.58 kΩ
21. 18.7 H; 18 V
23. 37.02°; 8 mA
25. 1.45 H; 376 μA
27. a. $275\angle 43.3°$;
    b. $182\text{ mA}\angle -43.3°$;
    c. $36.39\text{ V}\angle -43.3°$;
    d. $34.31\text{ V}\angle 46.7°$
29. 58.7 ms; 21.02 V
31. $\theta = 36°$
33. 75.175 V
35. 36.87°
37. 6.06 kHz

Chapter 16

1. 53 kΩ; 26.5 kΩ
3. 1.658 μF, 1.558 μF change
5. 1.11 mA
7. Less because of direct ratio
9. 1.48 V
11. 5.4 V; 70.8 mA
13. 184.5 Ω; 141.4 kΩ
15. 45°
17. 667 pF
19. a. 219.7 Ω at $-24.5°$
    b. 91 mA at 24.5°
    c. 18.2 V at 24.5°
    d. 8.28 V at $-65.5°$
21. 0.33 A $\angle 60°$
23. 407.8 Ω
25. 100 V; $\theta = 53.13°$
27. 0.923 kΩ $\angle -22.62°$
29. 6.26 kΩ $\angle -26.56°$
31. $Z \rightarrow R$, $Z \rightarrow 0$
33. 8.74 mA
35. 0.145 mA
37. 1 kHz

39.

```
10      REM CONVERT THE WYE CONTAINING RX TO A DELTA
15      PRINT , "F (KHZ)" , "ANGLE (DEG)" : PRINT
20      P1=3.14159 : A=90 : C=2E-8 : R=1E3 : R1=R/3 : S=1E4 : T=1E5
25      FOR F=S TO S*100 STEP T=10
30      IF A<25 OR ABS(A-25)<=.0001 THEN 45
35      X=1/(2*P1*F*C) : A=ATN(X/R1)*(180/P1)
40      PRINT , F*1E-3 , A : NEXT F
45      IF ABS(A-25)<=.0001 THEN 90
50      PRINT "CROSSOVER POINT" : S=F-(T/10)-(T/100) : T=T/10
55      FOR F=S TO S/100 STEP -T/10
60      IF A>25 OR ABS(A-25)<=.0001 THEN 75
65      X=1/(2*P1*F*C) : A=ATN(X/R1)*(180/P1)
70      PRINT , F*1E-3 , A : NEXT F
75      IF ABS(A-25)<=.0001 THEN 92
80      PRINT "CROSSOVER POINT" : S=F+(T/10)+(T/100) : T=T/10
85      GOTO 25
90      F1=(F-T/10)*1E-3 : GOTO 94
92      F1=(F+T/10)*1E-3
94      PRINT "FREQUENCY FOR 25 DEG PHASE SHIFT IS " F1 " KHZ"
96      END
```

```
                    F (KHZ)        ANGLE (DEG)

                     10             67.2723
                     20             50.0452
                     30             38.512
                     40             30.8301
                     50             25.5229
                     60             21.697
CROSSOVER POINT
                     59             22.0298
                     58             22.3725
                     57             22.7254
                     56             23.089
                     55             23.4637
                     54             23.8501
                     53             24.2487
                     52             24.6599
                     51             25.0845
CROSSOVER POINT
                     51.1           25.0414
                     51.2           24.9985
CROSSOVER POINT
                     51.19          25.0028
CROSSOVER POINT
                     51.191         25.0023
                     51.192         25.0019
                     51.193         25.0015
                     51.194         25.001
                     51.195         25.0006
                     51.196         25.0002
                     51.197         24.9998
CROSSOVER POINT
                     51.1969        24.9998
                     51.1968        24.9998
                     51.1967        24.9999
                     51.1966        24.9999
FREQUENCY FOR 25 DEG PHASE SHIFT IS 51.1966 KHZ
```

Chapter 17

1. $5.84\text{ mA}\angle 54.3^\circ$
3. $2.6\text{ k}\Omega$; $2.5\text{ k}\Omega$
5. $379\angle 11.1^\circ$; $\theta = 11.1^\circ$ leading
7. 5.89 mA
9. $20\ \Omega$
11. 120 pF
13. $23.15\text{ mA}\angle -64.4^\circ$
15. $13.33\text{ mA}\angle 36.87^\circ$; 6.67 mA
17. $10\text{ k}\Omega$, $5\text{ k}\Omega$
19. $G = 2.06\text{ mS}$;
    $B_C = +j1.19\text{ mS}$
21. $X_C = 2.87\ \Omega$
23. $3.54\text{ k}\Omega\angle -50.46^\circ$
25. $1.44\text{ k}\Omega\angle 28^\circ$
27. $1.788\text{ mA}\angle -6.56^\circ$
29. $712\text{ mA}\angle 45^\circ$;
    $1\text{ A}\angle 0^\circ$;
    $712\text{ mA}\angle -45^\circ$
31. 885 pF
33. 20
35. 400 V
37. 104 pF
39. $1\text{ k}\Omega$, 1.59 kHz
41. 0.0398×10^4 Hz
43. $C = 0.14$ pF

45.

```
10      DIM A(3),B(2) : P1=3.14159
15      READ R1,R2,R3,L,C,V
20      PRINT "F (KHZ)" , "Z(IN)" , "I(MA)" , "ANGLE(DEG)"
25      PRINT
30      F1=(1/(2*P1*SQR(L*C)))*SQR((L-(R3↑2)*C)/(L-(R2↑2)*C))
35      FOR F=F1-14E3 TO F1+26E3 STEP 2E3
40      X1=2*P1*F*L : X2=1/(2*P1*F*C)
45      D=(R2+R3)↑2+(X1-X2)↑2
50      N1=(R2*R3+X1*X2)*(R2+R3)+(R2*X1-R3*X2)*(X1-X2)
55      N2=(R2+R3)*(R2*X1-R3*X2)-(X1-X2)*(R2*R3+X1*X2)
60      A(1)=R1+(N1/D) : B(1)=N2/D : A(2)=SQR(A(1)↑2+B(1)↑2)
65      B(2)=ATN(B(1)/A(1)) : A(3)=V/A(2) : A=-B(2)
70      DATA 1E3 , 100 , 2E3 , 10E-3 , 1E-9 , 20
75      PRINT F*1E-3 , A(2) , A(3)*1E3 , A*(180/P1)
80      NEXT F
85      END
```

F (KHZ)	Z(IN)	I(MA)	ANGLE(DEG)
25.0044	4066.19	4.91861	−11.6519
27.0044	4269.07	4.68486	−11.2927
29.0044	4493.85	4.45052	−10.5777
31.0044	4740.04	4.21937	−9.45896
33.0044	5005.6	3.99552	−7.88604
35.0044	5286.15	3.78347	−5.81009
37.0044	5573.96	3.58812	−3.18964
39.0044	5857.14	3.41463	1.98468E−05
41.0044	6119.29	3.26836	3.75452
43.0044	6340.4	3.15438	8.02726
45.0044	6499.65	3.07709	12.72
47.0044	6579.78	3.03961	17.684
49.0044	6571.83	3.04329	22.7351
51.0044	6477.79	3.08747	27.6816
53.0044	6309.84	3.16965	32.355
55.0044	6086.5	3.28596	36.6318
57.0044	5827.96	3.43173	40.4411
59.0044	5552.33	3.60209	43.759
61.0044	5273.7	3.7924	46.5968
63.0044	5001.94	3.99845	48.9877
65.0044	4743.15	4.21661	50.9758

Chapter 18

1. 12 H
3. 0.612 henry
5. 5 mH, 0.408
7. 4
9. 5200
11. 51 mA
13. 0.375 A
15. 900 Ω
17. 833.3 Ω
19. 2.22 mA
21. 38.3 kΩ
23. 0.522 mA $\angle -47.2^\circ$
25. a. 5.374 mH;
    b. 33.77 Ω;
    c. $\angle 90^\circ$;
    d. 2 mA $\angle -90^\circ$;
    e. −1.35 mA $\angle -90^\circ$;
27. 6.8 H, 2.56 kΩ
29. 100 pF

31.

```
10        P1=3.14159 : L1=1E-2 : L2=L1 : K=.95 : F=1E4
15        PRINT "RL(KOHMS)" , "Z-IN(OHMS)" , "Z-ANGLE(DEG)"
20        PRINT
25        W=2*P1*F
30        M=K*SQR(L1*L2) : Z3=W*M : Z1=W*L1 : Z2=Z1
35        FOR R=100 TO 10100 STEP 250
40        D=R↑2+Z2↑2 : R1=(R*Z3↑2)/D : X1=(Z2*Z3↑2)/D
45        I1=Z1-X1
50        Z=SQR(R1↑2+I1↑2) : A=ATN(I1/R1)*(180/P1)
55        PRINT R*1E-3 , Z , A
60        NEXT R
65        END
```

RL(KOHMS)	Z-IN(OHMS)	Z-ANGLE(DEG)
.1	115.816	40.5354
.35	310.41	39.0478
.6	436.186	49.5092
.85	506.573	57.6507
1.1	546.432	63.4528
1.35	570.229	67.6401
1.6	585.268	70.7529
1.85	595.267	73.1376
2.1	602.208	75.0139
2.35	607.202	76.5244
2.6	610.907	77.7642
2.85	613.725	78.7988
3.1	615.917	79.6745
3.35	617.653	80.4249
3.6	619.051	81.0747
3.85	620.193	81.6428
4.1	621.137	82.1434
4.35	621.926	82.5879
4.6	622.593	82.9851
4.85	623.161	83.3422
5.1	623.648	83.6649
5.35	624.07	83.9579
5.6	624.437	84.2251
5.85	624.759	84.4697
6.1	625.043	84.6946
6.35	625.293	84.9019
6.6	625.517	85.0937
6.85	625.716	85.2717
7.1	625.895	85.4372
7.35	626.056	85.5916
7.6	626.202	85.7358
7.85	626.334	85.871
8.1	626.454	85.9979
8.35	626.563	86.1172
8.6	626.664	86.2296
8.85	626.755	86.3357
9.1	626.84	86.436
9.35	626.917	86.531
9.6	626.989	86.6211
9.85	627.055	86.7066
10.1	627.117	86.7878

Chapter 19

1. 1.59 kHz
3. 0.02 μF
5. a. 3.16 V $\angle -72°$;
   b. 4.47 V $\angle -63.4°$
7. 14.88 V $\angle 7.1°$
9. 7.8 kHz
11. 1.034 μF
13. $R = 21.2\ \Omega$
15. 8 Hz
17. 18.85 Ω
19. 7.23 kHz

21.

```
10   P1=3.14159 : C1=1E-7 : C2=5E-8 : R1=1E3 : R2=2E3 : V1=10
15   FOR F=1 TO 3E3 STEP 100
20   X1=1/(2*P1*F*C1) : X2=1/(2*P1*F*C2)
25   D1=2*X1*R2 : D2=X1↑2 : A1=-ATN(P1/2) : A2=-ATN(P1)
30   Y1=D1+D2 : A3=A1+A2
35   D3=2*R2*X2 : D4=R1↑2 : A4=ATN(P1/2)
40   Y2=D3+D4
45   Z1=Y1/R2 : B1=A3 : T1=Z1*COS(B1) : I1=Z1*SIN(B1)
50   Z2=Y1/X1 : B2=A3-A1 : T2=Z2*COS(B2) : I2=Z2*SIN(B2)
55   Z3=Y2/X2 : B3=A4-A1 : R3=Z3*COS(B3) : I3=Z3*SIN(B3)
60   Z4=Y2/R1 : B4=A4 : R4=Z4*COS(B4) : I4=Z4*SIN(B4)
65   H1=Z1*Z3 : G1=B1+B3
70   S1=T1+R3
75   S2=I1+I3
80   S3=SQR(S1↑2+S2↑2) : S4=ATN(S2/S1)
85   Z5=H1/S3 : A5=G1-S4
90   R5=Z5*COS(A5) : I5=Z5*SIN(A5)
95   H2=Z2*Z4 : G2=B2+B4
100  S5=T2+R4 : S6=I2+I3
105  S7=SQR(S5↑2+S6↑2) : S8=ATN(S6/S5)
110  Z6=H2/S7 : A6=G2-S8
115  V2=Z6*V1/(Z5+Z6)
120  PRINT F , V2
125  NEXT F
130  END
```

```
1       9.9853
101     8.90479
201     8.24124
301     7.76773
401     7.40728
501     7.13108
601     6.92819
701     6.79307
801     6.71962
901     6.69916
1001    6.7209
1101    6.77361
1201    6.84714
1301    6.93328
1401    7.02586
1501    7.12058
1601    7.21455
1701    7.30598
1801    7.39379
```

```
1901          7.47741
2001          7.5566
2101          7.63132
2201          7.70168
2301          7.76784
2401          7.83003
2501          7.88847
2601          7.9434
2701          7.99505
2801          8.04366
2901          8.08942
```

Chapter 20

1. 3.5 kΩ
3. 268 Ω, 292 Ω, $\Delta Z_{in} = 24\ \Omega$
5. $2.82\ \text{k}\Omega \angle -67.55°$, $\alpha = 0.196 \angle 101.32°$
   $2.68\ \text{k}\Omega \angle -63.43°$, $\alpha = 0.316 \angle 71.57°$
   $2.66\ \text{k}\Omega \angle -50.38°$, $\alpha = 0.6202 \angle 60.25°$
7. $158 \angle 32.44°$,
   $\alpha = 0.325 \angle -12.53°$
9. $Z_{11} = 61.2 + j108.17$,
   $Z_{12} = Z_{21} = -30.8 + j46.17$,
   $Z_{22} = 15.37 + j76.97$
11. $74.9\ \text{mA} \angle 99.97°$
13. $Z_{in} = Z_{11} = 1\ \text{k}\Omega$,
    $Z_o = Z_{22} = 1\ \text{k}\Omega$,
    $Z_{12} = Z_{21} = 800\ \Omega$
15. $Z_A = 889\ \Omega$, $Z_B = 2.66$ mH

INDEX